SYSTEMS ANALYSIS AND DESIGN

Marjorie Leeson

Associate Authors

Stanford H. Rowe
Manager, Computer Technology
Dow Corning Corporation

Sandra E. Poindexter
Northern Michigan University

SCIENCE RESEARCH ASSOCIATES, INC.
Chicago, Henley-on-Thames, Sydney, Toronto
A Subsidiary of IBM

Acquisitions Editor:	Michael Carrigg
Project Editor:	Byron Riggan
Cover Design:	Janet Bollow Associates
Cover Photograph:	Tim Jewett EKM - Nepenthe
Text Design and Composition:	Graphics West

Library of Congress Cataloging in Publication Data

Leeson, Marjorie.
  Systems analysis and design.

  Includes index.
  1. System analysis.   2. System design.   I. Rowe,
Stanford H.   II. Poindexter, Sandra E.   III. Title.
T57.6.L43 1985   001.64   84-27611
ISBN 0-574-21785-1

10 9 8 7 6 5 4 3 2 1

ABOUT THE AUTHORS

MARJORIE M. LEESON

From 1962-1983 Marj was the Chairperson for Computer Information Systems at Delta College, University Center, MI. Her responsibilities included designing and developing the CIS curriculum, working with advisory committees, articulating the program with industry and other colleges, developing course materials, coordinating the day and evening course offerings, and teaching courses which included Systems Analysis and Design. She was also an active member of the Computer Policy Committee which was responsible for developing long-range objectives regarding computer utilization, developing RFBs, selecting computer systems, and developing guidelines regarding the acquisition and use of microcomputer systems.

While employed at Delta, Marj also worked as an analyst and designed and implemented financial systems, developed inservice training seminars for industry, and was a consultant for several small organizations. She was appointed an advisor for the New Literacy project sponsored by the California Consortium for Community College Television and to the ACM committee responsible for developing and writing curriculum reports. She has also been active in DPMA and received several outstanding performance awards. She has received both local and national awards for her contribution to EDP and has been a frequent speaker at seminars, workshops, and conferences. Marj has authored several other textbooks and has written numerous articles.

Currently Marj is teaching courses for MiraCosta College and Webster University, writing, and involved in small system consulting. She received her Bachelor's degree from Western Michigan University and her Master's degree from the University of Michigan.

STANFORD H. ROWE

Stanford H. Rowe is the Manager of Computing Technology in the Systems and Information Management Department at Dow Corning Corporation in Midland, Michigan. Stan joined Dow Corning in 1966 and in 1969 was assigned to Brussels, Belgium as the Manager of Data Processing for Dow Corning in Europe. In 1972 he returned to Midland as Manager of Computer Operations and in 1975 was named Manager of Technical Support.

For the last eight years Stan has been involved, primarily from a direction-setting, planning, and budgeting perspective, in all phases of Information Systems activity in Dow Corning. Some of these activities include Distributed Processing, Office Systems, Interactive Computing, and Computer Finance and Leasing. His current responsibilities include management of software installation and maintenance for Dow Corning's large mainframe computers and for voice and data communication within the company.

Stan is a member of the Association of Computing Machinery and the Data Processing Management Association. He teaches several courses at Delta College including Systems Analysis and Design, Operating Systems, and Telecommunications. Stan recently received the Lynn Heatley award for Distinguished

Service to the college. He received his A.B. Degree from the University of Redlands and his M.S. from the University of Southern California.

Since Stan had used SYSTEMS ANALYSIS AND DESIGN as the text for his course and has such a diversified background in EDP, his help in defining content and reviewing the manuscript was extremely beneficial and resulted in a text that reflects the current state of the art.

SANDRA POINDEXTER

In 1980 Sandra Poindexter joined the faculty of Northern Michigan University, Marquette, Michigan and is currently teaching COBOL and Systems Analysis and Design. Sandra also monitors the systems analysis and design workshop projects and is an academic advisor for Data Processing Majors.

Prior to joining the faculty at Northern, Sandra was employed as a Senior Systems Analyst by the National Bank of Detroit. In that capacity Sandra worked on the development of a new comprehensive savings system and was the enhancement coordinator for the 24 Hour Banker system. She was also employed as a Systems Analyst by the American Natural Resources Company in Detroit, MI. Sandra aided geologists and environmental engineers in their study of coal gasification and mining projects. She also completed a feasibility study for the development of a land lease management system. While employed by the Farm Bureau Insurance Group of Michigan Sandra was a project leader responsible for the conversion of two insurance systems from Assembler to COBOL. The project involved eight people and over 4000 people/hours.

Sandra received her Master's Degree from Wayne State University and her Bachelor's Degree from Purdue University. In 1983 Sandra was awarded the Certificate of Data Processing (CDP).

In regard to this edition of SYSTEMS ANALYSIS AND DESIGN, Sandra reviewed the first edition and made suggestions regarding material that should be expanded, added, or deleted. In addition, she did the research for and wrote the North Central University, Westside Medical Plaza, and Provincial Art Studio case studies. Sandra also very carefully reviewed the manuscript and made many suggestions for improvements.

CONTENTS

Preface   ix

Listing of Frequently Used Acronyms   xv

SECTION I   INTRODUCTION TO COMPUTER SYSTEMS AND
TO SYSTEMS ANALYSIS AND DESIGN   1

Stanford Rowe, Manager of Computer Technology in the Systems and Information Management Department of Dow Corning Corporation talks about changes in EDP and the role played by the users.

1   An Overview   7
Introduction   8
A Payroll Example   8
The Role of the Analyst   9
Systems Development Life Cycle   11
Why Some Systems Do Not Meet Their
Objectives   17
A Basic Understanding of Today's
Technology   19
Computer Systems   25
Classifications of Computer Systems   30
Summary   32
Discussion Questions   32
Team or Individual Projects   33
Glossary of Words and Phrases   34
Study Guide   39

2   Management and the Computer Information
Services Department   45
Introduction   46
A Systems Approach to Management   46
Components of a Management System   48
Managing a System   49
Computer Information Services   54
The Computer Information Service Manager   57
Office Systems Coordinator   58
Information Center Manager   59

Database Administrator   59
The Systems Department   60
Programming Department   63
Computer Operations   65
Organization of a Small Data
Processing Department   68
Microcomputer Environments   69
The Computer Policy Committee   69
Discussion Questions   71
Team or Individual Projects   72
Glossary of Words and Phrases   73
Study Guide   75

SECTION II   TECHNOLOGIES, TOOLS, AND TALENTS
OF ANALYSTS   81

Gene Arnold, Computer Policy Chairperson, discusses the role of the computer policy committee. John Fuller, Registrar, talks about the advantages of an online registration system.

3   Project Management, Analysis, and
Design Tools   87
Introduction   88
A Systems Analysis and Design Project   88
Understanding the Problem   89
Functions of Data Flow Diagrams
and Systems Flowcharts   93
Other Tools and Techniques   106
Summary   109
Discussion Questions   109
Team or Individual Projects   109
Glossary of Words and Phrases   111
Study Guide   113

4   Online Transaction and Decision Support
Systems   117
Introduction   118
Batch Processing Systems   118

Online Transaction Processing Systems   119
Management Information Systems   127
Decision Support Systems   128
Fifth-Generation Computer Systems   129
Summary   130
Discussion Questions   130
Team or Individual Projects   131
Glossary of Words and Phrases   131
Study Guide   133

5   New Technologies Within the Office   135
Introduction   136
Reasons for Automation of the Office   136
The Modern Office   137
Office Automation   141
Summary   150
Discussion Questions   150
Team or Individual Projects   152
Glossary of Individual Projects   153
Study Guide   155

Section III   INITIAL AND DETAILED INVESTIGATION   159
At Bay Medical Center analysts make certain that systems are designed to meet the needs of the users.

6   The Initial Investigation   165
Why Investigations Are Made   166
The Request For an Initial Investigation   170
Steps in Initiating an Investigation   172
Guidelines or Standards   183
The Standards Manual   187
A Request is Made for Word Processing   188
Summary   190
Discussion Questions   198
Team or Individual Projects   199
Glossary of Words and Phrases   200
Study Guide   201

7   Feasibility Study   205
Introduction   206
Involvement of Management   206
Team Approach   206
Investigations By Companies Without a Systems Department   207
Beginning the Feasibility Study   210
Conducting a Feasibility Study   215

Planning the Presentation   229
Making the Presentation   233
Concluding the Feasibility Study   234
Summary   239
Discussion Questions   240
Team or Individual Projects   241
Glossary of Words and Phrases   242
Study Guide   243

SECTION IV   DESIGNING SYSTEMS   249
During all phases of the systems development life cycle, programmer/analysts at the Texas Farm Bureau work closely with the users. Microcomputers have been successfully integrated into transaction processing systems.

8   General Design Specifications   255
Introduction   256
Guidelines Used in Developing Specifications   257
Project Organization   265
Design Principles   270
Procedures Needed for the Word Processing System   270
Design Review   273
Preparing the General Design Phase Report   273
Selecting Hardware and Software   274
Establishing Microcomputer Guidelines   281
Summary   284
Discussion Questions   285
Team or Individual Projects   286
Glossary of Words and Phrases   287
Study Guide   289

9   Detailed Design: Input and Output   293
Input/Output Devices Available   295
Output Considerations   298
Conclusions Regarding Output Devices   303
Printed Reports   305
Distribution of Reports   308
Display Screens   311
Major Concerns Regarding Input   317
Verification Methods   321
Distributed Data Entry   323
Trends Regarding Data Entry Devices   324
Designing Source Documents   325

Codes **334**

Summary **336**

Discussion Questions **337**

Team or Individual Projects **339**

Glossary of Words and Phrases **340**

Study Guide **341**

10 Detailed Design: Files and Databases **345**

Introduction **346**

Uses of Files and Databases **346**

Logical and Physical Records **348**

Processing Data Stored on Magnetic
Tape or Disk **348**

Types of Storage Media **349**

ISAM Files **358**

VSAM Files **360**

File Access **363**

Databases **365**

Data Dictionaries **373**

Summary **377**

Discussion Questions **377**

Team and Individual Projects **378**

Glossary of Words and Phrases **379**

Study Guide **383**

11 Detailed Design: Programs and Procedures **387**

Introduction **388**

Segmenting the System into Procedures **388**

Factors to Consider in Segmenting a System **388**

Design Standards **392**

Factors that Must be Determined
for Each Procedure **393**

Procedures Required For All Systems **393**

Developing Programs **396**

Presentation of the Detailed Report
to Management **400**

Summary **406**

Discussion Questions **406**

Team or Individual Projects **407**

Glossary of Words and Phrases **409**

Study Guide **411**

SECTION V IMPLEMENTATION AND EVALUATION **415**

Documentation for the first phase of the TASCO
project included data flow diagrams, hierarchy
charts, detailed logic plans, documentation for
the operators, and user manuals. The new
system was run parallel with the old system
before it was turned over to Administration
Services.

12 Programming Considerations **421**

Four Stages of Growth in Electronic
Data Processing **422**

Development of Structured Design
and Programming Concepts **423**

Development and Implementing
Programs **425**

The Inventory Change Program **433**

Design Programming Standards **443**

Documentation Standards **448**

Increasing Programming Productivity **448**

Summary **452**

Discussion Questions **453**

Team or Individual Projects **454**

Glossary of Words and Phrases **455**

Study Guide **457**

13 Preparing for the New System **461**

Introduction **462**

File Conversion **462**

Testing Programs and Systems **463**

Training Personnel **465**

Preparing Audiovisual Materials **467**

Programmed Learning Materials **468**

Testing Training Materials **468**

Inservice Training Programs **469**

Using Documentation and Standards Manuals **470**

Converting to the New System **476**

Summary **482**

Discussion Questions **483**

Team or Individual Projects **484**

Glossary of Words and Phrases **485**

Study Guide **487**

**14 Implementation: Documentation and
Systems Audit** **491**

Introduction **492**

Functions Performed by Documentation **492**

Functions Served by Documentation **493**

Types of Documentation **495**

Internal Program Documentation **500**

Playscript Method of Writing Documentation **501**

Maintaining Documentation   **503**

Format and Completion of Documentation   **503**

Systems Audit   **506**

A Comprehensive EDP Audit   **511**

Summary   **512**

Discussion Questions   **512**

Team or Individual Projects   **513**

Glossary of Words and Phrases   **514**

Study Guide   **515**

Appendix I   Computer Products Incorporated Case Study   519

Appendix II   Program Flowcharting Symbols   557

Appendix III   Answers to the Checkpoint Questions   560

Index of Forms and Charts   579

Index   581

PREFACE

An analyst is recognized as a catalyst of change and is directly involved in the design and development of complex systems. In order to design and develop a system, it must be divided into procedures and the procedures subdivided into manageable tasks.

In the design of this educational system (the revised textbook and instructor's guide which includes a test bank and three independent case studies), the same steps were followed as those used in designing and implementing a revised information system. During the initial investigation, data was gathered by conducting a survey of users—faculty who taught from the first edition. The problem was identified: the first edition needed to be updated to include microcomputers, office automation, decision support systems, data flow diagrams, and online, transaction processing.

A more detailed investigation was made by involving the associate authors in analysing the data gathered and in further defining the problem by determining the content for the revised edition. It was evident that the revised edition had to emphasize transaction processing, the integration of microcomputers into the organization, and the coordination between office systems and information services. The detailed investigation indicated the need for learning materials that would meet the following objectives:

1. Provide information to make students aware of the total computer information processing environment and of the concepts involved in the top-down design of a system or a procedure. Current topics such as distributed data processing, online transaction processing, data flow diagrams, office automation, ergonomically designed workstations, and database management must be covered. A foundation for change must be created while preparing students for working with fifth generation systems and tomorrow's jobs.

2. Provide case studies and projects that will help students understand and do the tasks performed by an entry-level analyst, a programmer/analyst, a member of a project team, or a user. The experiences students will encounter on the job must be simulated.

3. Relate correct procedures and techniques to examples that are understandable to learners. In performing the assigned tasks, students will need to use previously mastered skills, such as effective communication.

4. Include a variety of projects and activities so that individual instructors can select those that will help their students meet course objectives. There is a vast array of projects, three independent case studies which are in the instructor's guide, and other learning activities from which to choose.

5. Emphasize the importance of interpersonal communication as it applies to systems analysis and design. The importance of having the end user

and management directly involved in *all* aspects of the investigation, design, implementation, and evaluation of the system is stressed.

6. Reinforce learning by providing checkpoint questions, discussion questions, projects, end-of-chapter glossaries, and study guides.

A general design was developed and a task force of competent, well informed, classroom instructors and representatives from industry reviewed the materials and made suggestions for changes. SRA provided the professional suport staff that completed the conversion—transforming the reviewed manuscript and other materials into an effective learning system.

After a system becomes operational, it must be evaluated. Now it is up to you to evaluate this educational system which includes the second edition of the text, independent case studies, and instructor's resource materials. Before doing so, the organization of the textbook and resource materials should be reviewed.

Organization

The test is divided into five major sections. Each section begins with a scenario which introduces you to an organization and its use of computers, to an analyst CIS manager or user, and to an application. The scenario is designed to show how theory is put into practice by different organizations and allow you to meet some of the information processing specialists.

Interspersed throughout each chapter are checkpoint questions that enable the learner to evaluate his or her comprehension of the proceding text material. Since the answers are given in the back of the textbook, the checkpoint questions also reinforce learning.

Each chapter is followed by discussion questions, team or individual projects, a glossary of words and phrases, and a study guide. Many of the discussion questions and projects are mini case studies which require students to use the material presented in the textbook to solve typical problems involving an organization's personnel and procedures.

Appendix I contains an on-going case study of Computer Products Incorporated (CPI). Although each chapter of the text has a corresponding chapter in the case study, the material has been placed in an appendix to make it more convenient for students.

Appendix II contains program flowcharting symbols.

Appendix III contains the answers to the checkpoint questions for all the chapters.

Content

Section I provides background material that is needed for the rest of the text. A review of EDP concepts and terminology is provided along with information regarding the role within the organization of the CIS department. Section II covers some of the tools and techniques used by analysts, online and decision support systems, and new technologies within the office. Section III identifies the steps required to investigate and define a problem and to provide a solution that is within established constraints. Section IV covers the design phase of the systems life development cycle. Section V illustrates what must be done to implement, test, and document a system.

Evaluation Guidelines

As you make your evaluation, check the material in the textbook to see if it is:

Realistic A wealth of practical, down-to-earth examples are presented to illustrate the roles played by management, the end user, analyst, programmer, and operations personnel in the development and implementation of systems and procedures. Many of the forms, guidelines, and standards can be used by students when they obtain positions related to the utilization of computers for information processing and retrieval.

Practical   Correct procedures and techniques are illustrated and stressed. However, because each phase of the systems life development cycle is divided into small, meaningful tasks, learners who are not data processing professionals can understand what must be done and successfully complete assigned tasks.

Flexible   Because of the variety of projects available, you can select the material that meets your learning objectives. For example, you might elect to omit all of Section II and stress the ongoing case study found in Appendix I. Another instructor might opt to cover the material in Section II, use the ongoing CPI case study for classroom discussion, and assign students one of the three independent case studies.

Related to learning activities   The ongoing CPI case study reinforces the material presented in the textbook. The case study also correlates the textbook material with realistic assignments that can be successfully completed by students.

Relevant   Timely topics such as office automation, decision support systems, online documentation, microcomputer guidelines, videodisk, networks, and many other topics are covered. Although transaction processing is emphasized, the need for batch processing is also identified.

Learner-oriented   The text is organized and designed to remove as many obstacles as possible from the learning process. Students successfully complete small tasks that make up the complex assignments given to analysts.

Systems Audit   You and your students will complete the systems audit and determine how well the objectives set forth on the first page of this preface are met. Students will perform the ongoing evaluation that will determine how effective the text is as part of a learning system.

Acknowledgments   In order to create an effective learning system the ability and talents of many people are involved. A very sincere thank-you is given to the people who were so gracious in providing assistance.

The following individuals provided data for the initial investigation regarding the revision. Many of the individuals listed below not only completed a four-page questionnaire regarding the topics that should be deleted, added, or expanded but also offered helpful comments regarding the strengths and weaknesses of the first edition. Because of their helpful comments, the weaknesses cited have been converted to strengths.

Joe Adamski	Grand Valley State College
Donald Alley	Kansas City Kansas Community College
Henry Altieri	Norwalk State Technical College
Ray Backstrom	Pima Community College
Willie Frank Baptiste	Westchester Business Institute

Edwin W. Basham	Kansas State University
Carl W. Blanton	Pima Community College
Agatha D. Briscoe	Hawaii Community College
Jack L. Burgess	Muskegon Business College
Donald Carpenter	Pikes Peak Community College
Melvin H. Carr	Adelphi Institute
Mario Cecchetti	Westmoreland County Community College
Joseph J. Cebula	Community College of Philadelphia
Paul A. Chase	Becker Jr. College
Evan E. Confrey	Greater New Haven State Technical College
Paul G. Duchow	Pasadena City College
Pat Eaton	Upsala College
Mary Jane Fedor	Community College of Allegheny County-South
Daniel F. Fitzgerald Jr.	Herkimer Community College
Frank Greene	Ventura College
Dennis C. Guster	St. Louis Community College at Meramec
L. Edward Hart	Carson-Newman College
Brooks Hartzell	Indiana Vocational Technical College
Ruby B. Holliday	Delgado Jr. College
James M. Hunter	Director of Inflo Center Warner Bros. Music
Lawrence A. Jadico	Spring Garden College
U.T. Johnson	Watterson College
Willard H. Keeling	Blue Ridge Community College
Virginia Knight	Western New England College
Irwin F. Kraus	MIS Technical Services Hallmark Cards, Inc.
Gerald R. Lamphere	Detroit College of Business
George Liaw	University of Lowell
Ruth Malmstrom	Somerset County College
Gerald Marquis	Walsh College of Accounting and Bus. Ad.
Ted J. McGoron	DP Manager Sawbrook Steel Casting Co.
John M. McKinney	University of Cincinnati
Edward M. McLaughlin	Luzerne County Community College
Donald Merkey	Hesser College
Duncan I. Meier III	North Dakota State School of Science
Larry Newcomer	Pennsylvania State University
David Newhall	North Shore Community College
Ron Norman	University of Arizona
Theresa M. Peterman	Davenport College of Business
Mark Probasco	Indiana Vocational Technical College
Larry Richards	Eastern Washington University
Joan M. Roberts	Front Range Community College
John J. Rooney	Hartwick College
Robert Rollins	Shippensberg University
Arline Sachs	Northern Virginia Community College
Dean J. Saluti	University of Massachusetts
Ira Slobodien	Heald College
Walter W. Smock	Rutgers, the State University
Susan L. Solomon	Eastern Washington University
Larry Stevens	University of Alaska
Alfred C. St. Onge	Springfield Technical Community College
Alexis N. Sommers	University of New Haven
Ludwig Slusky	University of Northern Colorado
Sandra Stalker	North Shore Community College

Shirley Tainow	Middlesex County College
Anthony Verstraete	Pennsylvania State University
Clinton Wallbank	Southern Ohio College
Frank White	Catonsville Community College
Helen W. Wolfe	Post College

Just as brainstorming by a project team helps to develop a better solution to a problem, the associate authors and the reviewers analyzed the results of the questionnaires and made additional recommendations that were incorporated into the design of the learning system. The reviewers and associate authors did an outstanding job of reading the manuscript and making suggestions for its improvements. The assistance of the people listed below is very much appreciated.

Donald Alley	Kansas City Community College
Rita Fillman	Purdue University
Gary Hanson	Brigham Young University
Ruby Holliday	Delgado College
Kate Merkling	Utah Tech
Sandra Poindexter	Northern Michigan University
Larry Richards	Eastern Washington University
Stanford H. Rowe	Computer Technology Dow Corning Corporation

The individuals interviewed for the scenarios contributed their time and effort to the success of the project. The organizations they represented also contributed suggestions and materials.

Gene Arnold, Computer Policy Chairperson	Delta College
John Fuller, Registrar	Delta College
Stanford Rowe, Manager of Computing Technology	Dow Corning Corporation
Shelby Tunmire, Mgr. Methods and Procedures	Texas Farm Bureau
Tammy Waugh, Programmer/Analyst	Bay Medical Center

Aletha Bateman, David McGrain, and Kandi Johnson took the photographs that were used in the scenarios. The photographs used in Chapter 5 were contributed by Steel Case Inc.

Michael Carrigg, Acquisitions Editor, coordinated the entire project from the initial design phase through implementation—the creation of a marketable system.

Byron Riggan Development Editor for SRA, coordinated the transformation of the manuscript into a comprehensive learning system. Byron's efforts contributed a great deal to the success of the project. His talent and patience are to be commended.

LIST OF FREQUENTLY USED ACRONYMS

AI	artificial intelligent
ASCII	American National Standards Code for Information Interchange
ANSI	American National Standards Institute
CIS	computer information services
COM	computer output to microfilm
CPI	Computer Products Incorporated
CPU	central processing unit
DFD	data flow diagrams
DPMA	Data Processing Management Association
DSS	data support system
EBCDIC	expanded binary-coded decimal interchange code
EDP	electronic data processing
FIFO	first in, first out
FMS	flexible manufacturing system
HIPO	hierarchy/input/output/processing
IOCS	input/output/control system
IRG	interrecord gap
ISAM	indexed-sequential access method
LIFO	last in, first out
JCL	job control language
MICR	magnetic ink character recognition
MIS	management information system
OCR	optical character recognition
OMR	optical mark recognition
PERT	program evaluation review technique
RFB	request for bid
UPC	universal product code
VDT	video display tube
VM	virtual memory
VSAM	virtual storage access method
VTOC	visual table of contents (developing logic plans)
VTOC	volume table of contents (disk storage)

Introduction to Computer Systems and to Systems Analysis and Design

Dow Corning

Dow Corning Corporation is an originator and world leader in the field of silicones and one of the major producers of semiconductor materials. The company was formed in 1943 by the Dow Chemical Company and Corning Glass Works to develop the commercial opportunities of this unique family of materials. Dow Corning products range from heat shields for spacecraft to surgical devices which include pacemakers, finger joints, and hydrocephalus shunts. These small rubber drain valves have saved nearly a half million children from brain damage or death resulting from an abnormal accumulation of fluid in the cranial cavities.

Corning operations literally span the globe. With approximately $800 million of annual sales, Dow Corning employs over 5000 people. Our host, Stanford Rowe, describes Dow Corning as a "small, big company."

EDP AT DOW CORNING

Figure I-1 illustrates the organization of the System and Information Management (SIM) Department at Dow Corning. The director of SIM has never been someone from within the department. The person selected, however, has always been from within Dow Corning since one of the basic requirements for the position is a broad prospective of the corporation and its objectives and goals.

Long range goals and objectives for SIM are determined by the System Management Board. The board meets once a month and its membership is made up of vice-presidents from the various functional areas. A second board, the System Planning and Review group, meets once every two weeks and reviews projects, determines priorities, and develops short-range plans. This

The award-winning corporate headquarters of Dow Corning is located in Midland, Michigan.

board is lead by the manager of Application Planning and Development and its membership is made up of individuals appointed by the vice-presidents of the various functional areas. Although the individuals on this board are from the functional areas they are "systems-wise"—knowledgeable in the informational needs of their areas and in data flow.

Application Planning and Development

Application Planning and Development is the department that does traditional analysis and design. Everyone who works within this group does both systems analysis and programming. A new employee hired by the department might start as a programmer and then advance to a programmer/analyst and could eventually become a project leader. Applications are developed by a project team that has from five to eight members. Most teams have one or more users who are assigned full-time to the project. If the user won't commit to putting someone on the team full-time, the project will not be assigned a priority and the application will not be developed.

Other than users, all team members do both analysis and programming. Senior team members do more analysis and less programming while the junior members do more programming and less analysis. The tools used by department members include PL/I (Programming Language One), IMS (Information Management System software), TELON (a frontend productivity tool), EASYTRIEVE (software used to generate reports), and SAS (Statistical Analysis System software). SAS is used to analyze data and generate special reports.

Within this department are portfolio analysts who specialize in the application used by functional areas such as marketing or research. These individuals maintain the application software for their particular functional area and may have been on the project team that developed the software.

Computer Services

Computer services includes the functions performed by the computer and peripheral operators, network operators, data control clerks, and schedulers. The people in this group run the computer and network. A computer operator typically monitors 18-20 batch jobs that run concurrently. At the same time 200-300 users may be using the IMS software and another 200-300 users may be writing programs, creating graphics, or

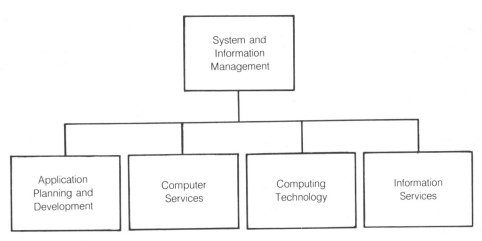

```
                    ┌──────────────────┐
                    │    System and    │
                    │   Information    │
                    │   Management     │
                    └──────────────────┘
```

FIGURE I-1   The System and Information Management Department at
Dow Corning Corporation.

utilizing the office automation software. The operator has the responsibility to see that the computer system functions efficiently and that maximum throughput is achieved.

Although data entry operators enter the data used by the batch system, it is considered to be an end-user function and is no longer a part of the services provided by SIM.

Dow Corning's Computer System

Both the IBM 3083-J, with 32 MB of memory, and the IBM 3083-BX, with 24 MB of memory, are monitored by the operator. During peak load periods the scheduler assists the operator. Attached to the computers are disk drives with nonremoveable disks that hold over 50 billion characters of data. Most of the reports are printed on an IBM 3800 printer which can produce anywhere from 10,000 to 20,000 lines per minute. A 1200 line per minute line printer and numerous letter-quality printers are also part of the hardware supported by the two computers.

It is interesting to note that Dow Corning's computers now have more main memory than the total amount of disk storage available for their IBM 360 third-generation computer which

Frank, the day shift computer operator, monitors as many as 18-20 batch jobs that are running concurrently with work that is initiated by network users.

LaDonna DeShone responds to a call from a manager in Canada who wants a job canceled from the queue.

3

Stanford Rowe, Manager of Computing Technology, frequently uses the terminal or the newly developed voice system to send messages to other network users at remote locations throughout the world.

was obtained in 1966. The 1984 figures also indicate that compared to Dow Corning's 1976 computer system the amount of computer power has increased by a factor of 90 and the performance per dollar of investment has increased by a factor of 15. Due to the increase in performance, statistics show that the cost of producing many of the reports is only 1/2 of the 1968 unit cost.

Uninterruptible Power Supply System Because so many functions are dependent on the availability of the mainframes, Dow Corning utilizes an Uninterruptible Power Supply (UPS) system. Under the UPS system the computers function 100% of the time on batteries that are under continuous charge from the public service company. In the event of a power failure, inhouse generators immediately take over the battery charging function. This eliminates the possibility of momentary current failure or of energy spikes or fluctuations that can be harmful to the computers and the peripheral equipment. If necessary, the batteries could sustain the computer operation for up to 20 minutes.

Telecommunications While traditional computer networks have terminals attached, Dow Corning is moving toward establishing a network with both big and small computers attached. Presently the network has 1,200 VDTs, 300 printers, and 80

remote job entry (RJE) stations. Tied into the network is an IBM 3033 computer located in Brussels, Belgium and also an IBM System/34 computer located in Canada.

The global network is controlled by an IBM 3725 telecommunications unit and is monitored by a telecommunication operator. Jobs are logged and monitored. If a hardware problem occurs, the problem is recorded and the telecommunication operator contacts a vendor who will solve the problem.

Computing Technology

Stanford Rowe, our host, has been with Dow Corning since 1966 and is currently the manager of the Computing Technology Department. Within Stan's department are two major areas of responsibility. The first is to provide technical support for the system software—maintaining the compilers, operating system, sort programs, utilities, and other major packages.

The other half of his responsibilities include voice and data telecommunications. Voice message capabilities were added about fifteen months ago and a separate computer was obtained to manage the system. Stan mentioned that when he was in Sydney and Hong Kong he could send voice messages to his staff in Midland and receive their answers as easily as if he were in his own office.

The librarian is responsible for maintaining the books, papers, and other documents stored in the technical, business, legal, and tax libraries.

Information Services

The Information Services Department is assigned two major functions—maintaining Dow Corning's libraries and end-user computing. The library was placed under Information Services as it is viewed as providing another type of information that should be managed.

The computerized library management system permits the librarian to perform an *any-word* search of the documents stored online. Online storage includes abstracts of many books, articles, and Dow Corning technical publications.

End-user computing includes interactive computing, and office automation. The end-user computing system is controlled by a host computer which handles both the interactive computing and office automation. The host computer supports standard VDTs, VDTs that provide colored displays, and IBM displaywriters for users who need word processing capabilities. However, the displaywriters will be replaced with regular IBM PCs that will be controlled by displaywriter software. The specifications for the equipment to be used for office automation and the PCs (personal computers) are developed by a subcommittee composed of individuals that are from SIM and users.

The major software package used for office automation is PROFS—PRofessional OFfice Systems. The software permits users to send messages, create documents, set up calendars, and schedule meetings. End users also use SAS, spreadsheet, and graphic software. At the present time there are 600 registered users of PROFS. To become a user an individual must obtain the permission of his or her department or system manager. Classes are provided to help users learn to use the terminals and the PROFS software. In addition there are 1,300 registered timesharing users who access the system by using TSO (Time-Sharing Option) software. New users will probably elect to use PROFS since it is supported by the "help desk" and is more user friendly and easier to learn than TSO.

It is anticipated that within two years every Dow Corning employee other than those who are directly involved in production will have his or her own terminal or PC. In addition some employees, such as technical support personnel and portfolio analysts, will also have a dial-up terminal in their home.

SUMMARY

Dow Corning has a policy requiring that any equipment used for electronic data processing functions or for office automation must have the capability of being tied into the telecommunications network. Although the PCs (microcomputers) may function as standalone computers and run word processing, spreadsheet,

Cynthia Studebaker operates the "help desk" and assists users who are having difficulty using PROFS.

or other software, the capability to function as part of the network must exist. Standards and guidelines for the acquisition and utilization of PCs are constantly being studied and revised.

Although both data processing and data entry functions are decentralized, Dow Corning has taken a strong stand on having centralized control and *one* corporate network. Functional areas perform their required computing tasks through the corporate network rather than building their own network.

In order to support the long-range objectives the computer system will need to be expanded, on a year-to-year basis, by upgrading the mainframes and adding more real memory, disk storage, printers, and terminals. Since SIM has a backlog of requests that would take from three to five years to implement, as each application is approved a decision is made regarding how much can be done by the users and what must be done by Application Planning and Development personnel.

By using SAS, spreadsheet software, and a database language such as SQL, end users can generate their own statistical and customized reports. Application Planning and Development personnel still develop the basic systems that are key to operations. These systems produce the standard and control reports. Control reports are necessary to safeguard the data and to protect its integrity. The systems also build fields and databases that can be used by the end users to produce statistical information and additional customized reports.

One of the measurable goals of SIM is to increase the productivity of the office worker by 20 percent. Another ongoing activity is the development and review of guidelines and standards that can be used in the analysis and design of systems.

When activity analysis occurs, jobs are restructured in order to utilize the computer more effectively.

Although Dow Corning has long been committed to the utilization of computers, networking, telecommunications, and centralized control, new applications must be cost justified and system audits are conducted to see if the goals and objectives for each new system have been met.

1

AN OVERVIEW

Looking ahead

After reading the text and completing the learning activities you will be able to:

- Identify the role of the analyst in systems analysis and design.
- Describe the major functions performed in each of the six phases in the systems development life cycle.
- Identify the difference between a measurable and nonmeasurable objective.
- Explain the reasons why a computerized system might not meet its objectives and might be considered unsuccessful.
- Identify the impact minicomputers had on computer utilization.
- Identify the impact microcomputers had on computer utilization and systems analysis and design.
- Identify the differences between online and batch applications.
- Describe the five components of a computer system.
- Identify the functions performed by the two major types of software.
- Describe why operating systems were developed and have become an important component of a computer system.
- Identify the three basic functions performed by operating systems.
- Identify the four major types of operating systems.
- Explain how the features incorporated into the operating system affect the design of a computerized system.
- Identify the features of the operating system that are of concern to an analyst.
- List the factors that are considered when determining whether a computer should be classified as a microcomputer, minicomputer, or mainframe.
- Identify the characteristics of the computer system utilized by Computer Products Incorporated.
- Define and utilize the words and phrases listed in the end-of-chapter glossary.

INTRODUCTION

As you read the brochure from your bank which describes the new bank-at-home service available to individuals who have **microcomputers***, you might wonder how such a complex application was designed. Who requested that the application be designed? Who was involved in its design and implementation? Were any individuals contacted to see whether they really wanted to use their personal computers to bank at home? How many people will take advantage of the service? What **hardware** and **software** is needed in order to implement the service? Will the revenue received from individuals who use the service offset the cost of its development and ongoing expenses?

A complex application such as the bank-at-home service is called a **system**. If a **top-down** approach is used to systems analysis and design, a complex system is divided into small, manageable parts. The system is first divided into subsystems. Each subsystem is divided into **procedures**, which are precise step-by-step methods of effecting a solution. Each procedure is made up of one or more **tasks**. Each task can be performed in a relatively short period of time. As you learn how computerized systems are designed and implemented, you will perform many of the small, but important, tasks that must be accomplished.

In the design and implementation of a system, the major focus must center on the needs of the users. The users are the individuals or departments that will prepare or input data and use the resulting output. Users must be directly involved in the design and implementation of a system. Although it is necessary to have the right hardware to implement the system, it is the design of the software that is critical to its success. It would be pointless to have state-of-the-art hardware and run programs designed for older, less sophisticated computers.

A PAYROLL EXAMPLE

Assume that the payroll manager wants a new payroll system. The payroll manager, the user, initiates a request for a study to determine whether a new payroll system is needed. If one is needed, early in the design process, the payroll system is divided into subsystems. Although the task analysis should produce the same results regarding the tasks needed to produce the desired results, there is no one correct solution regarding how a large, complex system should be divided into subsystems or procedures.

File Maintenance
Subsystem

The subsystem required to maintain the payroll master file is illustrated in Figure 1-1. Since the master file supplies so much of the data used in the payroll programs, the data must be current and correct. Therefore, it is necessary to add new records to the file, delete old records from the file, change data stored within records, and make corrections for past mistakes. Also because files are the life blood of a system, *files must be protected*.

The file backup procedure provides for executing a program that copies the payroll master file to a backup file. The tasks involved in executing the backup procedure might be to: mount a magnetic tape on a disk drive; initiate the program that will copy the master file records stored on disk to the backup file stored on tape; dismount the tape; label the tape; and store the tape in a vault.

Task Analysis

The tasks defined for the backup procedure might seem rather insignificant. However, a successful payroll system will be made up of thousands of small tasks organized into procedures that will be used to input data, process the data into

---

*Terms set in boldface are defined in the end-of-chapter glossary.

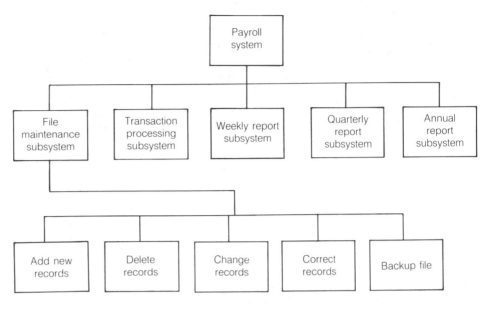

Backup Procedure: Tasks required
1. Mount tape
2. Initiate file copy program
3. Dismount tape
4. Label tape
5. Store tape in vault

FIGURE 1.1. Complex systems are divided into subsystem, procedures, and tasks

meaningful information, and either print or display the output. In designing a system, each required task must be identified. It is critical that each person perform his or her assigned tasks accurately. Everyone involved in a system—from the data entry operator to the payroll manager—must do their part accurately and when needed, if the payroll system is to be successful. Successful systems can have no weak links.

THE ROLE OF THE ANALYST

An **analyst** investigates a problem and then designs a solution. The analyst must determine the unique needs of management and devise a system that meets those needs. An analyst's job is not confined to working with computers and equipment related to **electronic data processing** (EDP) functions. Analysts work with many different people and many different types of systems. Their functions were defined long before computers were used to process data. When companies turned to computers to process the avalanche of data required to carry on an organization's activities, the need for systems work increased. The same basic techniques used for designing a simple procedure are used for designing a complex system such as the one that provides the bank-at-home service.

Analysts perform the required task analysis and explore methods that can be used to solve the problems identified. Since computers are used extensively in processing all kinds of data, in information retrieval, in office automation, and in different types of manufacturing systems, the role of the analyst has increased. The analyst's job is also more demanding and challenging than it has been in the past. A few years ago we would not have envisioned analysts working with senior citizens (users) to design robots that will allow elderly people to be more self-sufficient. Today that is just one of the expanding roles assumed by analysts.

The complexity of an analyst's job is partially due to the wide range of choices that now exist. When computers were first used in business and industry, few choices existed. There was a limited number of computers from which to choose. The available computers had a limited amount of control software, and most computer systems performed the same basic functions. Words such as **networking, databases**, and **teleprocessing** were not part of the basic vocabulary used to describe electronic data processing. Applications were batch oriented and there were few I/O devices from which to select. The computer processed one batch of data and then a new program was loaded onto its memory and the computer was given a new assignment. The computer and its I/O devices were securely tucked away within the data processing department. Few people had access to the computer. Also, few people wanted to work with or have access to the computer. Most users were not knowledgeable about what computers could, or could not, do. Therefore, users provided little input into the many decisions that must be made to create a system. Today most users are knowledgeable about what computers can do and what type of tasks computers should be assigned.

The analyst must be familiar with all of the choices that exist today. Can a microcomputer do the job or is a complex computer system that costs millions of dollars needed? What type of input and output devices can be used most effectively? Where should the data be captured and entered into the system? How can computers and online databases be protected from unauthorized use? Should the software be designed and written inhouse or are there software packages available that can process the data and provide the required output? As you progress through the text, you will learn about the techniques used to design complex systems that must meet the challenges of the future.

Many individuals in various occupations can profit by learning some of the basic techniques used by an analyst. Managers, accountants, auditors, and the users of computerized systems must understand how their roles relate to that of the analyst and to the systems that are designed. How well a computerized system serves management is often determined by how well the analyst studies the existing system and determines the tasks involved in processing the data.

Functions Performed by Analysts

Although the role of the analyst is constantly expanding, the functions performed by an analyst today are about the same as those performed when manual systems were used. The functions performed are:

- *To analyze existing systems and to design new ones or modify the existing systems*. The type of system designed will vary depending on the hardware and software available. The computer system (hardware and control software) being used may prevent the analyst from designing the ideal system. A payroll system designed for a microcomputer might be very different from one designed for a large computer capable of serving the needs of hundreds of users simultaneously.
- *To design forms used to collect data and to distribute information*. Since terminals and transaction processing are so widely used today, some forms are designed to provide part of the **audit trail** rather than to be used as a **source document**.
- *To analyze the distribution and use of reports*. One of the basic questions that must be answered is "Does the report provide management with the information needed?" The analyst also must determine whether a full report or an exception report is needed and if hardcopy (a printed

report) is needed or the information can be displayed on a **VDT (video display tube or CRT)**.

- *To administer the creation, use, and retention of forms (records management).* Today fewer forms are used and more information is stored in **online** or **offline files** in a format that makes it possible to process the information electronically. However, laws do require a clear audit trail that is supported by documents that must be retained for a specified period of time.
- *To participate in the measurement and simplification of work.* One of the primary objectives of computerizing a task or procedure is to make the worker more productive and to simplify the task being performed.
- *To be aware of the total environment of the worker in order to increase productivity and to ensure the safety of the worker.*
- *To make certain the design of the system is compatible with the organization's goals and objectives.*
- *To document all procedures according to the organization's standards.* Today there are **standards** developed preparing **documentation** for both manual and computerized procedures.

Expanding Role

Analysts must be concerned with an increasing number of factors such as the conditions under which employees work and the impact the worker's environment will have on his or her physical and mental health. Federal regulations, company policy, and union contracts are also considered in designing a system.

Many articles have been written on **ergonomics**—the study of the relationship between humans and machines. Questions such as how comfortable or how safe the person will be when working with machines must be answered. Analysts must be aware of the impact the systems they design will have on our life-style and the conditions under which meaningful work is performed.

Before exploring the techniques used to modify existing systems and to design a new system, it might be wise to determine what phases or steps are involved in analyzing the problem and designing a solution.

SYSTEMS DEVELOPMENT
LIFE CYCLE

The six phases in the **systems development life cycle** shown in Figure 1-2 on page 12 can be identified by different names. Also, *there are no definite rules regarding what must be included in each of the six phases.* Figure 1-3 on page 13 illustrates the names used in this text, other names used for an entire phase or portion of a phase, and the major activities performed in each phase. Although established guidelines and standards exist, each system study is unique. The terminology used in this text to identify the various phases of the systems development life cycle are initial investigation, feasibility study, general design, detailed design, implementation, and system audit. Since the system audit is ongoing and because at some point it will be necessary to redesign the system, the whole process is referred to as the systems development life cycle. Most systems have an anticipated life of from three to five years.

Figure 1-2 is an overview of the phases, or steps, involved in designing and implementing a new procedure or system. Analysts cannot design a task, procedure, or system unless they understand:

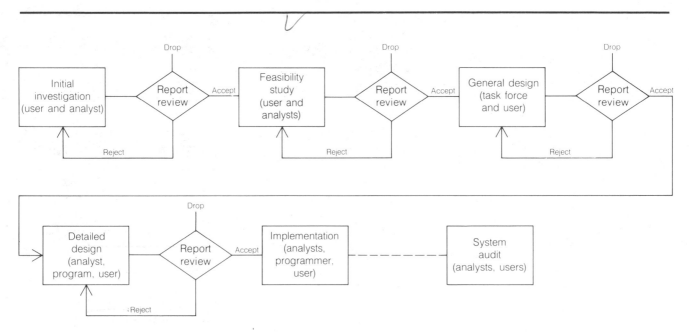

FIGURE 1.2. Phases in the systems development life cycle

- the organization and its personnel practices; and
- the characteristics of the computer system's hardware and **operating system**.

An operating system is the control software that makes it easier for us to communicate with computers and to implement complex applications.

Another major factor in determining the design of a new system is the commitment that the organization is willing to make in personnel, time, dollars, and equipment to implement the system. Often the ideal solution to a problem cannot be implemented due to constraints imposed by top management. Typical constraints are:

- no new personnel can be hired:
- the system must cost less than a given sum;
- existing hardware and control software must be used; and
- a limited amount of money can be spent to investigate the old system and to design and implement a new system.

Initial Investigation

An initial investigation is conducted to determine the scope and nature of the problem. The investigation may be made at the request of either the user or management. A limited amount of time is spent on the initial investigation. The report submitted upon completion of the investigation should provide a clear definition of the problem and an estimate of the time and resources needed to complete the feasibility study. If the report is acceptable, the project will be assigned a priority and the feasibility study scheduled. If the report is unacceptable, the project may be dropped or postponed, or further investigation scheduled.

As you examine Figure 1-2 you will note that each of the first four phases of the systems development life cycle results in a report that is reviewed to determine what action should be taken. Upon the recommendation of the committee or individual that reviews the report, one of the actions shown in Figure 1-4 will occur on page 14.

Text Name	Other Names	Major Functions
Initial Investigation	Preliminary Investigation	Determine nature and scope of the problem.
Feasibility Study	Detailed Investigation Situation Analysis Business Requirements New System Requirements	Study existing system Study alternatives Recommend course of action Prepare estimates of costs and benefits
General Design	System Design Functional Design	Design a system from the users' point of view Make a commitment to the type of system to be developed Make decisions regarding personnel, hardware, and software needed
Detailed Design	System Development Technical Design	Determine technical design of how the system will function on the computer Design databases, files, reports, VDT screens in detail
Implementation	Programming and Testing System Installation Cutover/Turnover	Develop detailed logic plans Write and test programs Convert files Develop operational procedures Train personnel Test system
System Audit	Evaluation Review	Compare results with objectives

FIGURE 1.3. Major functions performed in the six phases of the systems development life cycle

Feasibility Study

Since it is necessary to study the workflow, existing system, and alternatives, more people are usually involved in the feasibility study than in the initial investigation. The feasibility study report should detail the alternatives and provide preliminary cost and benefit estimates. The report should also estimate the time, personnel, and other resources needed to design, develop, and implement the recommendations.

If, after the feasibility study is completed, the recommendations indicate that only minor modifications to the existing system are needed, the recommendations might be accepted, the project assigned a priority, and the implementation scheduled without going through the general design or detailed design phase of the cycle. If the feasibility study recommends that specific application software be purchased, when the software is obtained, the implementation phase of the project can begin.

General Design

A design is created that clearly shows the users' view of the application. During this phase, as well as the two previous phases, users must be directly involved. In the design phase, the analyst's creativity, imagination, and initiative are used

Action taken	Reasons	Outcome
Next phase scheduled	Recommendations accepted for implementation	Detailed plan for implementating next phase developed
Recommendation and report rejected	Inadequate information Lack of resources Disagreement regarding recommendations Further study needed Technological changes have created additional solutions that should be studied	Problem and recommendations restudied
Project dropped	Minor problem is solved Lack of resources No apparent solution	Solution implemented Study postponed When technology changes the problem will be restudied

FIGURE 1.4. Actions that may occur when a report is presented

to their fullest. Most of the decisions regarding the final outcome of the project will be made at this time. A **task force** made up of people with varying talents is usually involved in the design of the application.

The task force should explore different possibilities before determining the design of the system. Based on the desired results, procedures are outlined for obtaining and processing the required input. What files or databases will be needed, how the variable data will be entered into the system, and the format and content of the reports and visual displays on VDTs must also be determined.

System flowcharts, which give an overview of the system and its procedures, will usually be developed. In addition, broad specifications that describe how the data is to be processed by the computer will be developed.

The final report for the general design phase will include the system flowcharts, input and output requirements, and the general specifications for the programs that will be needed to implement some of the procedures. Cost estimates made during the feasibility study may need to be revised since the solution to the problem has been defined more clearly. At this point, the projected cost for the system should be fairly accurate.

Detailed Design

Interaction among members of the task force, users, and representatives from management resulted in the general design specifications. Although the requirements were identified, a detailed study of how the design would work on the computer was not made. Technical creativity is needed to translate the general design specifications into detailed requirements that can be used in implementing the many tasks that make up the system.

For example, assume that for the system being designed, both text and graphics are to be retrieved from a storage medium and displayed. The general design specifications indicate that **videodisks** are to be used to store both graphics and text. Videodisks are plastic platters resembling phonograph records. Low-intensity laser beams are used to store data on the disks. The analyst must use ingenuity to adapt the newer storage medium to the application.

The final report should include **procedural flowcharts**, record and report layouts, and a realistic plan for implementing the design. Detailed information should be available regarding the resources (personnel, materials, machines, methods, money, and time) needed to implement the system. Earlier cost estimates may need to be revised. At this point, the projected cost should be very close to the actual cost of implementing and running the system.

If the detailed design is approved, no major changes should be made until the design is implemented and the system is **operational**. When a system is operational, the programs for both **transaction processing** and **batch jobs** are executed. If problems occur, the professional data processing staff is responsible for determining the cause of the problem and implementing a solution.

Implementation

Detailed logic plans must be developed for all programs before they can be written, tested, and documented. After each program and procedure is tested, the entire system is tested. The conversion to the new system is made according to a plan developed in the detailed design phase.

While the programmers are writing and testing programs, users can be documenting procedures and creating additional test data. There must be constant communication between the project team working on the implementation and the users for whom the system is designed. Many companies feel it is a good idea to have users document their own systems. This practice keeps the users directly involved. Since the documentation will be employed within the user's area, there may be a greater incentive to do a better job than when a member of the data processing staff is assigned to do the documentation.

The final report contains the documentation that tells what occurred during implementation of the system. Also included is the formal documentation that contains information regarding how the system is to be used. The cost of implementing the system should be compared with the estimates.

System Audit

When the implementation report is submitted, an evaluation should be made to determine whether the system meets the objectives stated in the general design report. Unfortunately, some companies stop at this point and do not conduct an ongoing evaluation of the system.

The actual system audit should occur about six months after the system is considered operational. During the six-month period prior to the system audit, any problems that occur should be documented. Flaws in the design, such as inadequate **editing** or **controls**, may also be detected. Editing routines are developed in order to determine the validity of the data being processed. Controls are built into the system to determine whether the output is valid. During the system audit, users may be able to suggest easy-to-implement improvements.

Termination of the Project

As indicated in Figure 1-2, the project can be dropped at any point prior to implementation. It is unlikely that it will be dropped after the general design phase is approved. At that point, realistic objectives and costs should have been developed. The major commitments to the system should also have been determined.

A project may be dropped if the benefits derived from the proposed system do not justify commitment of the needed resources. Sometimes the user's budget will not permit the continuation of the project if the costs are higher than the estimates made during the detailed investigation. In other cases, the resources needed are not available and the company cannot justify the additional costs. Perhaps until a new computer system is acquired, there is not enough CPU (central processing unit) time available to implement additional applications that require a large amount of computer resources.

Because ongoing operational costs cannot be justified in terms of benefits received, a system may be scrapped after it is operational. This may be the result of a poorly designed system or inaccurate operational cost estimates. As analysts, programmers, management, and users gain more experience in developing complex computerized applications, more accurate estimates can be made. Standards should also be developed that improve the quality of all phases of designing and implementing applications.

CHECKPOINT

1. Where is a word set in boldface type defined?
2. Where can you find the answers to checkpoint questions?*
3. What is the difference between a system, a procedure, and a task?
4. What is a top-down approach to systems analysis and design?
5. What is a batch system?
6. What information is graphically shown in a system flowchart?
7. In one sentence, describe the role played by an analyst.
8. Why is there a greater demand today for the services performed by an analyst than there was ten years ago?
9. Why is an analyst's job more complex today than it was ten years ago?
10. What are the six phases involved in the systems development life cycle?
11. What is ergonomics?
12. What is an operating system?
13. What are some of the reasons that make it impossible to design and implement an ideal solution to a problem?
14. What might occur when a phase-end final report is given to the computer policy committee?
15. Why might the recommendations of an analyst be rejected?
16. What are the primary objectives of the initial investigation?
17. Does the general design reflect the user's view of the application or the analyst's concept of the problem and its solution?
18. What is a task force and why it is wise to use one in the completion of the general design phase of the project?
19. How is data stored on a videodisk?
20. During the implementation phase of the project, what tasks might be accomplished by the users?
21. What is the purpose of the system audit?
22. If the system audit shows that it meets its stated objectives, is a system ever scrapped?

---

*Answers to checkpoint questions are listed at the end of the book, beginning on page 560. Often new material is presented in the answers that will increase your understanding of the information covered in the text.

WHY SOME SYSTEMS DO NOT MEET THEIR OBJECTIVES

Today a dollar buys about 20 percent more computer power and online storage than it did a year ago. Compared to the cost of processing data five years ago, the increase in cost/performance is remarkable. It is anticipated that in another five years a dollar will buy twice as much computer power as it buys today. Unfortunately the costs of designing new systems and entering data into existing systems are increasing due to the rising cost of labor. One of today's challenges is to increase the productivity of the workers and the quality of the products produced. By taking advantage of today's technology and past experience in designing systems, data processed by a new system should cost less than the same volume of data processed using previous methods, materials, and machines.

Measurable Objectives

In the early phases of the system design cycle, the objectives for the system should be clearly stated. Some will be measurable while others will not. During the system audit, the analyst can easily determine whether the measurable objectives have been met. For a new sales-order system, the following objectives are measurable:

- Orders will be processed on the day they are received.
- The number of complaints from customers will decrease by 50 percent or more.
- The cost per sales invoice will decrease from four dollars per invoice to less than two dollars per invoice.
- A comprehensive database will be developed that will contain more information than is available in the **master files** used with the present system.
- Since **terminals** and a **natural language** can be used to obtain information from the online database, managers will have additional sales information available upon demand.

Nonmeasurable Objectives

Some of the objectives listed for a proposed sales system might not be measurable. The nonmeasurable objectives, such as the ones listed below may be as important, or more important, than the measurable ones. Nonmeasurable objectives for a sales system might include:

- Improved service to the customer.
- The availability of more reliable information that can be used in the decision-making process.
- Since the data entry function will be distributed, fewer errors will be made in entering the data into the system.

Evaluating a System

Decreasing costs should not be the only criterion for evaluating a system. Intangible benefits such as providing better service to customers and having more relevant information might increase sales. The increase in revenue might more than offset the increased cost of a new sales system. Also it is difficult to compute the high cost of *not* having relevant, timely data. If all of your competitors can access information stored in online databases to answer questions posed by customers, how costly is it when you have to reply to a customer's question with ''I'll look it up in your file and call you back tomorrow''? Within a matter of seconds, some of your competitors may be able to answer most of the questions customers ask. The answer is no farther away than the nearest terminal—and that may be next to the phone on your competitors' desks.

Return on an Investment in Computer Systems

Many large companies commit anywhere from 2 to 4 percent (or more) of their gross sales to EDP. This expenditure is for computer systems (hardware and software) that are committed to far more than processing numeric data. It includes the use of computers to: process financial data, assist in the decision-making function, automate many of the functions performed within the modern office, retrieve information from databases, and assist in areas referred to as **computer assisted design/computer assisted manufacturing (CAD/CAM)**. Companies that commit resources to obtain computer systems expect to be able to show that their investment in computers and related equipment yields as high a return as it would if the resources were invested in any other phase of the business.

An Unsuccessful System

A system audit conducted six months to a year after the system is operational may show that the system does not meet its objectives. Assuming the five measurable objectives listed on page 17 were the only ones taken into account, the system would be considered unsuccessful if:

- Due to delays in workflow, the sales orders are not processed on the same day they are received.
- Based on the number of orders received, there is an increase in the number of complaints from customers.
- The cost per sales invoice increased rather than decreased.
- Less data about customers and products is available in the online databases than had been available in the master files.
- The terminals and natural language available are not being used to obtain additional sales information. Managers still depend on analysts and programmers rather than using the resources available.

The ongoing evaluation might also show that **data exceptions** occur when numeric data is not edited and nonnumeric data is found in a numeric field. The report might also indicate that operations personnel have trouble running some of the batch jobs.

In summarizing the types of complaints received from customers, it might be found that the wrong merchandise was shipped, orders were shipped to the wrong customers, and many items were **backordered**. Orders are placed in the backorder file when there is not enough merchandise on hand to fill a request. Although the analyst who designed the system might not feel accountable for the items backordered, the problem may be in the design of the inventory control system. During the design and implementation phases of the systems development life cycle, the impact of the sales system on the inventory control system should have been considered.

Reasons for Failure

There are many reasons why a new system may not measure up to its objectives. If problems are identified for an operational system, it may be possible to make changes that will correct the problems. In the past it was expected that it would be necessary to correct operational programs. When better design techniques are used, it should not be necessary to make corrections to operational programs. Analyst should profit from past mistakes and make certain that a new system is free of errors and meets its objectives before it is considered operational. Some of the reasons new systems are sometimes unsuccessful are summarized in Figure 1-5 on page 20.

Changing Failure into Success

The list of failures summarized in Figure 1-5 can be stated in a positive manner. If the list of suggestions illustrated in Figure 1-6 is followed, yesterday's failures may be turned into tomorrow's successes. During the system audit, the techniques used in studying the old system and designing the new one should be examined to determine whether existing standards regarding systems analysis and design were followed.

The list illustrated in Figure 1-5 could be expanded to include many more items, and each of the items could be expanded to include subtopics. Although there have been many advances in computer technology, the success of a new system is directly related to how carefully the old system is studied and the new one designed. Input regarding the needs of management must be obtained early enough to have a direct impact upon the design of the system.

An analyst cannot design a successful system alone. Users and management must be directly involved in the feasibility study and in the design phases of the new system. The analyst needs the assistance of many individuals within the company. Although the analyst must understand the capabilities of the hardware and software, additional knowledge and skills are needed to design and implement a successful system. It is also possible to develop a system that meets the objectives stated at the beginning of the project only to find that changes in environmental conditions require a change in objectives.

CHECKPOINT

23. Assume that you are designing a new payroll system. Identify one measurable objective and one nonmeasurable objective.

24. Based on your report to management, it is anticipated that the new sales-order system you are designing will cost more than the old system. When you give your verbal report to the computer policy committee, what reasons would you give for implementing a system that is more costly than the present system?

25. The system audit for the new sales-order system indicates that it should be considered unsuccessful. If you were the EDP manager, why would you want to determine how the system had been developed? What impact might your findings have on the development of future applications?

26. Is the computer and its I/O devices the most important factor in determining whether a system will be successful or not? Justify your answer.

A BASIC UNDERSTANDING OF TODAY'S TECHNOLOGY

Before learning more about systems concepts, how systems are studied and designed, and the role management, users, and analysts play in the development of a system, we should determine:

- Why computers are used so extensively in business and industry.
- What types of computer systems are available and what impact the type of system being utilized has on the design of the application.
- Why there is so much concern over office automation and the impact it will have on EDP.
- More of the basic terminology associated with systems analysis and design.

Causes of Unsuccessful Systems

- Users and management were not directly involved.

- Objectives were not clearly defined.

- A top-down approach to systems design was not used.

- Not enough time was spent in the feasibility and general design phases.

- The analysts and programmers were inexperienced.

- Standards regarding systems design were not followed.

- The workflow of the old system was not studied.

- The impact of the new system on existing subsystems was not studied.

- An old application was implemented, without being redesigned, on a new computer system.

- The computer system used was inadequate.

- The analysts and programmers did not have an in-depth understanding of the computer system.

- The new system was poorly documented.

- Personnel involved with the system were inadequately trained.

FIGURE 1.5. Why new systems are considered failures after they are implemented

Computer Utilization

When computers were first invented, no one envisioned the impact they would have on society and how extensively they would be used. Even Thomas Watson, one of the founders of International Business Machines Corporation (IBM), felt that only major universities and the government would utilize computers. Five years ago who would have predicted that soon the computer would be as commonplace in business and *in the home* as the telephone? Would you have predicted five years ago that architects would consider computer utilization an important factor in designing new office buildings and *homes*?

As computer technology advanced, the cost of computing and storing online data has steadily decreased while the computational ability of computers has increased. A few years ago only large corporations, major universities, and governmental agencies could afford their own computer systems. Today a computer is affordable by any company and by many individuals. As small "mom and pop" type companies change from manual methods to computerized data processing, there is a tremendous impact on the computer industry and the role played by the analyst.

Checklist for Successful Systems

- Direct involvement of users and management.

- Define objectives for each task, procedure, and system clearly.

- Use a top-down approach to system design. Determine objectives first for the system, then for the procedures, and then for tasks.

- Allow adequate time for the investigation and design phases of the system development life cycle.

- If experienced analysts and programmers are not available, hire a consultant or contract programmer. Investigate purchasing software that is already developed and tested.

- Develop standards that must be followed when systems are designed.

- Study the entire workflow related to the system.

- Integrate the design of subsystems into the development of major systems.

- When a new computer system is obtained, redesign the existing applications to take advantage of technological changes.

- Make certain the computer system that will be used to implement the design is adequate. There must be "left over" computer power that will take care of normal growth and unusual expansion.

- Make certain the analysts and programmers have an in-depth understanding of how the computer system functions. They may need to attend vendor schools or receive other types of on-the-job training.

- Document all phases of the project and the system.

- Provide adequate training for all individuals who will be directly involved with the new system.

FIGURE 1.6. Considerations that must be followed to make certain a new system will be successful

Traditional Applications   Traditionally, when a company acquired its first computer, the applications assigned to the new system were accounting oriented such as payroll, sales-order processing, inventory control, accounts receivable, accounts payable, and general ledger accounting. Although it was clearly established that computers could do far more than the recordkeeping functions, by the time those applications were up and running it was necessary due to company growth or a greater dependency on computers, to get a larger, more powerful computer system. The new computer was then assigned the same tasks as were run on the old computer.

Often the new computer had to be reprogrammed to accomplish the tasks previously assigned to the old system. Little time was spent on the design of the applications, and consequently, many systems performed inefficiently. Also, converting old systems to new computers left little time for the development of new applications. **Centralized data processing** was the norm and it was felt that the computer had to be used two or three shifts a day to be cost-justified. The applications typically assigned the computer usually did not challenge its computational ability. Few people within the organization had any direct contact with the computer and even fewer wanted to get directly involved with what was seemingly a complex, difficult-to-communicate-with, unfriendly machine.

Impact of Minicomputers   Until the powerful but relatively inexpensive **minicomputer** systems became available, it was difficult to cost-justify the use of a computer system for nonfinancial applications. Unless the tasks were repetitive and had a large volume of input, an application was considered unsuitable for computerization. Computer time was considered too costly for many applications.

When minicomputer systems were developed that could be programmed in **high-level programming languages**, it was no longer necessary to worry about how fully a computer was utilized. The cost per hour of computer time decreased to a fraction of the cost of using medium- or large-size computers. Often a minicomputer system was acquired and dedicated to performing one specific function. Other minicomputers were placed within users' departments, and some of the data processing functions were distributed. The benefits derived from using minicomputers for applications still had to be cost-justified. The benefits had to offset the costs of designing, implementing, and maintaining the new computerized application.

The minicomputers placed in the user's departments had to be able to communicate directly with the centralized corporate **mainframe**. The databases developed for the minicomputers were also accessed by the centralized mainframe. The need for such compatibility caused the whole area of standards regarding teleprocessing and networking of computers to be studied. Some of these problems are not yet resolved.

Attention was focused on the much-debated question of centralization versus decentralization as it relates to information processing. Large companies are now moving toward decentralization of equipment and personnel. There are many arguments in favor of retaining centralized control over the development and management of decentralized systems. Major companies have developed standards and guidelines regarding the use of microcomputers, minicomputers, and mainframes. Although the term mainframe is used to mean the actual central processing unit (CPU), when used in the previous sentence it should be defined as a large, centralized computer system that has a extensive online database and supports a wide range of applications. Many companies now use the term **information processing** to cover both the traditional data processing applications and office automation functions. Well-organized companies also insist that microcomputers, minicomputers, mainframes, and other related office-type equipment communicate with each other and share databases.

New Applications   Many new applications are now being developed. Computers are now used in CAD/CAM and well as in the development of **flexible manufacturing systems (FMS)**. In an FMS, equipment can be used to manufacture more than one product. The manufacturing systems can be programmed to alter their procedures to suit production requirements. Developments in the field of **robotics** have affected manufacturing processes and information processing. Today the entire environment of the factory and office may be controlled by a computer. Other computers monitor machines and keep track of their rate of production, downtime, and maintenance schedules. Still other computers are used for recording the cost of labor and raw materials used in production. Because of the low cost of today's microcomputers, new applications can be designed that require dedicated computers. The small computer system dedicated to a specialized task may gather and process the information. Condensed information or exceptions may then be transmitted to the corporate mainframe.

Just as space exploration has expanded our universe, low-cost, high-performance computer systems have expanded the horizons of the analysts. What once might have been considered as fantasy is now becoming reality. Perhaps the greatest challenge an analyst faces is keeping current in regard to new developments in hardware, software, and the utilization of computer systems.

Impact of Microcomputers   The early microcomputers were used for industrial applications. Hobbyists also delighted in obtaining a **microprocessor chip** and building a microcomputer system. It was not long before the manufacturers of microcomputer systems realized the potential of the mighty microcomputer. Small, compact, and as powerful as many of today's minicomputers, the microcomputer has invaded the home, small business establishments, educational institutions, large corporations, and governmental agencies. At the present time there are over 150 companies that manufacture or distribute microcomputer systems.

The widespread use of microcomputers by all kinds of people, not just computer specialists, has increased concern regarding computer crime and the privacy of information. The question of security regarding hardware, software, and information must be restudied. A recent survey summarized in *Computerworld* indicates that the majority of the corporations surveyed have developed standards regarding the acquisition and use of microcomputers. The results of the survey tend to discredit reports that there is an uncontrolled, widespread use of microcomputers within large companies that have a relatively long history of computer utilization.

In designing a new application, the analyst must determine what type of system is best suited for gathering and processing the data. Another major concern is how the summarized information can be communicated to the corporate mainframe—without duplication of effort. The key issue may be the development of standards and guidelines that require computer systems to be capable of communicating with each other.

When a Computer System Is First Used   When an organization or individual obtains its first computer, the machine is probably going to be assigned recordkeeping and **word processing** functions. The computerized recordkeeping applications should make it possible to process data faster, cheaper, and more accurately than by the previous method. When a computer is assigned the task of word processing, more productivity and higher quality output is expected.

The first applications assigned to a new computer system might still be payroll, sales-order processing, accounts receivable, inventory, and accounts payable. Data is entered into the system, processed, and put into a form that can be easily used by management. Once these applications are computerized and management becomes more aware of the computer's potential, other tasks can be computerized in order to provide additional information and assistance.

A major concern for either a new or veteran user is the *safety and integrity* of the input, output, and information stored in online files. Controls must be built into a computerized application that ensure the safety and integrity of the data.

If a small company or self-employed individual starts with a microcomputer or minicomputer system, it might be impossible to develop the type of system that is desired. The characteristics of the system's hardware and software determine the type of applications that can be implemented.

Establishing Goals for a Management Information System   Medium- and large-size corporations have had computerized sales, payroll, personnel, and inventory systems for a long time. As a result of defining their needs, knowing the capabilities of computer systems, and brainstorming ideas, many corporations have established goals pertaining to the development of a management information system (MIS). A true MIS depends on the availability of a large computer system and a comprehensive online database. While some of the computers described as "superminis" can support a fairly comprehensive MIS, at the present time, such complex applications are beyond the capabilities of many microcomputer systems. When integrated systems and databases are used, it is possible to retrieve more information than when separate systems are maintained. Many companies still have not achieved their goal of developing a comprehensive MIS.

Options Available   Would it be correct to say that only medium- or large-size companies can have some type of computer-based management system? Today small companies, as well as larger companies, have many options. Companies electing not to have their own inhouse computer systems may use terminals and take advantage of **timesharing** services. Many **online applications** are designed in such a way that transactions are entered directly onto the system for processing as soon as they occur. Immediate feedback is available and it is often possible to correct errors that occurred in entering the data.

Many of today's microcomputer systems permit multiple users to access the system. This makes it possible to develop online applications for microcomputers. One of the constraints that must be faced in developing the application is the amount of online storage available.

Other small companies may elect to send a portion of their data to a service bureau for processing. The source documents containing the data to be processed are sent in a batch to a service bureau. This permits a company or individual with a small amount of data to be processed to take advantage of the hardware and software capabilities of a large mainframe. The company or individual is only charged for the resources used—computer time, file space, and use of the printer. The data is processed in a batch mode and the source documents and resulting output are returned to the owner or manager of the company. Although the output is not as timely as it is when a timesharing system is used, many applications (such a payroll) are still well suited to being run in a batch mode. Since employees are paid on a weekly basis, the data can be submitted in a batch and processed. A number of different programs would be run in order to edit (check the validity of) the data, produce reports such as the payroll register and paycheck, and to update the company's payroll master file. In most situations, it is not essential that the data be processed on a day-to-day basis.

Informational Needs

Today for many companies, an online sales system is almost a necessity. As the sales data is captured, immediate feedback should be available regarding the status of the customer's account and the availability of the items ordered.

Many small companies have the same informational needs as larger companies. Today most companies can **cost-justify** their own inhouse computer system. An independent contractor who has two or three employees, a self-employed plumber, the proprietor of the corner grocery store, a dentist, or a family doctor can obtain a computer system that will meet their particular needs. The problem becomes one of selecting the computer system that best meets these needs. Both the hardware and software must be studied to make certain the system is capable of performing the desired tasks. Some analysts have specialized in helping small organizations determine their needs and select suitable computer systems. Today there is a great deal of prewritten software that is designed specifically for small computer systems.

The complex management information system of General Motors or a major university is bewildering to the individual who is just beginning to understand the capabilities of the microcomputer. It seems as if relevant, timely information is produced by merely pushing a button or entering verbal commands. A company such as General Motors, due to its size and the scope of its activities, requires a more complex system than does the sole proprietor of a small business. However, *it is just as important for a small company to have timely information as it is for a large corporation.*

One of the major problems of a large corporation is the lack of communication that often exists among its various components. Improving communications must be one of the major goals of any computerized system. Can you imagine what would happen to General Motors if our communication systems were to revert to the systems used during the Pony Express era?

Although the needs of a large corporation or major university might seem very complex, they are not that much different from those of the small entrepreneur. At the same time, each level of management has its own unique needs. The president of a large organization is only interested in summarized data and trends that are occurring. A supervisor in a plant wants very detailed information regarding the amount of labor and materials needed to produce products on a day-to-day basis.

CHECKPOINT

27. When computers were first used for recordkeeping functions, why was it felt that EDP should be centralized and that computers should be kept running for two or three shifts rather than one?

28. Today's EDP systems are often decentralized. Should control over the acquisition and utilization of the computer systems also be decentralized?

29. What types of concerns have been expressed due to the widespread use of microcomputers?

30. Within large corporations is the acquisition and utilization of microcomputers controlled?

31. Are the informational needs of a small company different from those of a large corporation?

32. Should the communication needs of an organization be considered when a new system is being designed?

COMPUTER SYSTEMS

A complete computer system is composed of hardware, software, procedures, data, and people. Hardware is the computer and its peripherals. Software consists of the control software (the operating system) and application programs used to process data. Procedures are the step-by-step methods used to effect a solution. The procedures are documented and describe how the data is entered, what output is produced, and how the output will be used and handled. While data is often defined as raw facts entered into the computer as input, it also includes the online information stored in files and databases. Individuals who design systems, write programs, operate mainframes, enter data into the system, and work with the output are key factors in the development and utilization of computerized systems.

Hardware Hardware consists of the central processing unit (CPU) and the required I/O devices. The CPU is responsible for the execution of instructions. Its main components are memory (referred to as primary or main memory), a central control unit, and an arithmetic/logic unit. Stored within the primary memory are instructions that make up programs. Also stored within the CPU's memory are one or more input records, constants used in doing the necessary calculations, and one or more output records. Work areas are also assigned for the results of the intermediate calculations. The central control unit is responsible for the execution of the instructions, and the arithmetic and logic unit performs the necessary calculations.

Software There are two major types of software. Application software consists of programs such as one that calculates an employee's earnings or prints his or her paycheck. After application specifications are developed for a system, an investigation should be made to determine whether a software package can be purchased or whether it is more advantageous to develop the software inhouse. Application software can sometimes be purchased from the vendor of the computer or a software company.

The second type of software is composed of complex programs the give the computer many of its basic abilities and also determines how easy it is to communicate with and to work with the system. Each of these programs has a unique function and makes up what is called the operating system. The operating system's software is usually managed by a **monitor program** or **supervisor program** which resides within the CPU.

Defining Operating Systems The term operating system has many definitions. The American National Standard Institute (ANSI) defines it as ''software which controls the execution of the computer programs and which may provide scheduling, debugging, input/output controls, accounting, compilation, storage assignments, data management, and related services.'' Sayers in his text *Operating System Survey* defines an operating system as ''a set of programs and routines which guide a computer in the performance of its tasks and assist the programs (or programmer) with certain supporting functions.''

Regardless of the definition used, most people consider the **utility programs, compilers**, and software that manage the libraries, job-to-job transition, and the I/O devices as the operating system. While some companies include the nucleus of the operating system in the cost of the computer, other vendors have priced it separately. Some vendors provide only one type of operating system, while other tailor the system to the needs of the user. Often an operating system for a mainframe or minicomputer is leased. The monthly charge includes the use and maintenance of the software. A company may start with only a nucleus and then, as its needs change, add additional software that supports data communications, management information systems, and online databases.

Operating System Functions Operating systems were developed to improve communications between the operator or programmer and the computer. In addition, the software made the computer more productive since setup time was decreased and more than one program could be executed concurrently. The functions performed by operating systems are:

- Job management. This function is responsible for managing the total computer system—both hardware and software. Communication between the computer and operator is provided as part of this function.

- System resource management. The resources of the system must be allocated to programs being executed. These resources include the memory of the computer, its computational ability, and its I/O devices.
- Data management. Routines within the operating system must provide for utilization of space on the **direct-access storage devices**, the allocation of the tape drives, the reading of cards, and the output of printed material. Usually the operating system controls all I/O operations.

Development of Operating Systems   Input/output control systems (IOCS) were developed in the early 1960s and became the forerunners of operating systems. IOCS commands are stored in an online library and are now considered part of the operating system. I/O control initiates, executes, monitors, and controls the transfer of data into and out of the computer. For the application programmer, the I/O control software eliminates most of the detail of handling input/output operations. Since separate I/O routines are not written by each application programmer, uniformity is achieved.

The handling of data stored on direct-access media is simpler because of the availability of control routines. Different access methods are supported by operating systems. The IOCS usually takes care of unblocking the physical records (the entire block) into logical records (a single record) that can be processed by the application program. Some computers require the use of **job control language (JCL)** or operation control language (OCL) which tells the computer where a file of records is stored. Other operating systems keep track of where files are stored and JCL statements regarding where files are stored are not needed. The use of JCL or OCL also provides flexibility and a means of communication between the computer and the operator.

Types of Operating Systems   *Serial batch.* In serial batch processing, jobs are executed in the sequence entered, one job at a time. Early mainframes all had serial batch operating systems. The first minicomputer and microcomputers were also serial batch systems. Today most mainframes and minicomputers as well as some of the microcomputers with more advanced operating systems support both multiprogramming and timesharing.

Multiprogramming. A **multiprogramming** system permits the computer to run two or more jobs concurrently. The computer's productivity is thus increased because the computational ability of the computer is used more efficiently. While one job is waiting for an I/O event, another job is using the central control unit to execute an instruction. Only one mathematical or logical operation can be executed at a time. Since I/O activities usually take longer than the mathematical or logical operations, a second program can be executing while the first program waits for input to be read into memory or for output to be sent to an output device. The functions incorporated into an operating system differ as do the number of jobs that can run concurrently.

Timesharing. Because of the difference in the time that it takes to transmit or receive data and the time that it takes to execute commands, many users can be served at one time by the CPU. By using terminals, users may access the computer from remote locations. Since data travels from the terminal to the computer at about the speed of light, a user in New York can get an answer from a computer in San Francisco within two seconds. Usually if the response time to the user is more than two or three seconds, the system is considered inadequate. Timesharing operating systems are more complex than batch systems because more I/O activities are required and the data and programs belonging to many different users have to be protected.

Transaction processing systems. Data is entered and processed as the transaction occurs and the user receives the results in time to influence the transaction. In transaction processing, invalid data can often be corrected before it is transmitted into the system for processing. Transaction processing (**realtime systems**) has the same characteristics as multiprogramming and timesharing systems. However, in a transaction processing system often one program serves many users. In a timesharing system, each user may call in his or her program from either a private or a common library and then enter data into the system for processing. Since the terminals used to transmit data into a transaction processing system may be remote from the computer, teleprocessing capabilities may also be required.

Many transaction processing systems can run batch jobs concurrently with the realtime applications. The earliest realtime applications that processed data in time to influence the transaction were airline and motel reservations systems and sales-order processing systems.

Maintaining an Operating System While the supervisor or monitor of the operating system resides within the memory of the central processing unit, the majority of the software is stored on either hard disk or **diskettes**. The supervisor or monitor is responsible for bringing into memory, when needed, the other components.

Each computer has its own operating system. When a new computer is developed the operating system may not provide all of the functions normally associated with a complex system. As improvements are made to the operating system, the additional routines are made available to the organizations and individuals who obtained the earlier versions of the software. A **systems programmer** is normally responsible for incorporating the new features into the existing operating system. If a number of changes are made, an entire new version is made available to individuals or firms who are renting or leasing the software.

Analysts, programmers, and computer operators must be informed when new features are added to the operating system. Analysts and programmers need to know about the new features so that applications can be designed and programmed to take advantage of the new capabilities.

What an Analyst Should Know about the Operating System An analyst must be knowledgeable about the various features of the operating system in order to design applications in the most efficient manner. The data processing manager must assume responsibility for seeing that the analysts and programmers are informed of any change made in the operating system.

The features incorporated into most mainframe and minicomputer operating systems with which the analyst must be familiar are:

- the **interrupt system** that provides default routines and error messages. For example, the analyst must know what will occur when nonnumeric data is found in a field defined as numeric or when a file is not closed.
- the library management and maintenance facilities. Application programs, the operating system programs, **functions**, and **subroutines** are all stored in online libraries. The analyst must know how programs and subroutines can be added, deleted, or changed.
- job control language (JCL) statements. JCL is used to provide instructions to the computer regarding the execution of programs. JCLs provide flexibility and better communication between the computer and the operator. By proper use of JCL statements, priorities can be   assigned and changes can be made in the way programs execute. Some computer

systems have **system commands** rather than JCL. However, system commands perform the same functions as JCL.

- the utility programs available. Utility programs provide functions such as sorting files, merging two files into one, or copying a master file to a sequential backup file.
- data-management facilities. Often the data-management facilities determine how records within files can be organized and accessed. In addition, software is provided that protects files from unauthorized users.
- scheduling and priority features. Some operating systems permit jobs to be scheduled in advance. When it is time for the job to be executed, the program will be automatically loaded into the computer from an online library and executed. Jobs are also assigned a priority which determines how fast the program can be executed.
- job accounting features. The cost of running programs and using other resources are prorated to the individual user or to a department. In determining the cost of an application, the analysts must understand the job accounting features and how costs are prorated.

There are many more features incorporated into the design of a comprehensive operating system. Some of the features are of direct concern to the operations staff, while other features are of little concern to anyone other than the systems programmer.

When software that provides the ability to create, maintain, and utilize databases is added to the system, an analyst must understand how the software interfaces with the operating system. Also, when teleprocessing capabilities are incorporated into the system, the software must interface with the operating system.

Microcomputer operating systems tend to be less complex than those designed for minicomputers and mainframes. For example, scheduling, job accounting, and a priority system usually are not provided. Newer and more complex microcomputer operating systems do provide for multiprogramming, teleprocessing, and interfacing the microcomputer with other computers.

CHECKPOINT

33. What are the five components of a computerized system?
34. What are the main components of a computer?
35. What are the two major types of software?
36. Why were operating systems developed?
37. What three basic functions are performed by the computer's operating system?
38. What did IOCS, the forerunner of operating systems, do for the application programmer?
39. What are the four types of operating systems?
40. What type of maintenance does the operating system require?
41. Although some individuals feel JCL and system commands are complex, what major function is provided by using JCL and system commands?
42. Why should an analyst be familiar with the components of the operating system being used by the computer for which applications are being designed?

CLASSIFICATIONS OF COMPUTER SYSTEMS

Unfortunately it is becoming increasingly difficult to provide a definition for the different types of computer systems. There is a tendency for each vendor to determine the category in which its computers belong. The class, or category, under which a computer is listed is generally determined by the amount of real memory that is included in the base price or can be added to the system, the **word length** used to store data and instructions, the amount of online file space that can be utilized, the speed with which calculations are performed, the capabilities of the I/O devices that can be utilized by the computer, and the complexity of its operating system.

Mainframes

The computer at the upper end of the scale—the mainframe—has more memory, executes instructions faster, and has faster I/O devices, more online storage, and a more complex operating system than does the computer at the opposite end of the scale—the microcomputer.

Mainframes vary in size and complexity from what might be termed large systems to "monsters." In contrast to microcomputer and minicomputer vendors, there are few companies that manufacture and distribute mainframes. There are even fewer companies that market the monsters—the computers that cost millions of dollars and are utilized by very large organizations that need a tremendous amount of data processed and must have the availability of vast amounts of online data. Most mainframes require special environmental considerations such as air conditioning and special wiring.

Large organizations will often use a large mainframe that is part of a network that includes microcomputers, minicomputers, and smaller mainframes. One of the decisions that must be made in designing a new application is "to what type of computer should the job be assigned?" Is it more efficient to use two or more smaller computers to accomplish the required tasks or can the tasks be accomplished more efficiently and cost effectively by using a mainframe? Mainframes, excluding the I/O and communication devices, are listed as costing anywhere from one hundred thousand to several million dollars.

Minicomputers

The term *minicomputer* was first introduced in the early 1960s. The early minis had a rather short wordlength (8 to 16 **bits**—8 bits are often called a **byte**) and were programmed in **assembler** or **machine language**. Usually they required no special environmental considerations such as air conditioning or wiring. Because they were rugged, the early minis were often used for industrial applications or in laboratories. Often minicomputers were preprogrammed to perform one function.

As the demand for low-cost computers increased, I/O devices were designed that were slower and had less capacity, but were far lower in cost than those used by mainframes. More comprehensive operating systems became available and minicomputers could be programmed by using a high-level programming language. Some software companies specialized in developing application programs for the minicomputers. By the middle of the 1970s, low-cost minicomputers made computers affordable to small- and medium-sized organizations.

The availability of minicomputer systems brought forth a new concept, **distributed data processing**, to the data processing community. Distributed data processing is the term used when small computer systems are acquired by individual departments to process their data. Under the distributed data processing concept, data is often captured and processed at the source and then transmitted in a condensed form into the centralized corporate database.

Minicomputers were also used to develop **turnkey systems**. The computer hardware is marketed as a unit with the operating system, application software, documentation for executing procedures, and the forms to be used as source documents and in printing reports.

An analyst or a consultant working with a small organization should be aware of the many possibilities that exist within the minicomputer marketplace. More firms market computers that are listed as superminicomputers, minicomputers, or small business systems than market mainframes. Prices for the central processing unit vary anywhere from $10,000 to $250,000.

Microcomputers

The microcomputer, the so-called "computer on a chip," was introduced in the early 1970s and was once again used for industrial applications. Microcomputers were preprogrammed and built into machines used in production and for providing office automation. In a microcomputer, the entire CPU is on a very small silicon chip. Most microcomputers have batch processing operating systems that permit one job to execute at a time. In contrast to mainframes, microcomputers have small memories and a limited amount of online storage. However, almost on a day-to-day basis, the memories and online storage available for microcomputers is increasing.

The first personal computer, or microcomputer system, was introduced in 1976. Since that time over 150 companies have entered the microcomputer system marketplace. Some microcomputers cost less than $100 while others are in excess of $10,000. The newer operating systems provide for multiple users and multiprogramming. Microcomputers are widely used as standalone systems or a part of a network of computers. Microcomputers have had a far greater impact on computer technology than did the early mainframes. For the first time, computer power is affordable to individuals and very small organizations.

For the analyst who must study the problem and design a solution, a whole array of new possibilities exists because of the availability of microcomputer systems. The mighty micro has focused the attention of vast segments of our population on the computer, its use and potential. Its availability has caused more people to become knowledgeable regarding how computers can be utilized. It has also caused management and users to become more demanding in regard to what they want and when they want the application implemented.

If analysts had become complacent because they felt they had mastered all aspects of designing new applications and understood the characteristics of the available computer systems, the advent of microcomputers and the integration of EDP and office automation has jarred them out of their complacency.

CHECKPOINT

43. What are the three major classifications of computers?
44. What factors are used in determining the class into which a computer should be placed?
45. Is there agreement among the individuals directly involved in the computer industry as to how each class of computers should be defined?
46. Which class of computers has had the greatest impact on the general public and in how computer applications are designed?
47. Depending on the operating system being used, can a microcomputer do all of the basic functions that are performed by a mainframe?

SUMMARY   Computers are changing the way data is processed, the way tasks within the office are performed, and the way goods are manufactured. When an organization first utilizes a computer, electronic processing of data occurs and facts are converted into meaningful information. As the organization expands and becomes more knowledgeable regarding computers, the goal of developing a more complex management information system may be established. Careful planning and a dedication to long-range objectives is needed to develop complex computerized systems.

Analysts are using techniques associated with system analysis and design to create a broad range of applications—from robotics to office automation—for computers. Small companies and individuals, who are for the first time obtaining a computer system, can profit from what has been learned in designing systems for large organizations.

When a request for a new or revised application is received, the first phase of the system development life cycle may begin. An initial investigation is made to determine the nature and scope of the problem. The next phase of the cycle is to do a feasibility study. If the recommendations listed in the feasibility study are approved, the general design phase is begun which determines, *from the user's point of view*, how the system should be designed. In the detailed design phase, the general design specifications are refined to include detailed specifications regarding how the computer and its I/O devices can be used to process the user's input and provide the desired output. In the implementation phase, files are converted or developed, programs are written, tested, and documented, and users and EDP personnel are trained to work with the system. The system audit determines whether the design objectives have been achieved and whether the actual costs are comparable to the budgeted costs. How the system was developed and implemented is also reviewed.

Analysts designing systems must be knowledgeable regarding the hardware and software available. Once a mainframe, minicomputer, or microcomputer is obtained, the analyst must study its operating system to determine the various functions available. An analyst must understand the needs of the user, regulations that might apply to the application, and the components of the computer system that will be used to implement the system.

Not all applications that are operational are considered successful. By determining why an operational system was unsuccessful, standards and guidelines for systems analysis and design can be revised or developed that will make it easier to develop more successful systems in the future.

Perhaps the most significant development in computing was the microprocessor which is a chip containing the entire CPU. The availability of low-cost microcomputers which utilize microprocessors has caused increased attention to focus on distributed data processing, new applications, and the development of standards for networking microcomputers, minicomputers, and mainframes into complex information systems. Due to the wide range of options available, the complexity and challenge of designing applications has increased.

As you progress through the text you will learn more about the roles of the user, management, the analyst, and other members of the EDP department's staff in the design, development, and implementation of tasks, procedures, and systems. Although the use of top-down designing and programming will be stressed, attention will be given to the many single tasks that make up a complex system.

DISCUSSION QUESTIONS   1.   Based on the information presented in the chapter, how important is it that users be directly involved in the design of a system? In what phases will the users be most involved? Will the users be involved during the implementation phase of the project?

2. Why should edit routines and controls be built into an application? As systems become more integrated and the output from one program becomes part of the databases or input for the next program, do you feel there will be more or less need for controls than when each programs was more of a separate entity?

3. Six months after the new sales-order entry system was operational, the system audit was completed. The report indicated that the system did not meet its objectives. What objectives might not have been met? What problems might have been noted regarding how the programs executed? What mistakes in developing the system might the analyst have made? How can the EDP department profit from analyzing how the system was developed?

4. Today when a company obtains its first computer system, the initial applications to be developed might be the automation of some of the recordkeeping functions. As the company expands and the users gain experience in working with computers, what other applications might they wish to implement? What criterion should they use in determining whether an application should be computerized?

5. Are the informational needs of a small company the same as those of a large organization?

6. Today we have microcomputers, minicomputers, and mainframes. Which type of computer system has had the most impact on our life-style and potentially on our concept of work? What problems must be faced regarding the utilization of microcomputers? What impact has the microcomputer had on the general public and their concept of what a computer is and what it can do?

7. A computer system consists of hardware, software, data, procedures, and people. Does hardware or software have the greatest impact on the way a system can be designed? What are the two major types of software and what does each do in regard to processing data? Why was control software developed? How do procedures relate to hardware, software, data, and people?

8. From your own experience (reading articles, material covered in other classes, working in a business environment, or making observations):
   a. Describe one application that should be run as a batch job.
   b. Tell why it is important that minicomputers and mainframes have multiprogramming capabilities.
   c. Indicate why it would be impossible to develop an online registration system if the computer didn't have timesharing capabilities.
   d. Tell why a charge sales system for a department store should be a transaction processing system. Indicate how the customer database and inventory database might be used in recording the sales transactions.

TEAM OR INDIVIDUAL PROJECTS

The following team or individual projects require a more in-depth study of material covered in the text. You may be asked by your instructor to work independently or as a member of a project team to prepare either a written or oral report.

Other projects and discussion questions that relate to the Computer Products Incorporated case study are found in Appendix I. Although each segment of the case should be studied along with the chapter to which it relates, the material is presented together in Appendix I to make it more convenient.

Team or Individual Projects

1. Utilize articles found in current publications such as *Computerworld, Office Administration and Automation,* or *Datamation* and prepare a report on one of the following topics:

   A. Management information systems. Include information regarding the type of hardware and software needed to implement the system, whether companies normally need to obtain additional control software, and the extent to which such systems are implemented and utilized.

   B. Problems created by the integration of microcomputers into applications involving office automation or electronic data processing.

   C. Distributed data processing. Include the type of equipment that is needed and the type of interface that should be available between the computer systems located at remote locations and the mainframe.

2. Interview an analyst or EDP manager and prepare a report on your findings regarding their views of how and when an application is considered either successful or unsuccessful.

GLOSSARY OF WORDS AND PHRASES

analyst Individuals who study existing systems and design new systems are called systems analysts. Analysts are concerned with the entire workflow, which often includes manual operations and computerized functions.

assembler A computer program that takes instructions written in nonmachine language and converts them into the machine language used by the computer.

audit trail A means of tracing transactions through the entire processing system. Federal law states that there must be a clear audit trail.

backordered When there is not enough merchandise to fill an order, the items that cannot be shipped are placed in a backorder file. When new merchandise arrives, the order is automatically filled.

batch job Data is accumulated and processed all at one time rather than being processed as the transactions occur. As example would be processing the entire day's sales transactions at one time.

bit A binary digit; the smallest storage device within the memory of the computer.

byte A group of eight bits within the memory of the computer or on an external storage device. Normally each byte of storage is addressable.

centralized data processing All of the data processed within an organization by computers is processed in one central location which is usually managed by a data processing department.

compiler A program used to translate source code written in a language such as COBOL into machine language commands. As a program is compiled, the compiler checks each source statement for errors.

computer assisted design/computer assisted manufacturing (CAD/CAM) The term is used to denote the use of computer-assisted automated methods for designing and manufacturing products.

controls Methods used to make certain the data is processed accurately.

cost-justified An analysis is made to determine whether the money saved in computerization of the tasks offsets the costs involved in using a computer system.

CRT See video display tube.

database A collection of data stored together to serve one or more applications. The use of a common database promotes the sharing of data and the development of standards.

data exceptions Nonnumeric data is found in a field defined as numeric. When this occurs, software provided as part of the operating system may cause an error message to be displayed and the job to be canceled.

direct-access storage devices Media such as magnetic disk, videodisks, and magnetic drums are used to store data that can be randomly accessed.

dial-up terminals Terminatls that have a modem and transmit data to the computer by using phone lines or a common carrier. A number is dialed to the computer. If all **ports** are occupied, a busy signal is received.

disk controller The interface between the disk drives and the computer. The controller coordinates the movement of data between the computer and the disk drives.

diskette A small disk made of a flexible material which is coated with iron oxide. The disks are often 3, 3½, 5¼, or 8 inches in diameter and are used for storing programs and data. Diskettes, or floppy disks as they are also called, are frequently used with minicomputers and microcomputers.

distributed data processing Each department or function may have its own computer system that is used for processing data. Often data is entered into the computer system and processed at the locations where the transactions occur.

documentation　Detailed information regarding the development, implementation, and operation of a system. Documentation provides information about the system and is also used to provide instruction.

editing　The validity of data to be processed is checked.

electronic data processing (EDP)　A method of processing data that requires the use of computers.

ergonomics　The word, which comes from *ergo* (work) and *nomics* (law or management), indicates the study of the relationship between humans and machines.

flexible manufacturing systems (FMS)　Equipment that can be used to manufacture more than one product. The manufacturing systems can be programmed to alter their procedures to suit production requirements.

frontend processor　An interface between terminals and the computers. The frontend processor, or controller, may only assemble the transmitted bits into characters and provide a parity check. Other frontend processors are minicomputers or microcomputers and assist the host computer with routine functions required in teleprocessing and in timesharing systems.

function　A prewritten routine that can be called into a program to provide the commands to do a particular task, such as finding the square root of a number.

hardware　The computer, I/O devices, and other offline machines, such as diskette recorders, are referred to as hardware.

hardwired terminals　Terminals that have a direct connection to the frontend processor. A port is dedicated to a hardwired terminal and immediate access to the computer can always be achieved.

high-level programming language　A powerful computer language. One instruction may generate numerous machine language commands. A program written in a high-level language is translated by a compiler or interpreter into machine language commands. The programmer need not understand how the machine language commands are executed.

information processing　The utilization of a computer to process data, retrieve data from databases, transmit data to distant locations, and format information into meaningful reports and displays.

interrupt system　Software that is part of the operating system is used to interrupt the execution of the application program's commands. Although some interrupts are normal (such as when the end of a file is reached), other interrupts occur when an error has been detected. Depending on the cause of the interrupt, an error-recovery routine may be invoked or the program may be canceled and an error message printed.

job control language　Statements take care of the job-to-job transition and communicate to the operating system what

options are to be executed. JCL was developed to provide additional flexibility and communication between users and the computer.

keypunch　A machine used to punch data onto a card. It has a keyboard much like that of a typewriter. However, the keypunch can be programmed to automatically shift from alphabetic mode to numeric mode, to duplicate fields of data, and to skip fields.

machine language　The commands the computer understands and can execute. Before a program written in a high-level language can be executed, the instructions must be translated into machine language commands.

mainframe　A large-size computer. The term is also used to denote any CPU.

master file　A file that contains constant and updated information. Often the data is used as input for many application programs.

megabyte　One million bytes (8 bit words) of memory.

microcomputer　A computer system that has its entire CPU on a single microprocessor chip. However, some of the more advanced microcomputers have coprocessors that are assigned some of the functions normally performed by the microprocessor chip.

microprocessor chip　An integrated circuit chip that contains the arithmetic, logic, and control units required for processing.

minicomputer　Although a standard definition is not available, most people associate the term with a small, rugged, inexpensive computer system that requires no special power or environmental conditions. Currently the minicomputer is midway (in terms of memory, speed, and online storage capacity) between a microcomputer and a mainframe.

monitor program　Software that resides within the central processing unit and coordinates all of the components of the operating system. A monitor is also called a supervisor or executive program.

multiprogramming　Two or more computer programs are executed concurrently within a single computer. However, only one command can be executed at a time within the computer's arithmetic/logic unit (ALU). The number of programs that can run concurrently is determined by the operating system. As more advanced computers are built, more than one central processing unit (CPU) will be located within the computer and more commands can execute concurrently.

natural language　An English-like language that makes it possible to obtain information in the form of reports or displays by either answering questions or by using key words to formulate sentences. Natural languages can also be used to create and maintain databases.

networking　Connecting two or more computers together so that I/O devices, programs, and databases can be shared.

offline files　The storage of data on a medium that is often stored in a vault or some type of fireproof container. An offline file can be removed from its storage location and placed on a device that is being controlled by the computer.

online application   The execution of a program that records data generated by transactions directly into the computer. Immediate feedback is available regarding exceptions that occur.

online master files   Files that contain updated or permanent information and are under the direct control of the computer. Online also denotes a method of processing that is directly controlled by the computer.

operating system   Software that controls the execution of computer programs. An operating system may provide for timesharing, scheduling of jobs, input/output control, library management, error routines, and many other services.

operational   An operational system is one that has been designed, tested, documented, and implemented.

ports   A device within a controller or frontend processor which permits a terminal to communicate with a computer. Depending on the system being used, there is often a limited number of ports available.

procedural flowchart   A graphic representation, diagram, or form that shows the flow of information throughout the organization, the locations at which tasks are performed, and the time that is required to perform each task.

procedures   A step-by-step method of solving a problem. Systems are usually divided into procedures. A typical payroll system might be divided into subsystems and the subsystems into procedures.

realtime system   Data is processed in time to provide immediate feedback and in time to influence the transaction.

robotics   The science of robot design and use. Robots are computer-controlled devices capable of performing many of the tasks performed by individuals.

software   Software is composed of many different types of programs such as application programs, utilities, compilers, and the components of the operating system.

source document   A form such as a sales-order form used to capture data pertaining to a transaction. Often the data recorded on the document must be transcribed into a machine readable format before it can be processed by the computer.

standard   A uniform method of accomplishing a task. Standards are available for developing systems and programs.

subroutines   Small programs that are compiled into machine language and stored in a library. Subroutines are executed by being called into other programs. System designers design and develop subroutines for special purposes and for a series of instructions that must be used in several programs.

supervisor program   See monitor.

system   An orderly means of accomplishing one or more procedures or tasks. An information system is usually composed of several subsystems. The term is also used to denote the computer.

system commands   An instruction used to tell the computer what task is to be performed. A typical example is LIST which

tells the computer to display the program on either a printer or a VDT.

system flowchart   A graphic representation that shows the movement of data throughout the system. Both manual and computerized procedures are shown. Sometimes a system flowchart is used to show the input and output required for a computer program. How the computer processes the data is not illustrated.

systems development life cycle   The phases, or steps, required to study a problem, design a solution, implement the system designed, and evaluate the solution. During the evaluation it may be determined that steps in the systems development life cycle need to be repeated.

systems programmer   An individual who maintains the operating system software. These individuals are highly skilled and must understand the architecture of the computer.

task   A single step within a procedure. For example, one task involved in printing paychecks is to put the checks in the printer.

task force   Two or more people, each with different talents, assigned to study a problem and determine the most effective solution.

teleprocessing   Remotely located terminals are used to receive data from and transmit data to a host computer. Additional software is needed as well as terminals, modems, communication lines, and a frontend processor.

terminal   A device remote from the computer that is used to input data into the computer or to receive data from the computer.

timesharing   The interactive use of a computer system by more than one user. The term usually implies that the users communicate with the computer through terminals.

top down   A method used to design a system. The system is divided into procedures and the procedures into tasks. Objectives are defined first for the system, then for the procedures, and finally for each task.

transaction processing   Data is entered into the system where and when transactions occur.

turnkey system   A complete processing system that includes the computer, I/O devices, operating system, application programs, and documentation. In order to use the system, all the user needs to do is to plug in the components, ''turn the key,'' and follow the instructions provided in the documentation.

utility programs   Frequently used programs that perform a task such as copying a file, sorting records stored in a file, or formatting the data stored within the computer.

verifier   A machine used to confirm the accuracy of the data punched into the card.

video display tube   (VDT) A screen similar to one used on a television that is used to display information and graphics. VDTs are also called CRTs (cathode ray tubes).

videodisk A plastic disk that is used for storage and retrieval of text and graphics. Data is recorded on the disk by using laser beams.

word length The length of a unit of primary storage that is addressable and stored in one location.

word processing Unformatted text, rather than formatted data, is stored and can be manipulated and used in the preparation of letters, memos, reports, and other types of printed materials.

STUDY GUIDE 1

Name _____

Class _____ Hour _____

A.   Indicate whether the following statements are true (T) or false (F). *Indicate in the margin if the statement is false, how it should be changed to make it true.*

_____   1.   A procedure is an orderly means of accomplishing one or more systems or tasks.

_____   2.   Usually there is a bottom-up approach to designing a system. The objectives are first defined for tasks, then procedures, and then for the total system.

_____   3.   Computerized systems are usually more successful if the users are not directly involved in the project.

_____   4.   The hardware used to implement a new system is more important than the type of software that is used.

_____   5.   When computers were first utilized, the applications for which they were used were usually recordkeeping functions.

_____   6.   A system flowchart shows the input required, the precise steps required for the computer to process the input, and the output produced as a result of processing the data.

_____   7.   An analyst works exclusively with tasks that are computerized.

_____   8.   The role of the analyst and the demand for his or her services is about the same as it was ten years ago.

_____   9.   Standards are only developed for tasks which are computerized.

_____   10.   When designing a new application, analysts are not concerned with ergonomics.

_____   11.   When an analyst is designing a job, the type of computer system that will be used in implementing the design is of little significance.

_____   12.   In solving most problems, only one approach can be used. Therefore, if two analysts independently designed solutions for the same problem, the solutions will be almost identical.

_____   13.   If the reviewing committee approves the recommendations presented at the end of the feasibility study, the analyst is guaranteed that the approved recommendations will be implemented.

_____   14.   Usually one or two analysts working as a team do the necessary tasks to complete the general design phase of the project.

_____   15.   A system is considered operational after the programs are written and tested. The documentation should be completed during the system audit phase of the project.

_____   16.   Once a system is operational, it will not be scrapped.

_____   17.   Today a dollar buys less computer power and online storage than it did a year ago.

_____   18.   For a new sales system, the objective "the quality of the service to the customer will be improved" would be measurable.

_____   19.   Unless the cost of the proposed system is less than the cost of the old system, the design should not be approved.

_____ 20. In designing a new sales-order system, the analyst should not be concerned with the type of systems used by the organization's competitors.

_____ 21. Large corporations that utilize computers in office and factory automation do not expect to get a return on their investment.

_____ 22. Knowing why a recently implemented system was not meeting the objectives that were stated during the design phase might help analysts create more successful systems in the future.

_____ 23. Partly because of the high cost of computers and the need to cost-justify new applications, EDP was decentralized.

_____ 24. Organizations that have well-defined standards often insist that microcomputers and minicomputers be able to communicate with the centralized mainframe and be able to share databases.

_____ 25. At the present time there is very little evidence of the impact computers are having on either factory or office automation.

_____ 26. Because the cost of computer power is decresing, computers are being used to perform many more functions than they did in the past.

_____ 27. A large organization that would like to implement an MIS should consider using microcomputers rather than the corporate mainframe.

_____ 28. The only choices available to a small company with three or four employees are to process data manually or obtain a microcomputer system.

_____ 29. It is difficult for a small company with relatively few employees to obtain the services of an analyst.

_____ 30. A small company's informational needs are less complex and very different from the information needs of a large organization.

_____ 31. To have a complete computer system, it is only necessary to obtain a CPU and whatever I/O devices are needed.

_____ 32. In selecting a computer, the major consideration is the hardware.

_____ 33. When obtaining a new computer system, it is always necessary to purchase the operating system.

_____ 34. Operating systems were developed to improve communications between the users and the computer and to make computers more productive.

_____ 35. Although operating system software provides many different types of services, such as managing the resources of the system, an analyst designing a new application need not be concerned about the functions performed by the computer's operating system.

_____ 36. An application programmer is usually assigned the task of maintaining the operating system.

_____ 37. In a multiprogramming system, several mathematical or logical operations are executed concurrently.

_____ 38. All users of a timesharing system share the use of the computer, programs, and a common database. Whatever is stored in online files is always available to all of the users.

_____ 39. In a transaction processing system, data is processed in time to influence the transaction.

_____ 40. Many operating systems provide for an interrupt system, a job control language or system commands, utility programs, scheduling of jobs, a job-accounting program, and a system of assigning priorities.

B.   Multiple choice. Record the letter of the **best** answer in the space provided.

_____   1.   A diagram that graphically illustrates the input and the resulting output is a
             a. system flowchart.   c. task analysis chart.
             b. procedural chart.   d. I/O chart.

_____   2.   An online payroll masterfile
             a. contains one record for each employee.
             b. contains constant and updated information.
             c. is under the direct control of the computer.
             d. all of the above answers are correct.

_____   3.   An analyst's role has become more complex due to
             a. the use of larger mainframes.
             b. less flexibility in the way an application can be designed.
             c. the use of standards in the design and development of systems.
             d. the availability of more alternatives regarding CPUs, I/O devices, and types of operating systems available.

_____   4.   In designing a new application, an analyst
             a. is only concerned with tasks that are computerized.
             b. only develops documentation and standards for procedures that are computerized.
             c. is not concerned with the measurement and simplification of work.
             d. is concerned with both manual and computerized procedures.

_____   5.   In the systems development life cycle
             a. the last phase is implementation.
             b. there are definite rules regarding what must be included in each phase.
             c. the users only participate in the feasibility phase.
             d. the initial investigation is conducted to determine the nature and scope of the problem.

_____   6.   Perhaps the most technical creativity—translating specifications into specific requirements that can be implemented in the system—occurs in the
             a. feasibility phase.   c. implementation phase.
             b. general design phase.   d. detailed design phase.

_____   7.   The design created during the general design phase should
             a. clearly focus on what the analyst feels is needed.
             b. show the users' view of the application.
             c. be very specific and detailed. Exact specifications are given for the input, files, and output.
             d. be developed entirely by system analysts since they have the most extensive knowledge of computer technology.

_____   8.   The detailed design phase report
             a. does not include information regarding the resources needed to implement the system.
             b. includes projected costs that are not expected to be very close to the actual cost of implementing the system.
             c. must include detailed information about the resources needed to implement the system.

_____ 9. The documentation for the system.
   a. is completed after the system is operational.
   b. should be completed before the system is considered operational.
   c. is only used as a reference if a problem occurs.
   d. is completed in its entirety by the analyst.

_____ 10. One of the new media used for storing text and graphics is
   a. diskettes.   c. magnetic tape.
   b. magnetic disks.   d. videodisks.

_____ 11. Due to availability of low-cost microcomputers and minicomputers
   a. there is a trend towards greater centralization.
   b. there is an increase in the amount of distributed data processing.
   c. there is no longer a need for mainframes.

_____ 12. It is anticipated that the impact of the microcomputer on computer technology will be
   a. less than the impact of mainframes.
   b. less than the impact caused by the development of minicomputers.
   c. greater than the impact caused by either mainframes or minicomputers.

_____ 13. Online applications are designed so that
   a. all of the data is processed in a batch.
   b. transactions are entered into the system as soon as they occur.
   c. there is little need for an online database.
   d. there is no longer need to do any batch processing.

_____ 14. A small organization with only a few employees
   a. does not need a computerized sales-order system.
   b. should only consider becoming part of a timesharing system.
   c. must have timely, relevant information in order to operate effectively in today's competitive environment.
   d. should not consider obtaining its own microcomputer system.

_____ 15. A complete computer system is composed of
   a. hardware, an operating system, procedures, data, and people.
   b. hardware, software, systems, data, and people.
   c. hardware, software, procedures, data, and people.
   d. hardware, software, procedures, files, and people.

_____ 16. The two major types of software are
   a. operating systems and utilities.
   b. operating systems and compilers.
   c. control software and application software.
   d. assemblers and application software.

_____ 17. The basic functions performed by the operating system are job management, system resource management, and
   a. file management.
   b. management of direct-access storage devices.
   c. data management.
   d. managing the total computer system.

_____ 18. A system that permits the computer to run two or more jobs concurrently is:
   a. realtime.   c. serial batch.
   b. timesharing.   d. multiprogramming.

_____ 19. When a transaction processing system is developed
   a. only one user at a time can access the computer.
   b. multiprogramming capabilities are not required.
   c. it is usually necessary to have teleprocessing capabilities.
   d. the computer can only be used for online applications.

_____ 20. The following is to provide instructions for the computer regarding how programs are to be executed:
   a. subroutines. b. functions. c. JCL. d. scheduling software.

_____ 21. Due to the widespread use of the microcomputer, analysts are more concerned with
   a. the security of data.
   b. computer crime.
   c. the lack of compatability between computer systems.
   d. all of the above factors.

_____ 22. A storage medium that uses laser beams to store data is:
   a. magnetic tape.   c. videodisk.
   b. magnetic disk.   d. diskettes

_____ 23. The estimated cost of the new sales-order system is more than the cost of the present system.
   a. Because of the increased cost, the new system must be dropped.
   b. The increase in cost is due to the higher cost of the hardware and software needed to implement the new design. Today we get less computing power for the dollar than we did five years ago.
   c. It may be possible to justify the implementation of the system on the basis of nonmeasurable or intangible benefits.
   d. If the system is well designed and provides management with additional information, management will not be concerned about the increase in cost or lack of return on their investment.

_____ 24. A report in *Computerworld* regarding the acquisition and use of microcomputers indicated
   a. each department within the organization selected its own hardware.
   b. that the microcomputers were not being interfaced with the mainframe.
   c. there were no guidelines concerning the acquisitions and use of microcomputers.
   d. that standards and guidelines were available requiring microcomputers be able to communicate with the mainframe and utilize the centralized database.

_____ 25. Because of improvements in all phases of computer technology (including teleprocessing),
   a. all procedures should be designed as online applications.
   b. there is no longer any need to edit data or to build in controls.
   c. interrupts will never occur due to data exceptions.
   d. online applications are more affordable. However, because of the input, processing, and output required, some programs should be executed in a batch mode.

2 MANAGEMENT AND THE COMPUTER INFORMATION SERVICES DEPARTMENT

Looking ahead

After reading the text and completing the learning activities you will be able to:

- Identify the key factors in the systems approach to management.
- Explain why a systems approach should be used in designing and implementing computerized applications.
- Identify the five components of a management system.
- Describe the four levels of management and identify the type of information each needs.
- Explain the five major functions performed by management.
- Explain how a *traditional* medium- or large-sized data processing department might be organized.
- Explain why it might be logical to include the departments managed by the office systems coordinator, database administrator, and information center manager as part of the computer information services department.
- Explain why the manager of the computer information services department must have interpersonal, managerial, and technical skills.
- Identify the major responsibilities of the computer information services manager, analysts, and programmers.
- Describe the five functional areas listed under operations.
- Explain how the role of an analyst might differ depending on whether the organization had a mainframe, minicomputer, or microcomputer system.
- Define and utilize the words and phrases listed in the end-of-chapter glossary.

INTRODUCTION   A creative analyst plays a vital role in the design and development of new systems and is often recognized as a catalyst for change. In designing a new system one of the most important things an analyst does is to work with people.

Since the analyst works with all levels of management, as well as with the subordinates who perform tasks at the operational level, it is necessary for the analyst to understand the role each plays within the organization. The analyst must also determine what information is needed at each level of management.

Although the analyst may be the catalyst, a far better system will be designed if a team approach is used and all of the functional areas within the data processing department are directly involved early in the project. Each area within data processing will play an important role in one or more phases of the project.

Equally as important as the early involvement of data processing personnel is the continued involvement of management and the user.

In managing either a systems design project or any other phase of business, the approach used is important. When a systems approach is used, the project is divided into subsystems or tasks small enough so that it is possible to measure the effectiveness of each.

A SYSTEMS APPROACH   As a result of the technological revolution, many industries have undergone ex-
TO MANAGEMENT   traordinary changes in products, methods, output, and productivity. Management must deal with a high product mortality rate since research and development constantly strive to improve products and methods. Management must keep up with all factors that influence the company's products and operations.

Businesses are larger and have more complex organizational structures than they had prior to World War II. There are more social and legal restrictions, and change is ever present. Although the future cannot be predicted, three things seem certain:

- Change will continue at an even more rapid pace;
- Change will demand improvements in management; and
- Management will need better information.

Management's ability to react swiftly to change will determine the profitability of the company's operations. Survival depends upon the quality of **information** available and management's ability to react to change. To react to change, management must have relevant information *when needed*

The complex organization of business today must be broken into manageable units. However, the parts or units (subsystems) must be related to each other and to the whole. The **goals** and **objectives** of the overall organization must be determined before they can be determined for the individual units. There must be communication between the various units. A systems approach to management must be used.

The systems approach to management is designed to utilize scientific analysis in a complex organization for (1) developing and managing subsystems such as cash flow, manpower systems, and manufacturing systems; and (2) designing information systems for decision making.

What Is a Systems Approach   When a complex problem is to be solved, major goals are determined. Then the
to Management?   problem is broken into manageable subsystems or components. A realistic approach is to:

- Determine the major goals;
- Determine the components, or subsystems, that make up the problem;
- Determine the relationships that exist among the components;
- Determine the objectives for each component;
- Determine the tasks that need to be performed for each of the components;
- Determine the objectives for each task; and
- Determine the trade-offs required if the resources needed to accomplish the objectives are not available.

The relationships that exist between the various components and tasks must be determined before implementing any task. Each task must be manageable in size and must have its own objectives.

Key Factors in a Systems Approach to Management

As illustrated in Figure 2-1, the systems approach is:

Creative. The design of a system depends on the originality and creativeness of the team members. Since many problems are unique and ill structured, *there is no formula to follow.* Therefore, the approach used to design a solution must be creative. Required information may not be available and alternatives must be formulated and barriers must be overcome to arrive at the best possible solution. The barriers that exist in an organization may be due to a lack of communication. A creative analyst must overcome the barriers and other problems.

A SYSTEMS APPROACH IS:

- CREATIVE
- ORGANIZED
- STRUCTURED
- ACTION-ORIENTED
- PRODUCTIVE
- BASED ON RELEVANT INFORMATION

FIGURE 2.1. Key factors in a systems approach to management

Organized. Since the systems approach is used for solving problems that require a large amount of resources, a team approach is often used. The team members have different areas of specialization. However, each must understand the systems approach and have an in-depth understanding of the problem. Management must be directly involved in the definition of the problem and have continued involvement during the design and implementation of the solution.

Structured. The framework, or structure, of the design is completed by using black boxes. The term *black box* is used when a function or task is shown but not explained. *What* must be done is illustrated by the black box but *how* it is done is not shown. The contents of the box remain a mystery until the entire framework is completed, procedures are established, and tasks are identified.

Action-oriented. Once the structure is completed, the detailed work begins on the black boxes. The work must be divided into manageable units that can be completed within a relatively short period of time. One week is an ideal period of time. Results are then measurable and progress can be charted.

Productive. When large projects are divided into small, manageable tasks, there is greater productivity than when a large segment of a project is assigned at one time. Since the systems approach requires the use of a planning chart which shows when each task is to be started and finished, the project manager can determine the productivity of his or her task force and can also chart the progress being made on the project.

Based on relevant information. Information must be available pertaining to the goals, objectives, practices, functions, interrelationships, attitudes, and other characteristics of the organization. Information is usually more available when top management is directly involved in the project.

Factors Contributing to Success

A systems approach to management is necessary to compete successfully in business. Therefore, management must use, and understand, the systems approach before an effective information system can be designed. In developing a comprehensive information system, the computer is only one of the tools used. Integrated systems developed by teams consisting of analysts, users, and management are the components of a comprehensive information system.

In designing a component of an information system, a systems approach must be used. *Goals are established, the structure is designed, and then the black boxes are filled in.*

COMPONENTS OF A MANAGEMENT SYSTEM

There must be a structure in order to make use of a systems approach to management. Computers have had, and will continue to have, an impact upon the organization of this structure. The structure consists of the individual, the formal organization, the informal organization, managerial leadership styles, and the physical setting.

The individual. The individual's personality, motives, and attitudes affect the organization. When a team approach is used to study a problem and arrive at a solution, the members of the team must be able to function effectively as a unit. Individual differences, likes, and dislikes must be put aside. The success of any organization requires effective management—individuals who are concerned with people and aware of the factors that make each person productive.

Formal organization. An organization chart shows authority, responsibility, accountability and the interrelationships that exist. The formal organization is further explained by the use of job descriptions, standards, and controls.

Informal organization. While the organization chart shows the relationships that should exist, an informal chart (if one existed) would show the actual relationships. While job descriptions usually show the authority, responsibility, accountability, and duties associated with a given position, an informal job description would tell what individuals actually do.

Leadership styles. The productivity of a unit is often directly related to the style of leadership used by management. When objectives are not met, it may be due to the leadership style used. While some individuals react favorably to a particular leadership style, others might have an adverse reaction and may be less productive. If possible, there should be a match between managers with given leadership styles and individuals that respond most productively to that style.

Physical setting. Since the environmental conditions under which individuals work have a direct impact upon productivity, analysts must be concerned with the conditions in which individuals perform their assigned tasks.

Analysts designing systems must be aware of the components of the management system. If the analyst fails to consider the physical setting or the informal organization, the effectiveness of the system will be affected. This would also be true if the analysts overlooks the individual in the design of the system. The best results will be obtained when the analyst understands and considers all of the components within the organizational structure.

CHECKPOINT

1. Why has it become necessary to use a systems approach to management?
2. What is meant by the term *systems approach?*
3. Why is a systems approach considered creative?
4. When a systems approach is used, major goals for the entire project are established before the subsystems are determined. Is it necessary to know the relationships that exist between the various subsystems?
5. Can an individual use a structured systems approach to solve a personal problem?
6. Can an architect use a structured systems approach in designing a house?

MANAGING A SYSTEM

Any business is a system that must be managed. Therefore, it must be divided into manageable components. Figure 2-2 shows a portion of a traditional organizational chart. At the top level, the division is by function: Industrial Relations, Marketing, Manufacturing, Finance, and Research and Development. Marketing is again divided by function: Advertising, Sales, and Research. Sales is divided by territory: Eastern and Western. Each division is divided by customer: Industrial, Government, Retail, and Manufacturer's Representative.

Divisions of Responsibility

The vice-presidents of the top-level functions are **line officers.** Line officers report directly to the chief administrative officer and work to formulate goals and policies. The product development team is responsible only to the president and is not part of line management. The members of the team will change as different projects are assigned.

Although not shown on the chart, manufacturing could be divided by products into such categories as Furniture, Appliances, and Floor Covering. Each of the product divisions could again be divided by functions such as Design, Purchasing, Producing, and Servicing.

The chart shows the formal organization. By studying it you should be able to determine how the business is divided into subsystems. It also shows the lines of authority and responsibility.

Should a company's organizational chart be the same today as it was ten years ago? If the only constant in today's business environment is change, the answer must be no. In a dynamic company the informal organization evolves to keep pace with change. There should be an ongoing evaluation to determine what changes should be made in the formal organization as a result of changes in technology, environmental conditions, governmental regulations, economic conditions, and society.

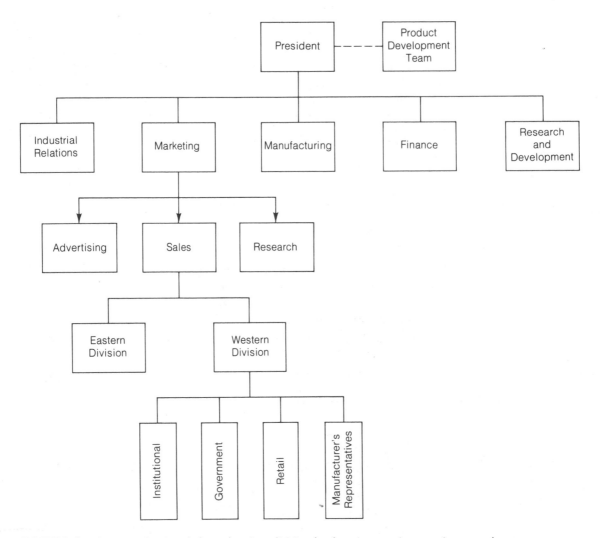

FIGURE 2.2. An organizational chart showing division by function, project, territory, and customer

Levels of Management

Figure 2-3 illustrates the four levels of management normally found in a medium- or large-size company. Each level has its own unique informational needs.

1. *Strategic.* Top management's functions are strategic. These functions include the establishment of goals, long-range planning, and new market and product development. Top management makes all major policy decisions.

2. *Tactical.* Managers in this group are responsible for allocating and controlling the resources necessary to accomplish the company's objectives. They must plan, delegate authority to the supervisory level, and measure the performance of those to whom they delegate authority.

3. *Supervisory.* The supervisory functions are short term and deal with day-by-day job scheduling, checking the results of operations, and taking necessary corrective actions.

4. *Functional.* At the operational level, routine production or clerical operations are performed. The employee at this level receives very little feedback. The supervisor at this level evaluates the performance of the employee as various tasks are performed.

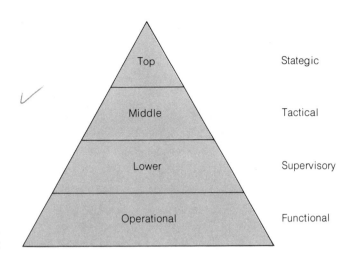

Top — Stategic

Middle — Tactical

Lower — Supervisory

Operational — Functional

FIGURE 2.3. The four levels of management

Data processing personnel involved in designing information systems must work with all four levels of management. The functional area cannot be overlooked since this is the origin of most of the data. The data must be entered into the system, sorted, merged with other data, calculated, condensed, and put into meaningful reports. Management must be able to get from the system *what* is needed, *when* it is needed, and *in a format* that meets existing needs.

In order to ensure the integrity and security of the data, many of the controls built into the system are implemented at the operational level. Under the distributed processing concept, much of the data will be inputted into the system from an operational area rather than from the centralized data processing department.

Information Requirements

Each level of management has specific information requirements. As indicated in Figure 2-4, top management must have a great deal of **external information** as well as a variety of **internal information** that shows the status of the company. **Exception reports** highlight problem areas, show the deviations from standards, or list items that need attention. Most of the reports are yearly or monthly ones that show trends rather than day-by-day operations.

Long-term trends can be determined by using historical data. Using this same data in various **mathematical models** will help management to determine what might occur in the future. In models, external information is also used. Although the information generated by using models is helpful, it is no better than the model and no more reliable than the input. It is often difficult to test a model to determine whether the correct relationship exists between the various components of the model.

The "what if" information is used along with historical data to determine what might occur when conditions change. What might happen if the cost of labor increases, the selling price of the product is increased, production increases and the selling price of the product is decreased, or there is a change in a government regulation?

Middle management's information is more detailed than top management's. Financial reports include more detail and are used to compare what actually happened with what was predicted. Their exception reports might list items that sold unusually well or items that were below the projected sales.

Management Level	Information Requirements

Top Management	External Information
	    Competitor actions
	    Government regulations
	    New markets
	    Availability of resources
	Internal Reports
	    Financial reports
	    Selected exception reports
	    Summarized historical data
	Long-term trends
	    "What if" considerations

Middle Management	Internal Reports
	    Financial reports
	    Exception reports
	    Historical data
	    Summarized operational reports
	Short-term trends
	    "What if" considerations

Lower Management
(supervisory and functional)
	Internal Reports
	    Detailed operational reports
	    Recent historical data
	    Exception reports

FIGURE 2.4. Information requirements of management

Lower management's information is far more detailed and is used in making immediate adjustments when the actual results deviate from the expected. Since EDP has been in use, detailed operational reports are often printed daily. The exception reports might include items that cost more than the projected standard to produce, customers who exceeded their credit limits, or items that need to be reordered.

Management Functions

Five functions—planning, organizing, staffing, directing, and controlling—make up the process by which a manager manages. As each function is defined, it will be related to an example involving the selection and implementation of procedures on a company's first computer.

Planning *Planning* is the thought that must precede any action. Alternatives must be developed. A course of action is then selected from the alternatives which will meet the objectives. Before the company obtains its computer it should determine objectives. What is it that the company hopes to accomplish by securing a computer? Next, alternatives are developed. What different types of systems are available that meet their needs and are within the price range they wish to pay? Decisions also need to be made regarding when the computer should arrive, what procedures will be developed, and who will be responsible for the computer center.

Organizing *Organizing* involves defining the tasks necessary to carry out the plan. In order to get the job done, the tasks must be divided among subordinates and authority must be assigned. In obtaining a computer, the individuals responsible must see that: the site is prepared; job descriptions are developed for the personnel who will be working with the system; systems are analyzed, designed, and implemented according to schedule; and the necessary supplies and related equipment are available when needed.

Staffing   *Staffing* involves determining the personnel needed and taking the necessary action. When new positions are being created, most companies first determine whether any of their own personnel have the required skills. It may be necessary to recruit, select, hire, place, and train the required data processing personnel.

Depending upon the types of computer selected, the applications to be implemented, and the size of the company, it might be necessary to hire a data processing manager, a systems analyst, a programmer, and an operator. After these individuals are recruited, selected, and hired, it will be necessary to provide an orientation to the company. Perhaps two of the company's clerk/typists have the necessary skills to be retrained as data entry clerks.

The job descriptions should identify the tasks to be performed. The job descriptions should also include the training and experience needed for the position, the anticipated career path, the title of the individuals to whom the person in the position is responsible, and who is subordinate to the person in the position.

Directing   *Directing* involves the guiding and supervising of subordinates. It is an essential leadership function. Various leadership styles are used in the motivation and coordination of subordinates. Subordinates must work toward the achievement of their objectives.

Today, when companies use what is called **management by objectives** (MBO), subordinates are evaluated on the basis of how well they achieve their objectives.

Perhaps the two most important things that a manager can do in motivating employees is to provide open, two-way communication and to keep employees informed regarding policy changes and other matters that directly influence working conditions. When open, two-way communication is provided, management understands the informal organization (what is occurring) as well as the formal organization.

In a small installation, the analyst, programmer, operator, and data entry clerks would be supervised by the data processing manager. Short-term objectives can easily be defined for the analyst. The programmer can also be given specific dates when tasks associated with the implementation of a procedure are to be completed. The objectives developed for the operator and data entry clerks would improve their performance and increase their skills.

Controlling   *Controlling* serves to identify major deviations from plans so that corrective action can be taken. In this function:

- Standards of performance are set that often involve time, money, materials, and methods.
- Actual performance is measured against the standard.
- Corrective action is taken when necessary to make sure objectives can be met.

In the case of a new department or a firm with little managerial experience, the development of realistic standards might be difficult. In some areas there may be industry-accepted standards that can be used as a basis for developing specific standards. When actual performance does not measure up to the standard, it is possible that the standard may be incorrect.

When computers were first used, there were no established guidelines or standards to evaluate the effectiveness of analysts, programmers, or operators. Tasks often took longer to complete than the time allocated and projects were seldom completed on time. Actual costs often exceeded budgeted costs.

Experienced data processing managers and analysts can develop realistic standards for most data processing functions. If the manager understands the procedure to be coded and knows the ability of a programmer, the manager can determine fairly accurately how long the coding, testing, and documentation should take.

When a deviation from standard occurs, someone is usually held accountable. Assume that a standard was established for entering sales data into the system. It was estimated that the work would take four hours per day and that the cost of labor would be $7.32 per hour. In actual practice, however, the work usually took five hours per day at a cost of $8.16 per hour. For what factors should the supervisor be held accountable? An investigation might show that because a new data entry clerk was assigned to do the work it has taken more than the estimated four hours. The supervisor would be expected to provide additional training to increase the productivity of the operator. The investigation might also show that the volume of data had increased and that the standard is no longer accurate. The time factor is an area of accountability for the supervisor while the increase in cost probably is not. Since the standard was developed, the cost of living increased and employees were given a raise.

CHECKPOINT

7. What is one of the most widely used methods of dividing the responsibility for managing a firm into subsystems?
8. What can you learn from studying an organization chart?
9. A company just went through a complete reorganization and new job descriptions were developed.
   a. At that point will the formal and informal organizations be similar?
   b. If the organization chart is not changed, will the two be similar ten years from now
10. What are the four levels of management?
11. How do the informational needs of top management differ from the needs of lower management?
12. What are the five basic functions performed by managers?
13. Which function is involved in each of the following?
    a. Authorizing James Green to write a program.
    b. Deciding which computer to order.
    c. Hiring the data processing manager.
    d. Determining that the costs allocated (time and labor) for entering data into the system are not exceeded.
    e. Defining the specific tasks necessary to carry out the plan and delegate the tasks to subordinates.

COMPUTER INFORMATION SERVICES

When punched-card equipment and computers were initially used in business, accounting or recordkeeping functions were the first to be automated. The machine room, as it was called, was located next to the accounting department. The responsibility for supervising the operation was given to the chief financial officer or to an accountant. As the role of computers in business expanded, data processing departments were developed and the role of the data processing (DP) manager emerged. The DP manager often reported to the vice-president of finance. Figure 2-5 illustrates a traditional organization of a medium- or large-size data processing department. In the illustration, the DP manager reports to the vice-president of finance.

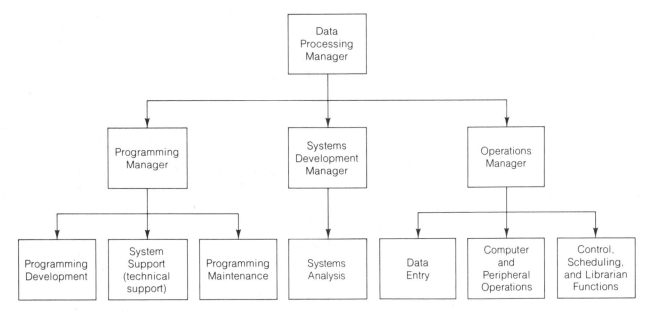

FIGURE 2.5. Traditional organization of a medium or large-size department

The Traditional Data Processing Department

The data processing department has always been considered a service department since its primary function is to provide advice and service to other departments. Although the department had its own budget, the costs of the hardware, software, personnel, and supplies were usually allocated to the users. The DP manager was accountable for keeping actual costs in line with estimated costs. Standards had to be developed so that estimates regarding the time, personnel, material, and equipment needed to complete projects could be accurately determined. Having projects completed within the budget and according to schedule involved a great deal of planning, organizing, and controlling.

Since the computer is used for more than accounting or recordkeeping functions, there were four things wrong with having the DP manager report to the vice-president of finance.

1. The data processing manager was subordinate to the individuals involved in the development of new applications.
2. Because the manager was not part of top management, he or she was sometimes not well informed regarding major changes in goals, objectives, or policy.
3. Financial applications often received a higher priority than warranted due to the placement of the department within the organization.
4. Eventually, more than 50 percent of the applications were nonfinancial. Many of the systems developed were not directly related to functions controlled by the vice-president of finance.

Because the role of the DP manager expanded, the placement of the department changed on the organization chart, the department's budget increased, and the department assumed a more descriptive name.

Computer Information Service Department

Today the computer is being used for engineering, manufacturing, environmental control, process control, numerical control, robotics, flexible manufacturing systems, and a host of other applications that deal with production. Within the **office** (the term is used to denote the communication and paper-handling functions and not necessarily an actual office), the computer is used in five major areas: electronic data processing, communications, **information retrieval, records management,** and word processing. In all five areas, the use of the computer can be cost-justified; the functional areas tend to merge together under a common title—information processing.

Because of the department's expanding role and the nature of its activities, the name Computer Information Services (CIS) is considered to be more appropriate. From now on we will refer to it as the CIS department. Figure 2-6 represents a much broader view of the role the computer plays within the corporation.

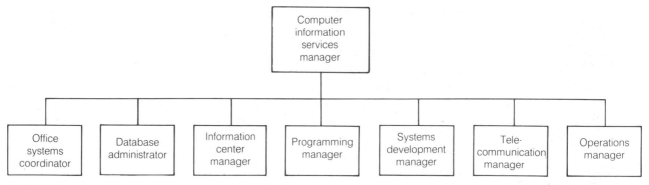

FIGURE 2.6. The computer information services department expands as more functional areas depend on computers for information handling

Today in some organizations, the CIS manager is a line officer. Computer information services is defined in a much broader context and includes: electronic processing of financial data (traditional recordkeeping functions), processing text, coordination of communication services, storage and retrieval of information, and possibly maintaining technical libraries.

The role of the CIS department is far more diversified and *must* provide leadership in the development of standards that will ensure compatibility between computers and *databases*. On the organization chart new functional areas are shown that match the expanding role of the department.

The role of the information center manager is to help the user by providing technical and educational support. As more microcomputers are obtained and used by other departments, the information center manager's role is to: assist in looking for appropriate software; help the user to use the software effectively; and determine whether the corporate guidelines in regard to the acquisition of microcomputers are being followed. Perhaps a more generic name for the function might be the "help desk."

Telecommunications becomes even more important as more online systems are developed. Also as **teleconferencing** and **telecommuting** (the home-to-office link) become more widely used, there must be a greater integration of EDP and office services. Having strong, centralized leadership for all computerized activities that involve telecommunications is essential.

Although the use of computers may be distributed due to the low cost of microcomputers and minicomputers, centralized management is necessary in order to avoid duplication and inefficiency. The database administrator must make certain that databases developed for one computer system, either a microcomputer or a mainframe, can be used by other computer systems within the organization.

THE COMPUTER INFORMATION SERVICE MANAGER

A successful manager must have interpersonal, managerial, and technical skills. The information manager must communicate effectively with all members of the organization (superiors, peers, and subordinates). To be effective, the manager must be able to work with and motivate others. An atmosphere in which employees can achieve their individual goals must be created.

Interpersonal Skills

The manager must be able to communicate affection and concern for the group. In times of stress or uncertainty, the manager should speak authoritatively, fairly, and with concern regarding the individuals involved. The manager must give ongoing guidance and feedback to groups and to individuals. The manager's words and actions must show an expectation of excellence.

The manager's words and actions must also bring security to the group. In communicating with others, the manager must use positive words and speak with vigor depicting drive, energy, vision, and unity. The manager must tolerate ambiguity and stress—and help others to do the same. Although sympathetic and concerned about others, the manager must be willing to take a stand on issues that affect individuals, the department, and the organization.

Managerial Skills

The CIS manager must be able to: define objectives based on the goals of the organization; understand the firm's organizational structure and the relationship of information services to other departments; utilize the tools of management such as MBO, simulation, and mathematical models; and plan, organize, staff, direct, and control the activities necessary to meet the department's and the organization's objectives.

Technical Skills

The manager must have an in-depth understanding of the computer system—both hardware and software—being used. It is necessary to keep up with technological changes. The manager should be knowledgeable in all areas of information processing and should be aware of the standards used in each area.

Technical versus Managerial Skills

Within an organization it is important that job descriptions show the career paths that are available. However, problems sometimes occur when programmers and analyst are promoted to managers. Unfortunately, excellent programmers or analysts may not make good managers. Often people trained in technical fields have not achieved the finesse and authority necessary for speaking, *listening* to, or directing a group. Being comfortable with others and communicating effectively are not inborn traits. The qualifications for a professional manager expand beyond strategic planning to include a vast array of ''people skills.'' Today communication skills and management success are tightly linked.

Major Responsibilities of the CIS Manager

The CIS manager should be responsible for:

- developing long-range plans for the utilization of computers for EDP, MIS, office automation, teleconferencing, and telecommuting.
- developing telecommunications standards so that information networks can be established within the organization and communication can occur with established, external networks.
- developing standards and procedures that ensure compatibility between: databases and computers; microcomputers and mainframes; and computers and other types of equipment designed for office automation.
- developing standards and procedures that will ensure the safety and integrity of information.
- developing standards for designing, implementing, testing, and documenting systems.
- developing and controlling the budget.
- working with the computer policy committee to establish priorities.
- scheduling requests for systems and programming work.
- determining the priorities for all operational procedures.
- communicating changes in procedures, hardware, or software to the department staff members and users who are affected.
- hiring, training, and retaining the department's employees.
- supervising and evaluating all CIS personnel.
- developing job descriptions and career path information for all positions within the department.
- providing opportunities for in-service training.

Each area of responsibility requires the CIS manager to understand what occurred within the organization, what is occurring in regard to technological changes and the needs of management, and what the future information and communication needs will be. The organization must respond to the changes occurring due to the utilization of computer technology.

OFFICE SYSTEMS COORDINATOR

The office systems coordinator must be knowledgeable regarding computer systems and procedures used to electronically process and retrieve information. Good communication and interpersonal skills will be needed along with an understanding of office procedures. Under the area designated as office systems will be:

- *word processing.* At one time most word processing applications used standalone equipment that could not interface with the corporate computer. Separate databases were maintained for EDP functions and for word processing.
- *electronic transfer of information.* As **PBX systems** are being expanded to include computerized communication services, more information may be transferred by using computers and online databases than is transferred verbally.
- *records management.* Within large organizations, the storage and retrieval of information are major functions. Since more information is stored on media from which it can be retrieved electronically, it is difficult to determine where EDP ends and records management begins.

- *teleconferencing.* The ability to transmit pictures, graphics, and information permits individuals in different locations to participate in a teleconference. Participants often feel as if they were all in the same conference room communicating with each other.

INFORMATION CENTER MANAGER

The information center is one of the functional areas within the CIS department. Its major function is to provide help for those who wish to become more directly involved in the utilization of computers. The information center has the responsibility for:

- Selecting and maintaining software which permits users to extract their own reports and to manage their own databases.
- Conducting in-service training seminars to provide instruction for those who wish to use the software.
- Providing some systems analysis support for users who would like to develop their own application software.
- Assisting users to develop and maintain their own databases. The databases must be accessible to the users and also must be compatible with the centralized corporate databases.

Information center employees must understand the characteristics of the computer systems being used. They must be very well informed regarding the database software and the languages available which permit users to extract meaningful reports from online databases.

DATABASE ADMINISTRATOR

Today organizations that develop transaction processing systems and provide for office automation must have comprehensive databases that serve numerous functional areas. While a file is a collection of records often used in a limited number of applications, a database is a shared resource used in many applications. The use of a common database promotes the sharing of data and the development of standards. If an administrator is not appointed, standards may not be developed. The **database administrator** (DBA) is responsible for the initial design of the database structure, monitoring the contents and structure of the database, developing and implementing recovery procedures, maintaining documentation, and evaluating proposed additions to the database software.

CHECKPOINT

14. Why should the data processing (computer information services) department be considered a service department?
15. What are the three functional areas within a traditional data processing department?
16. Why is the title computer information services department more descriptive that the title data processing department?
17. The computer information services manager must be knowledgeable regarding hardware, software, and procedures. What other types of skills must the manager have?
18. Why might an excellent analyst or programmer make a poor manager?

19. What functions are handled by the office systems coordinator?
20. What functions are controlled by the database administrator?
21. What is the major function of the information center?

THE SYSTEMS DEPARTMENT

Systems analysis is not a new field. Long before there were computers, analysts were engaged in many of the functions they now perform. The six major functional areas of systems work are:

- *Forms design and retention.* What format should be used for source documents and reports and how long should they be kept?
- *Work simplification.* Is there an easier, most cost-effective way of accomplishing the objectives that will provide more relevant and timely information?
- *Information management.* How should information be organized, retained, and retrieved? Should a medium such as videodisks be used that permit graphics, pictures, and text to be stored online?
- *Work measurement.* Under varying conditions, how long should it take to perform tasks, where should the tasks be performed, by whom should the tasks be performed, and what standards can be set?
- *Analysis and design of new procedures and systems.* Should the systems be designed inhouse by analysts working with users or is software already available that can accomplish the same objective? How can the new system be interlinked with other systems to contribute to the development of a more comprehensive MIS?
- *Ergonomics.* As new applications are designed, the productivity, health, and safety of the employees must be considered. The analyst must be considered with the total environment of the employee.

Often the role of the analyst is associated only with the analysis and design of new procedures and systems. As you read the job description illustrated in Figure 2-7 and study the qualifications needed to become an analyst, you will realize how much more an analyst does.

The Systems Analyst

While the analyst's primary job is to design systems, the individual is also involved in a wide range of other activities. The job description illustrated in Figure 2-7 is somewhat typical, although the ones used in industry vary. The responsibilities and duties differ from firm to firm depending upon the size of the department and how management views the role of the analyst. One analyst may be involved only in the investigation, design, and evaluation phases, while a second analyst may be actively involved in all phases of the project.

Within the analyst **job family** the descriptions can range as follows:

Manager	Is in full charge of all activities of the systems department.
Lead	Usually considered an assistant manager.
Senior	Works independently and is competent to work on all phases of the project.
A	Works under general supervision and can usually work independently on most phases.

computer products inc. POSITION DESCRIPTION 05/13/86

TITLE Systems Analyst	SECTION/DEPARTMENT Computer Information Service
REPORTS TO: Systems Development Manager	SUPERVISES: Analyst Trainees Programmer/Analysts

DESCRIPTION: An analyst gathers and analyzes information in order
to develop and modify information processing systems. He or she also
designs and specifies systems and methods for implementing the systems.
While evaluating operational systems, the analyst works closely with the
personnel in the area being investigated. After the fact finding is
documented and studied, a formal presentation of findings and recommendations is made.

RESPONSIBILITIES:

1. Define requirements for changing or developing systems.
2. Determine the cost-effectiveness of recommendations.
3. Guide systems development and implementation. The analyst will
   sometimes serve as project leader or be a member of the project
   team.
4. Ensure optimum control and security of data.
5. Utilize available resources efficiently.
6. Test all segments of a system to make certain requirements have
   been met.
7. Assist in the development of complete documentation for all systems.

DUTIES:

1. Assist in programs to educate management and users so that they
   fully understand the capabilities and requirements of their systems.
2. Develop, implement, enforce, and adhere to procedural standards.
3. Design systems and procedures.
4. Review and approve proposed system solutions.
5. Estimate the resources needed for proposed systems projects.
6. Perform resource planning and scheduling.
7. Define requirements for studies.
8. Perform interviews and pursue other methods of gathering data.
9. Apply current technology to solution of problems.
10. Remain current regarding technological changes in information
    processing and other directly related areas.
11. Define systems security and control procedures.
12. Develop testing and conversion plans.
13. Develop file structures and databases.
14. Design forms and reports.
15. Design data collection, processing, and control procedures.
16. Develop specifications for programmers and procedure writers.
17. Document results of design and testing.
18. Assist in systems testing and implementation.
19. Fulfill administrative reporting requirements.

WORKS CLOSELY WITH: Management, user's technical personnel, programmers,
operations personnel, and other analysts.

REQUIREMENTS:

1. Bachelor's degree (preferably in one of the computer sciences,
   accounting, statistics, or mathematics).
   Graduate education is desirable.
2. Training in administrative management, project control, supervisory
   techniques, and analysis and design techniques.
3. Experience in information processing with demonstrated project management
   responsibility.
4. Familiarity with the information needs of management.
5. Ability to effectively use verbal, graphic, and written skills.
6. Experience in working with mainframe and microcomputers.

FIGURE 2.7. A typical job description for a systems analyst

B	Works under direct supervision. Will sometimes need instructions or supervision.
C	Should only work under immediate supervision.
Trainee	A probationary employee who has no previous experience and will need immediate supervision and detailed instruction.

There are also job families for the analyst/programmer and programmer/analyst. If there are levels within a category such as analyst, each should be defined and an appropriate job description developed. Job descriptions vary in the detail provided regarding aptitudes, education, and the experience required to obtain the position. Some forms also include objective measures (how the person is evaluated), the career paths that can be pursued, and the salary ranges.

The Characteristics of a Successful Analyst

Basic Skills   In order to be successful, the analyst must keep technical skills current. This may mean learning to use new languages, software packages, and hardware. As new projects are assigned, new technical skills may need to be developed. In addition to technical skills, an analyst must be creative and have good interpersonal, communication, and conceptual skills.

Creative   In systems work there is seldom an exact pattern to follow. Therefore, the analyst must be creative in order to visualize what can be accomplished when numerous pieces are fit together to form a meaningful whole.

Interpersonal skills   An analyst must work effectively with others within the organization. In designing and implementing a system, an analyst will work with top management as well with as operational-level staff members.

Communication skills   An analyst must be able to use the English language correctly to communicate verbally and in writing. Clear, concise language understandable to the listener or to the reader is essential.

Conceptual skills   The analyst must see the relationships that exist between problems. Causes must be related to effects. Analysts identify problems and then determine the causes. A solution must then be found that stays within a given set of rules or parameters.

Communication and interpersonal skills can be developed by people who wish to succeed and realize the importance of working effectively with others. Teachers can teach and students can learn these skills. It is more difficult to teach someone to think conceptually since each problem encountered is different. Conceptual skills can be developed by working with an individual who is skilled in this area. Creativity may be more of a gift than a skill that can be developed.

Understanding Basic Motivation   The analyst who understands basic motivation, why people resist change, and how they resist change may be able to overcome some of the resistance. Most management books make reference to psychologist A.H. Maslow's *hierarchy of need* in terms of what motivates people. Maslow's categories, from the lowest priority to the highest are:

Need	*Example*
1. Physiological	Food, clothing, and shelter
2. Safety and security	Protection against danger and loss of job
3. Social	Being able to identify with individuals and groups
4. Egotistic	Recognition, status, and importance
5. Self-fulfillment	Realizing one's fullest potential in creativity and self-development

Lack of cooperation during the investigation and design phase of a systems project is often the result of one of the five basic needs not being met. For example, assume that during the investigation of a new sales system, the sales manager was not involved. Again during the design phase, the sales manager was not consulted or provided with any type of progress report. During the implementation phase, the sales manager may be very uncooperative because he feels rejected and is concerned about job security.

A successful analyst must understand human behavior, have the required skills, and also understand the organization, governmental regulations, and economic conditions. An analyst as a *change agent*, or catalyst for change must plan for and gain acceptance of organizational change.

PROGRAMMING DEPARTMENT

The organizational chart illustrated in Figure 2-5 indicates that there are three major divisions within the programming department: programming development, system support, and programming maintenance. The programming development group is responsible for developing, testing, and documenting new programs. The system support group maintain the operating system and other complex software packages such as the ones required for telecommunications and online databases. Maintenance programmers make modifications to application programs.

The Programmer

Within the programming job family there are descriptions for the manager, lead programmer, senior programmer, programmer A, programmer B, programmer C, and programmer trainee. Each level should have a detailed job description similar to the one illustrated in Figure 2-8.

Job descriptions should be updated as changes are made within the computer information services. Current descriptions for job classifications should be available during recruitment and hiring new employees. They should also be available to any member of the organization.

Although programmers do not work as closely with management and users as analysts do, they still need conceptual, interpersonal, and communication skills. A programmer must also have a certain amount of creativity.

Some individuals may understand the syntax of a language and may also be able to verbally define a problem. However, if they lack conceptual skills, they may not be able to write a program that will solve the problem.

Communication skills are necessary since programmers will work with others and be responsible for a major portion of the documentation. They must be able to accurately describe what the individuals who work with the system (users, data entry operators, and computer operators) must do. As part of a team they must be able to effectively communicate with other members of the team.

Programmer/Analysts

Exactly how the functions are divided within the CIS department depends on the volume of work, the size of the installation, and the number of people under the manager. As indicated on the organizational chart illustrated in Figure 2-5, the major functioning areas are systems development, programming, and operations. In a small department, systems development and programming may be combined in one area of responsibility. There is also a trend in some companies to merge the two areas so that the title of programmer/analyst is often found on career path information or job description forms.

computer products inc.   POSITION DESCRIPTION   05/13/86

TITLE Programmer	SECTION/DEPARTMENT Computer Information Services
REPORTS TO: Programming Manager	SUPERVISES: Programmer Trainees

DESCRIPTION: A programmer designs and tests program logic, codes programs, and prepares them for computer operations. Prepares logic plans and codes routines required to process data. (After the design process, a team walkthrough occurs before the program is coded. When the coding is completed, another walkthrough occurs to test the validity of the code.) Assists the systems analyst in establishing file requirements and processing specifications. Performs programming tasks according to established standards. Thoroughly tests the operation of programs.

RESPONSIBILITIES:

1. Develops accurate and efficient computer programs.
2. Maintains current knowledge of standard languages, coding methods, operations requirements, and the company's programming standards.
3. Tests programs completely.
4. Analyzes existing program logic to determine the best method for making required changes.
5. Designs change modules and adjustments to existing code to accomplish corrections or modifications.
6. Tests the program modifications.
7. Maintains a test database.
8. Participates in design and code walkthroughs.

DUTIES:

1. Review program specifications submitted by the analysts.
2. Determine personnel requirements, cost, and time for programming projects.
3. Review and approve proposed program logic.
4. Develop, implement, enforce, and adhere to procedural standards for programming.
5. Thoroughly test all programs before releasing them to operations.
6. Assist in programs to educate management and users in the capabilities of the computer system and in the procedural systems being developed.
7. Assist in the design of procedures and systems.
8. Analyze program specifications for completeness and conformity to standards.
9. Design program logic to meet specifications.
10. Conform to all programming standards.
11. Code programs in the language authorized.
12. Prepare test data.
13. Test programs.
14. Document programs according to installations standards.
15. Submit the necessary reports to management.

WORKS CLOSELY WITH: Analysts, programmers, operations personnel, users, and other programmers.

REQUIREMENTS:

1. Two-year associate degree in data processing or computer science; bachelor's degree preferred.
2. Technical training in programming.
3. Training in the use of job control languages.
4. Skills in accounting, statistics, mathematics, management, and communication are desirable.

FIGURE 2.8.   A typical job description for a programmer

CHECKPOINT 22. What are the six major functional areas involved in systems analysis and design?
23. What is a job family?
24. What are the differences between conceptual skills and interpersonal skills?
25. In regard to the hierarchy of need, which need has the lowest priority and which has the highest priority?
26. What is the egotistic need?
27. Why must an analyst continually update his or her skills?
28. How does a programmer's job differ from that of an analyst?

COMPUTER OPERATIONS

Within computer operations there are five major areas: scheduling, data control, data entry, computer and peripheral operations, and librarian functions. In some companies there will be someone from operations on the project team to design and implement a new system. In other companies operations personnel are not included in the planning and design phases. Companies that have been most successful in implementing new systems have been those that involved operations personnel in all phases of the analysis and design of a system.

When new systems are being considered for implementation on an existing mainframe, operations must confirm that there is time available to enter the data, process the data, produce the reports, and maintain the necessary files. Since operations personnel understand the problems that occurred in running the old system, they should be consulted regarding how the problems can be eliminated.

During the implementation phase, time must must be scheduled to build test files and to make test runs. The actual files will also need to be ready by the time the procedures become operational.

In some computer service areas, operations is perhaps the weakest link. Due to a combination of poor documentation, inadequately trained operators, and a lack of well-defined procedures, the **throughput** is far less than it should be. Throughput is the amount of data that is entered and processed, and results in displays, printed reports, or output stored in files. In addition, reports may be inaccurate or late.

Manager of Operations

The manager of operations has the responsibility for:

- making sure jobs are run according to a predetermined schedule.
- delivering good response time to users.
- implementing security measures that will provide safety for hardware, software, files, documents, and reports.
- hiring, training, and evaluating operations personnel.
- seeing that all equipment is properly maintained. Often preventive maintenance and other types of service are provided under a maintenance agreement with the vendor.
- making certain that all documentation used within operations is updated whenever operating procedures or programs are changed.
- maintaining a problem log that identifies the reasons execution-time errors occurred.

- maintaining an inventory of supplies such as forms, magnetic tapes, diskettes, and disks.

Data Control   The control function is vital for maintaining the integrity, security, and accuracy of data. While the analyst designs the controls, it is the data control section which will see that some of the controls are implemented. Certain controls are designed to make sure that the documents are processed according to schedule and then returned to the user. Others indicate that the data has been processed correctly. Some controls are initiated by the users, others by the data control clerk.

External Controls   Batch jobs often have more external controls than do online applications. In regard to the workflow for a batch card edit program, the following might occur. The time cards and transmittal forms are sent by the payroll department to data control. The transmittal form contains a document count (how many time cards and how many individuals will be paid), the total number of employees to be paid in each of the major categories, and the total number of regular hours and overtime hours for hourly employees. On the form the numbers of the first and last time cards are also recorded. The data entry operator makes certain all of the time cards are accounted for, logs in the time the documents were received, and sends the documents to the data entry department.

After the data entry department processes the documents by recording the data stored on the documents in the format described in the documentation, the source documents (time cards) are returned to data control along with the transmittal form. After machine readable input is processed, the computer output is sent to data control to be compared with the information recorded on the transmittal form. The data control clerk can determine whether the correct number of payroll records has been processed and whether the total number of regular and overtime hours processed agrees with the total on the form. If the printed totals do not agree with those printed on the computer output, the error must be found and corrected. It may be necessary to rerun the job and then check the new output with the transmittal form.

Internal Controls   When online systems are designed, building controls into procedures and programs is one of the most important aspects of the analyst's job. The importance of the control function in determining the integrity of data and information cannot be stressed enough. A large percentage of analysts' time should be spent in building controls into the system. One CIS manager indicated that from 40 to 45 percent of his programmers' time was spent in designing and coding editing and internal controls for programs.

A banking application might be an online system. Each teller starts with a given amount of cash. As the day progresses, cash is deposited by some customers while other customers withdraw cash. In addition, large quantities of cash may be "deposited" by the tellers. Since all transactions are entered by the teller into the computerized system, at the end of the day a printout can be obtained that summarizes the day's activities and also indicates how much cash on hand each teller should have. Although the control feature was carefully built into the system, the cash-in-drawer total printed on the report is a by-product of the transaction recording system. As more online systems are designed and implemented, more of the control functions will be built into the balancing procedures performed by the person who enters the data. In the illustration, the teller has the responsibility for checking the cash on hand with the printed balance.

Data Entry

Until recently, the data entry functions have been somewhat neglected by the analyst. Since costs associated with the data entry functions continue to increase, more attention is focused on how and where data should be entered into the system.

Centralized Data Entry In a centralized data entry department, the supervisor is responsible for the security of the documents, processing the documents according to schedule, the accuracy of the data recorded onto the input medium being used, and the proper maintenance of the equipment.

Distributed Data Entry Often the data entry function is distributed to the users' areas. Since the users are made responsible for *their* data, errors are decreased. When there is incomplete data on a document, the resources needed to complete the documents and to check what appears to be incorrect are readily available.

In designing new applications or modifying old ones, the analyst must determine the most cost-effective way of entering data into the system and where the data should enter the system. The method selected must provide for the accuracy and security of the data. When data is entered into the system at the source, additional controls need to be built into the system.

Scheduling

A scheduler must understand all areas of computer operations in order to build a realistic schedule. Determining priorities for the jobs that must be run starts with management and serves as a basis for building the day-by-day schedule. Time must be allocated for scheduled production runs, test runs, running those "must-have-it-now" jobs for top management, preventative maintenance on the equipment, and contingencies.

Before the implementation phase of a project is started, the analyst should determine how much time will be needed to execute the procedures and when the jobs will need to be run. Once the system is operational, the scheduler must fit the new procedures into the schedule.

Librarian Functions

Job descriptions for librarians vary. Usually the librarian is responsible for the proper handing, labeling, and storing disks and tapes. The librarian also must maintain a history of all tapes and disk packs. When the file stored on a reel of tape or a disk pack is no longer needed, the tape or pack is released. It is up to the analyst to determine how long each file should be retained.

The history indicates the file or files stored on the medium, the retention period for the data, the utilization of the file, and any problems that might have occurred when the file was being used.

Librarians may also be responsible for maintaining source decks, updating documentation, labeling and storing computer printouts, and updating standards manuals.

Computer and Peripheral Operations

The major function of computer operations is to achieve the maximum amount of throughput possible. Jobs should not **abort** (cancel) due to bad data, operator error, or malfunctions of the hardware or operating system. If a program does cancel due to a data exception (bad data), it should be noted in the error log and the problem referred to the manager. When a program cancels, it may be an indication that the data used in the program was not completely edited. For example, the programmer failed to edit a field that should contain only numeric data.

When character data was recorded in the field, the job canceled and the programmer had to be called to make the necessary change in the program.

When a job cancels due to an operator error, it could mean that the operator failed to follow the documentation or that the documentation was innaccurate or incomplete. A change could have been made in the program and the documentation might not have been updated to include the change. Operator errors can be costly. For example, one operator failed to follow backup procedures and then, due to a careless error, destroyed the file that he failed to backup. The company estimated that it cost more than $50,000 to rebuild the file.

To ensure maximum productivity of the computer system, operators must be well-trained, follow the documentation provided, run jobs according to schedule, know how to use the reference materials available, respond to error messages quickly, perform some of the routine maintenance on the equipment such as cleaning tape drives, and maintain a problem log which indicates why jobs aborted.

How well analysts and programmers do their jobs in regard to editing, controls, form design, record layouts, and documentation has a direct impact on operations. If an analyst or programmer is permitted to develop his or her own guidelines rather than following established standards, difficulties arise for operations personnel. The analyst should work closely with operations during the investigation, design, and implementation phases of the project.

In some CIS departments, part of the training provided for analysts and programmers includes spending time in operations so they understand the roles of the scheduler, data control clerk, librarian, and operator.

In the decentralized computer site, someone within the department will have to assume the responsibilities normally assigned to a computer operator, scheduler, data control clerk, or librarian. Although many of the newer minicomputers require very little operator intervention, it is still necessary to do some of the functions normally assigned to a computer operator.

ORGANIZATION OF A SMALL DATA PROCESSING DEPARTMENT

In a textbook, the small shop is often excluded from any discussion. This is unfortunate, since small shops far outnumber medium- and large-size computer centers. Individuals entering data processing should be aware of what is expected in both environments—the small shop and the large shop.

Figure 2-5 illustrates a typical organization for a fairly large computer center. What about a small shop that has one or two people? *The functions to be performed are the same as the ones performed in a large organization.* In the small shop, data control and scheduling are not eliminated. The tasks normally performed by the data control clerk and scheduler are performed by the operator or manager. It is not uncommon for the programmer to also perform tasks normally assigned to a computer operator.

In a two-person shop, an analyst would wear many hats—perhaps those of the computer information services, programmer, scheduler, data control clerk, and operator. An analyst in a small shop will find the job demanding, challenging, and rewarding. The person must be more resourceful than his or her counterpart in a large-shop environment.

People working in a small-shop environment have to be willing to perform a variety of tasks. If an accurate job description were available for the small-shop manager, the list of responsibilities and duties would be almost endless. The formal and informal job descriptions would vary. Since the small-shop manager must become directly involved in the design and implementation of systems, having current technical skills is critical. While more emphasis is placed on the manager's technical skills, less emphasis is placed on managerial and interpersonal skills.

Job descriptions might be developed for the analyst/programmer and programmer/operator. In a small shop, the operator may also perform some of the data entry and data control functions.

MICROCOMPUTER ENVIRONMENTS

Organizations that do not have a CIS department and only have microcomputers (no minicomputers or mainframes) are encountered more frequently than are the large or small shops. The person who operates the computer normally enters the data and performs the functions assigned to an operator. Since the manufacturers of the system have tried to create a "user friendly environment," most of the programs are internally documented and the operator is guided through the execution of the program.

Often the full potential of the computer is not utilized since no one within the organization fully understands the potential of the system. Also, some of the major software packages, such as the ones labeled database management software, **spreadsheets**, and word processing, require training. Employees using the computer systems may have a superficial knowledge of the software but have not taken the time to obtain an in-depth understanding that would enable them to use the software more effectively.

In the microcomputer environments, someone must perform many of the tasks assigned to an analyst. A feasibility study should be made to determine which functions can be computerized effectively, and a study must be made to determine what software is available that meets the needs identified. Training must be provided for the individuals who will work with the computerized systems.

Someone should be designated as the "computer expert." This individual will be responsible for coordinating the use of computer technology and for developing long-range plans. The individual may be the *entire* CIS department and assume the roles played in large organizations by the database administrator, analyst, programmer, and technical support staff.

Vendors, colleges, and universities are now beginning to address the needs of small organizations that obtain their first computer system—a standalone microcomputer.

THE COMPUTER POLICY COMMITTEE

Since the computer information services manager receives requests for system studies from many different areas, an advisory committee or computer policy committee is often established to help determine priorities. In large companies, there may be two computer policy committees.

The first committee is composed of top management. Because members of this committee have so many other commitments, they meet rather infrequently—perhaps only three or four times a year. They are concerned primarily with long-range planning, major goals, major expenditures, and overall design and effectiveness of systems. Since computer technology influences all functional areas within the organization, representatives from all areas should be on the committee.

The second committee is composed of middle-management personnel. Members of this committee meet more frequently. They are concerned with determining priorities, approving the design of a proposed system, and approving changes to existing systems. They are also directly involved in allocating and controlling the resources necessary to accomplish the objectives established for systems and procedures that involve the use of computers. In addition, the committee helps evaluate the reports produced during the various phases of systems analysis and design.

CHECKPOINT

29. What are the five functional areas within operations?
30. What is throughput?
31. What controls were built into the time card program? Who performed the control functions?
32. When online systems are developed, are controls eliminated?
33. Why have analysts been concerned about how, where, and when data should be entered into the system for processing?
34. Is the trend today to centralize or decentralize the data entry function?
35. What tasks are performed by a librarian?
36. What is the major function of computer operations?
37. What might occur if the job documentation is inaccurate or incomplete?
38. If you were an analyst working in a large CIS department, would your job description differ from the one prepared for someone working in a small organization in a shop that uses a rather small minicomputer system?
39. In regard to the utilization of the potential of the computer system, why might a microcomputer system not be used very effectively?
40. If a company has two computer policy committees, how do the two differ in terms of who is on the committee and the type of decisions made?

SUMMARY

When designing a system or procedure, the analyst must determine what type of information is needed for each level of management. Using a systems approach, the project can be divided into manageable tasks that have measureable objectives.

In order to design meaningful systems, the analyst must understand the organization and its information needs. It is also necessary to understand the capabilities and limitations of the hardware and software that will be used to implement the design.

During the project, many people with different talents and responsibilities will be involved. The analyst may be the catalyst, or change agent, but the changes will occur only if there is support and assistance from data processing personnel, management, and operational level personnel. Ultimately, it is the operational level personnel who will work with the system after it is implemented.

Each person within the CIS department has unique responsibilities and duties. In investigating the need for new systems, as well as for designing and implementing them, many skills are needed—technical, interpersonal, communication, and conceptual. Although creativity is needed, it is also necessary to adhere to standards.

After you learn more about the various phases of systems analysis and design, you will be asked to do projects that require a variety of skills and talents. Each task will be defined and have measureable objectives.

DISCUSSION QUESTIONS

1. Why do many companies feel it is necessary to use a systems approach to management?

2. When the systems approach is used, how is a complex problem solved?

3. What are the components of a management system?

4. Why might the formal organization of a company differ from the informal one?

5. The CIS manager has indicated that only top, middle, and lower management are to be consulted during any phase of the system study. Do you agree with this policy?

6. The CIS manager has stated that the staff is very busy and that as a result, only one sales report will be available. The report, which will be printed monthly, will be large and detailed. It will list all 10,000 products the company manufactures and under each product, it will list the top ten sales representatives. The report will include the sales figures for the month, previous month, year, previous year, and percentage increase or decrease for both the month and the year. Producing only one report rather than several will save computer time. Do you agree with this plan? Tell why you agree or disagree with the manager's plan.

7. The manager is concerned with the length of time that it takes to get systems up and running. In order to save time, numeric fields will no longer be edited since the data read in as input has already been verified by the data entry operator. The time cards will no longer be processed through data control since controls such as record counts or totals on regular and overtime hours will not be established. The manager feels this new policy will save both time and money. Tell why you agree or disagree with the manager.

8. Systems work is divided into what six major functional areas?

9. Describe the type of person you feel would make a good analyst. What educational background should the analyst have? What additional background would you recommend for an analyst who would like to work for a financial institution?

10. Mark Wilson would like to become an analyst but has a difficult time expressing himself. He has taken all of the required computer science courses and decided to take advanced PASCAL as an elective. What elective would you recommend that Mark take? Why would you make the recommendation?

11. Since Sue Novak understands the COBOL syntax, she can write an "A" COBOL test, providing it is an objective test. Sue can also verbally explain the problems that must be solved. However, when she writes a computer program in COBOL, the program never seems to work. What skill might Sue lack that is needed by both analysts and programmers?

12. A company placed an ad for a CIS manager. All but two candidates have been eliminated. There are 50 people in the CIS department and the department is still expanding. Study the qualifications and determine which of the two candidates you would hire and describe the basis on which you made the decision.

	Mr. A	*Ms. B*
College major	Accounting	Statistics
Minors	Computer science/ management	Economics/mathematics
Graduated	1973	1970
Work Experience	Sales/management/DP	All DP

Both individuals have good references. Ms B worked first in a small shop with microcomputers and then in a large organization with a mainframe. She progressed from operations to programming and then became an analyst. In some areas of DP, Mr. A is a little weak. However, he has done a great deal of programming and understands most of the concepts pertaining to operating systems, databases, and teleprocessing.

Team or Individual Projects

1. Study the following mini case and then determine what is wrong with the approach used. Relate your solution to the material presented in Chapters 1 and 2.

   Mini Case Study The president of Blaine International decided that a new inventory system was needed. He asked Frank Lopez, one of the analysts, to design and implement a new inventory system. Lopez started immediately on the design phase. After the new system was designed and implemented, Lopez called a meeting of the inventory control staff and explained the new system. The president was pleased that the system was up and running in such a short period of time. However, after it was operational, a number of things occurred that concerned the manager of the computer information services department.

   a. A number of abort reports were sent to Frank Lopez. However, Lopez did not reply to the memos that were attached and the jobs continued to abort.

   b. The manager of the inventory control department resigned. He indicated that he had an offer that was too good to pass up. He had been with the company for 23 years.

   c. The old inventory control system had a daily exception report which listed all items that were below the reorder point. The report was eliminated to save money. Lopez defended his position by stating that the monthly detailed report listing all items in stock would serve the same purpose.

   d. The sales department indicated that the items backordered had increased by 10 percent and that a number of old customers were no longer placing orders.

   e. The organization's auditor indicated that he felt there were not enough controls built into the system.

   f. All of the input entered into the inventory control system is keyed in by data entry operators located within the centralized data entry department. When entering the data into the input medium, the operators often complain that the source documents are incomplete and have to be returned for completion before the data can be entered into the system.

2. Using resources such as your textbook, Figures 2-7 and 2-8, management texts, articles in current publications, and occupational information reports, write a job description for the manager of the computer information systems department. It would also be helpful to secure information from local companies regarding their concept of the requirements for the manager of the CIS department. Be sure to include a description of the position, responsibilities, duties, and requirements.

3. Interview an analyst or programmer and obtain the person's opinions regarding the following topics:

   a. The advantages or disadvantages in being an analyst or programmer.
   b. Educational background of the person interviewed:
      (1) during college—major, minors, courses that were very helpful.

(2) after college—formal classes, workshops, conferences.
(3) in-service training provided by the company
  c. Employment history:
    (1) previous jobs.
    (2) career paths available.
  d. Responsibilities and duties of their present position.

GLOSSARY OF WORDS AND PHRASES

abort The job is canceled by the system's control software since an unrecoverable error occurred. Generally a message will be displayed on the operator's console.

database administrator An individual who coordinates the planning, development, and implementation of the company's databases.

exception reports Reports listing items that do not fall within an expected range, exceed the established standard, or need special consideration. An inventory exception report might only list items that need to be reordered.

external information Information available from various organizations, such as the government, professional organizations, research associates, and other companies. The organizations provide information regarding economic trends, new regulations, and research projects.

goals Broad, long-range statements of intent. Usually goals are not measurable.

information Results of processing data.

information retrieval The methods and procedures for recovering specific information from a storage medium.

internal information Information generated by a company as it processes its own data.

job family A group or cluster of jobs that relate to the same activity, such as that of programmer or analyst. The classifications of manager, lead, senior, A, B, C, and trainee have evolved. Each level has its own qualifying characteristics.

line officers Officers who make a direct contribution to the achievement of the firm's goals and objectives. The officers are involved in creating, financing, and distributing goods and services. Line officers usually are directly responsible to the president of the organization.

management by objectives (MBO) Specific objectives, achievable within a given period of time, such as six months or a year, are developed for each individual. The individual is then evaluated on the basis of how well the objectives are accomplished.

mathematical model An approximate description of a business problem written in mathematical language. After the model is established, it must be tested for validity.

objectives Objectives are derived from goals and are usually specific, short-range, and measurable statements of intent.

office The term is used to denote the communication and paper-handling functions of an organization and not necessarily an actual office.

PBX system An acronym for private branch exchange. Most medium- and large-size firms have their own inhouse PBX phone systems. Today, many are computerized and used as part of the telecommunication system.

records management Procedures designed to provide economy and efficiency in the creation, organization, maintainance, use, and disposition of records.

spreadsheets A display (either printed or on the CRT) that contains rows and columns of data. Spreadsheet software can cause the original data recorded in the rows and columns to be manipulated and new data to be created.

telecommunications The transfer of data from one place to another over some type of communication system.

telecommuting Using a terminal or a personal computer to communicate with others within the organization and for accessing the database and utilizing the necessary software so that work can be accomplished at home rather than by commuting to the office. By the end of the century, there may be 13 million telecommuters.

teleconferencing The communication capabilities of the computer are used to project images, graphics, pictures, and sound to another location. Some of the conference attendees may be located in one geographical location and the rest in a different location. Two or more groups can communicate with each other as effectively as if they were all located in the same conference room.

throughput The amount of data that can, within a given period, be entered into the system, processed, and outputted in final form.

STUDY GUIDE 2

Name _____

Class _____ Hour _____

A.   Indicate whether the following statements are true (T) or false (F). If false, indicate in the margin how the statement should be changed to make it true.

_____   1.   A far better system is developed when an analyst designs and implements the system before contacting the users.

_____   2.   A systems approach begins with developing objectives for each task that makes up the procedures.

_____   3.   The systems approach focuses on methods first and goals second.

_____   4.   When a team approach to systems analysis and design is used, each member of the team should have an identical background.

_____   5.   When a structured approach is used, the details of each task are identified before the design of the total project is completed.

_____   6.   A management system consists of only four components: the individual, the formal organization, leadership styles, and the physical settings in which the individual tasks are performed.

_____   7.   Staff officers report directly to the president of the organization and work to formulate goals and policies.

_____   8.   Top management's functions are tactical. Managers in this group are responsible for allocating and controlling the resources necessary to accomplish the objectives of the company.

_____   9.   Middle management's functions are short term and deal with day-by-day scheduling, checking the results of operations, and taking necessary corrective action.

_____   10.   Analysts only work with top management.

_____   11.   Top management needs more external information than is needed by middle management.

_____   12.   The five management functions are planning, organizing, staffing, directing, and comparing.

_____   13.   Organization charts are only used by top management.

_____   14.   When EDP departments were first developed, the manager usually reported directly to the chief financial officer.

_____   15.   Today over 70 percent of the new systems designed are for financial applications.

_____   16.   Within the office, the computer is only used for electronic data processing, information retrieval, records management, and word processing.

_____   17.   Teleconferencing permits members of the organization's staff to work at home. Either microcomputers or terminals can be used to access the mainframe.

_____   18.   Today the role of CIS is diversified and should provide leadership in the development of standards that will ensure compatibility between the various computers and databases used within the organization.

_____   19.   The CIS manager needs technical skills. Managerial and interpersonal skills are in the nice-to-have-but-not-essential category.

_____ 20. A good manager must be concerned with the individual and still be able to take a stand on issues that affect the individual, the department, and the organization.

_____ 21. There is very little correlation between communication skills and success in management.

_____ 22. One of the responsibilites of the CIS manager is to provide in-service training for staff, management, and users.

_____ 23. Electronic transfer of information is one of the areas included under office systems.

_____ 24. The information center's only responsibility is to develop software for users.

_____ 25. A datebase is a collection of records used in a limited number of applications.

_____ 26. The systems department is only concerned with the development of new applications.

_____ 27. According to the job description provided for an analyst, he or she is not responsible for determing the cost-effectiveness of a new system.

_____ 28. If analysts have good conceptual skills, they will see the relationships that exist between problems and will be able to relate the causes to the effects.

_____ 29. An analyst need only be able to communicate ideas and concepts verbally.

_____ 30. Conceptual and creative skills are easier to develop than are communication and interpersonal skills.

_____ 31. People often resist change because of the hierarchy of needs. Until the physiological need and the need for safety and security are achieved, the individual involved may resist change due to concerns over the future.

_____ 32. Programmers have more direct contact with management and users than do system analysts.

_____ 33. The three functional areas within many CIS departments considered traditional are systems design, programming, and data control.

_____ 34. Egotistic needs involve food, clothing, and shelter.

_____ 35. The five major areas within computer operations are scheduling, data entry, computer and peripheral operations, librarian functions, and documentation.

_____ 36. Operations personnel should not be directly involved in the design or implementation of a new application.

_____ 37. The manager of operations is responsible for seeing that security measures are taken to protect the hardware, software, and databases.

_____ 38. When online applications are developed, internal controls need not be built into the programs since the data is entered at the source by individuals such as bank tellers who have been trained in the use of the terminals.

_____ 39. Building controls into computerized procedures is one of the most important aspects of software design.

_____ 40. When the data entry function is distributed, there is often an increase in the number of errors made in entering the data.

B. Match the definition of the task with the individual who performs it.

a. analyst   c. programmer   e. data control clerk
b. CIS manager   d. data entry clerk f. operator

_____ 1. Completes the abort report.

_____ 2. Compares the actual cost of processing data with the amount budgeted.

_____ 3. Described as a catalyst for change.

_____ 4. Checks the totals on the transmittal form with the ones on the computer printout.

_____ 5. Transforms data recorded on a source document into a machine readable form.

_____ 6. Is responsible for developing long-range plans for the utilization of computers in EDP, MIS, and other applications.

_____ 7. Is responsible for changing or developing systems.

_____ 8. Should have an in-depth understanding of the language being used as the inhouse language, of coding methods, of operations requirements, and of the company's programming standards.

_____ 9. Is responsible for developing in-service training programs for users and members of the EDP staff.

_____ 10. Must be creative and have good conceptual, interpersonal, and communication skills.

_____ 11. Works with the computer policy committee to establish priorities.

_____ 12. Works in the area of forms design and retention, work simplification, information (records) management, and work measurement.

_____ 13. Must know how to use reference materials, respond to error messages, and run jobs according to schedule.

_____ 14. May be a line officer and responsible to the president of the organization.

_____ 15. During the design and implementation phases of an application, works very closely with the users.

C. Match the examples of five basic motivational needs.

a. physiological   c. social   e. self-fulfillment
b. safety and security d. egotistic

_____ 1. Recognition, status, and importance.

_____ 2. Realizing one's fullest potential in creativity and self-development.

_____ 3. Protection against danger and loss of a job.

_____ 4. Being able to identify with individuals and groups.

_____ 5. Food, clothing, shelter.

_____ 6. Promotion to lead analyst. A pay increase was not included with the promotion.

_____ 7. Job tenure is awarded. It is very unlikely that a tenured person will be let go.

_____ 8. A pay raise was received. This made it possible for the employee to take his daughter to the orthodontist.

_____ 9. One of the new programmers was asked to bowl on a team with some of the programmers who have been with the organization for a long time.

_____ 10. The analyst who designed an environmental control system was given an award for making an outstanding contribution to the company.

D.   Match the management function with the task described.
     a. planning   c. staffing   e. controlling
     b. organizing   d. directing

_____ 1. Standards are established regarding the length of time needed to complete a task.

_____ 2. Must precede any action.

_____ 3. Subordinates are supervised.

_____ 4. Tasks are divided among subordinates and authority is assigned.

_____ 5. A determination is made regarding the number of employees needed to implement the application.

_____ 6. Actual performance is measured against the standard.

_____ 7. An orientation is provided for the employees who will implement the new application.

_____ 8. Management by objectives is used as part of the evaluation of employees.

_____ 9. Objectives are determined regarding what the system should accomplish.

_____ 10. The site for the computer is prepared, job descriptions are developed, and the application is implemented according to schedule.

E.   Multiple choice. Place the letter of the best answer in the blank provided.

_____ 1. An officer of the company who makes a direct contribution to the achievement of the firm's goals and objectives.
     a. staff officer b. analyst c. line officer d. programmer

_____ 2. Provides the home-to-office link.
     a. teleconference b. telecommuting c. distributed data processing

_____ 3. One of the goals of operations is to
     a. eliminate the need for documentation.
     b. eliminate the need for external controls.
     c. increase throughput.
     d. decentralize data processing functions.

_____ 4. A display that contains rows and columns. The data can be manipulated in order to create new data.
     a. exception report b. spreadsheet c. external report

_____ 5. A systems approach is creative, organized, structured, action oriented, productive, and
     a. cost justified.
     b. management oriented.
     c. based on relevant information.
     d. more expensive than when other techniques are used.

_____ 6. Since many problems are complex, ill structured, and without a model that can be followed, the analyst must
     a. work effectively with others.   c. be organized.
     b. be creative.   d. follow instructions.

_____ 7. The components of a management system are the formal organization, informal organization, leadership style, physical setting, and
   a. the computer.   c. the individual.
   b. the line officers.   d. goals and objectives.

_____ 8. Often an organizational chart shows the division of responsibility by
   a. function. b. product. c. territory. d. a, b, and c.

_____ 9. Managers in this group are responsible for allocating and controlling the resources necessary to accomplish the objectives of the organization.
   a. top b. middle c. lower d. operational

_____ 10. Supervisory and functional management need
   a. less detailed information than top management.
   b. less detailed information than middle management.
   c. more detailed information than top or middle management.
   d. more detailed information than top management.

_____ 11. Has the greatest need for external information.
   a. top management
   b. middle management
   c. lower management

_____ 12. When data processing departments were first organized, the manager was usually responsible to
   a. the president.
   b. the vice-president of industrial relations.
   c. the vice-president of finance.

_____ 13. Within a traditional data processing department, the three functional areas are
   a. systems, programming, and operations.
   b. systems, programming, and technical support.
   c. programming, systems, and data control.

_____ 14. A term used to indicate concern over the effect machines and the environmental working conditions of the employee have on the employee's health and productivity.
   a. environment b. ergonmics c. work measurement

_____ 15. The ability to relate causes to effects and to design a solution that stays within given parameters.
   a. conceptual skills b. interpersonal skills c. communication

Technologies, Tools, and Talents of Analysts

Delta College

Delta College located in University Center, Michigan opened its doors to students in the fall of 1961. Since that time its enrollment has grown to 13,000 students enrolled in academic courses, the skilled trade programs, or in community education programs. Delta serves Bay, Saginaw, and Midland counties and is financed by tuition, property taxes, and state funds.

Like other colleges, Delta College is deeply committed to providing adequate computer facilities for its students and to using computers to provide management with information to manage the college in the most efficient manner. The trustees, President of the college, administrators, and faculty have long been aware that what was to have been the nuclear age may be recorded in history as the "computer age."

COMPUTING AT DELTA COLLEGE

Although the computer is often viewed as a mystic box that produces reports and other types of information, Delta recognizes that a successful system that meets the needs of the organization must have dedicated personnel, the right computer system to produce cost-effective results, and well-designed application software.

Prior to obtaining a computer, data was processed using electromechanical equipment. What fun it was to wire the boards for the 402 accounting machine! In 1963 an IBM 1620 computer was obtained that had 20KB of memory. Using punched card input, printing was still done on the 402 accounting machine. When the IBM 360/40 was obtained in 1968 with its 64KB of memory, high-speed printer, 2 tape drives, and

M. Gene Arnold, Chairperson of the Computer Policy Committee, is responsible for conducting needs analysis, developing request for bids, and determining how the project can be financed.

3 disk drives, it was looked upon as being the latest in computer technology. Guidelines were developed that enabled students, faculty, and administration to utilize the 360/40.

The Computer Policy Committee

At the November 10, 1969 Board of Trustees meeting an item under "Immediate Action" was the establishment of a Computer Policy Committee composed of first-line administrators as well as representatives from the faculty. The report requesting the appointment of a committee ended with the statement "To sum it up, we have made remarkable strides to date but as for the future all I can say is 'We ain't seen nothing yet!'."

In 1970 a Computer Policy Committee was appointed by President Donald J. Carlyon that included the first line officers of the college and five faculty members. The committee chairperson was Administrative Dean M. Gene Arnold. The committee's function was to:

1. Develop immediate, intermediate, and long-range goals for computer utilization.

2. Influence policies that govern the utilization of computer facilities and resources by community and campus requestors.

3. Encourage, develop, and promote the effective utilization of computers.

4. Identify, investigate, and report innovative applications with consideration as to their use at the College.

Project CITALA

In 1973 project CITALA (Computers in Teaching and Learning Activities) was funded by the Board of Trustees. The faculty and administrative members of the CITALA project studied interactive computing, timesharing applications, and online registration systems at selected colleges and universities. A great deal of data was collected and analyzed which justified the acquisition of a timesharing system.

Timesharing at Delta College

When a PDP/50 timesharing system was obtained in 1976 a new dimension in computing occurred. Central Administrative

Services, under the direction of Gene Arnold, established an institutional database which permitted the building of course schedules, online registration, and the generation of management reports. The college was also selected as a test site for S.I.G.I. (System of Interactive Guidance Information).

Inservice training programs were conducted for faculty, administrators, and staff members. The timesharing system was barely in place when surveys were conducted to determine the need for additional computing facilities and equipment. A great deal of time and effort was spent investigating the needs of the college and translating those needs into hardware and software requirements.

Once more specifications were developed, funds were obtained, and requests for bids were sent to vendors. As a result, two Prime 750 computers were acquired and many of the batch systems were redesigned as transaction-processing systems and more of the data entry functions were distributed. Each system had 4 MB of main memory, 2 disk drives with removable disk packs, a tape drive, and a printer. Up to 96 terminals can be attached to each computer.

John Fuller, Registrar of Delta college, studies an enrollment report. John is an active member of the Computer Policy Committee and was a member of the search team responsible for developing the specifications for the hardware and software needed to support an online registration system.

Each of Delta's Prime 750 computers has two 630 MB disk drives and a 1600 CPI tape drive which is used to off-load and back-up files. Although most of the payroll programs now are designed as transaction-processing jobs, Betty Nicholson, the operation's manager explains the new schedule for running the batch jobs to Jan Wegener.

New Applications and Requests

Although guidelines for the acquisition of microcomputers had been developed during the early 1980s, the Computer Policy Committee is currently establishing pilot studies that can be used to determine which applications are most cost effective when run on standalone microcomputer systems rather than being run on the mainframes. Also being investigated is high-speed printer and hard disk for storing files and software.

Although some word processing equipment is available, the committee.is studying various ways of implementing additional office automation and extending its word processing services to more users. It is considered essential that the two areas—EDP and office automation—be closely coordinated.

ONLINE REGISTRATION

During the initial and detailed investigation phase of the online registration systems life development cycle, John interviewed faculty, staff, and administrators who were concerned about

Ann Nitschke keys in the class number and the class list is retrieved from the file and displayed on the VDT. The operator need only key in the grade. Under study is a recommendation to ask Datatel to make the necessary changes in the software so that grades can be read by a scanner from the grade sheets rather than being keyed in by an operator.

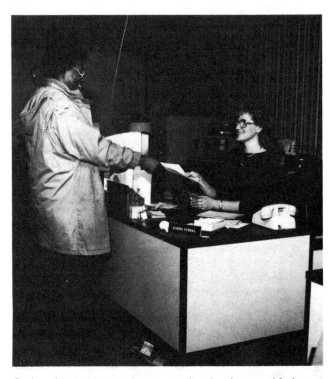

In the admissions center, a new student hands a simplified admission form to Kendra Kendall who will assign the student a number and enter the data needed to create an additional record in the online database.

As students register, immediate feedback regarding the availability of classes is provided and both the online databases are updated.

the registration system and reports that would be available. Other colleges were visited and vendors made presentations on both hardware and software. Both the hardware and software that was eventually obtained was demonstrated. All of the data was compiled and analyzed in order to prepare a report for the Board of Trustees and to prepare a request for bids.

In the feasibility study report it was recommended that the Datatel "Colleague" software package be obtained. The software is currently used by 6 community colleges and 14 four-year colleges. At the present time the system is operational and the college's analysts and the users of the system are determining what changes should be recommended to the consortium of users. When the system was implemented, a new database was established rather than converting the information stored in the old database to the new system.

Adding Records to the Online Database

Since the database stored on a removable disk packs is the basis for the transaction-processing programs, adhoc requests for information and reports, and for hatch-processing jobs, it is important that new student records are added to the system. Within the admissions center, operators can also change records and add additional data to existing records. Periodically the database is backed up on magnetic tape.

Registration

At the current time, most students enrolling in courses must come to the campus to enroll. Students attending the off-campus

In working with a student, Donald Bieri retrieves the student's record from the online database.

centers may elect to mail in their registration forms. As each student enrolls, the computer determines if space is available in the classes requested. If so, the student is added to the class list and one is subtracted from the "seats available." If the class is already filled, the student's name and number is added to a "wait list." Each student's schedule is printed and forwarded to either financial aids or to the cashier's office. Students have until 4:00 p.m. of the following work day to pay for their courses. At that time, the unpaid for schedules are reviewed. For each student who has not paid the computer asks "Do you want to cancel this schedule?". If the operator responds with "Y***", the class schedule is removed from the database and one seat is added to each of the classes that made up the student's schedule. By requiring the addition of three asterisks to the response, it is unlikely that a schedule will be deleted by mistake.

ADVANTAGES OF THE ONLINE SYSTEM

According to Registrar John Fuller the major advantages of the online system based on a comprehensive database are:

- More accurate class lists. Under the old system the "no shows" were often as high as 15%.
- More timely and accurate information is available. Since only 1/2-1 percent of the schedules are canceled, department chairpersons and deans have better information to use in adjusting the master schedule.
- More accurate control over the collection of tuition and fees.
- Students can register faster.
- More automated procedures are utilized. For example, when a student drops a class the amount of the refund is calculated and a check for that amount is printed.
- Tighter control over registration. Because of the controls built into the system, there are fewer opportunities for human errors to occur.
- Adhoc reports can be generated upon request.

A proposal has been submitted to convert the present registration system to a telephone registration system. The proposal includes a design for the location of the hardware, the design of the individual workstations, and the estimated cost of additional hardware, and the additional ongoing costs that would occur. The proposal also indicates that a commitment will be needed for computer time and for modifications in the Colleague software. Dedicated lines will be needed for the VDTs used in the telephone registration system.

Under the proposed telephone registration system, students will be able to register by phone rather than coming to the campus. According to Fuller, the system would be even more efficient than the present one and scheduling problems could be more easily resolved as they occurred.

CONCLUSIONS

The online registration system at Delta college serves to illustrate why the title systems life development cycle is used for the investigation, implementation, and evaluation of a project. Users and EDP professionals gather data, analyze the data, and develop a proposal. If the proposal is accepted and the system is implemented, an ongoing study is made to see if changes are required to solve problems or to make the system more efficient and cost effective.

3 PROJECT MANAGEMENT, ANALYSIS, AND DESIGN TOOLS

Looking ahead

After reading the text and completing the learning activities you will be able to:

- Identify the characteristics of a project.
- Construct a simple Gantt chart or PERT chart.
- Identify the symbols used for constructing data flow diagrams.
- Develop a data flow diagram.
- Identify the symbols used for constructing system flowcharts.
- Develop a system flowchart.
- Develop key questions and exit criteria for the tasks described.
- Define and utilize the words and phrases listed in the end-of-chapter glossary.

INTRODUCTION   A **project** is usually extensive and involves multiple, interrelated tasks. For a contractor, building a house is a project. The project is divided into subsystems such as site preparation, building the foundation, framing, and so forth. Each of the subsystems can be divided into tasks that take anywhere from a few hours to several days. An experienced contractor knows how long each task should take and also knows the relationships that exist between the tasks that must be performed. Obviously the roof cannot be started before the framers complete their part of the construction.

As work on the project procedes, the contractor can compare what is in progress and finished with the projected schedule of when each task should have been started and completed. Costs are also maintained for labor, materials, and overhead for each phase of the project. A large construction company will have many projects underway and will have assigned each project to a manager. The manager is expected to give progress reports to top management on the status of each project. Top management needs to know whether a project is on, ahead of, or behind the projected schedule. They must also know whether the project is over or under the projected budget.

In the example given you can see the importance of planning. The plan is used for scheduling men and materials and for evaluating the progress being made on the project. An experienced architect in charge of a project would know how to proceed. Perhaps the key factor in developing the plan for a project is the blueprint. The project manager can also refer to records maintained on similar projects. Although each custom-built house is different, after a little experience the project manager knows exactly how long it should take to complete each task. The tasks remain the same—prepare the lot, complete the foundation, and so forth.

A SYSTEMS ANALYSIS AND   In systems analysis and design a blueprint, or plan, for a project must be
DESIGN PROJECT   developed. Due to the following characteristics of projects, a ready-made plan is not available:

- Each project is unique. Although a contractor may build several houses that are alike, in systems analysis and design, each project is different. However, with experience, an analyst can see the similarities that exist between projects.
- Each project has a definite beginning and end. The project manager determines the first task and the last task to be completed.
- Different skills and talents may be needed to complete the tasks that make up each project.

Since each project is different and a blueprint for each is not available, how does an analyst proceed? What have experienced analysts learned that can be passed on to the neophyte? Within the industry and individual organizations, there are standards regarding the tools and techniques that can be used. A standard is a uniform method of accomplishing a task. However, there are different opinions regarding the best way to show the flow of data throughout the organization and the relationships that exist between the procedures that make up informational systems. This chapter is designed to give you an overview of some of the basic tools used in planning projects and in showing the flow of data throughout the system. Some of the topics presented will be illustrated in other parts of the text by using additional examples.

UNDERSTANDING THE PROBLEM

Before the project leader and members of the project team can create a plan, they must have a *complete understanding of the problem.* Assume that you have been asked to do the analysis and design for a billing system for a physical fitness center which is operated under the direction of the city. Members are charged membership fees and fees for the use of the facilities. The pro shop carries a limited inventory of equipment that is purchased by the members. Members must charge all items and are billed at the end of the month for use of the facilities, equipment, and for their membership fees.

An analyst must have a complete understanding of the problem in order to design a system that gives management the type of information that is needed. This requires an understanding of far more than the present billing system and the problems that occur when it is used. The analyst must understand the computer system that will be used and the needs of management.

Developing a Plan

After the initial meeting regarding what data should be gathered and how this should be accomplished, a list of required tasks can be developed. Once the list is completed, a plan can be developed that shows: the length of time that each task should take; when each task should be completed; and what tasks can be completed concurrently.

The initial plan will cover only the first part of the project which is to gather data, analyze the data, and submit a recommendation for completing the project. If a well-developed plan is available, at any point an evaluation can be made to determine the status of the project. It should be possible to determine the tasks that are completed and the ones that are in progress. The projected schedule of events should be compared with what is occurring.

Once the tasks are defined and a tentative plan developed, there are many ways of presenting the plan. Some analysts prefer to use a **Gantt chart** which shows when each task is to be started and finished. As the work progresses, the chart can be used to show which tasks are finished and which ones are in progress. Other analysts prefer to develop PERT charts. PERT is an acronym for **Program Evaluation Review Technique**. The PERT chart shows the sequence in which the tasks must be executed and which tasks can be executed at the same time. A mathematical formula can be used to determine the shortest period of time in which a project can be completed.

Figure 3-1 on page 90 illustrates the tasks that must be performed in the initial phases of the billing project. The time estimated to complete the task is also provided. The code number is used to show the tasks that can be completed at the same time and the sequence in which the tasks will be executed. In the illustration, the first three items coded with a 1 can be completed at the same time. All three items must be completed before item 4, which is coded with a 2, can begin. If five analysts prepared a plan for initial investigation, each plan probably would be different. Each analyst would have a logical explanation for the sequence that the tasks should be done in and the time needed to complete the tasks. The initial phase of the billing system project is used as an illustration to present the concepts as simply as possible.

Figure 3-2 is a Gantt chart that is constructed using the data presented in Figure 3-1. As the tasks are begun, a second color or different type of line can be used to indicate that the task is in progress. A third color or third type of line can be used to indicate the completion of the task. At any time by referring to the Gantt chart, the status of the initial phase of the project can be determined.

Figure 3-3 illustrates a PERT chart. Note that the PERT chart by itself would be meaningless. The information that describes the event must be provided in

Initial Investigation for the Billing System

Task	Time (days)	Code
1. Study the documentation that is available for the present system. The documentation should explain the procedures involved and provide illustrations of the input forms needed, file structures used for the master and transaction files, and the reports produced as output.	1	1
2. Obtain and study organizational charts that illustrate how the management of the physical fitness center relates to other departments within the organization.	1/2	1
3. Obtain and study the policy and procedure manuals that provide information regarding policies that are related to the operation of the physical fitness center.	1/2	1
4. Interview the manager of the physical fitness center to determine his views regarding the problems that exist.	1/2	2
5. Observe the way the data is collected and processed. Determine how the reports produced as output are used.	1	3
6. Contact managers of other municipal centers to determine the type of systems they have. Also obtain information regarding policies, fee structures, and how pro shop accounts are handled.	2	4
7. Analyze the data.	2	5
8. Brainstorm the alternatives available.	1/2	5
9. Formulate recommedations regarding the next phase of the study and the alternatives that are available.	1	6
10. Prepare and present written and verbal reports.	1	7

FIGURE 3.1. Tasks required to complete the initial investigation of the fitness center's billing system

order to give meaning to the chart. More detailed information regarding Gantt and PERT charts will be provided in later chapters.

There are a number of different types of planning books and charts that can be used. Some charts are designed to mount on the wall and have magnetic symbols that can be used to show the status of various tasks. Planning books can be used to record when each task is to begin and end. Usually there is space available for additional notes.

The type of chart of planning aid is not as important as learning how to divide a project or any part of a project into small, manageable tasks that can be completed in a short period of time. Once the plan is constructed, the tasks can be assigned and the progress of each individual charted.

CHECKPOINT   1. What characteristics are associated with a project?

2. What is a standard?

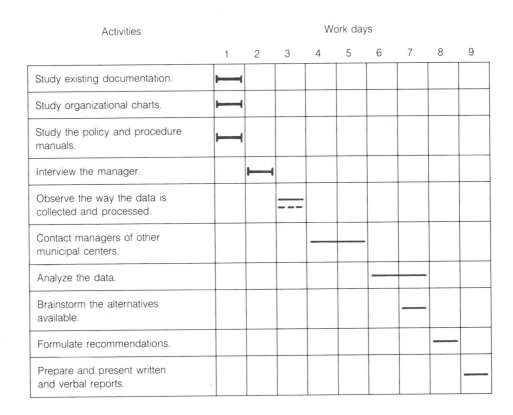

Activities	Work days								
	1	2	3	4	5	6	7	8	9
Study existing documentation.	⊢─⊣								
Study organizational charts.	⊢─⊣								
Study the policy and procedure manuals.	⊢─⊣								
Interview the manager.		⊢─⊣							
Observe the way the data is collected and processed.			– – –						
Contact managers of other municipal centers.					───				
Analyze the data.						───			
Brainstorm the alternatives available.							───		
Formulate recommendations.								───	
Prepare and present written and verbal reports.									───

FIGURE 3.2. Gantt chart for the initial investigation of the fitness center's billing system

─────── Scheduled events

– – – – – Events in progress

⊢──────⊣ Events completed

3. What must be done before a plan can be developed?
4. What information can be obtained by referring to a Gantt chart?
5. What information can be obtained by referring to a PERT chart?

Gathering Data

Present System   Once a plan is developed, execution of the tasks can begin. In the planning phase it should have been determined what data regarding the present system is needed and how the data can be obtained.

The parameters that define the total project cannot be determined until all facts relating to the current system are gathered. Although each of these methods will be covered in depth, the basic fact-finding techniques used by analysts are studying available materials such as annual reports, minutes of meetings, organization charts, job descriptions, procedural or standards manuals that describe how each task is to be accomplished, source documents used to capture data,

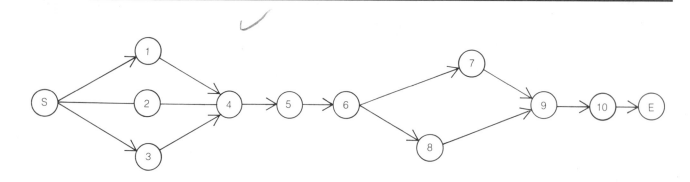

Event	Description
S	Start initial investigation phase.
1	Study existing documents.
2	Study organizational charts.
3	Study the policy and procedure manuals.
4	Interview the manager.
5	Observe the way data is collected and processed.
6	Contact managers of other municipal centers.
7	Analyze the data.
8	Brainstorm the alternatives available.
9	Formulate and Present recommendations.
10	Prepare and present verbal reports

FIGURE 3.3.   PERT chart for the initial investigation of the fitness center's billing system

detailed reports, exception reports, and information available by making ad hoc requests into the system. In the fact-gathering stage, observations are made and inteviews conducted with individuals directly involved with the system. As much information as possible should be gathered regarding the present system.

Alternatives Available   Once information regarding the present system is obtained and studied, information regarding alternate methods should be gathered. What must be done in this phase of the project is determined by the type of system that is to be designed. Other key factors are the constraints that are imposed upon the project. Management may indicate that there are no funds for new equipment or additional personnel. In some projects, it is necessary to investigate the possibility of obtaining new hardware or software.

In brainstorming the alternatives available, all possibilities should be explored. Available literature is studied, vendors may make presentations, and visits may be made to organizations that are using the hardware or software being investigated.

Presenting Information   Once data regarding the system is gathered, it has to be assembled for presentation. Because they are more easily understood, graphic presentations are preferred. Some analysts prefer to use system flowcharts that illustrate the movement of the data throughout the organization, while others prefer to use **data flow diagrams (DFD)**. Data flow diagrams show how data flows through a system and is transformed into information. However, a DFD is not concerned with the physical form of the data (disk, tape, cards, main storage), the resources (programs or individuals) that will operate on the data, or the control logic that will be used in processing the data. A systems flowchart uses different symbols to indicate the different storage media (tape, disk, cards), clerical activities, programs, and subprograms used in the transformation of data.

FUNCTIONS OF DATA FLOW DIAGRAMS AND SYSTEM FLOWCHARTS

Data flow diagrams (DFD) and system flowcharts can be referred to as models. Data flow diagrams are often developed during the feasibility phase of a project to show the **logical model** of a proposed system. System flowcharts provide more detail and are often used during the general design phase of the study. There are no set rules regarding how or when DFDs or system flowcharts can be used. Noted are the similarities between system flowcharts and DFDs:

- Both can be used to provide an understanding of the present system, identify the changes in user requirements, and illustrate the design of a new system.
- Uniform symbols are available for developing models. Since a limited number of symbols are available, some explanation must be recorded on the chart or diagram.
- Either can be used to show the movement of data through the organization and the steps needed to transform the data into information.
- Both show the relationships that exist between procedures. For example, the output from the time card edit procedure becomes the input for the procedure that produces the payroll register.
- Additional documentation is required to support the diagrams or charts.
- Both can be done at several levels—at one level only an overview is provided, while at a lower level a great deal of detail is included.
- Few rules are available that must be followed in the construction of the diagrams and charts. The amount of detail added to the diagrams and charts varies.
- The models can either be physical or logical. A **physical model** shows the processing activities in sequence and illustrates all changes in data and files and the outputs produced. Emphasis is placed on documents, people, and forms. The logical model shows what the system does and the emphasis is placed upon the data and the processes.
- The characteristics of the physical and logical models can be combined into one diagram or chart.
- The distance that a document travels or the time required to complete a task are not shown on either model.
- With practice, both models are easy to draw and use.

Data Flow Diagrams

Figure 3-4 illustrates the symbols used in the construction of a data flow diagram. Note that two symbols are given for process and for external entity. In the illustrations provided, the circle and square will be used for processes and external entities.

The levels associated with a DFD are:

Context	An overview of the scope of the system as a means of reviewing an existing system or providing an overview of a proposed system. While the inputs provided and outputs received by each entity are shown, it does not show subsystems that make up the system.
Diagram O	The major processes or subsystems are shown along with the data flow, entities involved, and data storage.
Detailed	Major processes or subsystems can be diagrammed individually and in more detail. Often the diagrams are very large. While considered too bulky to be used as part of the

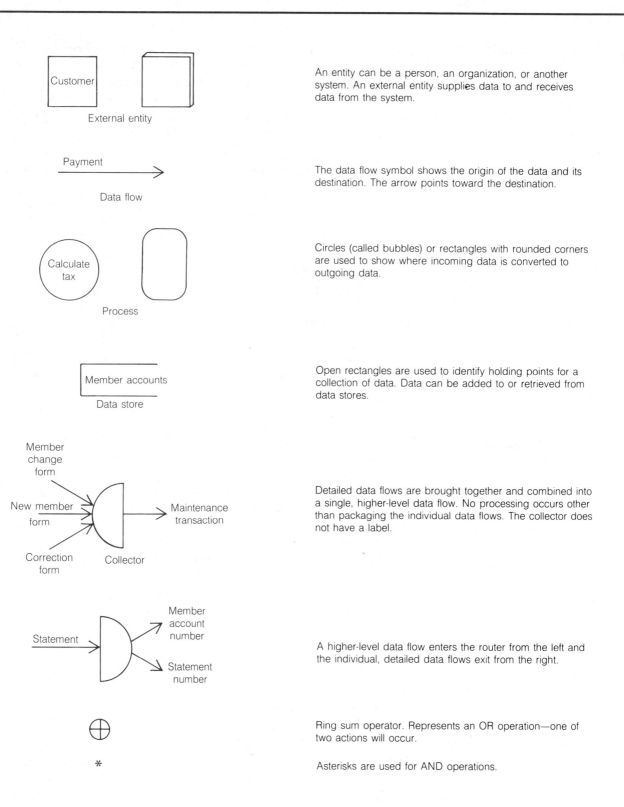

An entity can be a person, an organization, or another system. An external entity supplies data to and receives data from the system.

The data flow symbol shows the origin of the data and its destination. The arrow points toward the destination.

Circles (called bubbles) or rectangles with rounded corners are used to show where incoming data is converted to outgoing data.

Open rectangles are used to identify holding points for a collection of data. Data can be added to or retrieved from data stores.

Detailed data flows are brought together and combined into a single, higher-level data flow. No processing occurs other than packaging the individual data flows. The collector does not have a label.

A higher-level data flow enters the router from the left and the individual, detailed data flows exit from the right.

Ring sum operator. Represents an OR operation—one of two actions will occur.

Asterisks are used for AND operations.

FIGURE 3.4   Data flow diagram symbols—meaning and uses

documentation, the diagrams are an excellent reference for the individuals directly involved with the project.

Steps in Developing a Diagram O   The steps that should be followed in developing a Diagram O are:

1. Determine the data stores currently in use. Usually the data stores will include both transaction and master files. Records stored within the **transaction file** contain current information used for a limited number of applications during the current fiscal period.
2. List the major events that occur within the system. An analyst learns to look for acceptance of major input to the system (payment of dues), production of major outputs (statements are prepared), and functions that occur on a regular basis (monthly or annual reports).
3. Draw a segment of the complete data flow diagram for each event.
4. Assemble the segments into a single data flow diagram.

Guidelines in Constructing Data Flow Diagrams   In constructing the segments and then the single data flow diagram, the following guidelines should be observed.

- Use bubbles to show processing. Processing occurs when data is changed.
- Begin and/or end data flow at a processing bubble.
- Show only the flow of data. *Do not* show controls or how the computer program processes the data. Controls are routines built into a procedure in order to determine whether data is entered or processed correctly.
- Place a diagonal slash in the lower right-hand corner of a symbol that is repeated.
- Use an asterisk (*) when two data streams are required in order to produce the output or when two outputs are produced from one process.
- Use the ring-sum operator. ($\oplus$) when one of two data streams will be followed. The ring-sum operator denotes an either/or situation.
- Keep data flow diagrams simple.
- Ignore control logic. You should not be concerned with processing loops or decisions made during processing.
- Ignore the initialization (starting up) and termination (ending) of a system.
- Omit error routines that would lead from a bubble to a point outside the scope of the data flow diagram.
- Label data elements carefully.

Illustrations of Data Flow Diagrams   Figure 3-5 illustrates a context diagram of the billing system now being used for the physical fitness organization. Note that detail is not included regarding the procedures or data stores that are needed. The context diagram shows the entities, inputs, and outputs that make up the system.

Figure 3-6 is a Diagram O of the member billing system. Note that the individual procedures that make up the subsystems are identified in terms of individuals involved, input required, and output produced. By studying the chart, the importance of transaction and master files becomes apparent. The chart does

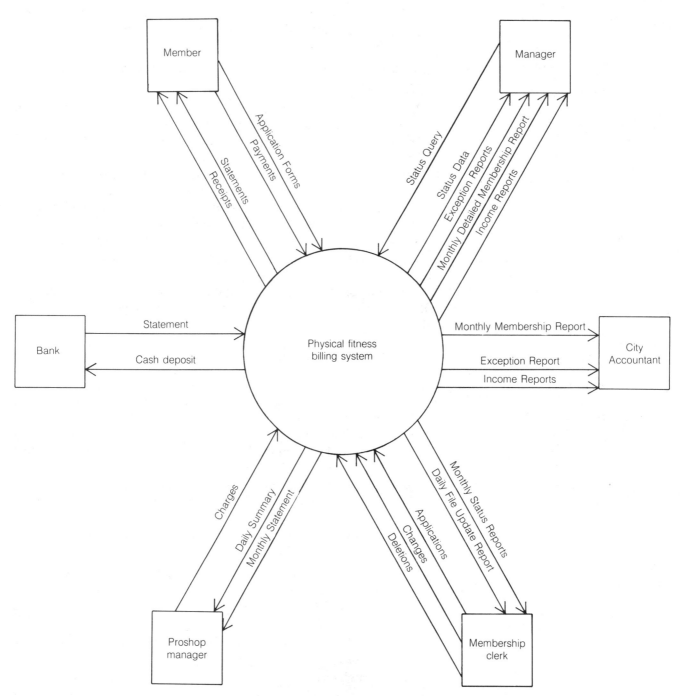

FIGURE 3.5. A context data flow diagram of the physical fitness billing system

not show editing, controls, or routines followed where there is a deviation. The chart does show five subsystems:

1. The change subsystem illustrates how changes are made to the membership master file.
2. A query subsystem is available that permits the manager to make queries into the system in order to extract information.

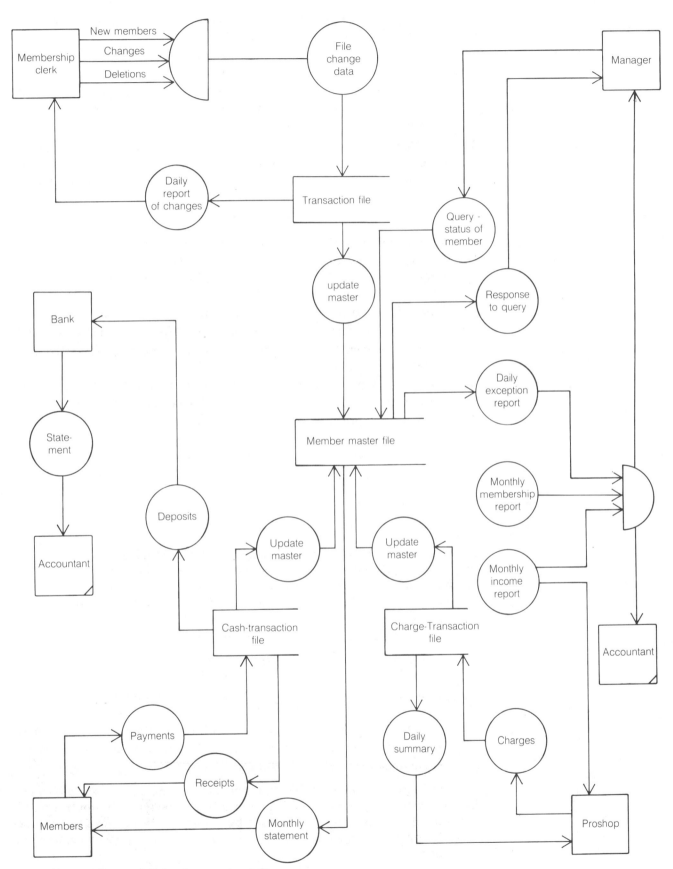

FIGURE 3.6.   A Diagram O for the member billing system

3. The membership billing subsystem shows how charges made in the pro shop affect the membership file.
4. The cash receipts subsystem illustrates how payments received from the members are recorded and the money is deposited.
5. The report subsystem indicates that daily exception reports, monthly membership reports, and monthly income reports are generated by information stored in the master file.

Figure 3-7 is a detailed data flow diagram which shows the physical characteristics of the membership change subsystem. Although more detail is included, the routines required to handle the rejected forms and records are not illustrated.

CHECKPOINT

6. What fact-finding techniques are utilized by analysts?
7. By studying a detailed data flow diagram or a Diagram O, what information can be obtained? What information regarding a system is not shown on a data flow diagram?
8. What information can be obtained by studying a context data flow diagram?
9. What are the differences between the physical model and a logical model?
10. Are the characteristics of a physical model and a logical model ever combined into one diagram?
11. What four steps should be followed in developing a data flow diagram?
12. Are the start-up procedures for a system illustrated on a data flow diagram?
13. What symbols are used for an external entity, data flow, process, and data store?
14. For what purpose is an asterisk used on a data flow diagram?
15. For what purpose is a ring-sum operator used on a data flow diagram?

System Flowcharts

A system flowchart shows the movement of data through an entire system or an individual procedure. It does not show how the computer processes the data. Analysts use system flowcharts to graphically show the relationships that exist between the procedures and the data. Management, auditors, and users of the system find that system flowcharts are a valuable source of information.

By studying a system flowchart it should be possible to determine:

- the starting and ending point for the system.
- the input media used for each procedure.
- the output produced as a result of processing input or using information stored in existing files.
- the major processing requirements such as sorting or updating files.
- the major decisions and conditions that control processing.
- the major controls built into the system.

The flowchart provides a graphic representation of what occurs as input comes into the system, is processed, and then results in some type of output. A working copy of the flowchart may be used by the analyst throughout the project. A final copy of the system flowchart is often included in the documentation for the project.

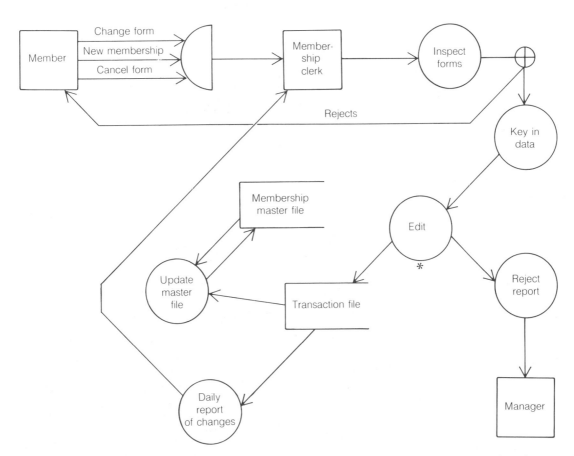

FIGURE 3.7. Detailed data flow diagram for the physical characteristics of the membership change subsystem

System Flowchart Symbols   Figure 3-8 on page 100 shows the symbols recommended by **American National Standards Institute (ANSI)**. While how the input is processed by the computer is not illustrated, flowcharts do provide detail regarding the input, output, or storage medium used and also identify the programs used to process the data. Control is stressed and there is a tendency to include more detail than is shown on data flow diagrams.

Often **exception routines** are also illustrated. An exception routine is used to handle abnormal conditions. For example, in the billing system the cashier makes out a deposit slip that shows the total deposit. When the payments are entered into the computerized system, the total credits to the members' records should equal the sum of the deposit. If the two do not agree, an exception routine is followed.

Guidelines in Developing System Flowcharts   In developing a system flowchart, the following guidelines should be used:

- Before starting the flowchart, break the system into subsystems. For each subsystem determine the output required, input required, and major controls and decisions needed.
- Use the standard ANSI symbols.
- Include a limited amount of detail within the symbol.

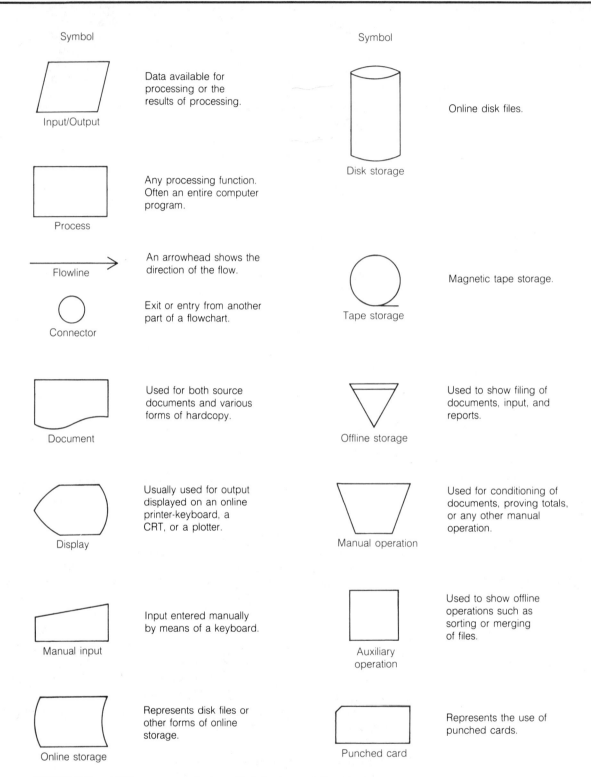

Symbol

Input/Output
Data available for processing or the results of processing.

Process
Any processing function. Often an entire computer program.

Flowline
An arrowhead shows the direction of the flow.

Connector
Exit or entry from another part of a flowchart.

Document
Used for both source documents and various forms of hardcopy.

Display
Usually used for output displayed on an online printer-keyboard, a CRT, or a plotter.

Manual input
Input entered manually by means of a keyboard.

Online storage
Represents disk files or other forms of online storage.

Symbol

Disk storage
Online disk files.

Tape storage
Magnetic tape storage.

Offline storage
Used to show filing of documents, input, and reports.

Manual operation
Used for conditioning of documents, proving totals, or any other manual operation.

Auxiliary operation
Used to show offline operations such as sorting or merging of files.

Punched card
Represents the use of punched cards.

FIGURE 3.8.   ANSI recommended symbols for system flowcharts

- Develop the flowchart from left to right and from top to bottom.
- The data flow is usually horizontal rather than vertical.
- Determine how the flowchart will be used in order to include the proper amount of detail.

Illustrations of System Flowcharts Figure 3-9 on page 102 illustrates a conceptual system flowchart that provides an overview of the member billing system. By studying the diagram you can see what will probably be developed as subsystems. The first subsystem will take care of what are considered normal changes to the master file—new members are added, address, phone number, and other changes are made, and the records of inactive members are deleted from the file. The second subsystem records in the master file the charges made by members in the pro shop. The third subsystem records the cash paid by members. The fourth subsystem creates daily and monthly reports. The fifth subsystem shows the queries that can be made by the manager.

The conceptual system flowchart does not show the external entities involved in the transactions that cause the transformation of data. Also controls, exception routines, and the handling of source documents and reports are not shown.

Figure 3-10 on page 103 illustrates a detailed system flowchart for the member billing system. By looking at the flowchart you can see that there are three daily procedures that create transaction files which are used to update the master file. When a file is updated, records are retrieved, new data is added to the records, and then the records are recorded back on the storage medium. When files are updated, the flowline has arrows at both ends. This indicates that the records stored in the file are read as input and written back into the file as output.

Unless comments are added to a system flowchart, a time delay that might be built into the system is not shown. For example, before the member master file is updated for the changes that occur, the summary of changes and error report is studied. If necessary, changes are made in the source data and the file change and edit program rerun to create a new change transaction file.

The billing system is primarily a batch system. An analyst assigned to design a new member billing system might elect to develop an online system. However, the exception, membership, income, and billing reports would still be produced as the result of running a batch program.

CHECKPOINT 16. Does a system flowchart show how the computer processes data?

17. Other than an analyst, who might refer to a system flowchart?

18. What is an exception routine? Would you anticipate that on a daily basis, all of the exception routines would be executed?

19. How is the updating of a master file shown on a systems flowchart?

20. In developing a system flowchart, does one identify the input, output, and storage media?

21. Are there more or fewer standard symbols used in constructing a system flowchart than are used for data flow diagrams?

22. What guidelines are followed in constructing a system flowchart?

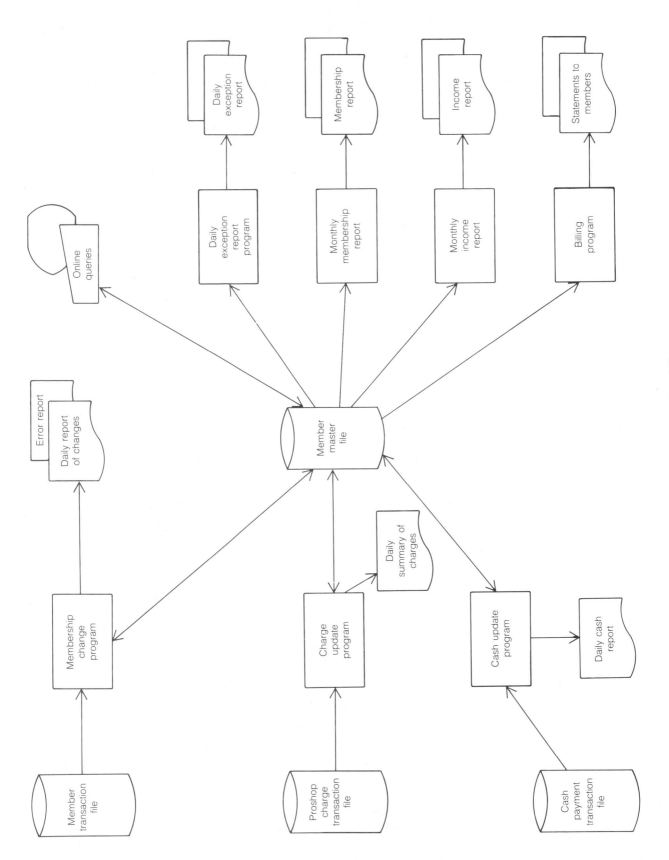

FIGURE 3.9. Concept system flowchart that provides an overview of the member billing system

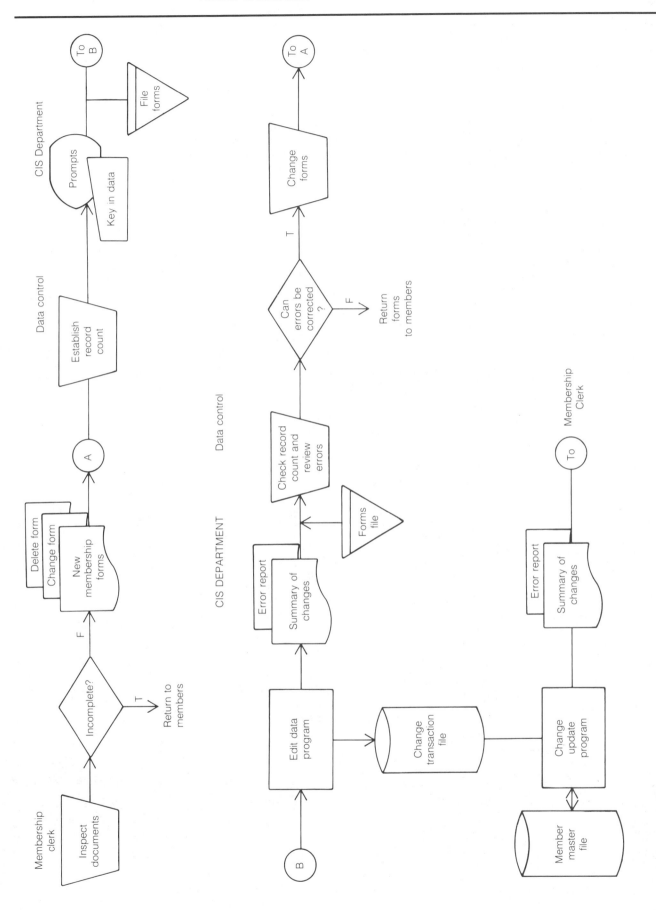

FIGURE 3.10. System flowchart for the member billing system   Part I

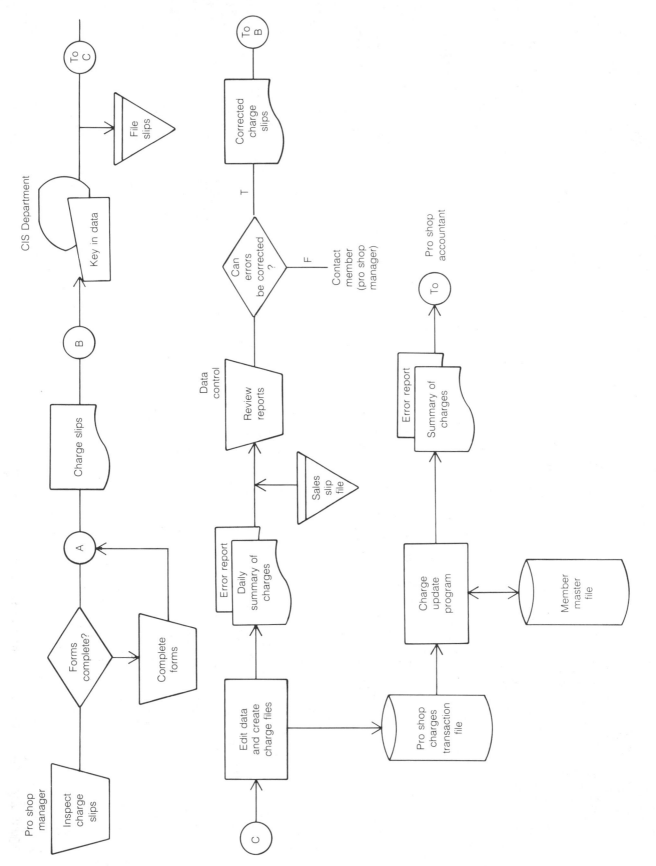

FIGURE 3.10. System flowchart for the member billing system   Part 2

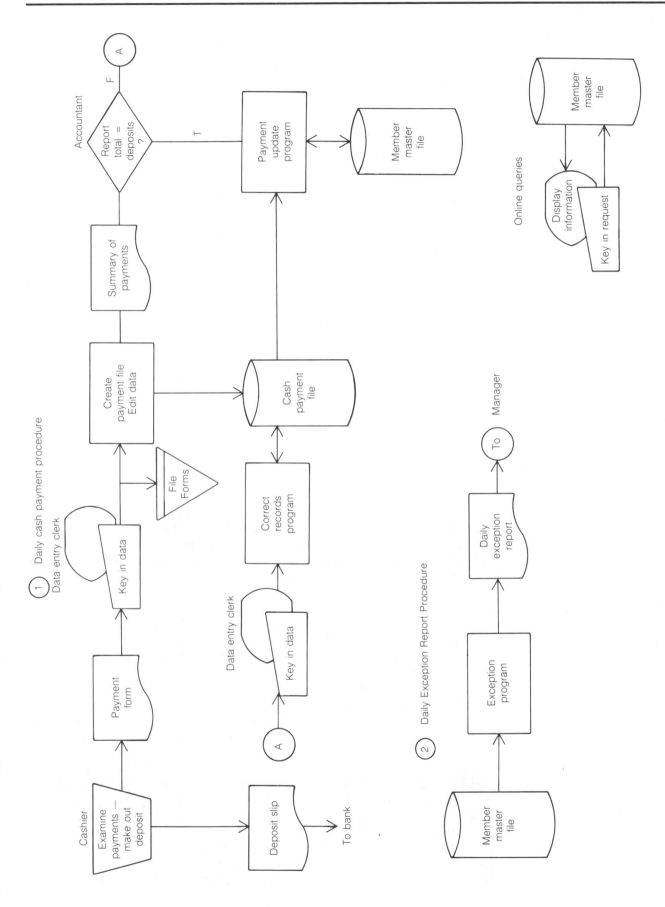

FIGURE 3.10. System flowchart for the member billing system Part 3

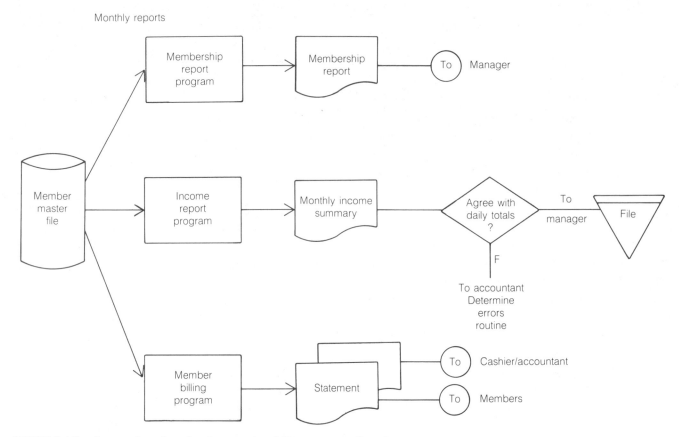

FIGURE 3.10.   System flowchart for the member billing system   Part 4

OTHER TOOLS AND TECHNIQUES

Once all of the facts are gathered and presented in a logical manner, the data must be analyzed in order to determine the scope and nature of the problem, solutions available, and the solution within the constraints specified that provides management with the information required. The solution should also be cost-effective.

Key Questions and Exit Criteria

In analyzing data, an analyst must rely on his or her experience and training as well as on the opinions and insights of others. As an analyst gains experience, analyzing data and developing a solution become easier. There are always parts of a system that are similar to other systems that have been developed.

Analysis becomes easier if for each phase or task of a project a **key question** is determined and **exit criteria** are established. The key question is the major question that must be answered before the task can be started. The exit criteria shows what must be done before the task is considered finished.

An analyst starts with the highest level and then breaks each phase into subsystems and tasks. Figure 3-11 illustrates how key questions and exit criteria can be established for the major phases of the systems development life cycle of a system. In using key questions and exit criteria, the analyst works from the objectives (outcome) toward the tasks that must be completed to accomplish the objectives.

Key questions and exit criteria can be used effectively for almost any task. For example, assume that the task is to design a paycheck or to pack to go to a Data Processing Management Association (DPMA) meeting in San Francisco.

Phase	Key Question	Exit Criteria
Initial investigation	What is the nature and scope of the problem?	Problem definition and objectives for the project.

Feasibility study	Is there a solution that will be cost effective and provide management with more current information?	Cost/benefit analysis
		Objectives for the system
		Logical model of the system.
		  Data flow diagram or
		  system flowchart
		  Data dictionary

General design	What must be done to accomplish the objectives?	System flowchart or
		  data flow diagram
		More specific cost/benefit
		  analysis

Detailed design	How should the system be implemented?	Hardware specifications Program specifications Program logic plans Initial test plan Implementation schedule Detailed cost estimates
Implementation	What tasks must be performed to make the system operational?	Code, test, and document the programs. Install hardware Develop operating procedures Develop security procedures Develop audit procedures Train employees
System audit	Is performance according to established criteria or are revisions necessary?	Ongoing evaluation using established criteria.

FIGURE 3.11. Use of key questions and exit criteria

Task	Key question	Exit criteria
Design paycheck	What information must be printed on the check?	Design a check with the required information that is acceptable to management, auditors, and the bank.
Pack for DPMA meeting	What type of clothing is appropriate for the meeting in San Francisco?	Pack suitcase to go to DPMA meeting.

In the examples, a list of additional questions can easily be formatted and a list of tasks that must be performed to meet the objective (the exit criteria) can also be formulated. Analysts must learn to think in terms of required outcome, key issues, and the steps needed to meet the established criteria.

Data Dictionaries

Data dictionaries can be developed and maintained manually or by using software. A data dictionary contains data pertaining to each data item within a file or database. A data item is the smallest storage unit within a database and is also called a *field*. Usually the size of the data item, the type of data stored in the field, the name used to identify the data item, and the programs that use the data item are identified. Since the foundation of any system is the data stores, data dictionaries are developed early in the analysis and design phase of the project.

Throughout the project modifications will be made to the data dictionaries. In the chapter on Detailed Design: File and Databases, an example of a data dictionary is provided. The dictionary serves as a tool throughout the development of the system and also becomes part of the final documentation. Maintaining the dictionary helps an analyst to visualize the impact each program has on the data items stored within a file or database.

Procedural Flowcharts

Procedural flowcharts are sometimes used that show the movement of documents, the time that each document remains in a given location, the distance the document travels, and the tasks that are performed at each location. Procedural flowcharts are used effectively in evaluating manual tasks to determine the cause of delays. The charts are used less frequently with the analysis of online transaction systems. Since data is entered when and where transactions occur, there is less movement of data and fewer delays.

Logic Plans

After an alternative is selected, the detailed design of the system begins. The system is divided into procedures and the procedures into tasks. Often each procedure requires one or more programs. Before the program is coded, a detailed logic plan must be developed. Usually the program is broken into segments which are called *modules*. For each module, a step-by-step logic plan is developed. There are many different formats that can be used to develop a logic plan. The important factor, however, is that a step-by-step plan is available before the source code is written.

A separate chapter, Programming Considerations, illustrates some of the ways in which logic plans are developed. A detailed logic plan might be compared to a road map. An experienced traveler would not start out from New York City for Los Angeles without a good road map; an experienced analyst or programmer should not attempt to write a program until a detailed logic plan is developed.

Use of Forms

Analysts often design forms that are used as source documents or for external reports. In addition, there are a number of standardized forms that can be obtained from an office supply company that make the work of an analyst easier. For example, forms are available, such as flowcharting paper, for use in developing logic plans. A template that has the standardized symbols should also be used.

When forms are used, data is presented in a uniform manner. Also when a form heading is completed, the required identifying information regarding the project is available. For example, the form used for job descriptions (illustrated in Figure 2-8 on page 64) has a heading which identifies the position and captions for DESCRIPTION, RESPONSIBILITIES, DUTIES, WORKS CLOSELY WITH, and REQUIREMENTS. By using the form, the same type of data is made available for each position.

SUMMARY

Since each system is different, a standard "cookbook" receipe for developing a system is not available. However, once the tasks that must be accomplished are identified, a plan can be developed. Two excellent planning devices are Gantt and PERT charts. A plan provides a guide for the course of action which must be followed and also a means of measuring progress.

After the plan is developed, data is gathered and analyzed. During this phase, existing data flow diagrams or system flowcharts may be used to illustrate the present system. If none are available, DFDs or system flowcharts may be developed to illustrate the present system. When new systems are developed, data flow diagrams or system flowcharts may be used to illustrate the proposed system and are then put in final form and used as part of the documentation.

There are accepted techniques used in gathering data. However, in analyzing data, the analyst must rely on ingenuity and experience. After the design of the system has been determined and accepted, any one of several techniques may be used for developing the logic for the programs that will be used to process the data.

An analyst should be aware of the forms that are available. Some forms are standardized, such as flowcharting forms, and available from office supply companies. Other forms are developed within the company and comply with its requirements.

DISCUSSION QUESTIONS

1. What characteristics are associated with a project?
2. What must be done before a Gantt or PERT chart can be developed?
3. What additional information is available when a Gantt chart is used that is not normally shown on a PERT chart?
4. What steps should be followed in developing a Diagram O?
5. What can be learned from referring to a system flowchart?
6. What guidelines should be followed in developing a detailed system flowchart?
7. In developing the logic of a program, is there more than one accepted method?
10. Why should an analyst learn what forms are available and use forms for doing tasks such as writing job descriptions?

Team or Individual Projects

1. **Mini payroll case study**

The payroll department has requested that a system study be done to convert the present batch oriented system to an online system. The analyst assigned to the project investigated and found the payroll system consisted of five major subsystems. Each of the subsystems is described in some detail. The system as described has some major shortcomings and is also incomplete. However, the intent is to provide data that can be used to develop data flow diagrams and systems flowcharts. (You will be asked to look at the case again as you learn more about systems analysis and design.)

Subsystems:

Master file maintenance. The forms are made out by a payroll clerk and then transmitted to a data entry clerk. The payroll system is a centralized batch system. Forms are available for adding new employees to the file, changing any of the data stored in an employee's master file record, and deleting employees from the file. The data is keyed in by an operator and edited, and the records without errors are stored in the change transaction file. An exception routine is followed for records that have errors. In some cases, the operator is able to make the necessary corrections in order that the record may be stored in the file. In other cases, the data needed to make the correction is not available and the data entered will be listed on the error report. The change transaction file is used to update the payroll master file. Normally the procedures that make up the file maintenance subsystem are executed prior to the procedure that calculates the employees' earnings. Other than editing the data for errors such as wrong codes and nonnumeric data stored in numeric files, there are no controls built into the file maintenance procedure.

Edit current payroll data.   Time sheets are submitted by each department head and each has the employee's name, identification number (used to retrieve their record from the master file), department for which they worked, regular hours, and overtime hours. A payroll clerk does an audit of the sheets to see whether all the data is recorded on the sheet. The total number of employees to be paid is established.

The time sheets are sent to the computer information services and the data is keyed in by an operator. As the data is keyed in, the payroll master file is online because some of the data in the file is needed to edit the data recorded on the time sheet. Unrecoverable errors are listed on an error report and returned along with the time sheet to the payroll department for correction. The good records are stored in a time transaction file. After the corrections are made, the same procedure is repeated. However, this time, the new records are appended on to the time transaction file. The procedure must be repeated until all the data for all employees to be paid is stored in the transaction file. Usually only one or two errors are detected.

Calculate pay procedure.   A series of steps are needed to complete the procedure. The time transaction file is sorted by employee number. A utility program is used for this purpose. The reason for sorting the file is so that the report will be in sequence by employee number. The time transaction file and the payroll master file are used as input. As a result of processing the data, a payroll register is printed and a more complete payroll transaction file is created. The total number of employees to be paid is printed on the last page of the payroll register. This total must agree with the one developed by the payroll clerk. If the totals do not agree, the error must be found and the necessary correction made. This often involves adding additional records to the time transaction file, re-sorting the file, and rerunning the payroll register program. The payroll department manager checks the totals.

Print paychecks.   The payroll transaction file is used to print the paychecks. The totals printed after all the checks are printed must agree with the totals on the payroll register. If the two do not agree (and they always have), it would be a software problem that would need to be solved by the individual who developed the system.

Update master.   After the paychecks are printed and the totals agree, the payroll transaction file is used to update the payroll master file. The employee's number is used to retrieve his record from the payroll master, the current earnings are added to the year-to-date totals, and the record is rewritten into the file.

Directions for project 1

    a. Review the material in the text related to the development of DFDs and system flowcharts.

    b. Construct a context data flow diagram of the payroll system.

    c. Construct a Diagram O for the payroll system by:

        (1) determining the data stores needed for the system.

        (2) listing the major events that occur within each subsystem. (Look for major inputs into the system.)

        (3) drawing a segment of the DFD for each major event.

        (4) assembling the segments into a single data flow diagram.

    d. Develop a detailed system flowchart for the payroll system.

2. For each of the following tasks, develop key questions and appropriate exit criteria. In addition, list the subtasks that you feel will need to be performed in order to meet the exit criteria and indicate what resources you might use. The following tasks are to be used:

    a. Develop a training manual that can be used in training personnel to use the existing payroll system.

    b. Design a payroll register. If you are not certain what information is printed on the payroll register (or journal), what resources could you use to learn more about payroll registers?

    c. Develop a Gantt chart that might be used to illustrate the plan to be following in developing a new payroll system.

    d. Plan a one-week sight-seeing trip to New York City.

GLOSSARY OF WORDS AND PHRASES

American National Standards Institute (ANSI) An organization that coordinates the establishment of standards for products as well as programming languages and flowcharting symbols.

data dictionary A file maintained by software or a chart kept manually that contains information relating to the data items within a file or databases. The size of the field, type of data stored within the field, name of the field, programs in which the data item is used, and other information are provided.

data flow diagram (DFD) A graphic representation that shows data movement, processing functions performed on data, entities involved, and the files used in performing the tasks.

exception routines Statements incorporated into a program that handle deviations from normal processing such as an invalid code or alphabetic data in a numeric field. Usually an error message is printed or displayed and the record that contains the invalid data is bypassed.

exit criteria Events or activities that must be completed before the task is considered completed.

Gantt chart A graphic representation of a plan that shows scheduled, in-progress, and completed events. The chart also shows which events can be performed concurrently.

key question The major questions that must be answered before a procedure or task is started.

logical model A model showing what the system does. The emphasis is placed on the data and the processes.

physical model A model which shows the processing activities in sequence, changes in data and files, and the output produced.

Program Evaluation Review Technique (PERT) A planning and analysis tool that illustrates the relationships between the tasks to be performed. When a PERT chart is constructed, a mathematical formula can be used to determine the shortest period of time in which the project can be completed.

project A series of multiple, interrelated tasks that have a predetermined beginning and end. Because every project is unique, each requires different skills and talents.

transaction file The records stored within the transaction file contain current information. Usually the file is used for a limited number of applications during the current fiscal period.

STUDY GUIDE 3

Name _____

Class _____ Hour _____

A. Indicate whether the following statements are true (T) or false (F) in the space provided. If false, indicate in the margin how the statement should be changed to make it true.

_____ 1. A project is usually simple and involves only one task.

_____ 2. Developing a plan for systems work is important as it provides a means for scheduling the personnel that are needed.

_____ 3. Since the skills and talents needed to complete a project are always the same, projects tend to be simple.

_____ 4. The plan for a project should show the tasks that need to be completed, how long each task should take, and the tasks that can be completed concurrently.

_____ 5. PERT is an acronym for Program Evaluation Review Technique.

_____ 6. A Gantt chart can be used to show scheduled events, events in progress, and completed events.

_____ 7. During the fact-finding phase of a project, all facts are obtained by observation and by interviewing key individuals directly involved with the system being studied.

_____ 8. Since all projects are similar, a "cookbook" approach can be used to find a solution to the problem.

_____ 9. Constraints imposed regarding the design of a new system might include using existing equipment and personnel.

_____ 10. Data flow diagrams show how data flows through a system and is transformed into information.

_____ 11. A data flow diagram shows the physical device on which data is stored.

_____ 12. A data flow diagram can be used to illustrate the design of a new system.

_____ 13. A context diagram provides a great deal of detail about the present or proposed system.

_____ 14. Detailed data flow diagrams are sometimes included as part of the documentation for a system.

_____ 15. A logical model shows the processing activities in sequence and illustrates all changes in data and files and the output produced.

_____ 16. The symbol used to indicate a data store is a square.

_____ 17. Before constructing a data flow diagram, the analyst should determine the data stores in use, the major events that occur within the system, and functions that occur on a regular basis.

_____ 18. In a data flow diagram, a rectangle is used to show processing.

_____ 19. When a ring-sum operator is used, it indicates that one data stream is always followed.

_____ 20. In developing a data flow diagram, special routines are included for initiating and termination of the system.

_____ 21. Labels are not included on a data flow diagram.

_____ 22. A square is used to depict an entity.

_____ 23. An entity is always a person.

_____ 24. A system flowchart illustrates how data is processed by a computer.

_____ 25. Uniform symbols are not used in constructing a system flowchart.

_____ 26. Usually a system flowchart is developed from right to left and from top to bottom.

_____ 27. External entities are always shown on a system flowchart.

_____ 28. System flowcharts show the time that is required to complete a task.

_____ 29. The billing system that was shown in a flowchart is primarily an online transaction system.

_____ 30. The only advantage in using forms is that uniform headings are developed.

B.   Multiple choice. Record the letter of the correct answer in the blank provided.

      a. data dictionary   c. Gantt chart   e. procedural flowchart
      b. data flow diagram d. PERT chart   f. system flowchart

_____ 1. Illustrates graphically work scheduled, in progress, and completed.

_____ 2. Provides information regarding the distance documents travel and the length of time that documents remain at each location.

_____ 3. A square is used to represent an entity such as an individual or another computerized system.

_____ 4. An asterisk is used to denote an AND condition.

_____ 5. When it is used, the minimum amount of time needed to complete the project can be determined.

_____ 6. Can be used to find out what programs use a data item such as gross pay or federal income tax deductions.

_____ 7. Does not show the exception routines that are not considered part of normal processing.

_____ 8. A circle is used to represent a process.

_____ 9. The major emphasis of the chart is to show the processes that transform data.

_____ 10. Symbols are available which show the medium on which the data stores are maintained.

C.   Multiple choice. Record the letter of the best answer in the space provided.

_____ 1. An organization that coordinates the establishment of standards and is responsible for establishing standardized flowcharting symbols.
    a. DPMA b. PERT c. ANSI d. ACM

_____ 2. More detail is usually provided by constructing
    a. a physical model. b. a logical model. c. an exception routine.

_____ 3. A file that contains a limited amount of current data that is used during a given fiscal period is a
    a. master file. b. transaction file. c. database.

_____ 4. A series of multiple, interrelated tasks that have a predetermined beginning and end.
    a. project b. system c. procedure

_____ 5. In system analysis and design you often start with
   a. a detailed list of all the tasks that must be performed.
   b. the desired outcome or objective.
   c. the key question and the exit criteria.

_____ 6. When used on data flow diagram, an either/or situation is illustrated by using
   a. an asterisk. b. a ring-sum operator. c. a diamond.

_____ 7. When an entity is repeated, the symbol used is
   a. an asterisk. b. a ring-sum operator. c. a diagonal slash.

_____ 8. A Gantt chart can be used to
   a. show work in progress.
   b. determine the status of the project.
   c. show scheduled events.
   d. determine all of the above.

_____ 9. In regard to DFDs and system flowcharts
   a. there are few rules relating to their development and to the detail that should be included.
   b. the rules and guidelines for their development are very specific.
   c. all analysts will include the same amount of detail.

_____ 10. Updating a master file is illustrated by using
   a. a special symbol.
   b. a flowline with two arrows.
   c. a circle within a rectangle.

4 ONLINE TRANSACTION AND DECISION SUPPORT SYSTEMS

Looking ahead

After reading the text and completing the learning activities you will be able to:

- List the four types of electronic data processing systems and identify the major characteristic of each.
- Identify the characteristics of a traditional batch processing system.
- Identify the characteristics of an online transaction system.
- Determine what changes occur when a batch system is converted to an online transaction system.
- Indicate what information is displayed on a menu screen.
- Describe the type of hardware and software needed to support a management information system.
- Describe the type of hardware and software needed for a decision support system.
- Define and utilize the words and phrases listed in the end-of-chapter glossary.

INTRODUCTION

As is the case with many other aspects of information processing, there is no standard definition or distinction between the type of systems that exists. In this chapter, four types of computerized data processing systems will be discussed. The discussion starts with a brief review of the simplest type of computer data processing system and then advances to the most complex. The systems to be discussed are:

1. **batch processing systems.**
2. **online transaction processing systems.**
3. **management information systems (MIS).**
4. **decision support systems (DSS).**

Today over half of our working population consists of **knowledge workers** who rely on information to perform their jobs; the amount of new data is doubling every 20 months. If a company is to survive, the computer must do far more than process data and create historical information that can be used for the preparation of reports and statements. However, some companies are still at this minimum level of sophistication. Data is accumulated, entered in a batch, and then processed. In batch processing, the master files may be stored offline until needed.

The computer used to execute the batch program may be a small microcomputer capable of executing only one program at a time. Or it may be a huge mainframe capable of executing many programs at the same time. Many individuals and small businesses are at the threshold—processing in a batch environment but ready to develop more complex systems.

BATCH PROCESSING SYSTEMS

A program stored in the memory of the computer processes the data and converts the raw data into meaningful information. Until sophisticated operating systems were developed, jobs ran in a batch environment. Today, microcomputers normally process data in a batch environment. In a single **batch processing system**, each job is scheduled to run in a given sequence. After the first job is completed, the computer notifies the operator that a second job can be loaded into its memory and executed. One by one, each job is executed.

The flowchart example used in Chapter 3 of the member billing system depicted a series of batch jobs. Figure 4-1 illustrates a typical batch job for a payroll system. Cards are used to enter the employee's number, regular hours, overtime hours, and pay code. Usually one operator punches the data into the cards and a second operator verifies the cards to make certain the data was recorded correctly. In the illustration, the master file records stored on disk will be used to edit the data punched in the time cards. The primary purpose of the program is to edit the data and to establish a transaction file. After the job is completed, the error report will be studied and the necessary corrections made. Until the input is entered and the message NO ERRORS IN RUN appears, the edit program will continue to be run. After the operator sees the message NO ERRORS IN RUN, the next job in sequence will be loaded into the memory of the computer and executed.

The key ingredient to a successful batch system is the availability of data stored in online master files. Each system, such as payroll or accounts receivable, will have its own master file. Within the file will be stored a collection of **records**. A record contains information pertaining to one person or object. Records are divided into **fields**. Each field contains one type of data such as an employee's name or social security number.

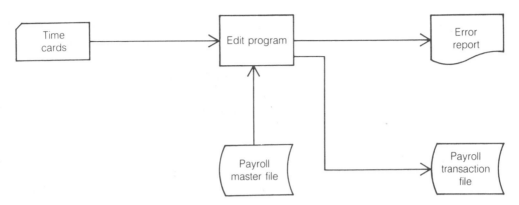

FIGURE 4.1. In batch processing, data is accumulated and then processed in a batch

Batch systems were often developed in a nonintegrated manner. Also the output from one program was evaluated and then used as input to a second program which might be part of another system. Seldom were files updated until the initial edit program was run and the output evaluated. Once the data was stored in either a master or transaction file, numerous reports could be obtained. Because each program required a limited amount of new data, it could easily be proved that processing the data electronically was faster, more accurate, and cheaper than manual processing. The multiple use of data stored electronically was the key to success. This factor becomes increasingly important since the cost of keying in new data is increasing while most costs associated with EDP are decreasing.

CHECKPOINT

1. What percentage of our population is classified as knowledge workers?
2. What is a knowledge worker?
3. What is the major characteristic of a batch processing system?
4. Why is processing data in a batch environment usually less expensive than processing the data manually?

ONLINE TRANSACTION PROCESSING SYSTEMS

The creation of **online transaction systems** dates back to early airline and hotel reservation, sales and inventory, and work-in-process systems. Online systems require the master file or database to be online whenever transactions are occurring. Since these systems have been in existence in large organizations for many years, analysts and management have had an opportunity to gain expertise in developing and managing online databases. More progressive organizations also developed guidelines and standards for online systems. Management was also forced to develop solutions for key issues such as the privacy and security of information and the rapid increase in the number of computer crimes.

Although some of the conditions listed may also apply to batch systems, when online transaction systems are developed, the following usually occurs:

- The computer becomes more "user friendly." This is accomplished by providing "menus" and more online documentation. Figure 4-2 illustrates the main menu that might be displayed for the member billing system discussed in the previous chapter.
- More data must be online. In an online sales transaction system, the customer master, inventory, and backorder files or databases must all be online. In working with a small microcomputer system one of the contraints that may affect the design of the application is the amount of storage space that can be online.
- Although files may be created that serve as a backup or an audit trail, transaction files used to update master files are no longer needed. As the transaction occurs, the variable data is edited and used to update the online master files.
- Fewer reports are printed. Since the validation of all transactions occurs as the data enters the system, most error reports are eliminated. Management is often asked to determine which reports they actually need. Since ad hoc requests can be made by using the available query language, some of the detailed reports should be eliminated. Reports that summarize data, make projections, and contrast actual data with budgeted items should not be eliminated.
- Data entry and data processing become more distributed.
- Fewer source documents are required. There also tends to be less movement of existing documents throughout the system.
- While more individuals access the data stored in online files, fewer individuals are directly involved in the handling and processing of source data.
- More controls are designed into the system.
- More user involvement is required.
- Users become less dependent on the EDP department for the maintenance of databases and the generation of ad hoc reports.
- The physical model of the application changes far more than the logical model.

Changes Made in the Physical Fitness Billing System

In order that you may contrast a batch system with an online transaction system, the member billing system discussed in Chapter 3 is used as an illustration. If the system were designed for a very small organization such as a single-operation health club, it could be designed to run on a minicomputer system. As long as there was a sufficient amount of online storage, the computer system would only need to support three or four terminals. The design could also be changed so that it would run on microcomputers that are networked into a system and supported by a centrally located printer and online database stored on hard disk.

The design illustrated is intended for implementation on the city's host computer. Workstations consisting of a keyboard, VDT, and a small, letter-quality printer are available in the manager's office, pro shop, cashier's office, and membership clerk's office. The database is physically located in the city's computer center.

The operating system provides for recording all transactions entering the system in a transaction-backup file. On a regular basis (every few hours) all databases are backed up. When a database is backed up, a copy is made that could be used in the event an original database is no longer available or for some reason

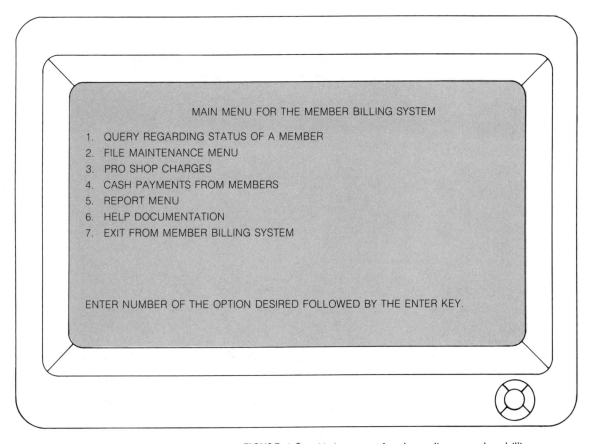

MAIN MENU FOR THE MEMBER BILLING SYSTEM

1. QUERY REGARDING STATUS OF A MEMBER
2. FILE MAINTENANCE MENU
3. PRO SHOP CHARGES
4. CASH PAYMENTS FROM MEMBERS
5. REPORT MENU
6. HELP DOCUMENTATION
7. EXIT FROM MEMBER BILLING SYSTEM

ENTER NUMBER OF THE OPTION DESIRED FOLLOWED BY THE ENTER KEY.

FIGURE 4.2. Main menu for the online member billing system

was updated incorrectly. Although seldom used, backup files are created in order to provide additional security.

Physical Fitness Billing System Procedures

The system flowchart on pages 122-123 shows the revised procedures. In contrasting it with Figure 3-10 on page 104 you will see that the data is entered when and where transactions occur. Either **formatted screens** are used or prompts are displayed that tell the operator what data must be entered. Figure 4-4 on page 124 shows how one of the screens used for entering the data necessary to create a record for a new member might be formatted. The slashes (/) on the screen indicate the number of characters that can be entered. When the screen is displayed, the cursor will appear at the point where the first / appears for the number field. Each time a character is entered, a / will be replaced on the screen by the character. When the enter key (sometimes shown as <CR>) is used, the data is transmitted into the computer. Before <CR> is used, the data is in the terminal's **buffer**. After all 10 fields of data are keyed in, the operator can review the data. If a field of data was keyed in incorrectly, the operator presses the number of the field. The cursor will return to the location where the correction is needed. Screen design will be covered in more depth in Chapter 9 Detailed Design: Input and Output.

Since the system flowchart shows a fair amount of detail, there is no way to indicate that all of the online procedures can occur concurrently. From one workstation, file maintenance could be occurring and at the same time the charge and cash payments procedures could also be executing.

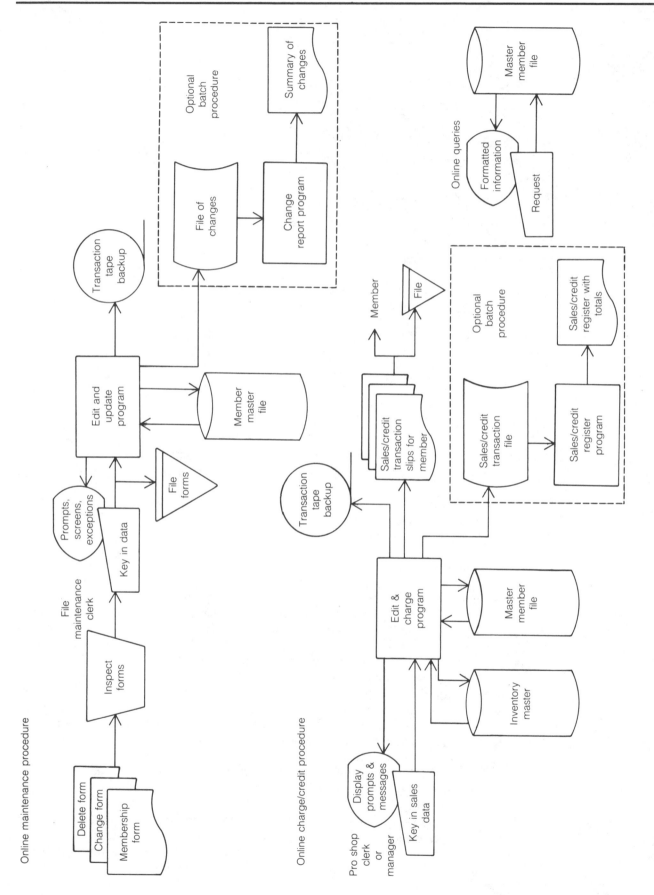

FIGURE 4.3. System flowchart for the online member billing system   Part 1

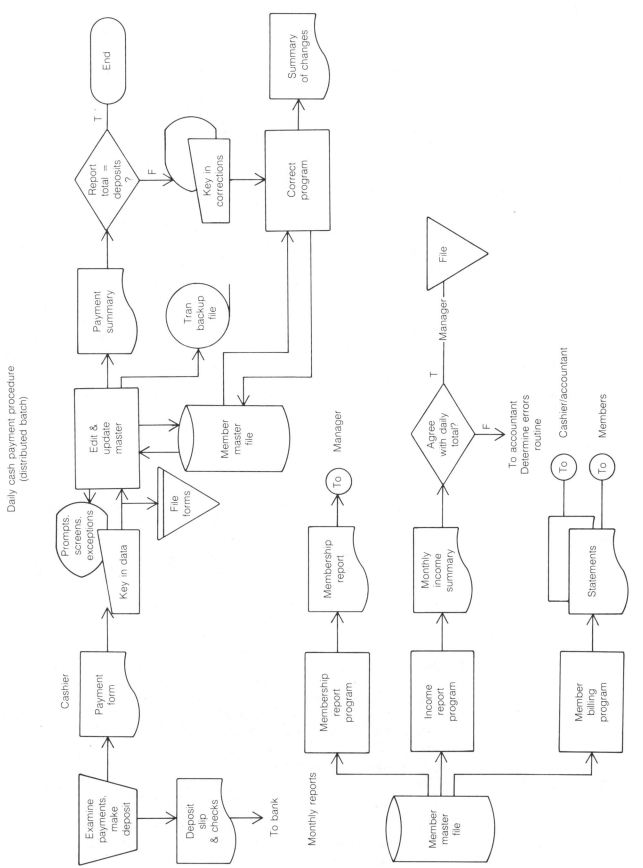

FIGURE 4.3. System flowchart for the member billing system Part 2

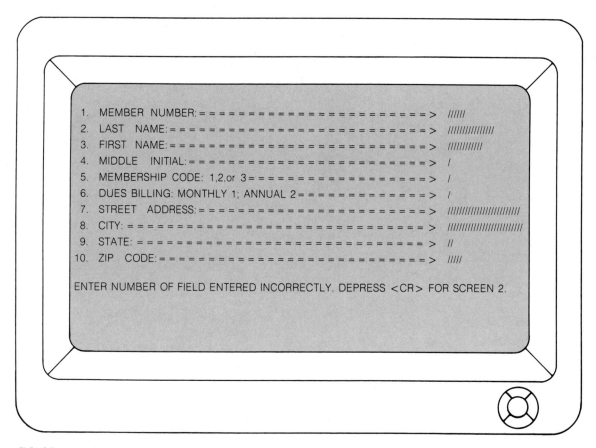

FIGURE 4.4. Screen 1 for entering data for a new member

In the documentation for the member billing system, each procedure will be described and the tasks performed by each individual will be detailed. The detailed logic plans, developed prior to writing the programs that process the data and print reports, will show how the data is edited and processed. All exception routines followed when invalid or incomplete data is entered are also explained in the documentation.

If the detailed logic plans were available for both systems, you would find that the new system had more editing and exception routines. This is possible because the member master file is always online when data is being entered.

File Maintenance Procedure Whenever any of the programs that are part of the member billing system are to be run, the main menu will be displayed. If an operator or manager enters a 2, the file maintenance menu illustrated in Figure 4-5 will be displayed. The maintenance menu displays six selections: help, add new records, change data stored in existing records, delete records, print a report, or return to the main menu. A brief description of what occurs when each of the six options is selected follows:

Option

1 The help option causes a ''help'' menu to be displayed. Depending on the option selected from the help menu, documentation for one of the other five options will be displayed. After the requested

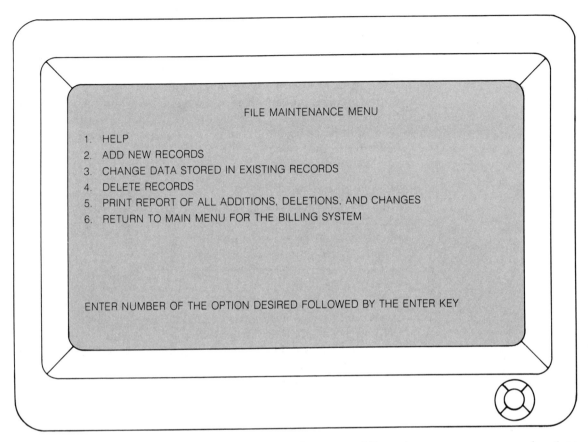

FILE MAINTENANCE MENU

1. HELP
2. ADD NEW RECORDS
3. CHANGE DATA STORED IN EXISTING RECORDS
4. DELETE RECORDS
5. PRINT REPORT OF ALL ADDITIONS, DELETIONS, AND CHANGES
6. RETURN TO MAIN MENU FOR THE BILLING SYSTEM

ENTER NUMBER OF THE OPTION DESIRED FOLLOWED BY THE ENTER KEY

FIGURE 4.5. Menu of file maintenance programs and options

documentation is displayed, control returns to the maintenance menu.

2   The data entered by the operator is edited and confirmed. After all of the data is entered for a new member, the member's record is added to the member master file.

3   The operator enters the member's number and a record is retrieved from the member master file. The name of the member is displayed and the operator is asked if the name matches the one on the charge form. If so, a formatted screen appears and the operator fills in the required data pertaining to the fields to be changed in the member's record. Some of the entered data is edited by the program. After all data is keyed in, the operator is asked to visually check to see that all data was keyed in correctly. The original record that contains additional or new information is then written on disk.

4   The operator enters the member's number and a record is retrieved from the member master file. The name of the member is displayed and the operator is asked if the name matches the one of the delete form. If so, a delete **code** will be placed in the   delete field of the record. The record is rewritten. When the file is reorganized, the records with delete codes will not be included in the reorganized file.

5   A batch program is called in that prints a daily sales transaction report. The totals printed are used as part of an audit and control procedure.

6   Control returns to the main billing system menu.

Indicated as an optional batch procedure in Figure 4-3, a file of the changes made to the member master file is created. After all of the online changes have been made, the clerk may request a summary of changes report. Unless the report is used to conduct a final audit to determine whether the changes were made correctly, the report is probably filed and never used. Many analysts would recommend that a file-change audit procedure be developed. If an error is detected during the audit procedure, the change procedure can be used to make the necessary correction to the master file.

Charge/Credit Procedure   When the system went online, the procedure used for sales and items returned (credits) changed a great deal. Sales slips are no longer made out manually and then audited. Since the charge/credit procedure has only two options, a separate menu is not displayed.

When a sale occurs or credit is granted, the member's record is retrieved by inserting the member's membership card into a slot at the top of the cash register terminal. Once the sales clerk or manager has verified that the right record was retrieved, a code is used to indicate whether the transaction is a credit or a charge. When the new system was initiated, a policy was established that required all purchases made in the pro shop to be charged.

When the item number is entered, the item's record is retrieved from the inventory master file. If the correct item was retrieved, the operator keys in the quantity. The price and description of the product are retrieved from the master file record.

The printer is also part of the cash register terminal and is used for printing sales and credit slips. One copy of the form is given to the member and two are retained and filed. In the evaluation of the system, it was noted that members were impressed with the new system. Fewer errors were made and it took less time to complete transactions. When the old system was used, the sales or credit slips were sometimes misplaced and batch totals did not agree. With the new system, as soon as a sale is made, the master files are updated.

Daily Cash Payments Procedure   As you compare Figure 3-10 on page 000 with 4-3 there appears to be little difference between the two procedures. However, the second illustration shows that the cashier will be keying in the data. Since the member master file is updated as the data is entered, the need for a cash payment file is eliminated. More than one batch of checks can be processed in a day.

Daily Exception Report Procedure   Since each program edited the data entered and displayed error messages regarding the exceptions that occurred, it is not necessary to print a separate report. Also since data is entered when and where (pro shop) the transactions occur and by individuals who are responsible for the transactions (cashier), errors are less likely to occur. When an error is detected by the program processing the data, the operator is given an opportunity to make the necessary correction.

Monthly Reports and Online Queries   The monthly reports are batch procedures that are executed the last day of the month after all of the daily online procedures are run. When the system was revised, more data was added to the online database (member master file). Since more information is available, more online queries are made.

CHECKPOINT

5. What has occurred in regard to EDP due to the development of online transaction processing systems?
6. When the physical fitness billing system was redesigned, what major changes were made?
7. What is a menu?
8. What is a formatted screen?
9. What procedures in the physical fitness billing system could be classified as transaction processing systems?
10. What procedures in the physical fitness billing system were still designed as batch jobs?
11. Why might fewer reports be printed if the operating system and database software support online queries?

Management Information Systems

A management information system (MIS) provides information needed to manage a business effectively. The outputs of the electronic data processing system become an integral part of a management information system. As illustrated in Figure 4-6, an MIS is dependent on the availability of comprehensive databases rather than master files. A database is often defined as a collection of files that are interrelated. Also a database usually serves many applications in contrast to files whose use was generally limited to one type of application.

An MIS must be capable of providing management with the proper level of detail. At the operational level, management must have the details of the transactions.

Middle management is usually more concerned with knowing the exceptions that occurred. For example, if a company has 500,000 receivables, a large percentage of the accounts are of little concern to management. These companies or individuals pay their accounts on time and do not ask for special considerations. Management is, however, concerned with the exceptions—the accounts that are past due, inactive, or require special attention.

Management is equally concerned when the cost of producing a product or service exceeds the budgeted amount. If the cost of material, labor, or overhead is greater than anticipated, management must be aware of the problem and create a solution before the problem compounds. In more traditional batch systems, production costs were usually only available on a month-to-month rather than a day-to-day basis.

Top management must be aware of the exceptions that occur on a day-to-day basis. In addition, it must be possible to retrieve whatever additional information is needed. As illustrated in Figure 4-6, batch jobs, transaction processing, and the online retrieval of information using a query language can occur simultaneously. Companies that require managers to submit written requests to the CIS department for information and then require them to wait until the reports are generated may not survive in this competitive society.

Although microcomputers do have database software, it takes a larger computer system to support a true MIS. A massive amount of information must be online. An online database under the direct control of the computer can be thought of as a gigantic tank of information that is used as a resource for all types of applications. The information is used to assist in the managing of the organization's resources—people, money, materials, methods, and markets.

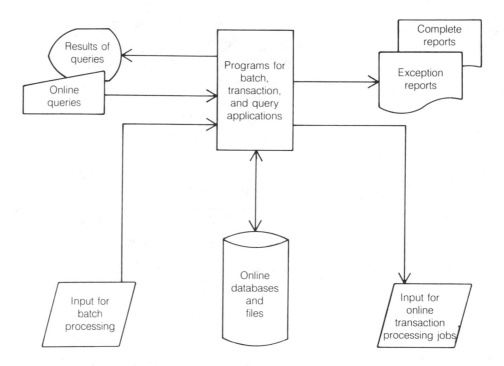

FIGURE 4.6. The key factor in development management information systems is availability of a comprehensive database

While batch systems provide detailed information, much of which is historical in nature, an MIS system also provides selective information requested by management or produced automatically as the result of exceptions that occured in the execution of tasks or procedures. Individuals who need information stored in a database must be able to use a terminal to make an English-like query to obtain the required facts or figures.

DECISION SUPPORT SYSTEMS

EDP and MIS provide management with existing information about the organization. The information is used to make the day-to-day decisions necessary to carry on the many functions of the organization. Some organizations may extend the use of the computer to include decision support systems (DDS). Figure 4-7 illustrates a decision support system. At an annual Association of Systems Management conference, Joseph F. Bode, director of Decision Support Services for Firestone Tire & Rubber Company, defined DSS as a means of marrying analytical tools with computer technology. A DSS provides the means for modeling, planning, and establishing long-term goals.

A DSS must integrate the capabilities of the computer to:

- create databases
- extract data from the databases
- create graphics
- perform functions associated with office automation
- provide written reports
- provide voice output
- display information
- perform analytical functions

When a dialogue is carried on between the user and the computer, not only the, "what if?" questions but also the "what's best?" questions should be

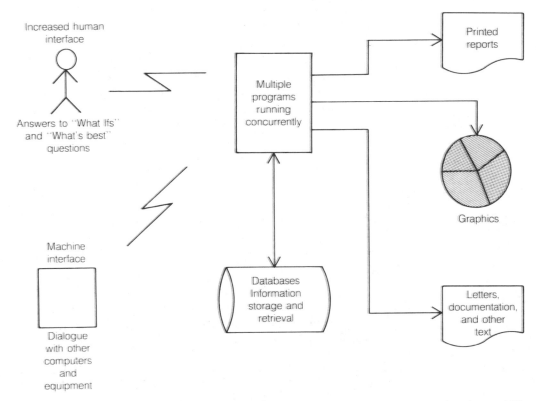

FIGURE 4.7. A decision support system has more components and is far more complex than a MIS

answered. Perhaps the most important ingredient of a DSS is a user-friendly human interface—an easy-to-use language that permits managers and their support staff to use the computer's capabilities in the decision-making function. A successful DSS must be easy to create, easy to understand, and easy to use, and must be user controlled.

A true DSS integrates into a meaningful whole a number of functions that were once thought of as separate entities. Large computers and microcomputers are linked together into a network that permits users of the typical office functions to interface with individuals and databases created as part of EDP or an MIS system. Only when the barriers—such as the lack of standardization and the lack of communication between EDP and office automation personnel—are removed can DSS continue to advance in sophistication. As computer technology—hardware and software—continues to advance and computers have **artificial intelligence (AI)**, more creative DSS can be developed. Artificial intelligence is the computer's ability to simulate human functions such as reasoning, learning, and self-improvement.

FIFTH-GENERATION COMPUTER SYSTEMS

Most computer scientists feel artificial intelligence will be a reality in the early 1990s when a new generation of computers will be introduced. The new generation will provide: easier ways of carrying on an interactive dialogue with the computer; better methods of database interrogation; easier-to-use natural languages; automatic programming; more effective use of handwritten documents; and more computer-aided learning programs. Although many of the features listed are now available, the fifth-generation systems will be more user friendly, faster, and have the ability to store more information online. It is also predicted that the new generation will be able to process data a thousand times faster than to-day's computers. The entire architecture of the computer and the way it processes data will also change.

The way knowledge workers (52 percent of today's work force) perform their jobs will change. Greater attention will be focused on subjects such as the concept of work, working, and the work environment. It is anticipated that as many as 20 percent of the knowledge workers will be employed in "electronic cottages" and telecommute with the organization's database and personnel. The motivating force behind all of this is the computer and its relatively untapped capabilities.

CHECKPOINT

12. What are the differences between a file and a database?
13. What appear to be the major differences between a batch oriented system and an MIS?
14. What is considered one of the most important ingredients for a successful DSS?
15. In a large organization with a great deal of hardware and software, what might prevent an effective DSS from being developed?
16. What is artificial intelligence?

SUMMARY

Until more sophisticated operating systems and better online storage media became available, most EDP was done in a batch environment. Online transaction systems can be achieved by using mainframes, minicomputers, or microcomputers. Many organizations now use transaction processing rather than batch processing methods. As new computers are installed and old systems redesigned, each procedure must be evaluated to determine whether transaction processing or batch processing should be used.

As organizations expand and analysts became more knowledgeable regarding computers, the goal may be to establish a more complex management information system. The goals for such a system must be established by top management. The next level of sophistication is to merge computer technology with analytical tools such as financial modeling and simulation in order to develop a decision support system. To develop MISs and DSSs requires careful planning, a dedication to long-range objectives, and a complex computer system.

When fifth-generation computer systems are available, an entire new era of computing will begin. True artificial intelligence may be achieved and the entire computing environment may change. Analysts being trained today must have the necessary skills to design complex, integrated systems for fifth-generation computers with capabilities that may be beyond our wildest imagination.

DISCUSSION QUESTIONS

1. What major changes are made when a batch system is converted to a transaction or online system? What type of procedures or programs are likely to change to online procedures and which ones are most apt to remain as batch jobs?
2. Is there more or less involvement of users when online systems are developed and used than when batch systems are developed and used? Explain why you answered the question as you did.

3. Although there are no definite boundaries between MIS and DSS, what are the major characteristics of each?
4. Why will DSS become more sophisticated when artificial intelligence becomes a reality?
5. Why is having a comprehensive, online database becoming increasingly important? Is the information stored in the database accessible to knowledge workers? Why might some knowledge workers work in their electronic cottages in the future?

Team or Individual Projects

1. Mini payroll case study

   Review the material presented in Chapter 3 regarding the case study and the data flow diagrams and system flowcharts you developed for the existing system. Also review the changes that were made in the physical fitness billing system when it was converted to an online system. You are now to make the necessary changes to convert the existing payroll system into an online system. At each of the major cost centers (departments), workstations are available that have keyboards, VDTs, and a small printer.

   Directions for project 1
   a. Make the necessary changes to convert the payroll system to an online system. Consider carefully where data should enter the system and who should enter the data. Also determine which subsystems would still be batch jobs and those that would be changed to transaction processing.
   b. Construct either a Diagram 0 for the payroll system or a detailed system flowchart. Although there could be a great deal added to the system, work only with the existing subsystems and do not add additional features to the system.

2. Utilize articles found in current publications such as *Computerworld*, *Office Administration and Automation*, or *Datamation* and prepare a report on one of the following topics:
   a. Management information systems.
   b. Decision supports systems.
      (For either of the two topics, include information regarding the type of hardware and software needed to implement the system, whether the companies using them normally need to obtain additional control software, and the extent to which such systems are implemented and utilized.)
   c. Problems created by the integration of microcomputers into applications involving electronic data processing.
   d. Distributed data processing. Include the type of equipment that is needed and the type of interface that should be available between the computer systems located at remote locations and the mainframe.

GLOSSARY OF WORDS AND PHRASES

artificial intelligence The capability of a computer to perform human-like functions such as reasoning, learning, and self-improvement.

batch processing system Data is accumulated and then processed in a "batch" rather than being processed as transactions occur.

buffer A temporary storage area used for data or instructions.

code Characters or digits used to condense data. For example, an account number is used rather than an individual's name

decision support system (DSS)   The resources of the computer are used in the decision-making process and questions such as "what if?" and "what's best?" are answered. Analytical tools are merged with computer technology.

field   A location within a record or the memory of the computer used to store a unit of information such as an account number, name, or address.

formatted screen   A display on a VDT which provides prompts for the operator who enters data. Sometimes a formatted screen is similar to a form that would be filled in manually.

knowledge worker   An individual who does not produce goods or services and must rely on information to perform his or her job.

management information system (MIS)   A computerized system that has a substantial amount of online data, numerous controls, and a good deal of feedback. Many different software packages are developed which make it possible to develop a comprehensive MIS that has as its foundation a well-defined database from which information can be extracted.

online transaction processing system   Data is entered into the system when and where the transaction occurs. The data is processed in time to influence the process. Online transaction processing systems are also called realtime systems.

record   Data pertaining to one subject or object. Within a payroll file, each employee has his or her own record.

STUDY GUIDE 4

Name _____

Class _____ Hour _____

A. Indicate whether the following statements are true (T) or false (F) in the space provided. If false, indicate in the margin how the statement should be changed to make it true.

_____ 1. Since over half of our working population are knowledge workers who rely on information to perform their jobs, computers must do more than process data and output the processed information.

_____ 2. An inventory file updated once a day is considered current.

_____ 3. When online systems are developed, fewer reports may be printed.

_____ 4. Online systems are centralized rather than distributed.

_____ 5. Because of the complexity of online systems, users are less involved.

_____ 6. Online systems can only be developed for large mainframes.

_____ 7. One of the options available on most menus is to have documentation regarding the other available options displayed on the VDT.

_____ 8. In the physical fitness billing system, workstations were only placed in the manager's office and the pro shop.

_____ 9. Most terminals or microcomputers are designed so that as data is keyed in, it is available immediately to the CPU for processing.

_____ 10. In the online system developed for the fitness center, only one procedure could be executed at a time.

_____ 11. When a formatted screen is displayed, the size of the field of data to be entered is not shown.

_____ 12. Since the person responsible for the accuracy of the data also keys in the data, the possibility of data entry errors is increased.

_____ 13. In order to have a successful MIS, there must be a comprehensive, online database available.

_____ 14. A decision support system only provides answers for the "what if?" questions.

_____ 15. If the new generation of computers that is expected to be available in the early 1990s has true artificial intelligence, analysts and the way systems are designed will not be affected.

B. Multiple choice. Put the best answer in the space provided.

_____ 1. A temporary storage device.
  a. modem b. buffer c. file d. database

_____ 2. Online system are usually
  a. user friendly.
  b. designed to include more online documentation.
  c. menu driven.
  d. distributed.
  e. all of the above.

_____ 3. When a menu is displayed, which two options are usually included?
  a. Options to call in one of several programs and an option to end the session.

b. An option to display documentation and options to call in one of several programs.

c. Options to display a formatted screen and to end the session.

d. Options to display documentation and to end the session.

____ 4. When the physical fitness billing system was converted to an online system, which subsystem (or procedure) changed the most?

a. file maintenance   b. charge/credit   c. cash payments

____ 5. When a batch oriented system is redesigned,

a. each procedure should be evaluated to see whether transaction or batch processing should be used.

b. all procedures should be redesigned as transaction processing applications.

c. only procedures that produce monthly reports should be batch jobs.

d. fewer terminals and workstations will be needed.

____ 6. The type of system which provides management with the kind of information that can be used in formulating goals, objectives, and plans.

a. batch processing system   c. decision support system
b. management information system   d. teleprocessing

____ 7. When an MIS is used

a. the database is stored offline.

b. management receives selective information as requested and day-by-day information regarding exceptions that are occurring.

c. batch jobs are never run.

d. it is unnecessary to use terminals.

____ 8. The most important ingredient in a decision support system is

a. the master file

b. a database.

c. the human interface—an easy-to-use, user-friendly language that permits users to employ the computer's capabilities in the decision-making functions.

d. artificial intelligence.

____ 9. In the member billing system, screens provided

a. prompts which indicated the field size.

b. documentation.

c. menus.

d. all of the items listed under a, b, and c.

____ 10. In the revised member billing system,

a. all data was processed in an online transaction mode.

b. all of the data entry functions were centralized.

c. fewer controls were included.

d. some data was processed in a batch mode while other data was processed when the transactions occurred.

5 NEW TECHNOLOGIES WITHIN THE OFFICE

Looking ahead

After reading the text and completing the learning activities you will be able to:

- Identify the reasons why so many organizations are involved in office automation (OA).
- Describe the differences that exist between a traditional office and a modern, automated office.
- Describe a well-designed workstation.
- Explain why word processing is more cost-effective than using a typewriter to create a letter or a report.
- Explain why word processing and other office automation systems should be compatible with electronic data processing systems.
- Describe the advantages of using electronic mail over other methods of sending letters, reports, or other documents.
- Identify the differences between facsimile systems and electronic mail.
- Identify the rationale that would be used to justify the development of a teleconference facility.
- Explain why the cost of scheduling software could easily be justified.
- Give the rationale for involving the computer policy committee and analysts in the development of standards and guidelines for office automation.
- Identify the reasons why the organization within the office will change as more office automation occurs.
- Describe the similarities and differences between EDP and OA.
- Define and utilize the words and phrases listed in the end-of-chapter glossary.

INTRODUCTION

Analysts are concerned with new technologies and how they affect the modern office. Analysts should be involved in the creation of an up-to-date office from the time it is on the drawing board until the equipment, furniture, and personnel are all in place. What distinguishes a modern office from a traditional one? What role does the analyst play? Are there two distinct areas: data processing, which includes MIS and DSS, and office automation?

REASON FOR AUTOMATION OF THE OFFICE

Although office automation (OA) is still in its infancy, there is a wide range of equipment and procedures that are designed to solve communication problems and bottlenecks. The complexity of the solutions, the multiple sources of products, and lack of standardization have given management problems. In companies where two paths—often parallel—were followed in developing decision support systems and an automated office, additional problems were sometimes created.

A number of studies have indicated that while the average amount of capital behind a factory worker was $50,000, as little as $200 was needed to support an office worker. Studies also indicate that office support personnel were away from their desks and unproductive a large percentage of the time. While away from their desks, support personnel were retrieving materials from files or placing material in files, copying materials, obtaining information from other workers, and performing other tasks. The design of the office frequently was not conducive to productivity. The copy machine might be located farther from those who use it and closer to people who should be in an environment that offers as few interruptions as possible. Some of the specific problems cited included:

- Executives and highly paid support personnel are still using pencils and pushing paper. This is neither cost-effective nor productive.
- The cost of letters, memos, and typewritten documents has tripled in recent years.
- The amount of information needed by executives has increased and more information must be online to support MIS and DSS. Day-, hour-, or minute-old information is no longer considered current or reliable.
- The cost of misfiling information averages about $60 a misfile.
- Often several attempts are made before a busy executive can be reached by phone. Secretaries play "telephone tag" in an effort to get two executives on the line at the same time.
- The skills of many of today's support personnel are not utilized effectively.
- Fifty phone calls or more may be needed before a meeting can be arranged between four or five executives.
- Diversification of products and services as well as the complexity and size of today's organizations have caused district and regional offices to be distributed throughout the United States or the world.
- The high cost of travel and more restricted budgets have caused attention to be focused on alternatives to conferences that require widely disbursed executives to meet in what may not be a central location.
- Over half of the work force are knowledge workers that depend on up-to-the-minute information that can be retrieved and formatted according to their specifications. Knowledge workers are information handlers and do not produce goods or services.

Obviously the separation of what was once known as electronic data processing and the early concepts of office automation is becoming less distinct. Both rely on computer and communication technologies, and perhaps the most important single factor in the success of each is the availability of comprehensive, *up-to-the-minute databases.*

Decision makers and knowledge workers expect information to be current. In many situations a different decision would be made if the information was as it had been a day, an hour, or a minute earlier. For example, at 10:04 A.M. a company has 1,000 of item X on hand and can fill orders. At 10:05 an order for 1,500 is placed and customers now must be told the item is backordered. Or at 12:00 a customer has unused credit of $10,000 and at 12:01 an order for $15,000 is received from the customer. Although the examples illustrated are simple ones, they serve to demonstrate the importance of absolutely current, reliable information. As strategic decisions are made, the importance of up-to-the-minute information increases rather than decreases.

Regardless of who is responsible for office automation and online systems, an integrated information system must ultimately prevail.

THE MODERN OFFICE

Office automation projects require a top-level commitment and the direct involvement of architects, analysts, and top management. When a commitment is made to utilize computer technology to streamline office procedures, three types of problems are created: space must be provided for new hardware, communication equipment, and peripheral equipment; additional cables and wiring must be provided for the extra equipment; and provisions must also be made for the comfort of the people using the new equipment. For these reasons, the office of today and tomorrow look different than the traditional office, and more attention must be given to the physical environment in which personnel and equipment are expected to coexist.

At Paradyne Corporation, a computer firm in Largo, Florida, a high-density, open plan work environment provides every employee with ample work and storage space while at the same time maintaining much needed privacy. Many other features listed under the caption "modern office" in Figure 5-1 on page 139 are included in the design created to meet the programmers unique requirements. (Photo Courtesy Steelcase, Inc.)

Figure 5-1 indicates some of the ways the modern office differs from the traditional office. In studying Figure 5-1 you must realize that the table indicates trends that are occurring. Large organizations are rapidly replacing typewriters with VDTs and standalone equipment with equipment that can be part of a network. However, in small organizations with limited needs and capital, the equipment used may change very little. Many of the concepts of OA may be implemented by using standalone microcomputers and software designed for word processing, database management, scheduling, and so forth.

Regardless of the size of the organization or the office, the rate of change is increasing. Just as EDP has not been able to solve problems created by poor management, OA will not solve all of the problems that exist within the office. Also as some old problems are solved, new problems, such as the vulnerability of online data, are created.

Creating Workstations

In a modern office, modular workstations are designed with desk tops that move up and down, ergonomic chairs designed to prevent fatigue, and separate task lighting to prevent eye strain. Walls and furniture are often coated with sound-absorbing material and individually controlled thermostats allow workers to regulate the temperature within their own work area. Above the workstation may be a white-noise emitter that creates a nonirritating sound which covers up the voices of other workers. The use of plants, color, and innovative designs are all intended to create an atmosphere that helps to decrease fatigue and increase productivity.

Wall systems must be created that permit workstations to be redesigned around different types of equipment. Traditional permanent walls are too inflexible to cope with the changing technology and needs of the modern office. Panel systems consisting of partitions hung on posts can be combined in different ways to create different shapes, sizes, and styles of offices. Components that can provide work surfaces, files, drawers, shelves, lights, computer terminals and

New tasks often require new tools. A team of computer programmers developing a new software package need adequate worksurfaces for printouts, access to a glare-free VDT screen, readily available storage space for reference materials, a table for informal discussions, and comfortable, ergonomic seating. Ten or fifteen years ago, two desks in a corner might have been considered sufficient. Today workstations are designed to meet the unique requirements of the worker.   (Photo courtesy Steelcase, Inc.)

Traditional Office	Modern Office
Equipment:	
Typewriters	VDTs and letter-quality printers
Calculators and adding machines	Personal computers and terminals connected to mainframes
Copy machines	Documents retrieved from online files additional copies can be printed
Equipment that handles mail such as postage meters, envelop stuffers, bursters, decollators, and so forth.	Communication and EDP equipment that provides for networks, electronic mail, and **facsimile systems**
File cabinets	More data stored on microfilm, magnetic disk and tape, and videodisks
Office environment:	
Permanent walls	Panel systems
Fluorescent room lighting	Task lighting
Nonadjustable desks and chairs	Ergonomic workstations
Centralized air conditioning and heat	Individually controlled areas
Little attention to accoustics and noise problems	Concern regarding noise level and privacy
Frequently drab and uninteresting	More effective use of open areas, color, and live plants
Limited concern over the individual	Greater concern over the welfare and conditions under which an individual works
Conference rooms with limited equipment	Teleconference rooms
Job descriptions for personnel:	
Secretary, stenographer, clerks	New terminology used for support personnel that indicates changes in assigned tasks and areas of responsibility
Security Limited	Greater concern as equipment is more available and more information is online

FIGURE 5.1. Changes occurring within the office

keyboard platforms, coat hooks, and many more items can be obtained for panel systems. The National Office Products Association (NOPA) predicts that by 1988 half of the interiors of new offices will be panel systems.

In designing or remodeling offices, consideration must be given to acoustics, lighting, noise level, electrical outlets needed, and communications system needed. The communication system must make it possible to link personal computers, mainframes, and other types of equipment needed in the automated office into a network. A typical workstation should be equipped with electrical lines for machines and lighting, cables for data transmission, and phone lines for modem and voice transmissions. In many workstations there are two phones—one for the employee and one for the equipment.

From employees working on VDTs, there have been an increasing number of complaints regarding blurred vision, eye discomfort, and muscle, joint, and tendon pain. Analysts are beginning to attach a greater importance to the environment in which employees work and to furniture and equipment that can be adjusted to the needs of the individual. For example, many analysts feel it is essential that VDTs and keyboards are freestanding and adjustable.

Increase in VDTs

Today on an average there is one VDT for every 20 office workers (5 percent); by 1990 it is predicted that there will be one VDT for every two or three office workers (33 to 50 percent). Other predictions indicate that 80 percent of all office workers will have a VDT. Already in the office of Touche Ross & Company, one of the "Big Eight" accounting firms, each partner has a VDT and a personal computer.

Along with the number of VDTs in the office will be an increase in the number of "peripherals." A typical workstation will have a VDT, keyboard, and printer. Many will also have optical recognition equipment that can read and record typed or printed data. (In the past, optical readers were large, expensive, and found only in centralized computer centers.) Depending on the nature of the work performed, the workstation may also be equipped with a multitude of I/O devices such as graphic tablets, mice, joy sticks, and plotters that create colored graphics and charts.

Wider use of VDTs will cause a decline in the amount of printed reports used in the office. There will also be fewer filing cabinets, bookshelves, and wastebaskets.

CHECKPOINT

1. At what point should analysts become involved in the design of an automated office?
2. What three problems must be solved when a commitment is made to automating an existing office?
3. Why has management had problems designing and developing office automation?
4. What percent of today's work force is classified as knowledge workers who depend on up-to-date information in performing their jobs?
5. Each day all sales are recorded in a file. After all sales are recorded, the accounts receivable file is updated to reflect changes that occurred due to sales on account. Would you consider the accounts receivable file an up-to-the-minute database? If you were an analyst, what changes would you recommend?

6. Has the environment of office workers changed as the amount of automation has increased?

7. How does an office constructed with wall systems differ from one that has permanent walls?

8. By 1990 how common will VDTs be in the office of medium- and large-size organizations?

9. What impact will the increased use of VDTs have on the amount of reports printed?

OFFICE AUTOMATION

Office automation software is available as a package from many of the minicomputer and mainframe vendors. Software for many of the functions described under OA is also available for microcomputers. Effective OA is dependent on up-to-date databases, well-designed workstations, well-written software, and computers that have enough memory to provide multiprogramming and timesharing. Depending on the organization, teleprocessing capabilities might also be necessary.

OA will cause organizational changes in the office. There will be two categories of labor: executive knowledge workers (also called **principals**) who supervise subordinates, and support personnel. Executives will gain the most from OA. The amount of time they spend on paper handling, writing, searching for information, proofreading, and dictation will decrease.

Support personnel will feel the first impact of OA. The skills they need will change and there will be less demand for typing, stenographic skills, filing, indexing, and manually retrieving hardcopy from files. Traditional secretarial skills will be replaced with computer-based technical skills.

Integrated into systems that are already developed and available will be new ones that rely more on voice-based capabilities which include voice response, voice recognition, and voice mail.

The computer terminal is taking on growing importance in the executive office. Once a mere novelty, the VDT is now a tool that executives depend on for access to corporate records, distribution of electronic mail, and market forecasting. (Photo courtesy Steelcase Inc.)

Word Processing   Word processing (WP) was first used by large organizations. In many organizations **stenographic pools** were first developed. One or more stenographers worked in the pool and their services were utilized by many individuals within a department or the organization. As automation occurred, standalone word processing equipment was obtained. The equipment could only be used for word processing. One of the earliest types of word processing systems used magnetic cards. These cards (also called *mag cards*) were about the size of the standard punched card and were used for storing text material. Mag card equipment was controlled by a microprocessor that had been programmed to perform certain functions. The files stored on the magnetic cards could not be used by the EDP equipment. The two areas, word processing and EDP, were viewed as separate entities.

Use of standalone word processing equipment became popular in smaller organizations such as law offices, medical offices, insurance offices, and so forth. Both the hardware and software used for word processing improved.

Computer vendors realized the potential of the market and developed software for multipurpose computers. Although many of the vendors who provided dedicated word processing equipment maintained that the word processing software run on multipurpose computers did not provide the same functions, this is not always true. Excellent word processing software is also available for most microcomputers. One of the major applications for which microcomputers are used is word processing.

A trend within large organizations is to:

- either utilize the corporate mainframe for both EDP and word processing; or
- develop guidelines for the acquisition of standalone systems which made it mandatory that word processing equipment become part of the information network and have the ability to utilize the existing databases.

Many organizations seem to prefer multifunction systems that can do both word processing and EDP rather than dedicated systems.

Disadvantage of Standalone Equipment   Dedicated standalone equipment can usually only be used for word processing. Often the files are stored on eight-inch diskettes which means that a limited amount of data can be online at one time. The per-station cost is often higher than the cost of using a terminal attached to the mainframe. Often the files created by using dedicated word processing equipment cannot be used for EDP. Unless the files can be used for both types of applications, there will be duplication of effort in creating and maintaining name and address files and other specialized files. Sometimes the diskettes have to be formatted by the vendor and are much higher in price than other eight-inch diskettes.

An invalid concept is that word processing should only be used when text is created to produce more than one document. It can be proved that a stenographer, typist, or someone in in management is more productive when WP equipment is used rather than a standard or electronic typewriter. Text can be changed; therefore, rekeying a letter or document when a minor error occurs is unnecessary. Once it is recognized that many individuals within the organization should have access to word processing equipment, it seems likely that multiple-purpose workstations will be developed. In large organizations, either microcomputers that can be part of a network or programmable terminals will probably be installed. In small companies, microcomputers will be used. The microcomputers can also be used for other applications.

Advantages of Word Processing When text is being created, word processing software permits words or characters to be inserted in the text or deleted from the text. There are commands that permit "cut and paste" operations. Words, sentences, or paragraphs can be moved to new locations within the text material. Searches can be done to find a word or phrase. For example, the text could be searched and all occurrences of "Chicago" would be changed to "New York."

Most word processing software is menu driven and help menus are provided. Once the codes are learned, the operator can elect to eliminate the menus that are displayed along with the text that is being entered or edited. Commands are available for controlling the cursor and scrolling, editing data, formatting data, searching text, saving files, and printing. Depending on the printer being used, the size of the characters printed can be controlled.

Once text is created, it is stored in files. Most systems provide for a backup file. When a change is made and the file is stored back on the disk or diskette, the backup file remains unchanged. The next time the text is to be edited, the original file is copied to backup. The changes made during the second editing session will only be made to the original file. In order to protect the backup file, it is often impossible to edit text stored in the file. If anything happens to the original file, a command is available that will cause the backup file to be used to create a new file.

While the full capabilities of word processing require mastering a large number of commands, most people can learn to use word processing to create memos, letters, and other types of text with very little effort. It is important that the word processing software selected is easy to learn and to use. Today, the person who wants to send a message keys in the text. Studies have shown it is more cost-effective to have the principal key in the text rather than dictating the message to a secretary who will then do the necessary keyboarding.

Hardware and Software for Effective Word Processing When mainframes are used, it is recommended that **programmable** or **intelligent terminals** (also called smart terminals) are used. Stored within the memory of the terminal will be part of the WP software and the text being created or edited. When programmable terminals are used, the mainframe is not required to execute most of the WP commands. The mainframe's computational and logical ability can be used for more complex functions. If nonprogrammable terminals are used and the mainframe is carrying a heavy workload, the WP operator may experience slow responses to requests for text or for the execution of commands.

When microcomputers are used for word processing, the computer should have enough memory and software so that new text can be entered while the text previously entered or edited is printing. If hard disks are used rather than diskettes, more files can be stored online. Functions such as copying files, storing text, searching for a string of characters, and so forth will execute faster than when diskettes are used. The additional cost of hard disk should be offset by the increase in productivity of the operator. Since most microcomputers have only one hard disk, text files that are not in use can be copied on to diskettes. When the text is to be used, the file can be transferred back to hard disk.

Letter-quality printers which have both uppercase and lowercase letters should be available for word processing. Impact printers should have a changeable print mechanism (sometimes called a *daisywheel* or *thimble*) so that different print styles are available. Often multistroke carbon ribbons are used that produce high-quality printing. It should be possible to adjust the printer for different sizes of characters and for the amount of space between lines. Most of the printing features are controlled by WP software.

Analysts' Role in Word Processing   Analysts, along with representatives from top management, should be directly involved in establishing standards regarding the acquisition of WP hardware and software. Files created or edited using WP systems should be accessible to EDP; files created or edited using EDP systems should be accessible to WP. Recent improvements in software for microcomputers make it possible to use the same files for EDP or WP applications.

The project team should also establish criteria for the evaluation of the hardware and software used for WP applications. Some word processing software permits the text files to be sorted and also supports some mathematical operations. Other WP software only provides basic WP functions.

The project team should also make certain other related functions such as checking the spelling of text material are available. The spelling check can be online or a separate operation performed after all the text is entered. When spelling software is used, two types of dictionaries are supported. The main dictionary is supplied by the vendor and contains frequently used words. The second type of dictionary is developed by the user. Words can be added to or deleted from either

Analysts should be directly involved in the design of the word processing workstations. An open feeling should be created even in a high density office. Ergonomically designed seating should be provided to ensure comfortable eye-to-screen and hand-to-keyboard relationships.   (Photo courtesy Steelcase Inc.)

type of dictionary. The spelling software checks the words in the text against the ones in the dictionaries. If a word cannot be found, it is "flagged". The operator can either ignore the word, fix the word, or cause the word to be added to one of the dictionaries.

Software is also available that will check text created under a WP program for grammatical errors. When one of the basic rules of grammar is violated, the error is flagged. The operator can either fix the error or tell the system to ignore the problem. Spelling and grammar-checking software are not intended to replace the need for good communication skills. What is often detected is the result of typographical errors that sometimes are overlooked when a document is proofread.

Some Typical Examples   Let's assume that you wish to send collection letters to everyone who has a past-due account. Using the accounts receivable database, a file of all individuals with past-due accounts can be created. Each record would contain the firm or individual's name and address as well as the due date and the amount past due. The text, or letter, is created using WP. When the letter is created, a code is used to indicate where the data stored in the second file is to be inserted. The word processing mail-merge facility is used to work with the two files: the one supplying the body of the letter and the one supplying the names and addresses. Why not print the letters using EDP? This could have been done. However, word processing software is already available that provides the desired functions.

An attorney might have a file of paragraphs that are typically included in a will. Whenever variable data has to be included within the text material, a "stop" is programmed. When drafting a new will, the attorney indicates which paragraphs are to be included. The operator keys in the codes needed to obtain the necessary paragraphs. When the "stop" command is executed by the computer, the variable data is keyed in by the operator. The entire text, with the form paragraphs and variable data, is recorded in a file. If the variable data was entered incorrectly, the material can be edited and reprinted.

Programmers and analysts have found that WP can reduce the amount of time needed to maintain documentation. The documentation is stored in files and is essentially in pages. When one section of documentation needs changing, the required changes are made by editing the data. The pages that have been corrected can be reprinted.

WP is not only cost-effective but is essential if error-free documents are to be produced in a cost-effective manner. Erasures or white out is not permitted on many types of legal documents.

CHECKPOINT   10.   What hardware and software must available to support office automation?

11.   When word processing was first used, was it integrated with data processing?

12.   When did mainframe and minicomputers manufacturer decide to develop word processing software?

13.   What trend regarding word processing seems to be occurring within large organizations?

14.   How would you respond to the statement that "word processing should only be used when the text material is to be used for more than one document?"

15.   What major functions are supported by word processing software?

16. What is the purpose of the backup file that is created by word processing software?

17. Why should programmable terminals be used when a mainframe is used for word processing?

18. When microcomputers are used for word processing, what advantages are provided when hard disks rather than diskettes are used?

19. What role should the analyst play in the development of a word processing system?

20. Most word processing software also supports the use of programs that provide what three functions?

21. Every month you wish to send an individualized letter to 500 clients. Each letter is to have the client's name and address as well as a personalized salutation. What is the most efficient way to produce the individualized letters?

Electronic Mail

Interoffice Communication Electronic mail can be used as either interoffice mail or as a means of communicating with business associates at remote locations. Within the office, the organization's timesharing or office automation software handles the electronic mail. Assume that you are the sales manager and you wish to send a message to Jim Green who handles accounts receivable. In order to send a message, you must:

- Sign on your system. Usually an **account number** and a **password** must be used.
- Key in the message.
- Use the proper commands to transmit the message into the account number of the accountant.

When Jim signs on the system, he is notified that he has mail. When he gives the proper response, the message is displayed on his VDT. Most software gives Jim four options: the message can be printed and then deleted; the message can be deleted; the message can be indexed and kept in the online file; and a response can be returned to the sender.

Electronic mail has many advantages over interoffice mail. Perhaps the major advantage is that it costs less. Some companies require the person who wants to send the message to key in the message. This is much less costly than when an executive calls in a secretary, dictates the message, and then approves a printed copy of the message. Keying in the message saves the executive time over former methods and also permits his or her secretary to do more meaningful tasks. There was some resistance to this approach for sending memos. However, once people become comfortable with their terminal and gain experience in using the keyboard, they are supportive of the system.

Another advantage of using electronic mail is that the sender is usually notified that the message has been received and read. When Jim uses the commands necessary to have the memo displayed, a message is returned to the sender, indicating the time that it was read. The sender can have a printed log of the messages sent and time each message was received.

Using electronic mail in place of typed or written memos is another way of taming the "paper tiger." The need for communication coupled with the availability of high-speed printers has caused an explosion in the number of printed memos, letters, reports, and other documents. Some executives are literally buried under an avalanche of paper. When electronic mail is used, very few of the messages will be printed. The person receiving the message makes the decision as to how the message should be handled.

One company has made electronic mail easier to use and more cost-effective by using voice input and voice output. The person who wants to send the message designates to whom the message is to be sent. His or her voice input is digitized, sent over telecommunication lines to its destination, and stored in a file. When the person receives the message, it is converted from the code used to store the message in the file to voice output.

Electronic Mail Services   There are electronic mail services to which individuals or organizations can subscribe. Any subscriber can transmit mail to any other subscriber. Each subscriber must have either a computer or a terminal with a **modem**. A modem (modulator/demodulator) takes care of converting data to and from sound signals or other types of signals that the device used to transmit the message needs. Approximately 80 percent of electronic mail messages are transmitted over telephone communication systems. The organization providing the service must have networking and telecommunication capabilities. However, for an individual subscriber, a microcomputer, disk or diskette storage, a printer, and a modem will suffice. Using a commercial electronic mail service for sending memos, letters, reports, and other documents provides the following advantages:

- instantaneous service. The information is received within a second or two after being sent.
- printed copy can be obtained, *when needed*, by both the sender and the receiver.
- more reliable service. Traditional mail is sometimes lost or sent to the wrong location.
- less expensive than many of the overnight or next-day services that are available. Also when the information is already in a file, the cost is usually less to use electronic mail than to print the document and perform the additional tasks needed to mail the letter or package.
- audit trails can be established regarding what was sent, when it was sent, and when the documents were received.

Trends in the Use of Electronic Mail   Large organizations with computer networks can provide electronic mail service for all members of their network. While development of timesharing systems and MIS was the major thrust of many organizations in the '70s, development of telecommunication capabilities, networks, and DSSs are the major goals for the '80s.

Effective use of electronic mail requires the establishment of standards, guidelines, and policies. Some companies have established the following policies:

- Individuals who are considered users of the system must sign on and check for mail at least twice a day.
- Users of the system must key in short memos and responses to messages they receive.
- Files must be purged periodically to remove messages that are no longer needed. If a message is to be retained longer than a stated period of time, a printed copy should be obtained. Some operating systems are designed to delete any files that have not been used within a given period of time. However, when the information is deleted from disk storage, it is usually

written on magnetic tape. Other companies have codes whose use will cause a file to be deleted at the end of the day or at the end of a given period of time; or the file is retained until the owner of the file indicates the information should be deleted. There must be some type of law, such as Murphy's, that states "The more file space you have, the more you use." Unless policies are established and enforced regarding file storage, more online file space must be obtained than is needed. Also more time is required for the computer to search for an retrieve information stored in online files.

Managing Calendars and Scheduling Meetings

Scheduling meetings between three or four individuals can be very time-consuming. One study indicated that often more than 50 phone calls are necessary to schedule such a meeting. A computerized system requires that each individual's schedule is entered into the computer. Visualize this as being done manually on a weekly calendar that has each day divided into half-hour blocks.

When a meeting is to be scheduled, the principal (member of the executive team) or a member of his or her staff calls in the computer program that schedules meetings. The names or identifying codes of the individuals who are to attend the meeting and the dates that are acceptable are entered. Within a second or two, the date and time of the meeting is usually displayed. The operator then asks the computer to schedule a place for the meeting and also enters the purpose of the meeting. The conference room schedule is checked to determine where the meeting can be held. The time, place, and purpose of the meeting is entered into each individual's schedule.

In the event that it is impossible to schedule a meeting time that everyone has free, the computer will pick a time when most members can attend. The one or two members who cannot attend are notified of the meeting and provided an opportunity to adjust their schedules. As with many other types of well-designed computerized applications, there should always be an opportunity for human intervention based on new information priorities.

Since most computers have a timing device that keeps track of the date and the time, many additional features can be added to calendar software. Printouts that include graphics can be obtained. If the computer system being used has voice output capabilities, verbal messages and reminders can be provided. If voice output is not available, various sound signals can be used that will serve as reminders.

Teleconferences

In many industries it is necessary to price goods and services to be competitive with other firms. Therefore, budgets are reviewed. One budget item that is often decreased is the amount allocated for travel and conferences. Teleconference rooms are designed to include large monitors for receiving colored visual images and sound. Equipment is also required to transmit images and sound. A teleconference might be scheduled between two people in Chicago, one person in New York, and four people in San Francisco. The picture and voice of the person speaking will be transmitted as well as other visual images of slides, photographs, graphics, and so forth. At the present time, the cost of the equipment as well as the per hour cost of transmitting visual images and verbal messages is high. As with other types of services, as the demand increases, the cost should decrease.

The advantages provided when teleconferences rather than regular conferences are scheduled are:

- The principals are better prepared. Since a limited time is allocated for the conference, materials are sent out in advance and each person is asked to review the material and develop responses to key issues and questions.
- Travel time is saved. Although the flight from New York to San Francisco is approximately five hours, additional time is required to get to and from airports. Also participants must either stay overnight or return home late at night.
- Participants are less fatigued. Individuals who travel and fight the battle of getting to and from airports, experience delays causes by cancellations, bad weather, or computers being down, and those who must adjust to time differences often experience increased stress and fatigue.

The first time or two an individual participates in a teleconference he or she is somewhat apprehensive. However, most individuals relax and tend to forget they are "on camera" and feel they are having a one-to-one conversation with another person.

Facsimile Systems

Organization that must transmit a great deal of printed material to distant locations may elect to use a **facsimile (fax) system**. Newer technologies may cause these systems to become obsolete. A fax system can be visualized as two copy machine connected by a telephone line. A person in New York elects to send a letter to someone in Dallas. The necessary telephone connection is made, the letter inserted in the copier in New York, and almost instantaneously the copy machine in Dallas produces a hardcopy of the letter. Fax transmission is not a new technology. Its use has been cost-justified by organizations that need to transmit a great deal of hardcopy to branch offices or subsidiaries in distant locations. Major users tend to use dedicated telephone lines for the transmission of the images.

Fax systems can be integrated with computerized systems. The computer may perform functions such as maintaining records regarding the transmission and receipt of information and scheduling the transmission of the information. When fax systems are used, illustrations or graphics can be transmitted as easily as typewritten or handwritten information.

Other Types of Office Automation

Assume that you are a busy executive. Computerized telephone directories and answering systems are available to assist you. A computerized answering service uses voice output to tell a caller that you are not available. The caller can leave a message and a number. Some systems assign priorities to the messages, sort the messages according to priority, and print a log of the messages. At your convenience, you can review the messages and determine which ones you wish to return. Other messages can be turned over to your support personnel for answering or can be postponed.

There is an increased awareness that telephone systems should be linked more closely with the organization's computer system. Numerous benefits will be achieved. There is also an increased awareness that voice transmission is natural and user friendly. Equipment is available that can convert voice transmission to hardcopy that will provide the necessary backup for transaction and decisions.

CHECKPOINT

22. How does the computer know that you are authorized to use a terminal and gain access to certain files?

23. When you use electronic mail to send a message to another individual, how do you know he or she received the message?

24. What advantages are there in electing to use electronic mail rather than the regular United States Postal Service?

25. What is a modem?

26. Why should an organization have some type of system for deleting files from random access storage devices?

27. What must occur before a meeting can be scheduled by a computer?

28. How can teleconference facilities and equipment be justified?

29. How does a use of a facsimile system for sending a letter differ from using electronic mail?

30. How can a computerized telephone system be of benefit to an executive?

31. Although a great deal has been written and said about OA, to what extent has its used increased the productivity of office personnel?

SUMMARY

Sophisticated transaction processing systems and the advancements discussed under office automation require the same type of systems. Computers are required with enough memory and an operating system that provides for multiprogramming, timesharing, and often, telecommunication capabilities. Also necessary are terminals or microcomputers, online random access storage devices, printers, and a medium that can be used to back up transactions and files.

Transaction processing systems and office automation can be achieved by using large mainframes, minicomputers, or microcomputers. Mainframe office automation software provides for the integration of EDP and OA functions. Often the "software package" provides for word processing, maintaining calendars, developing schedules, electronic transfer of information, telephone directories, automatic answering and response to telephone calls, and other office-related services.

Although the hardware and software are available and a great deal is written about OA, surveys indicate that in the past few years productivity within the office has increased very little. More organizations must develop long-range plans for the development of comprehensive information systems that will include traditional EDP as well as improved visual and verbal communication that take advantage of the technology used for word processing, electronic mail, facsimile transmission, and teleconferences.

Analysts, management, and the committee established to determine goals pertaining to the use of computers, must establish guidelines and standards that provide for the integration of EDP and OA.

DISCUSSION QUESTIONS

1. In what ways does the physical appearance of a modern, automated office differ from that of a traditional one? Why is an analyst concerned with the physical aspects of an office?

2. What are the similarities and differences between the equipment and control software needed to support MIS or DSS and OA?

At Continental Western Life Insurance Company in West Des Moines, Iowa, decorative acoustical ceiling, carpeting, and tackable acoustical panels keep sound at a comfortable level. Creating more attractive and comfortable workstations has done much to reduce fatigue and increase productivity. (Photo courtesy Steelcase Inc.)

3. What type of activities or functions are available when most word processing software is used? How does the spelling software check for errors? Will using the spelling or grammar-checking software detect all of the errors made by the person entering the text?

4. What action must be taken and what normally occurs when you use electronic mail facilities to send a memo to someone within your organization?

5. Assume you live in Boston and find that you communicate on a regular basis with someone in Dallas. What equipment would you need to subscribe to one of the electronic mail services? What equipment would your friend need to communicate with you?

6. Assuming that it costs between 75 cents and $1 to send a letter using electronic mail. To send it by U.S. air mail costs 37 cents. What advantages are there in using an electronic mail service?

7. How does sending a letter using a facsimile system differ from using electronic mail?

8. You have a friend who is a lawyer and he has indicated to you that using word processing would cost more than having documents typed. What rationale would you use in trying to explain to him the cost-effectiveness of using word processing? What other benefits would you describe to him and what type of equipment might you recommend that he obtain? At present he does not utilize computers for either data processing or any form of office automation.

9. What advantages are there in having your secretary use a computerized scheduling system to schedule a meeting rather than calling all of the principals who are to attend the meeting?

10. Why should analysts and the computer policy committee be involved in developing standards and guidelines for OA? What are the projections for the market for OA equipment and software as compared to EDP?

Team or Individual Projects

1. Prepare a report on the electronic mail systems that are available from subscription services such as Telenet's Telemail, Tymnet's OnTyme II, The Source, or CompuServe. In your report:
   a. Describe the equipment that must be used by subscribers.
   b. Indicate the cost of subscribing to the service.
   c. Describe the features that are provided by three of the major subscription services.
   d. From your review of articles pertaining to OA and electronic mail, determine what types of individuals and organizations are subscribing to electronic mail systems.

2. Prepare a report on a standalone word processing system. In your report describe:
   a. The hardware that is available from the vendor.
   b. The cost of the system which should include the required software.
   c. The features that are incorporated into the software. For example, when financial reports are included are column totals and crossfootings available? Is the spelling-check software included, available as an optional item, or not provided?
   d. The other applications, if any, for which the equipment can be used.
   e. The systems, if any, that can be interfaced with the standalone system.
   f. The training program that is available from the vendor. Include an evaluation of the manuals available.
   g. If the system uses eight-inch diskettes, contrast the cost of obtaining the diskettes from the vendor with the cost of obtaining them through an office supply company or one that specializes in the sale of computer-related products.

3. Prepare a report describing word processing software available for a microcomputer. In your report:
   a. Describe two or three of the computer systems that can be used to run the software.
   b. Determine the cost of the word processing system—hardware and software.

c. Identify the major features incorporated in the WP software.
d. Indicate how the software can be obtained and describe the manuals and training materials supplied with the software.

GLOSSARY OF WORDS AND PHRASES

account number A unique number assigned to an individual or to a group and that is used to gain access to a computer system. After entering the account number, the operator enters a password which is checked by the operating system software.

facsimile (fax) system A system designed to transmit complete images from one location to another. The image may be a picture, chart, graph, drawing, or page of text.

intelligent terminal A terminal that is programmable. See programmable terminal.

modem A MODulator/DEModulator used to translate data back and forth from characters the computer can understand to the type of data that can be transmitted over a communication system.

password A unique word used by an operator to gain access to a computer. Often the user enters an account number, then a password. The system's software checks to make certain the password identifies the owner of the account.

principal A term used to describe an executive or someone in management who supervises support personnel such as executive secretaries, stenographers, or other personnel.

programmable terminal A terminal that has sufficient memory to store a program and a limited amount of data. A programmable terminal may have stored in its memory a program that performs some of the word processing functions or one that edits the data entered from the keyboard.

stenographic pool A group of stenographers who work for a number of different executives. Each member of the pool can do work for any of the executives.

STUDY GUIDE 5

Name _____

Class _____ Hour _____

A. Indicate whether the following statements are true (T) or false (F) in the space provided. If false, indicate in the margin how the statement should be changed to make it true.

_____ 1. In large organizations, typewriters are being replaced with VDTs.

_____ 2. One of the problems created as more systems are changed to online is the increased vulnerability of data.

_____ 3. The National Office Products Association (NOPA) has indicated that the trend is to develop offices with permanent walls rather than wall systems.

_____ 4. A typical workstation needs electrical lines, cables for data transmission, and phone lines.

_____ 5. It is predicted that by 1990 there will be one VDT for every five office workers.

_____ 6. Since more VDTs will be available, the amount of printed reports and documents is predicted to increase.

_____ 7. In developing an automated office, management faces the problems of complexity and lack of standardization.

_____ 8. Today executives and support personnel need less information to make decisions than they did in the past.

_____ 9. Currently the skills of many support personnel are not utilized effectively.

_____ 10. Approximately 25 percent of today's work force are knowledge workers who do not produce goods or services but depend on information in performing assigned tasks.

_____ 11. The separation of what is referred to as EDP and OA is becoming more distinct.

_____ 12. Many large organizations have guidelines for the acquisition of stand-lone word processing systems which make it mandatory that the equipment obtained can become part of the information network.

_____ 13. Word processing is only cost-effective when a document is created and printed several times.

_____ 14. Corrections are harder to make when word processing is used than when text is typed on a typewriter.

_____ 15. Response time to WP commands is usually faster when nonprogrammable terminals are used.

_____ 16. When microcomputers are used for word processing, it is unlikely that the use of hard disks rather than diskettes can be cost-justified.

_____ 17. When microcomputers are used for word processing, the files created cannot be used for the more typical EDP functions.

_____ 18. When software that checks spelling is used, words unique to a particular field are not available in the dictionaries.

_____ 19. When terminals are accessible to many individuals, there is no way of determining that only authorized users are signing on the system and accessing a user's files.

_____ 20. When electronic mail systems are used, principals never key in their own messages. It is considered more efficient to have their secretaries key in the data.

_____ 21. When online systems are developed, a policy must be established regarding when and how files will be deleted from online databases.

_____ 22. In order to have the computer schedule a meeting, each principal's schedule must be part of the system.

_____ 23. When applications are computerized, there is never any need for human intervention or decision making.

_____ 24. Teleconferencing may be a solution to the increased cost of travel.

_____ 25. A teleconference that lasts two hours is far less productive than a conference where all of the participants are in the same room.

_____ 26. Facsimile systems transmit data that is stored in an online file.

_____ 27. Facsimile systems always involve the use of the computer.

_____ 28. Some computerized answering services assign priorities to the messages, sort the messages by priority, and print a log of the messages.

_____ 29. Since most large organizations have modern offices and have implemented office procedures that utilize techniques considered as part of OA, surveys indicate that in the past few years office workers have become 30 percent more productive.

_____ 30. When electronic mail is received, it can only be displayed on the VDT or printed as a report.

B.   Multiple choice. Put the letter of the best answer in the space provided.

_____ 1. A device that converts data to and from signals that communication carriers can use to characters that can be understood by computers is called a

     a. terminal.   b. buffer.   c. modem.   d. microcomputer.

_____ 2. Required in order to sign on a timesharing system:

     a. identification code   c. account number
     b. password   d. account number and password

_____ 3. A temporary storage device.

     a. modem   b. buffer   c. file   d. database

_____ 4. In creating workstations, an analyst is concerned with:

     a. the equipment.   c. the physical environment of the worker
     b. the software   d. equipment, software, and
       needed.   environment.

_____ 5. One study showed that while $50,000 in capital was needed to support a factory worker, as little as _____ supported an officer worker in a large organization.

     a. $1,000   b. $2,000   c. $5,000   d. $200

_____ 6. Office automation systems were developed to

     a. solve problems such as the increasing cost of letters and memos.
     b. make principals and support personnel more productive.
     c. increase communications between people within the organization.
     d. accomplish all of the above.

_____ 7. Which of the following statements does not describe what is occurring when workstations are developed?
   a. Panels systems are used to provide flexibility.
   b. Furniture is adjustable to provide maximum comfort.
   c. More attention is devoted to making workstations and the surrounding areas attractive.
   d. Little concern is shown regarding the comfort of the individual worker.

_____ 8. Word processing was first developed for
   a. mainframes b. standalone systems. c. microcomputers.

_____ 9. The feature which permits a file containing a form letter to be combined with a file that contains names, addresses, and amounts is sometimes referred to as
   a. spelling-check software.   c. mail-merge software.
   b. grammar-checking software.   d. insert software.

_____ 10. When word processing is dependent on the mainframe, intelligent terminals should be used because
   a. the operator can program the terminals.
   b. they cost less than "dumb" terminals.
   c. all of the text to be edited is stored within the memory of the terminal.
   d. some of the software and text is stored within the memory of the terminal and fewer demands are made on the host computer.

_____ 11. Analysts and programmers often use word processing to
   a. enter source code into a program.
   b. create and maintain documentation.
   c. develop system flowcharts.
   d. develop detailed logic plans.

_____ 12. Electronic mail can be used
   a. only for interoffice communications.
   b. only to communicate with customers who subscribe to an electronic mail service.
   c. only to communicate with members of the organization who are remote from the corporate headquarters.
   d. for interoffice communications and communicating with individuals or firms that subscribe to an electronic mail service.

_____ 13. Statistics indicate that
   a. it is less cost-effective to have highly paid principals key in their own memos.
   b. a principal should key in all of his or her reports and other documents.
   c. it is cost-effective to have principals key in their own short memos and responses to messages.
   d. support personnel should key in all of the text material that is to be stored in a file and then transmitted.

_____ 14. When electronic mail is received, the message
   a. can only be displayed on a VDT.
   b. can be displayed or printed and then is automatically deleted from the file.
   c. can be displayed and/or printed, retained in the file, or deleted from the file.

        d. is never logged on the sender's system as being received.

_____ 15. Teleconference systems provide for
        a. voice transmission only.
        b. transmission of voice and visual displays such as graphics, text, and pictures.
        c. one-way communication only.
        d. visual transmissions only.

_____ 16. Teleconferencing
        a. is always cheaper than having an actual conference.
        b. only reduces travel time and expenses.
        c. is often more productive because the participants are better prepared than when they attend a conference.
        d. creates more stress and strain on the participants than does traveling a long distance to attend a conference.

_____ 17. Analysts, management, and the computer policy committee
        a. should be directly involved in the development of OA.
        b. should only serve as consultants in the development of OA.
        c. should not be concerned if the OA system's files are not compatible with those created under EDP.
        d. are not concerned with OA since different equipment is used than is obtained for EDP.

_____ 18. Perhaps the single most important item in both OA and EDP is the availability of
        a. VDTs that display colored graphics.
        b. well-designed workstations.
        c. comprehensive, current databases that support the random retrieval of records and queries.
        d. letter-quality printers.

_____ 19. Rather than stating that a company has EDP and OA it might be better to indicate that
        a. computers are used for many different applications including flexible manufacturing systems and CAD/CAM.
        b. a comprehensive information and communication system has been developed.
        c. word processing and DSS are available.
        d. no new applications will need to be developed.

_____ 20. In regard to EDP and OA
        a. the same standards and guidelines apply to both areas.
        b. less hardware and software will be needed to support OA than is now used for EDP.
        c. guidelines and standards must be established that provide for the integration of the two areas.
        d. analysts need to be knowledgeable only in regard to EDP.

Initial and Detailed Investigation

Bay Medical Center

The merger of General and Mercy Hospitals in 1972 to form the Bay Medical Center consolidated community health services to improve efficiency and reduce the cost of hospital service.

The medical center provides a broad range of specialized and general services for both inpatients and outpatients. Major programs and services are: an intensive care unit, radiation therapy, radio-isotope services, inhalation services, psychiatric services, emergency and ambulance services, migrant workers' clinic, home care programs, and outpatient services.

The Board of Directors of Bay Medical Center, the physicians within the community, the 1500 employees and 1000 volunteers who make up the Medical Care family have shown they are ready to meet today's challenges by helping to build the 25 million dollar, 342 bed medical-surgical addition to the original General Hospital. The Bay Medical Center Family pledged both funds and support for the addition.

EDP AT BAY MEDICAL

Bay Medical has utilized timesharing services such as MacDonald Douglas and batch-processing services supplied by local organizations such as Professional Online Corporation. At the present time within the hospital there is a variety of stand-alone word processing equipment and approximately 40 micro-computer systems. In the past as each new application was presented, the decision was made regarding what hardware should be obtained.

The computer utilization committee is working on a master plan that will define the specifications for additional equipment. When a needs analysis indicates a new system is required, the necessary form showing the equipment called for, benefits from

Bay Medical offers a broad range of specialized and general services for both inpatient and outpatient care.

the system, and cost savings is submitted to Jim Meister, head of the Systems Department, for his approval and then to the Vice President of the functional area that made the request. If the projected budget for the project exceeds a given amount, the proposal must be referred to the computer utilization committee for its approval. If the project is approved, the detailed design developed by the project team must be approved by Meister, the Vice-President of the functional area making the request, and the computer utilization committee.

At Bay Medical responsibility regarding the utilization of computers is divided between the Systems Department and the Data Processing Department.

THE SYSTEMS DEPARTMENT

The Systems Department is responsible for developing new systems and modifying existing systems. Jim Meister's staff consists of a program coordinator, three analysts, and two clerks. At the present time most of the computerized systems being developed will be run on microcomputers. Jim believes that users should be directly involved in the design of systems and

that all of the activities involved in the study should be well documented. He has developed a Project Development Methodology Manual, which details what must be done.

The Project Development Methodology Manual

In the introduction to the Project Development Methodology Manual it states "If this methodology is closely followed and users are competent in their field, the resulting system development efforts should produce *on time, within budget, user accepted* systems that do what they are supposed to do and require little maintenance immediately after becoming operational."

The manual is very specific and provides rules, policies, procedures, and techniques that must be used: to develop new manual or automated systems; to convert from manual to automated processing; and to change either manual or automated systems. The manual covers all phases of the systems development life cycle—from an "idea" to its implementation.

The manual recommends the use of system flowcharts, Gantt planning charts, HIPO charts, and extensive documentation, training, and testing. Brainstorming and walk through sessions have become a way of life for the analysts at Bay medical.

During a brainstorming session Pat Zaplitny, the project leader for the Physical Medicine System, explains his concept of the proposed system to Tammy Waugh and Ann Knight.

The Analysts at Bay Medical

The three analysts at Bay Medical have different areas of responsibility. However, they frequently work as a project team to develop major applications. One such application was the physical medicine system.

Tammy Waugh

Tammy Waugh received an Associate Degree in Data Processing from Delta College in 1980 and a B.S. degree with a major in Data Processing from Ferris State College in 1982. Since that time she has been employed either as a programmer or an analyst.

At Bay Medical Tammy is the project leader for the payroll system and does the required program maintenance, redesigns input screens, and corrects any problems that occur. Although her payroll work is done for a main frame, the new applications she has developed are designed for microcomputers.

Tammy indicated that what she likes the most about her job is working with people. She feels that one of the real challenges is to determine what the user really needs. She believes a successful analyst must have a good technical background and be able to communicate effectively. According to Tammy "you must develop a logical way of thinking and problems must be broken into steps—step by step the problem must be solved."

Since analysts spend a great deal of time working with users, she feels it is critical that they learn to explain procedures in terms users will understand. Tammy realizes that one of the disadvantages of being an analyst is that sometimes you must work closely with people who do not understand computers and who do not want to work with a computer.

Pat Zaplitny

Pat, the project leader for the physical medicine system, received his B.S. degree in Computer Systems Engineering from Western Michigan University in 1976. Before joining the Bay Medical staff in March of 1983, Pat worked primarily on large computer systems and with COBOL and CICS. Pat feels it is important for analysts to have a solid background in computer-related courses and that they learn to ask the right questions. Analysts must learn to ask "Why" and learn how to continue questioning when they receive "Because this is the way we do it" for an answer.

As an analyst Pat enjoys working with users and learning more about the organization. He does feel, however, that working with first-time users can be difficult, but rewarding. He feels that the type of information he is learning may qualify him for a managerial position.

Each analyst at Bay Medical has a modular office equipped with both a microcomputer and a terminal.

Data stored on diskettes is read and transferred to the System/36's hard disk for processing. The system/36 at Bay medical is dedicated to the processing of financial data.

Ann Knight

Ann graduated in 1974 with a Bachelor of Arts in Education from Michigan State University. Although trained as a teacher, Ann elected to obtain an Associate Degree in Computer Information Systems. Since November of 1982 Ann has been employed by Bay Medical. Ann has worked on a variety of systems which include Asset and Information Management, Nursing Statistics, and a Pharmacy Inventory System. She has been instrumental in the development and presentation of software training programs for employees and is participating in a hospital-wide study of its information requirements.

In developing software and in training employees Ann has worked with IMS-8000 and IBM PC and XT microcomputers, the CP/M operating system, COBOL and PL/I programming languages, dBASEII and other software designed for microcomputers which includes WordStar, DataStar, MailMerge, and ScreenGen.

Ann feels that being an analyst is interesting as the work load is always changing and an opportunity is provided to learn about many other areas. She feels the job is challenging because there is always more to learn.

THE DATA PROCESSING DEPARTMENT

The first step towards developing their own data center was to obtain an IBM System/36 and establish a data processing department. The computer obtained in 1984 is used to run a financial package that was tailored to the specific needs of the center by an independent software house. Prior to that time, financial data processing applications were processed on time-sharing systems. All of the other inhouse applications were run on microcomputers. The department's six employees are directly involved in data entry and in operating the computer.

THE PHYSICAL MEDICINE SYSTEM

The physical medical system is designed to maintain schedules for the physiotherapists and their patients. One of the outputs from the system, the charges to the patients, will become an input into the hospital's financial system.

The request for the system was made in May of 1982. After the initial investigation, the project was shelved until February of 1984. At that time a needs analysis was conducted which was approved by the committee in May of 1984. The three analysts worked from June through September on the detailed investigation and design phases of the project. During the detailed design phase, users approved the input forms, screen designs, and report formats. In addition, program specifications were developed.

From November 1, 1984 until January 15, 1985 all three analysts were committed to writing the required programs in Microsoft COBOL for the IBM PC and XT. However, in the future, applications will probably be programmed using the C language.

By April of 1985 the final testing, documentation, and training required to implement the system was completed. According to the policy set forth in the manual, the training was completed after the system had been completely tested.

Entering Data

A microcomputer within the physiotherapy department is used to enter the physiotherapists' schedules as well as the information pertaining to the treatment of the patients. The therapists record the data on the source document which was designed by the analysts. When the operator loads the program used to record the data, formatted screens provide the necessary prompts.

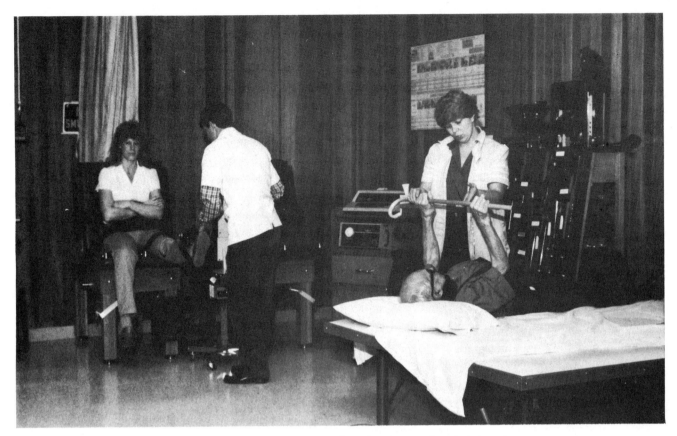

The data for the Physical Medical System is generated when the physiotherapists work with their patients.

Brenda Walters keys in data recorded on the source documents by the physiotherapists.

After the data is entered, a variety of reports can be printed on the letter-quality printer. The reports will provide the department with a great deal of information that was not available until the system was implemented. One of the printed reports details the charges that are to be made to the patients' accounts.

Entering the Data into the Financial System

Until the IBM PC and XT located in the physiotherapy department is networked with the IBM System/36, the billing information will need to be keyed in by an operator. Before the report is printed, the charges will be sorted by procedure code so there will be less variable data to enter. The screens are designed to look as much like the report from which the data is entered as possible. The data entered by the operator will be processed by the IBM System/36.

Eventually either a dedicated line will be used to transmit the data from the XT in the physiotherapy department directly into the System/36 or the billing information will be recorded on a 5¼″ diskette and taken to the data processing department for processing. However, in order to do that, an additional program will need to be written for the System/36.

Sue Ogden, or one of the other data processing department employees, will key in the charges listed on the report printed by one of the programs that make up the Physical Medical System.

SUMMARY

Until the master plan for computer utilization is developed, it is uncertain what will occur. Systems Department manager, Jim Meister, seems to feel that perhaps the direction to take is to obtain a mainframe which would be the "host" computer to microcomputers located within the users' departments.

Plans are well underway to provide additional education regarding computers and how they are utilized in a hospital environment to Bay Medical Center's employees.

Innovative, cost effective systems have been developed at Bay Medical Center for microcomputers. If the procedures outlined in the Project Development Methodology manual continue to be followed, user accepted systems that are completed on time and within the budget will continue to be developed.

6 THE INITIAL INVESTIGATION

Looking Ahead

After reading the text and completing the learning activities you will be able to:

- List the objectives of the initial investigation.
- List the steps required to initiate an investigation.
- Explain why requests are made for a systems study.
- Identify the purposes of each of the three sections of the request form.
- Explain why the computer policy committee should process the requests for system studies.
- List the tasks involved in the initial investigation.
- Determine the techniques used to gather data.
- Explain what should be done by the interviewer prior to the interview, during the interview, and after the interview.
- Identify the responsibilities of the person being interviewed.
- List the information and exhibits that should be included in the initial investigation report.
- Identify how a standards manual might be used.
- Explain how a standards manual should be formatted, written, and maintained.
- Define and utilize the words and phrases listed in the end-of-chapter glossary.

INTRODUCTION The purpose of an initial investigation is to determine whether a problem exists and, if one does exist, to ascertain its true nature and scope. No two initial investigations are exactly alike. Also there is no clear-cut distinction between what must be done in an initial investigation and what must be done in the feasibility study. Usually, limited **resources** (time, money, personnel, and materials) are committed to the initial investigation.

Although other techniques may be used to investigate the problem, a large percentage of the data is gathered by interviewing top or middle management. Usually only one analyst is assigned to the initial investigation.

Guidelines should exist for conducting the study and for writing the report. However, the length of time needed to complete the initial investigation depends upon the detail required. Figure 6-1 illustrates the objectives of the initial investigation.

Objectives of Initial Investigation

- Accurately define the problem.

- Determine the scope of the problem.

- Determine resources needed to complete the feasibility study.

- Estimate personnel, time, and cost needed to complete the project.

- Determine priority to be assigned to the project.

- Accurately estimate the resources needed to modify the existing system.

- When only minor problems with existing system are identified, accurately estimate resources needed to make the necessary modifications.

FIGURE 6.1. Objectives of the initial investigation

WHY INVESTIGATIONS ARE MADE Initial investigation are made for the following reasons:

1. Modifications are needed in an existing system that no longer meets the needs of the company.
2. A new system is needed to solve a problem that previously had not been identified.
3. Management believes an existing system could be redesigned so that it could be more cost-effective.
4. The ongoing system audit of an existing system indicates a variety of problems such as lack of controls and security of information.
5. New technological advances have occurred that should be investigated.

The request for a systems study can come from management, from the user, or from within the systems group. All requests should be handled in a uniform manner. A form should be used and a procedure should be established for processing the request.

A User Makes a Request More requests for systems studies come from users than from any other source. Depending upon the type of application, after a design is implemented, it may have a useful life of five years or less. Technological advancements, growth in companies, and changing business conditions are three of the major reasons an existing system may be considered obsolete. A user may request that the present system be reviewed to determine whether it is still adequate. Requests are also received to have additional features build into existing systems.

Frequently a request is made to convert a batch system to an online system. When the system was designed, an online transaction system could not be justified. With larger CPUs that can handle more jobs concurrently, more online file space, better database management software, and better communication systems, existing batch systems should be reassessed. It may also be possible to develop an online system that utilizes low-cost microcomputers; data can be recorded when and where the transactions occur.

Requests are also made to decentralize existing systems. As users become more familiar with hardware, software, and computerized procedures, some desire to become more involved. Microcomputers or minicomputers can be used to process a great deal of data within a department; summarized data can be transmitted to the corporate database.

There may also be problems with the present system. For example, the statements produced by the accounts receivable subsystem may often be printed after the due date. In addition, the wrong customer is sometimes charged for a sale. The accounts receivable manager may be concerned about the integrity of the entire system.

Management Makes a Request

Inadequate Information   The request for a systems study may be made by management for a number of reasons. One of the major reasons is that there is not enough information available for making decisions. For example, each semester when the class schedule is being determined, the academic dean's staff studies the enrollment trends over the past five years. A computer program consolidates the information into one report. For each course, the number of sections, total students enrolled each semester, and the percentage of increase or decrease in enrollment from the base year (first year) to the last year (fifth year) is shown. The report provides a great deal of information regarding enrollment trends. However, there is no way of knowing how many students wanted a class that was already filled or how many students could not enroll in a course because of conflicts.

An analyst might be assigned to make an initial investigation. By interviewing top-level administrators who are directly involved with scheduling, the analyst might find out that a great many additional problems need to be solved. The analyst might feel there is not enough information available to effectively plan, organize, and control the registration and scheduling procedures.

The analyst's report might indicate that a full-scale investigation is warranted which involves studying the entire area of student registration and accounting. Since a great deal of the information that is needed to make effective decisions is not in the database, the scope of the problem is broader than the request for a study had indicated. The report might also indicate what additional information should be in the databases and how simulation could be used. The justification for the additional information might list the reports that would be available and the type of online queries that could be made. Using historical data, enrollment trends, and a mathematical model, a suggested academic schedule could be generated.

Problems With an Existing System   When a number of complaints are received regarding a system, management might request an investigation. For example, a vice-president might request a study because of customer complaints about receiving the wrong merchandise or having items placed on backorder.

The analyst assigned to the investigation might find there are far more problems than those listed in the request. Because the other problems need to be identified, justification for a feasibility study will be included in the report. The initial investigation report should include an estimate of the number of **person/months** needed to complete the feasibility study. A person/month is the time equal to one person working one month (160 hours). Two individuals working two 40 weeks would equal one person/month.

Management Hears About an Innovative New System   While attending a conference, some of the line officers of a company hear about a new system being implemented by a company similar to theirs. The officers might request a study to determine whether it would be feasible to consider implementing a similar system on their equipment.   However, the initial investigation might show that the company's system is inadequate.

Since the only way a similar system could be implemented would be to obtain new hardware and software, the computer policy committee might decide not to pursue the project. In the future when there are additional requests for new services which could be used to justify upgrading the present system or securing a new system, the initial investigation report could be reexamined.

Brainstorming Computer Technology   Management might hold a brainstorm session that centers around the topic of computer utilization. Often the session is held in a remote location where there will be a limited number of interruptions. During the session, ideas are presented. Only what might be done is considered—not how it will be done or how much it will cost to implement the idea. After all the ideas are reviewed, some might result in a request for an initial investigation.

The initial investigation report should provide the rationale for either pursuing or dropping the idea. The report should also indicate whether resources are available to implement the suggestion.

The Systems Department Initiates the Request

The systems department might initiate the request for a systems study to determine the feasibility of designing a new application that is now possible because of a new capability recently added to the existing computer system.

For example, perhaps timesharing software was added to your college's operating system. One of the first applications often designed for a new timesharing system is an online registration system. Your college's systems department may have submitted a request to investigate the cost-effectiveness of an online registration system.

There might be two subsystems, such as payroll and personnel, that should be integrated into one system. This might now be possible because of the expanded capabilities of the system's database software. It may now be possible to design a database that would serve the combined needs of payroll and personnel.

When the systems department makes the request, systems personnel may have to sell the user on its merits. Actually, when each phase of the study ends, there must be a certain amount of selling to get management and the user to accept the recommendations presented.

When the request comes from the systems department, the proposed system is usually one that should be computerized. Analysts may become more excited about a system they propose than about one proposed by management. Their proposal may present more of a challenge than some of the more routine requests, such as adding further exception reports to an existing system or finding the bugs in an operational system.

External Forces Require a Change

New government regulations may make a change necessary. For example, new unemployment compensation regulations may require that additional information be available on request. The change may make it necessary to alter reporting procedures, the format of data being processed, the procedures used in processing data, the file structure, and the reports produced. Additional reports may also be needed.

How many procedures and programs would need to be altered if a change to the metric system is mandated? Although the same techniques that we will cover should be used to investigate the problem, the project differs from many since it could encompass all computerized systems within an organization. The initial investigation would determine the scope of the problem and provide recommendations regarding the resources needed to complete the feasibility study.

Since required changes often need to be completed within a given period of time, requests that result from new regulations, from economic or environmental changes, or from action taken by other companies are assigned a higher priority than other types of requests.

Unique or New Applications

Although many of the requests are for modification or redesign of an existing system, other requests may be for new systems. For example, as the result of an increased awareness of the need to conserve energy and due to increasing cost, many organizations have designed or purchased a computerized system for controlling the environments of plants and offices. The systems are cost-justified on the basis of the savings in fuel. The decrease in the cost of fuel may more than offset the cost of designing, implementing, and running the system.

A company with a number of branch offices might initiate a study to determine the most effective way of providing a communication system. In the initial investigation, the analyst might elect to determine the type of system now used, the cost of the system, and the problems that have been cited regarding its use. At the present time, most of the communications that occur between the branches and the corporate office are either by mail or by telephone. The corporate office has a separate mail department and employs a number of people who are responsible for sorting and distributing the mail. If the initial investigation recommended that a feasibility study be undertaken, various alternatives would be examined. During the feasibility study, the relative costs, advantages, and disadvantages of the alternatives would be studied.

The project team assigned to the communication project might elect to compare facsimile systems, the cost of a teletypewriter service such as Teletex, voice mail systems, and the merits of establishing a corporate communication network. The project team would find that newer facsimile systems, such as the Xerox Telecopier 495-I system, can integrate facsimile images with information stored in databases and office automation networks. They would also discover that Teletex is based on an international standard, is much faster than the older Telex, and that incoming messages can be received while outgoing messages are prepared. The estimated costs for the systems being studied might be more than the cost of the present conventional mail system. If this is the case, the value of having a faster, more reliable system would need to be weighed against the additional cost.

CHECKPOINT
1. What are the stated objectives for the initial investigation?
2. What are five reasons that an initial investigation may be made?
3. Why are users more likely to submit a request for a systems study than are managers or analysts?
4. What are some of the reasons that management might request a systems study?
5. In conducting an initial investigation, will the same procedures always be followed? Provide the rationale for your answer.
6. Is there a clear-cut distinction regarding what is to be done in the initial investigation phase and the feasibility phase of a study?

THE REQUEST FOR AN INITIAL INVESTIGATION

Standards are not always developed regarding the way requests should be submitted. Sometimes a phone call, memo, or conversation in the hall between the manager and the user is considered as a request. Some company policies merely state that the request must be in writing but do not specify the format of the request.

The use of a form, such as the one illustrated in Figure 6-2, is desirable for the following reasons:

1. Uniformity is achieved.
2. More complete information regarding the request is submitted.
3. It is easier to fill out a form than it is to write a memo.
4. Confirmation of the receipt and disposition of the request will be sent to the person requesting the study.
5. Misunderstanding can be avoided.
6. A copy of the form provides a basis for preparing a monthly or an annual report concerning the type of requests received and the disposition of each request. This type of information is often needed when an attempt is being made to justify upgrading an existing computer system or securing a new system.

Figure 6-2 illustrates how the form might have been made out by the payroll manager requesting a modification to the system in order to meet the requirements imposed by changes in the state unemployment compensation reporting procedures.

The Request Form

The primary objective of using a form rather than a memo is to obtain all of the information that is needed. The basic rule to follow in designing a form is: *Keep it simple*. The form must be easy to complete. A form should be a means of assisting the person making the request—not a means of discouraging requests for studies.

The Request for Systems Analysis form permits the individual making out the form to supply part of the required information by checking the appropriate square. This makes it easier to fill out and also provides uniform answers. The section in the **standards manual** that deals with requests for systems analysis explains when each of the boxes should be checked.

Adequate space must be left for answers. Have you ever tried to fill out a form that allows only a half inch for recording your full name? The lines left

$C^P I$ **computer products inc.**   REQUEST FOR SYSTEMS ANALYSIS

Section I

DATE SUBMITTED: _January 11, 1986_   REQUEST FOR: [X] MODIFICATION OF SYSTEM

REQUEST IS: [X] IMMEDIATE   [ ] SHORT-RANGE   [ ] REDESIGN OF SYSTEM

[ ] INTERMEDIATE   [ ] LONG-RANGE   [ ] NEW SYSTEM

SUBMITTED BY:_ J. Arnold, Payroll Department Manager_   _Payroll_
                              (Name)   (Department)

NATURE OF REQUEST:_The current payroll system produces biweekly reports. It_
is now necessary to have weekly reports that show the regular hours, overtime

hours, and earnings of all employees.

REASONS FOR MAKING REQUEST:_Change in state unemployment regulations._

SUPPORTING DOCUMENTS ATTACHED:_Publication explaining the new regulation._

DIRECTIONS: Complete Section I and submit it to the data processing manager. Please attach supportive documents such as new federal regulations, changes in corporate policy, or abend reports. If a new system is being requested, attach any additional information you might have that will help to explain the project.

Section II (to be completed by the data processing manager or an analyst)

MODIFICATIONS APPEAR TO BE: [X] MINOR   [ ] MAJOR   [ ] EXTENSIVE

IMPLEMENTATION MAY REQUIRE ADDITIONAL: [ ] SOFTWARE   [ ] HARDWARE   [ ] PERSONNEL

COMMITMENT OF RESOURCES WOULD BE: [X] MINOR   [ ] MAJOR   [ ] EXTENSIVE

INITIAL INVESTIGATION COMPLETED BY: _____Mike Wackerly_____
                                                      (Name)
PROJECT NUMBER:_1049_____   _____January 18, 1986_____
                                                      (Date)

Section III (to be completed after the committee has determined the priority and disposition
          of the request)

                                                                    PRIORITY [1]

[X] PRELIMINARY INVESTIGATION   Assigned to:_Mike Wackerly_____
    APPROVED
                                            Tentative starting date:_February 1, 1986_

[ ] UNSCHEDULED BECAUSE _____

DATE REVIEWED:_January 20, 1986_____

Form No.: DP 001
10/15/84

FIGURE 6.2.   Request for a modification in an existing system

for recording the reasons for making the request should be two spaces apart. Sometimes the lines are one and a half spaces apart, which makes it difficult for the typist. If the machine being used does not have a one-and-a-half line space adjustment, the paper must be manually adjusted each time a line is typed.

Each form should have a unique number that can be used in referring to the form. The date the design of the form was accepted should also be printed on the form. This becomes important when forms are revised. A manager may find that he or she has two different forms for requesting a systems study. Which one should be used? If both the original form and the revised version of the form are dated, it is easy to decide which one should be used.

Providing Directions

The directions on the form should be stated in nontechnical terms and must be clear and concise. After the rough copy of a form is completed, it is a good idea to have two or three people (ones who would normally fill in the form) try using the form. Do they understand how to record the data? Do they all make the same mistake? If so, the directions are probably not clear. After forms have been printed and used for a period of time, it may be necessary to make revisions because more or less data is required.

STEPS IN INITIATING AN INVESTIGATION

Figure 6-3 summarizes the steps required to initiate an investigation. However, not all requests are approved and some are delayed due to lack of resources. Some people might question using Form DP 001, which is shown in Figure 6-2, to request minor modifications to an existing system. It is true that a separate form could be designed. Although minor changes might not be reviewed by the committee, it is critical that even minor changes are reviewed by *more than one person*. The CIS manager may be authorized to approve requests for modifications that will involve less than 15 or 20 person/hours. Often what is considered a minor modification will affect several programs and procedures. If the computer information services manager or analyst reviews the request for a minor modification, very little time will be needed to process the request. *Written authorization to modify any operational program or database should be mandatory*.

Steps in Initiating an Investigation

- Complete the standard request form.

- Submit with the form supporting materials that provide rationale for modification of an existing system or for implementing a new application.

- Estimate resources needed to complete the initial investigation. An analyst or the CIS manager will determine the resources needed.

- Determine action to be taken. The computer policy committee bases the decision on resources available and corporate goals and objectives.

- Assign priority.

- Schedule initial investigation.

FIGURE 6.3. Steps in initiating an investigation

Establishing Priorities   Many companies do not have a computer policy committee or an advisory committee to determine policies and priorities. There are advantages and disadvantages in having a committee review *all* requests. Since there are often more requests for services than can be scheduled, establishing priorities is a critical part of the process. If only the CIS manager makes the decisions regarding which projects will be investigated, only certain types of requests may be approved. The manager may not be aware of his or her personal bias regarding the types of projects that will be approved.

The major disadvantage in having the committee review all requests is that some high-priority investigations may be delayed until after the committee meets. A memo could be written indicating the need for immediate action, attached to the request form, and sent through interoffice mail to the members of the committee. The memo would also ask the committee members to approve the request so that immediate action could be taken.

When a computer policy committee is not organized to assist in reviewing requests and assigning priorities, the CIS manager should establish a project review committee made up of staff members.

Computer Policy
Membership
and Meetings

Ideally, there should be two committees. One committee should be composed of top-management personnel and the second committee composed of middle-management personnel. If there is only one committee and if time constraints of top management allow for participation, top-management personnel should actively participate in the committee. Realistically, however, most members will be from middle management.

While some committee members might be directly involved in the CIS area, other members should be from different areas such as manufacturing or marketing. Members of the committee must have a clear understanding of the corporate organization and its goals, objectives, and policies. Although technical knowledge regarding hardware and software is not required, all committee members must be aware of the state of the art regarding applications. The committee *must* meet on a regular basis. Committee members *must* view their function on the committee as being important to the growth and development of their organization.

CHECKPOINT

7. Why should a form be used to either initiate a modification to an existing system or computerize a new application?
8. What key factors should be considered in designing a form?
9. What are the two most important functions performed by the computer policy committee?
10. What might occur if the CIS manager approves or disapproves all requests and also assigns the priorities?
11. What might be considered the major disadvantage of having all requests approved by the computer policy committee?

CONDUCTING THE INITIAL
INVESTIGATION

The technique used to complete the initial investigation will differ depending on the nature of the problem. Figure 6-4 illustrates the tasks required to complete the initial investigation.

Tasks Involved in the Initial Investigation

- Gather data.

- Analyze data to determine the nature and scope of the problem.

- Document the investigation.

- Prepare a report that includes a summary of the investigation, its findings, and its recommendations.

- Estimate resources needed to complete the feasibility study.

FIGURE 6.4. Tasks involved in the initial investigation

Although a limited amount of time should be spent on the initial investigation, enough data must be collected to determine the type of resources needed to complete the project. A recommendation must also be made regarding the merits of the request. During an initial investigation, data is usually obtained by studying documents, conducting interviews, and observing the way data is processed.

Studying Documents

Before conducting interviews and making observations, it is important to learn as much as possible about the problem or procedure being investigated. In all phases of systems analysis and design, internal and external documents will be studied. Figure 6-5 illustrates the type of documents that are included in the two categories.

There is a great deal of internal and external information that can be used in the investigation, design, and implementation phases of the project. An analyst must be familiar with the data that is available. An analyst must also remain current regarding changes in computer technology and the area in which the investigation is being made.

Conducting Interviews

Several key factors must be observed if an interview is to be successful. The analyst must determine who should be interviewed. If specific answers are required regarding how data is handled, the subordinate directly involved in the process is interviewed. Since a broader view of the problem is desired, individuals in middle-management positions will usually be interviewed. Within any given category, it is not easy to determine just who should be interviewed. If two people within the organization can provide the information, the one who will be most responsive should be selected. A good analyst can usually find out which of the two should be interviewed.

The individual to be interviewed must be contacted for an appointment. Although the appointment can be made by phone, confirmation must be made in writing. The memo should confirm any decision made by phone (date, time, and place of the meeting) and also list the points that will be covered during the interview.

Assume an analyst has been assigned to investigate the request from the counseling staff which stated ''an online information retrieval system that contains complete information for each student currently enrolled is desired.'' The

Documents Studied

Internal documents

- Organization charts, goals, objectives, and policy statements.

- Minutes of meetings, letters, memos, and reports.

- Formal reports prepared for the board of directors, stockholders, and line officers.

- Documentation regarding the previous phases of the project.

- Final documentation submitted prior to a system becoming operational.

External documents

- Technical publications.

- Magazines and reports published by professional organizations and research groups.

- Reports on new governmental regulations.

- Reports on economic trends.

- Publications regarding topics of interest such as in-depth reports on communications, office automation, or robotics.

FIGURE 6.5. Types of documentation studied

reasons for making the request had been given as ''the proposed information retrieval system would permit counselors to do a much better job of assisting students with their problems and class schedules. The system would provide more information than is currently available and would save the counselors a good deal of time.'' Figure 6-6 on page 176 illustrates how the memo would be written that confirms the appointment.

In regarding the memo, notice that the meeting is scheduled to be held in the analyst's office. Usually this is not done. Jon Arnold, the analyst, has interviewed the director of student services on prior occasions in the director's office. During those interviews there were numerous interruptions that made it difficult to concentrate on the topics being discussed. Arnold feels more can be accomplished if the two meet in a conference room so they will not be interrupted.

The Interview

Before John Woodcock arrives, Arnold should make certain that all materials he will need are well organized. He may have listed some additional questions he would like to ask. A form such as the one illustrated in Figure 6-7 and 6-8 might have been prepared.

When Woodcock arrives, Arnold should be ready to start the interview. During the interview he should give the director his undivided attention. While it is important to make a few notes on the interviews form regarding the answers to specific questions, this should not distract from the attention given to the interviewee. Usually a tape recorder is not used since many people feel inhibited when their conversation is being recorded.

OfficeMemo

TO: John Woodcock, Director of Student   DATE: Jan. 15, 1985
       Services
FROM: Jon Arnold, Senior Analyst
RE: Requested Information Retrieval System

Your proposal regarding the development of an online information
retrieval system sounds exciting and I am looking forward to
hearing more about it on Tuesday, January 26, at 10:00 a.m. Since
we will be meeting in conference room 416, you may wish to bring
your material regarding the proposed system.

At our meeting, I would like to discuss the following points:

1. What type of information do you feel should be included
   in each student's record?

2. Are you concerned only with academic information or do
   you want data such as a student's medical history,
   vocational interests, extracurricular activities, test
   scores, and family background included?

3. Is a visual display of the student's record sufficient
   or do you also wish to obtain a hardcopy of some of
   the records?

4. Where do you feel the terminals should be located that
   would give you access to the information?

5. Why do you feel this system would save time and enable
   counselors to do a better job?

6. Who else, besides counselors, should have access to the
   information?

7. Would you still maintain your hardcopy files?

8. Do you know counselors who have worked with an online
   information system who could be contacted regarding the
   advantages of their system?

If you have any questions regarding our meeting, please let me know.

JA/ld

FIGURE 6.6.   A memo confirming an appointment

C^PI computer products inc.   INTERVIEW FORM   Page 1 of 2

PERSON INTERVIEWED: _John Woodcock_____ TITLE: _Student Service Director_

TOPIC: _Proposed Information Retrieval Systems_____

INTERVIEWED BY: _Jon Arnold_____

DATE: _January 26, 1986___ TIME: _10:00 a.m.___ PLACE: _Room 416___

FOLLOW-UP MEMO SENT ON: _January 27, 1986_____

RESPONSE RECEIVED FROM FOLLOW-UP MEMO: _____

The topics to be covered should be listed below. Whenever possible checklists should be used **to** indicate the possible responses.

1. A printout of the student's master file should be available in order to determine what data is already in each student's record.

   What information is desired that is not currently in the student master file? I = very important; N = nice to have; U = not important

   ____High school record ____Family background ____Work experience

   ____Recommendations ____Medical history ____Recommendations

   ____Test scores ____Talents and ____Possible career
                          interests   choices

   Other:_____

2. Equipment needed:

   ____CRT with display only ____CRT and printer ____CRT and remote
                                                       printer

3. Location of terminals:

   ____Each counselor's office ____Centrally located--shared by
                                     three or four counselors

4. Who, other than counselors, should be able to use the system?

   What about the inclusion of privileged information that might be in a student's record?

5. Why will counselors be able to work more effectively with students?

6. Why will the system save the couselors time?

7. What about your hardcopy files?

   ____Maintained in present form ____Discontinued ____Modified

8. How would the additional data be entered into the file?

   ____Centralized ____Distributed

Form No. DP002
10/15/84

FIGURE 6.7.   Interview form used by an analyst

Page 2 of 2

INTERVIEW OF: _John Woodcock_

INTERVIEWED BY: _Jon Arnold_ DATE: _January 26, 1986_

9. Will the new system increase or decrease the costs associated with maintaining the necessary data in the files?

10. Sources of additional information:

Individuals to contact:

Publications:

Studies:

FIGURE 6 8. Second page of the interview form

As soon as the interview is finished, Jon Arnold should record any additional information on the form he has used in conducting the interview. Using the form will enable Arnold to formulate questions faster and keep the interview directed toward its objective (to gather specific data). The form will also make it easier to record the decisions made.

If the additional data is not added to the form soon after the interview, some of the pertinent facts may be forgotten. Arnold should try to keep his schedule open for a period of time immediately following the interview.

Follow-up memo   A memo, such as the one illustrated in Figure 6-9 should be written indicating the decisions reached during the interview. By not indicating there is anything wrong with the statements made on the memo, the individual receiving the memo is tacitly confirming the validity of the information. If there is a difference in opinion regarding the agreements reached, the difference should be resolved as soon as possible.

Responsibility of the Interviewer   Since a large percentage of the data gathered during both the initial investigation and the detailed investigation is from interviews, the responsibilities of both the interviewer and the interviewee should be reviewed.

The interviewer is responsible for conducting the interview and for making certain the required data is obtained. A successful interviewer works hard at preparing for the interview and in determining the types of questions that should be asked to obtain the necessary data.

Interviewing is a skill that can be developed. There are a number of different types of questions that can be used to obtain data. The interviewer must select the type that will be most effective. The primary questions can be formulated in advance and should be phrased in the most effective way.

Direct	Direct questions are explicit and require specific replies. How is the variable data entered into the system? How many terminals do you have?
Open	An open question provides alternatives from which to pick. Why do you think your department can cost-justify the use of a distributed data entry system?
Closed	Alternative answers are provided in order to narrow the possible responses. Do you feel that most of the problems related to the sales order system are the result of inadequate documentation or of inadequately trained sales personnel?
Leading	The way the question is phrased implies or encourages a specific answer. Your department probably would not be able to cost-justify a new sales-order entry system?

During the interview the interviewer will use one of two types of questions to follow up on the response of the interviewee. One type is the *mirror*, or reflective, question; the other type is the *probe*. A reflective question is intended to get an individual to expand further on the subject while a probe question is used when an in-depth response is imperative.

Mirror	In answering to the interviewee's response of ''one of the major problems concerning the order entry system is the number

Office Memo

TO: John Woodcock, Student Services Director   DATE: January 27, 1986

FROM: Jon Arnold, Senior Analyst

RE: Summary of January 26, 1986 Meeting Regarding an
        Online Retrieval System

Our recent meeting was both enjoyable and productive. Your proposed information retrieval system has a great deal of merit and I will continue investigating the feasibility of developing such a system.

During our meeting we arrived at the following conclusions:

1. It will be necessary to have additional data added to each student's record: high school record, test scores, family background, medical history, and career preferences.

    Although it would be desirable to have data pertaining to recommendations made by faculty, recommendations of former employees, and information regarding the extracurricular activities of each student, it might not be possible to have it included.

2. Each counselor should have a terminal with a CRT and a printer. If the costs of implementing the system are excessively high, four counselors could share a centrally located terminal.

3. Since priviledged information will be stored in the student records, only counselors and student service personnel should be able to use the system.

4. Because of having immediate access to the data pertaining to the student being counseled, time will be saved. Having data that is not currently available will assist the counselors in working more closely with students.

5. Hardcopy files, similar to the ones you now use, will also be maintained.

6. The additional data should be entered into the files by student service personnel. This would probably require the hiring of additional personnel.

7. No cost benefits are expected from the system.

The four individuals listed below will be contacted in order to obtain information regarding their online systems and student databases.

John Phillips, Dir. of Student Services   Mary Sue Black, Dir. of Student Services
Lake Forest College   Foothills College
Oak Park, Illinois   Golden, Colorado

Dennis Green, Dir. of Inst. Research   Alison Gentry, Dir. of Student Services
Camelback College   Goldenwest College
Greenbrier, Arizona   Palo Alto, California

You will receive a report of the preliminary investigation as soon as it is finished. If you have any additional information that you feel should be incorporated into the report, please let me know.

FIGURE 6.9. Follow-up memo summarizing the interview

of errors recorded on the sales order," the interviewer might ask, "If you feel there are too may errors on the sales-order form, how do you feel the errors could be reduced?"

Probe   In answer to the same response, the interviewer might ask, "Please explain the type of errors that are found on the sales order and explain why you feel each type of error occurred."

A summary of the responsibilities of the interviewer is provided in Figure 6-10. As in many other phases of systems work, a plan must be developed and executed.

Responsibility of the Person Being Interviewed   The interviewee must also carefully plan for the interview. The purpose of the interview, as well as what is expected by the interviewer, must be evaluated. Any background materials and supporting data that would be helpful in the interview should be gathered so that the interviewee will be able to make adequate responses to the questions.

The interviewee should take time to respond thoroughly and consistently to all of the primary and secondary questions. A good interviewee will not just sit back and let the interviewer carry the entire interview; instead, he or she will feel free to ask questions of the interviewer.

Above all, the interviewee should be a good listener. Attentive listening will enable the individual to respond appropriately to all questions. If both the interviewer and the interviewee can adapt to new courses of action that must be pursued due to the answers to the primary questions, the interview will be conducted effectively enough to accomplish its objectives.

When the written confirmation of the interviewee is received from the interviewer, the interviewee should review it carefully to determine whether it is accurate.

Documenting the Interview   When the initial investigation report is submitted, all three items (confirmation of the interview, the interview form, and the follow-up memo) will be kept as part of the documentation. The interview form is considered as a working document for the analyst and need not be given to the person being interviewed. Preparing the form in advance of the interview helps the person conducting the interview to organize his or her thoughts and do any research that is necessary. The analyst may wish to have a copy of the blank form available to use during the interview as a worksheet. The information recorded on the worksheet can be put in more complete form and typed on the original form that will be included in the documentation. Regardless of how carefully the analyst prepares, there will always be additional pertinent data that will come up during the interview.

Observations   Observations regarding how data is processed are made more frequently during a feasibility study than during an initial investigation. Before making an observation, the documentation pertaining to the workflow and the procedures to be observed should be studied. The analyst may be trying to determine the difference between the formal and informal method of accomplishing tasks. Partly as a matter of courtesy, the supervisors of the individuals who will be observed should be contacted. The observations should be summarized and included as part of the documentation for the investigation.

Summary of the Interviewer's Responsibilities

1. Determine the objectives of the interview.

2. Determine who must be interviewed in order to obtain the required data.

3. Learn as much as possible about the individuals who will be interviewed.

4. Secure the interviewee's cooperation.

5. Pick the time and the place for the interview. Make sure that unnecessary interruptions can be avoided.

6. Do not request an interview until you know a good deal about the subject.

7. Contact the person to be interviewed and arrange the time and place for the interview.

8. Send written confirmation of the time, place, and purpose of the interview. The confirmation should include the purpose of the meeting, specific points that will be covered during the meeting, and a list of materials that might be requested.

9. Develop a plan or a procedure.

10. Prepare a list of primary questions that will be asked during the interview.

11. Be on time for the interview. Do not keep the interviewee waiting.

12. Make sure you are not interrupted by unimportant phone calls or requests for information from subordinates.

13. During the interview DO:

   restate the purpose of the interview;
   observe not only what is said but how it is said;
   be attentive to the person being interviewed—
      good posture and direct eye contact help to
      convey the attitude of interest and respect;
   respect your interviewee's time;
   make a record of what occurred.

14. During the interview DON'T:

   reveal your doubts or disagreements—you are there
      to gather information;
   try to impress the interviewee with your knowledge
      of the subject;
   let the interviewee get off the subject;
   convey the impression that you are not interested
      in the interviewee's responses.

15. As soon as possible after the interview, intepret and evaluate the results of the interview.

16. Send the interviewee written confirmation of any conclusions that were reached during the interview.

FIGURE 6.10. A summary of the interviewer's responsibilities

CHECKPOINT

12. What steps are required to initiate an investigation?
13. What tasks are involved in the initial investigation?
14. What two major types of documents are studied during the various phases of a project?
15. What are the three major purposes of the memo sent to confirm an interview that has been arranged over the phone? (You should be able to determine the answer by reviewing Figure 6-6.)
16. Why is it important to find out as much as possible about the person you are to interview?
17. What are the two major reasons for completing an interview form prior to the interview?
18. During the interview should the analyst take extension notes or use a tape recorder?
19. What is the purpose of the follow-up memo that is sent to the person interviewed?
20. Besides conducting interviews and studying documents, what other technique can be used to gather data?

GUIDELINES OR STANDARDS

Guidelines or standards should be developed for the use of the request form. Many companies don't have written guidelines stating how requests are to be handled. If guidelines are not available, there may be a difference between the formal method (what is detailed in the manual) and the informal method (what actually occurs). Figure 6-11 illustrates the ways in which a well-written, updated manual can be used. Unless procedures are developed for maintaining manuals, the information becomes obsolete and of little value. Often the question ''Who writes and maintains a standards manual?'' is not answered in the guidelines. Figures 6-12 through 6-14 illustrate the section of the standards manual that explains how requests for studies are to be made and processed.

Use of Standards Manuals

- For training new employees.

- As a reference in determining how to perform a task.

- By analysts in studying systems and procedures.

- By auditors in determining what controls are build into a system or procedure.

- By individuals who need information regarding the organization of a firm.

FIGURE 6.11. Uses of standards manuals

C^P_I **computer products inc.** COMPUTER INFORMATION SERVICES STANDARDS MANUAL

SECTION: SYSTEMS STUDY	SECTION	CHAPTER	SUBJECT
CHAPTER: INITIAL INVESTIGATIONS	1	1	1
SUBJECT: REQUESTS FOR SYSTEMS ANALYSIS	Date: 4/15/85		Page: 1

1. All requests for an analysis of a present or proposed system should be submitted on Form DP 001.

2. Section 1 of the form must be completed by the person making the request.

3. The following guidelines should be followed in completing Section 1.

   A. DATE SUBMITTED: Enter the current date as mmm dd, yyyy.

   B. REQUEST FOR: Check one of the three reasons.

      MODIFICATION OF SYSTEMS -- Check if the request is for changes in existing programs, one or two additional reports, correction of one or two programs, additional editing, or the inclusion of additional controls.

      REDESIGN OF SYSTEMS -- Check if a complete study of the existing system is requested in order to implement a number of additional features, delete numerous reports or procedures, change the processing method from batch to online, or implement a more cost-effective system.

      NEW SYSTEM -- Check if the application or system to be studied is not at the present time a computerized system.

   C. REQUEST IS: Check either immediate, short-range, intermediate, or long-range.

      IMMEDIATE   - Modifications must be made within the next two months.

      SHORT-RANGE - Modifications must be made within the next six months.

      INTERMEDIATE - Modifications should be made, or the new system designed, within the next two years.

      LONG-RANGE   - Modifications should be made, or the new system designed, within the next five years.

   D. SUBMITTED BY: The form should be signed by the person initiating the request.

   E. NATURE OF REQUEST: Explain as clearly as possible exactly what you would like to have done. These are the type of statements that should be used.

      Example:

      The sales department should like to have a word processing system that could be used to send personalized letters to customers who request information or who register complaints.

      An online information retrieval system that contains complete information for each student currently enrolled is desired.

FIGURE 6.12. Data processing standards manual (page 1)

$c^P I$ **computer products inc.** COMPUTER INFORMATION SERVICES STANDARDS MANUAL

SECTION: SYSTEMS STUDY	SECTION	CHAPTER	SUBJECT
CHAPTER: INITIAL INVESTIGATIONS	1	1	1
SUBJECT: REQUESTS FOR SYSTEMS ANALYSIS	Date: 4/15/85		Page: 2

F. REASONS FOR MAKING THE REQUEST: State as clearly as possible the benefits you anticipate receiving from the proposed system or procedure. If the request is for the modification of your present system, indicate what problems will be eliminated if the project is implemented. The reasons should be stated as follows:

Examples:

The proposed word processing system would print individualized responses to either questions or complaints from customers. Substantial savings in secretarial costs should be realized.

The proposed information retrieval system would permit counselors to do a better job of assisting students with their problems and class schedules. The system would provide more information than is currently available and would save the counselors time.

G. SUPPORTING DOCUMENTS ATTACHED: The following examples indicate the type of documents that should be attached:

(1) A statement of the new policy or regulation that makes it necessary to change existing data processing procedures.

(2) Abend reports that illustrate the type of execution problems operators experience in running the programs.

(3) A description of a system that is similar to the one being requested. This could be an article, paper, report, or a brochure prepared by a vendor.

4. Section II is to be completed by the computer information services manager or an analyst. It may be necessary to review the supporting documents, review documentation of an existing system, or to contact the person making the request. The information will only be used to estimate the length of time that will be needed to complete the initial investigation and to estimate the scope of the project being requested.

MODIFICATIONS APPEAR TO BE: ☐ MINOR ☐ MAJOR ☐ EXTENSIVE

Minor modifications are ones that can be made in two person/ months or less.

Major modifications may take from three to six person/months.

Extensive modifications will require more than six person/months.

IMPLEMENTATION MAY REQUIRE ADDITIONAL: ☐ SOFTWARE ☐ HARDWARE ☐ PERSONNEL

Check software only if it will be necessary to incorporate an additional software package, such as CICS, into the operating system.

FIGURE 6.13. Data processing standards manual (page 2)

$c^{P}I$ **computer products inc.** COMPUTER INFORMATION SERVICES STANDARDS MANUAL

SECTION: SYSTEMS STUDY	SECTION	CHAPTER	SUBJECT
CHAPTER: INITIAL INVESTIGATIONS	1	1	1
SUBJECT: REQUESTS FOR SYSTEMS ANALYSIS	Date: 4/15/85		Page: 3

COMMITMENT OF RESOURCES TO THE PROJECT WOULD BE: ☐ MINOR ☐ MAJOR ☐ EXTENSIVE

A minor commitment would involve from one to two person/months to implement the required changes. Very little time would be needed for preparing or entering data. A minimal amount (one hour per week or less) of additional CPU time would be needed.

A major commitment would involve from three to six person/months to design and implement the changes in an existing system or to develop a new system. A substantial amount of time would be needed for preparing or entering data. Each week from one to two hours of CPU time would be needed.

An extensive commitment would involve six person/months or more to design and implement. Additional personnel would usually be required either to prepare or input data. Once the system becomes operational, more than two hours per week of CPU time would be needed.

5. After Sections I and II are completed, the request will be scheduled for review by the computer policy committee. The computer information services manager or the analyst who completed Part II should be prepared to answer questions.

If APPROVED is checked, an analyst will be assigned to complete the initial investigation. The starting date should be recorded on the form and the length of time needed to complete the initial investigation should be estimated.

If UNSCHEDULED BECAUSE is checked, the reasons should be recorded in the blank provided. If the proposal seems to have merit but an analyst is not available to complete the initial investigation this should be noted. The proposal will be reconsidered as soon as some of the projects that have higher priorities are completed.

6. PRIORITY: A 1 to 5 scale is used. The highest rating is a 1; the lowest is a 5.

7. After all three sections are completed, the processing of the three-part form will be completed as follows:

A. Part 1 (the white copy) will be filed under "Requests for Systems Analysis--Processed".

B. Part 2 (the yellow copy) will be returned to the individual who initiated the request.

C. Part 3 (the blue copy) will either be given to the analyst assigned to complete the initial investigation or will be filed under "Requests for Systems Analysis--Pending."

If the initial investigation was approved, the blue copy will become the part of the report submitted when the investigation is completed.

FIGURE 6.14. Data processing standards manual (page 3)

If a company has a technical writing staff, the person responsible for developing the procedure, guidelines, or standards submits a rough copy to a technical writer. When technical writers are not available, the final copy must be prepared by the individuals responsible for developing the guidelines or standards. The operations supervisor, programmers, and analysts may be required to write portions of the data processing manual. Maintaining the currency of the manual is much easier when word processing or a good **text editor** is used. The standards manual might state that word processing or the text editor must be used in the creation and maintenance of documentation.

Throughout the text, forms, memos, sections of the standards manual, and letters will be used as illustrations. *The illustrations may contain information that you will need in completing assignments.*

Before you read the following sections of the text regarding the request form and standards manual, study Figures 6-12 through 6-14.

THE STANDARDS MANUAL

Writing Style   Someone needs to actually write the standards manual. How should it be written? Many business communication students learn the "five Cs" of good letter writing. The same five Cs apply to the style of writing used in a standards manual. The writing should be:

Clear	People using the manual must understand what is being said. Whenever possible, technical terms should be avoided.
Concise	The material must be brief and to the point. The people using the manual are generally busy and do not have time to read unnecessary details.
Correct	We are judged by the way we write and speak. Since the analyst or programmer writing the material has no way of knowing who might use the manual, it is wise to use correct, standard English rather than slang or buzz words.
Complete	All pertinent information must be included. After reading the section in the standards manual on requests for systems analysis, the reader should find it unnecessary to ask someone how to fill in Form DP 001.
Courteous	When requesting information or giving directions, it never hurts to be courteous. Although it adds one word, how much nicer "please enter the date" sounds than "enter the date."

Besides the necessary technical skills, an analyst must have communications, interpersonal, and conceptual skills. As a programmer, analyst, or supervisor you will be writing directions, completing documentation, writing sections of the standards manual and writing memos. For this reason, keep in mind the five Cs.

Format

A standard format, such as the one illustrated in Figure 6-12, should be used in preparing forms. Each form should be consistent in the way the section, chapter, and subject are recorded. In the sample form, the date, 10/15/85, is when the guidelines were approved. Other than the heading, margins, and indexing method used, there may not be a great deal of consistency in the way the material is presented since different people will be writing the material. Also, the nature of the subject matter determines how the information should be presented. It must be easy to find specific information such as what type of supporting documents should be attached to Form DP 001.

Definition of Terms

In studying the Requests for Systems Analysis section of the standards manual, you should be aware that each section of Form DP 001 is explained. All of the terms that could be checked are defined. You need not be concerned with the actual definitions used in this form since standard definitions for terms such as short-range, intermediate-range, minor, or major are not available.

Whenever terms without standard definitions are used, a definition must be provided. When the terms are listed as items that could be checked, it is even more important that all people using the form interpret the terms in the same way.

In some areas of the standards manual it would be impossible to list all of the variations that could occur. For example, only three items are listed under the caption SUPPORTING DOCUMENT ATTACHED. Under NATURE OF REQUEST, the examples are intended to show that a short, simple statement is all that is required.

A REQUEST IS MADE FOR WORD PROCESSING

The sales department manager, Glen Damuth, sent a request to Dennis Paulson, the CIS manager, for a study to determine the feasibility of using word processing for sending personalized letters to customers. The majority of the letters received from customers ask for information or make complaints regarding products or the type of service received.

One day during lunch Paulson explained to Damuth the capabilities of their office automation software which included an excellent word processing program. Paulson also explained how the existing accounts receivable database could be used to provide most of the names and addresses. Since Damuth didn't seem to know whether a few customers made repeated queries or complaints, Paulson indicated it might be possible to maintain statistical data regarding requests for information and letters of complaint. Paulson remarked that it was unfortunate that word processing or any of the other features included as part of the office automation software were used so seldom at Computer Products Incorporated.

A few days after having lunch with Paulson, Damuth read an article in his *Office Administration and Automation* magazine which indicated that word processing, data processing, electronic mail, and information retrieval systems could be integrated. The article seemed to answer the question regarding compatibility between databases or files created using different software. Also while attending a conference, he heard several managers discuss the advantages, disadvantages, and problems encountered while implementing integrated word processing systems. From listening to the other managers' conversations, Damuth decided that if the resources were available at Computer Products Incorporated, the advantages obtained would certainly offset the problems encountered in making the necessary changes.

Summary of the Initial Investigation

Damuth submitted a request to Paulson for a study. The request submitted is shown in Figure 6-15. The request was assigned to Kay Walzcak, a programmer/analyst, who reviewed the proposal and secured additional information to complete Section II of the request form.

The proposal was submitted to the computer policy committee. Walzcak indicated that the resources were available to implement the change. She also indicated that substantial saving in secretarial costs should be realized.

After a brief discussion, the committee assigned a priority of 2 to the request and the initial investigation was scheduled to begin on May 10. Walzcak decided that she should review any documentation that was available regarding how complaints and letters of inquiry are handled. She also wanted to study the

c^{P}I   computer products inc.   REQUEST FOR SYSTEMS ANALYSIS

Section I

DATE SUBMITTED: __March 15, 1986__   REQUEST FOR: [X] MODIFICATION OF SYSTEM

REQUEST IS: [X] IMMEDIATE   [ ] SHORT-RANGE   [ ] REDESIGN OF SYSTEM

[ ] INTERMEDIATE [ ] LONG-RANGE   [ ] NEW SYSTEM

SUBMITTED BY: __Glen Damuth, Sales Manager__   __Sales Department__
　　　　　　　　　　　　(Name)　　　　　　　　　　　　　　　　　　　(Department)

NATURE OF REQUEST: __The sales department would like to use the word processing__

__system to send personalized letters to customers who request information or who__

__register complaints.__

REASONS FOR MAKING REQUEST: __1. Reduce secretarial costs for the tasks__

__identified.__

__2. Improve the image of Computer Products Incorporated.__

SUPPORTING DOCUMENTS ATTACHED: __An article -- "Doing Old Things in New Ways"__

DIRECTIONS: Complete Section I and submit it to the data processing manager. Please attach supportive documents such as new federal regulations, changes in corporate policy, or abend reports. If a new system is being requested, attach any additional information you might have that will help to explain the project.

Section II (to be completed by the data processing manager or an analyst)

MODIFICATIONS APPEAR TO BE: [X] MINOR   [ ] MAJOR   [ ] EXTENSIVE

IMPLEMENTATION MAY REQUIRE ADDITIONAL: [ ] SOFTWARE [X] Terminal HARDWARE [ ] PERSONNEL

COMMITMENT OF RESOURCES WOULD BE: [X] MINOR   [ ] MAJOR   [ ] EXTENSIVE

INITIAL INVESTIGATION COMPLETED BY: __Kay Walzcak__

PROJECT NUMBER: __1051__   __May 1, 1986__ (Name)
　　　　　　　　　　　　　　　　　　　(Date)

Section III (to be completed after the committee has determined the priority and disposition of the request)

                                                                                    PRIORITY [2]

[ ] PRELIMINARY INVESTIGATION   Assigned to: __Kay Walzcak__
　　　APPROVED
　　　　　　　　　　　　　　　　　　Tentative starting date: __May 10, 1986__

[ ] UNSCHEDULED BECAUSE _____

_____

_____

DATE REVIEWED: __April 20, 1986__

FIGURE 6.15.   Exhibit1: request for a word processing system for the sales department

documentation for the customer database to find out what information was available. Walzcak was already familiar with the office automation software available on CPI's mainframe. She wanted to interview Damuth and also obtain permission to observe how the correspondence was now being handled. The job descriptions of the employees directly involved with processing the letters were also reviewed.

The results of the interviews, observations, and review of the available documentation were studied and analyzed. A brief cover statement was prepared for the initial investigation documentation. The materials included in the documentation file are referred to as *exhibits* and are illustrated in Figures 6-16 through 6-22.

Reports Provided

A written report, illustrated in Figures 6-21 and 6-22, was prepared and submitted to the computer policy committee. Visual aids were prepared to make the oral presentation clearer.

A verbal report regarding the investigation was made by Walzcak. All of the computer policy committee's questions were answered, the recommendations approved, and a feasibility study was scheduled.

CHECKPOINT

21. What factors should be considered in designing a standards manual
22. In what way will the information standards manual be used?
23. Why should standards manuals be kept in a looseleaf notebook and each page dated?
24. What prompted Glen Damuth to submit a request for a systems study?
25. Why was the office supervisor interviewed?
26. Why were the job descriptions and current pay scales for the individuals directly involved in processing complaints and requests for information studied?
27. Why should a standard format such as the one illustrated in Figures 6-21 and 6-22 be used in preparing the initial investigation report?
28. Why is it important to know how many letters are processed each day and the length of time that it takes to process each letter?
29. From reading the memos, do you feel the sales manager and office manager are cooperative? Give the rationale for your answer.
30. Why are the tasks involved in processing both the incoming and outgoing letters identified?
31. Does the final report indicate that the objectives of the initial investigation have been fulfilled? Explain your answer.
32. Why should the initial investigation be documented?

SUMMARY

A form should be used for submitting a request for a system study. All requests should be processed in a uniform manner. The request should be reviewed by a committee which will either accept the proposal and assign a priority, reject the proposal, or ask that additional information be submitted. When a proposal is rejected, the individual who submitted the form is told why.

PROJECT NUMBER: 1051

INVESTIGATED BY: Kay Walzcak

DATE COMPLETED: June 25, 1986

NATURE OF INVESTIGATION:

Determine the feasibility of using a word processing system
for the sales department. The system would be used for handling
requests for information and for processing letters of complaint.

MAJOR OBJECTIVES OF NEW SYSTEM:

1. Process requests for information and complaints on the
   same day as received.

2. Reduce the costs of handling the responses.

3. Enable the secretaries to utilize their time in other
   areas.

4. Reduce the total number of personnel hours needed to
   process the responses.

METHODS USED TO INVESTIGATE THE PROBLEM:

1. Interviewed Glen Damuth, who originated the request.
   See Exhibits 2, 3, 4, and 5 (Figures 6-17 through 6-20).

2. Observed the actual processing of the letters to determine
   the types of tasks that were required and to estimate the
   time needed for processing the responses to the incoming
   mail. The office supervisor, Mary Green, was contacted
   regarding the observations. See Exhibit 5 (Figure 6-20).

3. Reviewed documentation pertaining to the accounts
   receivable database to determine what data is already
   in the system.

4. Reviewed relevant job classifications of the sales department
   staff to determine the various classifications and pay
   scales.

CONCLUSIONS:

1. The proposed system could be expanded to include additional
   activities that fall into the word processing category.
   However, the sales manager would like to limit the system to
   handling requests for information and letters of complaint.
   The cost-effectiveness of the system will then be studied
   to determine whether additional applications should be added.

2. There is sufficient time on our present system to process
   the letters.

3. A detailed investigation should be completed as soon as
   possible.

FIGURE 6.16. Cover statement for the initial investigation documentation

OfficeMemo

TO: Glen Damuth, Sales Manager   DATE: May 24, 1986

FROM: Kay Walzcak, Analyst

RE: Meeting Regarding Proposed Word Processing
       System

At our meeting on May 28 at 9:30 a.m. I would like to explore with
you some of the ways word processing might be used within your department. Personalized letters can be sent to customers who make inquiries
or complaints. In order to complete my investigation, I will need to
obtain answers to the following questions:

1. Why do you feel your present system is inadequate?

2. Approximately how many requests for information and
   complaints do you process each month?

3. Are there approximately the same number each month?

4. What procedure do you now use to process the requests
   for information and letters of complaint?

5. Do the letters received fall into different categories?

6. What percentage of the information used in the letters
   you send out might be considered as constant?

7. Where would the variable data required to personalize
   the letters be entered into the system?

8. When would be a good time to observe how your staff
   processes the letters to be sent out?

The article, "Doing Old Things in New Ways", which you attached to
your request form, was very interesting. We may also wish to discuss
some of the other points covered in the article.

I am looking forward to our meeting and am sure it will be very
productive.

KW/ld

FIGURE 6.17. Exhibit 2: memo outlining the topics to be covered during the interview

$C^P I$ computer products inc. INTERVIEW FORM

PERSON INTERVIEWED: Glen Damuth   TITLE: Sales Manager

TOPIC: Proposed word processing system for generating letters that will

be used to answer requests for information and letters of complaint.

INTERVIEWED BY: Kay Walzcak

DATE: May 28, 1986   TIME: 9:30 a.m.   PLACE: Room 416

FOLLOW-UP MEMO SENT ON: June 1, 1986

RESPONSE RECEIVED FROM FOLLOW-UP MEMO:

The topics to be covered should be listed below. Whenever possible
checklists should be used to indicate possible responses.

1. How many requests for information and complaints are processed
   each month? 2000 - 2200

2. Are there approximately the same number per month? X Yes ___No
   Number varies from day to day but each month about the same.

3. Procedure now used to answer the letters?
   a. General category of each request or complaint determined.
   b. Letter coded to determine what standard paragraphs can be
      used in answering the letter.
   c. Individualized paragraphs dictated to a secretary.
   d. Letter and envelope individually typed.
   e. Letters signed, inserted in envelope, and mailed.

4. Do the letters fall into different categories?

   Yes. Approximately 80 percent of the text is from precomposed
   paragraphs.

5. What percentage of the information used in the letters is
   constant? 80 percent

6. Why is the present system inadequate?

   If an unusally large amount of letters are received in one day,
   the extra work load cannot be handled with existing staff. This
   means it may be four or five days before things get back to normal.
   One of the objectives of the new system would be to answer all
   letters on the day they are received.

   High cost. Additional staff needed to handle the responses.

7. How would the variable data be entered into the system? Terminals

   Install terminal(s) in sales department. ___ yes ___ no

8. When can processing of letters be observed? Anytime--talk to

   office supervisor

Form No: DP 003
10/16/83

FIGURE 6.18. Exhibit 3: interview form used for the word - processing project

OfficeMemo

TO: Glen Damuth, Sales Manager   DATE: June 1, 1986

FROM: Kay Walzcak, Analyst

RE: Conclusions Reached During our May 28 Meeting

Since you were so well organized, our meeting on May 28 was very productive. The problems with the present system were identified and we reached the following conclusions regarding the proposed system:

1. Prewritten paragraphs would be used for about 80 percent of the text material.

2. The remaining 20 percent would be entered through a terminal.

3. Someone from the sales department would be trained to enter the account numbers, codes, and variable data. A new job description will need to be developed.

4. The terminals used for entering the data and the letter-quality printer will be located in the sales department.

5. Unless a more efficient system is designed, additional staff will be required to process the letters.

6. If the proposed system is implemented, there will be a redefinition of the tasks involved in processing the responses to incoming letters. Less secretarial time will be needed. In addition, there should be a reduction in the total amount of time needed to process the letters.

7. Continuous form envelopes will be used. The addresses will be printed from information stored in the accounts receivable database.

Within the next day or two I will contact your office supervisor, Mary Green, to indicate when I will be making my observations. When I talked to her in your office, she seemed very receptive to the proposal and indicated she would supply any additional information needed.

If you have any questions regarding the conclusions listed, please let me know.

After the initial investigation you will receive a copy of the report. Please feel free to contact me should you have any questions.

KW/ld

FIGURE 6.19. Exhibit 4: memo detailing the conclusions reached during the interview

PROJECT NUMBER: 1051

INVESTIGATED BY: Kay Walzcak

DATE COMPLETED: June 25, 1986

OBSERVATION REPORT: June 4, 1986

    Date of observation: Mary Green, Supervisor, Sales Department

    Contact:

OBJECTIVES:

1. Determine how the letters needed to answer requests for information or complaints are processed.

2. Determine the classification of the staff who perform the various tasks.

3. Determine the approximate length of time needed to process each letter.

TASKS IDENTIFIED:

1. The mail is opened and distributed by a clerk.

2. The requests for information and complaint letters are given to an assistant to the sales manager who:

    a. Codes the letters and indicates what form paragraphs can be used. A few of the letters may need to be referred to the sales manager.

    b. Makes phone calls and checks in the files to secure any additional information that is needed.

    c. After the coding is complete and the additional data is gathered, the assistant dictates the variable data to a secretary. Sixty-five percent of the letters consist of from two to five typewritten lines of variable data. Numerous interruptions occur while the executive assistant is trying to dictate.

3. Individual letters and envelopes are typed by the secretary who took the dictation. Each letter contains from 22 to 28 lines of text. Since the secretaries also have numerous interruptions, an average of 4.5 letters and envelopes are typed each hour by each secretary.

4. The letters are signed by the executive assistant. If she is not available, the letters may remain on her desk until the next morning.

5. A clerk folds the letters and inserts them into their respective envelopes.

6. The letters are placed in the outgoing mail, which is picked up at either 9:00 a.m. or 3:30 p.m. Mail is sent out from the company at 11:00 a.m. and 5:00 p.m.

Note: Several secretaries are involved in taking the dictation and typing the letters. One secretary may be give 20 to 25 letters and then will start working on those while the assistant dictates to a second secretary.

FIGURE 6.20. Exhibit 5: report of observation

computer products inc.   INITIAL INVESTIGATION FINAL REPORT

PROJECT NUMBER: 1051

REPORT PREPARED BY: Kay Walzcak

DATE: June 25, 1986

NATURE OF THE INVESTIGATION: Page 1 of 2

Determine the feasibility of using a word processing system for the sales department. The system would be used for handling requests for information and for processing letters of complaint.

MAJOR OBJECTIVES OF THE PROPOSED SYSTEM:

1. Process requests for information and complaint letters on the same day received.

2. Reduce the costs of handling the requests for information and complaint letters.

3. Enable secretaries to better utilize their time.

4. Reduce the total amount of person/hours needed to process the responses.

PROBLEMS IDENTIFIED WITH THE PRESENT SYSTEM:

1. Responses to requests for information and complaint letters are sometimes delayed as much as four or five days. This creates additional customer dissatisfaction.

2. High cost.

3. Inefficient utilization of secretaries' time.

DOCUMENTATION AVAILABLE:

1. Methods used to investigate the problem.

2. Exhibits include:

    a. Interview confirmation memo.

    b. Interview form used for recording answers to the points covered during the interview.

    c. Memo containing agreements arrived at during the interview.

    d. Report of observations made regarding the tasks involved in processing the letters.

FIGURE 6.21. Exhibit 6: preliminary investigation final report (page 1)

C P I computer products inc. INITIAL INVESTIGATION FINAL REPORT

PROJECT NUMBER: 1051

Page 2 of 2

SUMMARY OF FINDINGS:

1. The proposed system should result in decreasing the cost of processing letters and increasing the quality of the work produced. Secretaries' time will also be better utilized.

2. Additional terminals and letter-quality printers will be required.

3. The software will be designed so that one printer can be used to print the letters and a second printer the envelopes. As the files are created that contain the letters and addresses, the operator will queue the files to the printers.

4. If the project is approved and implemented, it will serve as a pilot study to determine whether the same technique can be used in other areas to reduce costs.

5. The findings up to this point indicate that a substantial amount of time and money can be saved if the existing procedure for processing the letters is redesigned.

ESTIMATED COST OF COMPLETING A FEASIBILITY STUDY:

Analyst	1/2 person/month	$2,000.00
Secretary	1/2 person/month	1,125.00
Materials and Supplies		300.00
		$3,425.00

RECOMMENDATIONS:

It is recommended that an analyst be assigned to complete the feasibility study. There is a good deal of evidence that the new system will be cost-effective and result in substantial savings.

FIGURE 6.22. Exhibit 7: preliminary investigation final report (page 2)

The purpose of the initial investigation is to define the problem and determine its scope. The analyst conducting the investigation must gather data, analyze the data, formulate conclusions, and make recommendations to the committee.

Facts are gathered by studying internal and external documents, interviewing middle-management personnel, and by observation. In the standards manual, forms should be illustrated and procedures detailed that should be used in interviewing or for making observations.

The initial investigation should be fully documented. The analyst making the study will provide both a written and an oral report to the computer policy committee. The analyst's recommendations may either be accepted or rejected. The report can result in: (1) a priority being assigned and the feasibility study scheduled; (2) the analyst being asked to continue the investigation and resubmit his or her report; or (3) the project being dropped completely or put on the shelf for an indefinite time.

DISCUSSION QUESTIONS

1. You are the new manager of a CIS Department. At the present time very few forms are available. What rationale would you use in presenting to your staff members forms for requesting systems studies, developing standards manual, and conducting interviews?

2. You have just hired Jim Brown as a technical writer. What instructions, or advice, would you give Jim regarding the development of standards manuals? How should the manual be written and how should the materials be organized?

3. The president of the company for whom you work has indicated he is not in favor of establishing a computer policy committee. What rationale would you use to support your request for the establishment of a computer policy committee? Indicate who should be on the committee, how often they should meet, and what authority they should have.

4. Bruce Gary, an analyst whom you just hired, is preparing to conduct his first interview regarding a request for a new environmental control system. He seems to be at a loss regarding what internal and external information might be studied and how he should prepare for his interview with the plant manager who has requested the study. What recommendations would you give Bruce?

5. One of the department's analysts made an appointment for an interview, developed a form to be used in conducting the interview, and conducted the interview. In the final report, the results of the interview were included. The sales manager who had requested the study indicated that the recommendations included in the final report were not in accord with those he made to the analyst during the interview. What step in the interview procedure seems to have been omitted? What advice would you give to the analyst?

6. Why should all of the activities undertaken during the initial investigation be documented?

7. You are the analyst assigned to do the initial investigation requested by Glen Damuth. You would like to interview his assistant, Gwen Fox, who handles all of the office procedures. Gwen has indicated several times that "there is a great deal of evidence that indicates constant use of VDTs is harmful to the operator's health" and "if her typists are forced to use VDTs she will resign." What would you do prior to your interview with Gwen Fox?

8. Why should standards be developed regarding how an initial investigation report should be submitted?

9. When an oral and written report is submitted to the computer policy committee, what three courses of action might be taken? What rationale might be given by the committee for each of the three?

10. The payroll department manager has submitted a request for modification of the present system. The new union contract for hourly employees indicates that double time, rather than time and a half, is to be paid for hours that exceed 35 per week. The new contract also indicated that double time must be paid whenever an employee works more than seven hours per day. The union contract was signed on July 6 and the changes are to go into effect on August 1. The payroll manager has called you and indicated that the changes must be made ''now'' and normal procedures should not be followed. As the CIS manager you have been trying to convey to the computer policy committee the importance of their role and how what appears to be a minor change can affect the entire system. How would you respond to the payroll manager and how would you handle the problem?

**Team or Individual
Projects**

1. Use a form similar to the one on page 184 and write the section of the standards manual that provides the guidelines for preparing for, conducting, and following up an interview. Make certain the guidelines are clear so that anyone who has not read the chapter understands exactly what must be done.

2. Refer back to Discussion Question 7. You are the analyst assigned to investigate the possibility of implementing an interoffice electronic mail system that would substantially reduce the time needed to dictate and transcribe memos. In addition, the amount of interoffice mail would be decreased. You need to interview Gwen Fox and are aware of her attitude regarding the use of VDTs. You have an appointment for July 15 at 10:00 A.M. Since you have been told how confusing it is to work in her office, you would like to meet in your office. You need to find out how many memos are sent, how much time is needed to dictate the memos, and the time needed to transcribe the memos.

   Before meeting with Gwen you want to find out more about the use of electronic mail and what new job descriptions are created when more routine office procedures are automated. A professional, one-day in-service conference on electronic mail and other phases of office automation will be presented on July 10 at your local civic center. A variety of equipment will be demonstrated along with a hands-on workshop.

   Above all you want Gwen's support and do not wish to create a hostile situation. You feel Gwen needs to be better informed and would like to have her attend the conference with you. The conference is sponsored by the Data Processing Management Association.

   Directions:

   a. Indicate the reference materials that you might review prior to your meeting with Gwen.
   b. Using a memo form similar to the one illustrated in the text, write a memo to Gwen confirming your appointment, indicating the topics you would like to discuss, and inviting her to attend the conference with you. Indicate what attachments you would include with the memo.

GLOSSARY OF WORDS AND PHRASES

person/month   A length of time equal to one individual working full-time for one month. Two individuals working two weeks would equal one person/month.

resources   Personnel, money, machines, materials, and methods needed to develop a system or to complete a project.

standards manual   A systems manual or a procedural manual. Guidelines, policies, and procedures are detailed.

text editor   A program used to change data stored in files. Most text editors also permit new records to be added to the file and obsolete records to be deleted.

STUDY GUIDE 6

Name _____

Class _____ Hour _____

A. Indicate whether the following statements are true (T) or false (F) in the space provided. If false indicate in the margin what is wrong with the statement.

_____ 1. The initial investigation should define the problem and determine its scope.

_____ 2. A request for a systems study is always made because of a situation that exists within a company, such as the lack of relevant data.

_____ 3. During a brainstorming session all ideas being discussed should be evaluated to determine whether they can be cost-justified.

_____ 4. Once a system is designed and implemented, it probably will not be modified for at least ten years.

_____ 5. The systems department might initiate a request for a study since they feel one of the systems has become obsolete.

_____ 6. The same tasks are always performed in conducting an initial investigation.

_____ 7. There is a clear-cut division between what should be done in the initial investigation and what should be done in the feasibility study.

_____ 8. Since the CIS manager is very busy, it is considered good practice to have requests for a systems study made verbally.

_____ 9. One of the major problems regarding the use of standards manuals is that when modifications are made to the system, the manual is not updated.

_____ 10. All parts of the Request for Systems Analysis form are made out by the person requesting the study.

_____ 11. In order to save the computer policy committee time, the CIS manager should approve most of the requests for systems studies.

_____ 12. One of the major tasks of the computer policy committee is to determine standards and guidelines for the operation of the computer center.

_____ 13. Any term that is used in the standards manual that does not have an accepted standard definition should be defined.

_____ 14. Since EDP professionals will use the standards manual, technical terms should be used in describing procedures.

_____ 15. If a well-written standards manual is available, an analyst need not make observations to see how tasks are performed.

_____ 16. A technical writer should always write the standards manual.

_____ 17. One of the disadvantages in using forms is that it takes more time to complete the form than to write a memo.

_____ 18. Checklists should not be used on forms.

_____ 19. The style of writing used in a standards manual is unimportant.

_____ 20. Everyone on the computer policy committee should be directly involved in EDP.

_____ 21. Minor modifications made to an existing system need not be approved or documented.

_____ 22. If a recommendation is made not to do a feasibility study, the documentation for the initial study need not be kept.

_____ 23. During the initial investigation, only individuals who perform the tasks being investigated should be interviewed.

_____ 24. You should not tell the person to be interviewed what topics will be discussed.

_____ 25. One of the objectives of the initial investigation is to determine the resources needed to complete the feasibility study.

B.   Multiple choice. Record the letter of the correct answer in the space provided.

_____ 1. The interview form is considered
   a. part of the documentation.   c. a working document.
   b. unimportant.   d. both a and c.

_____ 2. A summary of the decisions made in the follow-up memo is considered accurate
   a. only if the person interviewed responds to the memo.
   b. only if the person interviewed sends a memo stating the summary is correct.
   c. if the person interviewed does not respond to the memo.

_____ 3. More will be accomplished during an interview if the person being interviewed
   a. is notified regarding the type of questions that will be asked.
   b. finds and organizes materials that might be needed.
   c. notifies his or her secretary that there should be no interruptions.
   d. all of the above answers are correct.

_____ 4. Before making an observation,
   a. it is not necessary to notify the individual's supervisor.
   b. no advance preparation is needed.
   c. the available documentation should be studied.
   d. none of the above answers is correct.

_____ 5. During the initial study, usually only
   a. the individuals performing the tasks are interviewed
   b. top management is interviewed.
   c. top- or middle-management personnel are interviewed.

_____ 6. The report of the initial investigation
   a. is always given verbally.
   b. is only presented in written form.
   c. is often given in writing and verbally.

_____ 7. The cover statement for the initial investigation report should include the nature of the investigtion, the objectives of the new or revised system, the methods used to investigate the problem, and
   a. the exact cost of the proposed system.
   b. the estimated cost of the feasibility study.
   c. conclusions reached as the result of the investigation.

_____ 8. The major objective of the proposed word processing system is to
   a. save time by making the responses less personal. Only form letters will be used.
   b. increase the productivity of secretaries and lower the cost of sending responses to customers.

c. achieve the objectives stated in a and b.

d. allow Mary Green to be more directly involved.

_____ 9. The final report for the word processing initial investigation included the nature of the investigation, problems identified with the present system, documentation of findings, and

a. recommendations regarding the type of system to be developed.

b. estimated cost of the new systems.

c. recommendations regarding the feasibility study and the estimated cost of the study.

d. job descriptions for word processing technicians.

_____ 10. Included with the final report were

a. articles regarding office automation.

b. reports on new legislation regarding the work environment of terminal operators.

c. memos requesting interviews and the follow-up memos.

d. articles regarding health hazards associated with continued use of VDTs.

7 FEASIBILITY STUDY

Looking Ahead

After reading the text and completing the learning activities you will be able to:

- Determine the objectives for a feasibility study.
- Identify the advantages of assigning a feasibility study to a project team rather than to an analyst.
- Identify the situations in which a consultant might be obtained and the major advantages and disadvantages of using the services of a consultant.
- Identify the alternatives available in conducting a feasibility study for a small company that does not have an inhouse analyst.
- Identify the objectives that are often cited when a manual system is to be computerized or an existing computerized system is redesigned.
- List the steps required to complete a feasibility study.
- Identify the various ways in which data is gathered and determine when each might be used.
- Explain why guidelines are usually available for procedures used to gather data but seldom available to analyze data.
- Determine the various costs that must be estimated for a proposed system.
- Explain why guidelines must be developed and followed for estimating costs.
- List and explain the steps that must be taken in preparing for an oral presentation.
- Identify the various types of visual aids that might be used and indicate when each might be used in making a presentation.
- List the steps that were taken to conclude the feasibility study for the word processing project.
- Identify the ten major headings that were used in developing the Feasibility Study Final Report.
- Explain how the estimated cost of the proposed word processing system was determined.
- Define and utilize the words and phrases listed in the end-of-chapter glossary.

INTRODUCTION

While the techniques used in conducting an initial investigation and a feasibility study are very similar, the objectives are not. The objectives of the initial investigation are to define the problem and to determine its true nature and scope. The investigation must be done in enough depth so that the length of time and resources needed to do a detailed investigation can be determined with a reasonable amount of accuracy.

The major objectives of the feasibility study are to further define the problem and to determine the best way to solve it. The study will be conducted in more depth and the individuals who perform the tasks studied will be directly involved.

Several alternatives may be presented as possible solutions to the problem. From the choices available, the analyst or project team should determine which provides the best solution. The final report for the feasibility study should indicate the time and resources needed to complete the design and implementation phases of the study. Estimates for the design phase should be fairly close to the actual costs. However, the implementation and operational expense estimates might be as much as 50 percent more or less than the actual costs. Estimated costs made by experienced analysts will be closer to the projected costs than those made by inexperienced analysts or project teams.

INVOLVEMENT OF
MANAGEMENT

In both the initial investigation and feasibility study, users and management must be directly involved. Management should provide support for the project, answer questions about the present system, make materials and personnel available, and suggest solutions for some of the problems identified.

Without the full support of management, needed information might be difficult to obtain. Top management may provide a letter or memo that serves as authorization to conduct the study and to obtain materials. The memo should indicate why the study is being made and also ask middle- and lower-management personnel to cooperate. The success of most projects depends on the team receiving the full cooperation of middle and lower management.

The analyst can learn a great deal by studying the standards manual, reports, and other printed materials. All pertinent material must be available upon request. Since some of the materials might be classified, authorization from top management will be needed.

Key people must be available to be interviewed, to participate in brainstorming sessions, and to answer questions. This may become a problem during the feasibility study when the present system is studied in-depth. As soon as the study is begun, projections should be made regarding the type of support that is needed from personnel not directly involved in the study. These people should be contacted as far in advance as possible so that they may think about the problem, find materials they wish to use, and organize their materials.

When EDP first became available, it seemed to overwhelm individuals who were not EDP professionals. Often management accepted whatever the analyst said was the best solution and seldom questioned why projects were late and over budget. Today management is more knowledgeable about electronic processing of data and the retrieval of information. Middle- and lower- management personnel want to be more involved and have more direct control over computerized procedures.

TEAM APPROACH

In a very small company there may only be one analyst. However, for major projects in a medium- or large-size organization, more than one analyst should be directly involved in the feasibility study. One person working alone may have tunnel vision and see only the solutions that have been used before and may not see the vast horizons that exist today. A team can do a considerable amount of brainstorming and will probably come up with a better solution than one developed by an analyst working alone. A good, lively discussion can bring into focus a wealth of ideas.

Depending upon the complexity of the project, a project team might consist of from two to five members. A three-person team might consists of a senior analyst, a programmer/analyst, and a representative from the functional area being studied. During the course of an investigation, EDP personnel such as the operations manager, documentation specialist, communications specialist, and database manager will work with the committee when their expertise is needed.

This phase of the project charts the course for the rest of the project. Alternatives overlooked during this phase of the project may never be considered. Although the computer policy committee will review the alternatives presented by the project team, the best solution might not be among those presented. During this phase, it is extremely important that all feasible alternatives be pursued and that management, support personnel, and users have an opportunity to contribute their ideas.

INVESTIGATIONS BY COMPANIES WITHOUT A SYSTEMS DEPARTMENT

Regardless of the size of the company, the same procedures should be followed to determine the best alternative for a new or revised system. On a very informal basis, the manager of a small company may have conducted an initial investigation. It is doubtful that the manager called it an initial investigation; however, data was gathered and analyzed and problems were identified. Since the manager works closely with the staff that performs the various tasks, the manager may be more aware of problems that exist than is the management of a large organization.

What should a company without inhouse talent do? Certainly there is a vast array of hardware and software that might help to solve some of their problems. Should they attempt to carry out a feasibility study that might involve the selection of hardware and software? Today there are several alternatives that should be considered.

Use of Independent Consultants

Many independent consulting firms specialize in performing feasibility studies. There are also consulting firms that specialize in working with small companies that will probably use either microcomputer or minicomputer systems. Consultants proceed in the same manner as do inhouse project teams or analysts and must be knowledgeable about the type of business the client has and the type of equipment that can adequately meet the needs identified.

Large organizations also use consultants. When there are problems caused by internal politics or conflicts between functional areas within the company, it may be wise to hire a consultant who will present an impartial report. Also a technical problem may develop that requires the ability of a highly trained individual who has in-depth experience in a specialized area. At other times, a consultant may be hired because the company wants an unbiased opinion from an outside source.

Before consultants are hired, their qualifications should be carefully checked. All too often, someone with a limited background in systems analysis and design, programming, and computer technology may attempt to do small-system consulting. One company hired a free-lance consultant who said he would design and implement a complete financial system for their microcomputer. One year after the system was installed, the company had not received any monthly income and expense statements and had no way of knowing whether they were making or losing money. The company had to revert to their manual system to determine their profit and tax liability.

There must be a clear understanding *in writing* that indicates: the scope of the study—what actually will be done; the length of time the study will take; the type of report that will be received; and the total cost of the study. During and after the study there should be no surprises. It would be very discouraging to find that the report identified the major problems areas but provided no solutions. The advantages and disadvantages of hiring consultants are illustrated in Figures 7-1 and 7-2.

Advantages of Hiring Consultants

- The consultant is prepared by training, background, and previous experience to counsel on special problems, to conduct investigations, and to assist in designing new systems.

- The consultant may have worked with companies that are similar to the client's.

- Many pitfalls can be avoided. An experienced analyst should be able to identify the hardware and software needed to handle the client's work.

- The consultant has no preconceived ideas regarding the organization, its problems, or possible solutions.

FIGURE 7.1. Advantages of hiring an outside consultant

Disadvantages of Hiring Consultants

- Each organization is unique and an outside consultant may not understand its goals and objectives.

- Employees may not be informed regarding the objectives of the study. Due to fear of what might occur and resentment of an outsider making recommendations, employees may not cooperate with the consultant.

- Exploring new technology is a learning experience. Since the consultant investigates the use of new technologies, the client's staff does not benefit from this aspect of the study.

- Foot-in-the-door syndrome might develop. The consultant may be retained to perform tasks that should be done by the company's personnel.

FIGURE 7.2. Disadvantages of hiring consultants

In addition to the disadvantages listed in 7-2, management might feel that the consultant's fee is high. However, consider how expensive it would be to solve problems created when wrong decisions are made because the individuals making the decisions lacked training, knowledge, and experience. Management of a small company would probably find that the advantages outweigh the disadvantages.

Other Alternatives

Staff Personnel   Although an analyst is not employed by the organization, management may feel the feasibility study should be conducted by a member of their own staff. The individual selected to do the study should be released from his or her usual responsibilities and given the opportunity to become totally committed to the study. Total commitment is good in theory but may not be possible because key personnel who might be assigned to the study are also needed to solve day-to-day problems.

The individual selected will probably be able to identify the problems that exist with the present system but may overlook some of the possible solutions. The study may also take longer than if a qualified consultant was hired. Funds should be provided for travel, seminars, and workshops. The individual assigned to the study will need to investigate more external sources of information than would a trained consultant.

A consultant could be hired to work with the person selected to do the study. Critical problems may arise that require the expertise of a professional. Can non-data-processing personnel develop criteria to use in the selection of a computer? Without a good background in EDP, it would be difficult to develop hardware and software specifications.

Vendors   Some vendors are willing to conduct an investigation to determine hardware and software requirements. While this does produce an alternate method for the small company that is currently using mostly manual methods, it will not solve some of the other problems that exist. The vendors may look primarily at what would be done by a computer and not take into consideration other problems such as a lack of standards, a lack of documentation, a staff overtrained for the positions, or problems that result because the staff is not adequately trained.

The person representing the vendor may be primarily a sales representative and may not have much formal training in systems analysis and design. Such individuals may attempt to make all small companies fit into the same mold.

Hire a Small-Shop Specialist   If the company is growing and *a new position can be supported*, a small-shop specialist should be hired. If financial systems are to be implemented, the individual must have a background in finance, accounting, and marketing. In addition, the small-shop specialist should be familiar with electronic data processing procedures, systems analysis and design, minicomputer and microcomputer systems available, operating systems, and software packages that are available for small business systems. The small-shop specialist may do jobs normally assigned to the computer operator, maintenance programmer, analyst, documentation specialist, and database manager. His or her job description may also include assignments in other areas such as accounting or marketing.

It is difficult to find a true small-shop specialist that has all of the necessary skills and who wants to work in the small-shop environment. However, the diversity of the position adds to its challenge and enjoyment. The individual hired is also in a position to learn a great deal about the organization and probably would be considered a candidate for other managerial positions.

CHECKPOINT

1. What are the objectives of the feasibility study?
2. In what ways should top management be involved in the feasibility study?
3. What should be done before a consultant is hired?
4. Why might a large organization with an inhouse systems department hire an outside consultant?
5. What are the disadvantages of hiring an outside consultant?
6. The president of a small organization that does not have a systems department feels that some of the manual applications should be redesigned to take advantage of recent advancements in computer technology. If he elects not to hire a consultant, what other alternatives are available?

BEGINNING THE FEASIBILITY STUDY

The starting point for the feasibility study is to review the documentation and the report prepared by the individual who did the initial investigation. Even if the analyst who did the initial investigation is one of the team, the material should be reviewed. A considerable length of time might have elapsed since the first phase of the project ended.

Although the team might be tempted to start immediately on the project, a plan *must* be developed. However, before a plan can be developed, specific objectives must be developed and the procedures required to accomplished the objectives should be listed.

Developing Primary and Specific Objectives

Users and management must be directly involved in determining both the primary and the specific objectives. Primary objectives are broad, major objectives that spell out the need for a new system or the need to revise an existing system. The primary objective for the sales system might be to increase customer satisfaction, decrease the length of time needed to process orders, and provide more timely and reliable information.

Specific objectives are more definitive and will differ for each system being studied. Many of the objectives included in Figure 7-3, might well be among the primary objectives for any new or revised system. Each of the objectives specified in Figure 7-3 will be discussed briefly.

Unless centralized databases are already operational, establishing such databases should be a major objective. A master plan, based on intermediate or long-range objectives, should be developed for creating databases. When individual systems are developed, the long-range objectives for establishing databases should be followed.

At the functional level of management, some of the routine decisions can easily be automated. For example, exception reports can indicate when to reorder supplies or when to send out form letters to customers who have past-due accounts. Exception reports should detail any deviation from standards so that immediate attention can be given to the problem. For instance, in developing a new system for a college, provision should be made so that during registration a list of closed classes is printed upon request. Immediate attention can be given to the problem.

Many clerical functions can be automated. A secretary may spend hours on the telephone trying to arrange a meeting between several key executives.

Typical Primary Objectives

- A centralized database will be developed.

- Routine operating decision making will be automated.

- Clerical functions will be automated and clerical work by nonclerical staff will be eliminated.

- Control will be by exception.

- Unnecessary control procedures will be eliminated.

- Unnecessary decision points that delay workflow and th production of reports will be eliminated.

- A minimum of paperwork should be produced.

- The output produced by the system should be in a usable format.

- Information should be disseminated in time for it to be useful.

- The system should be flexible enough to allow expansion or refining without doing extensive analysis or programming.

- Methods must be built into the system that can be used to evaluate the effectiveness of the information used for nonroutine decision making.

- Work procedures should be standardized.

FIGURE 7.3. Objectives often included as primary objectives for a feasibility study

If each executive's schedule is in a common database, a printout can be obtained within seconds that indicates when the meeting should be held. The time is also blocked out on each of the executive's schedules and each executive is notified of the meeting. If some of the key people have unresolvable conflicts, the reasons for the conflicts will be listed and the time when the majority of the people can attend the meeting will be selected. Also clerical functions performed by highly trained personnel should be eliminated. For example, in the case study, well-qualified secretaries are typing form letters.

Unnecessary decision points that delay **workflow** and the production of reports should be eliminated. For example, the existing sales system of a company delays the printing of all sales invoices until the credit manager determines the disposition of orders for customers who have exceeded their credit limit. The decision point should be removed so that orders from customers who have not exceeded their credit limit can be processed immediately. The exception report can be sent to the credit manager along with the orders that cannot be processed. If the credit manager decides to extend the credit limit for some of the customers, the credit limit in their master file records can be adjusted. The rejected sales orders can then be processed along with the following day's batch.

Unnecessary control procedures should be eliminated. Although control procedures are necessary, there may be an overkill. Excessive controls delay the dissemination of information. Before any control procedure is eliminated, there should be an evaluation made to determine whether the procedure is needed. It is better to have more controls than are needed than to eliminate those controls necessary to ensure the integrity of information.

Existing reports should be analyzed to determine whether they are being used or there is unnecessary duplication. All output produced by the system should be in usable form. Very little interpretation, analysis, summarizing, reconciliation, or rerecording of the information should be necessary. Information should also be disseminated in time for it to be useful. To illustrate, a monthly inventory reorder report is of little value. An inventory exception report listing all items that should be reordered should be printed on a daily basis.

The system must be flexible so that it can be expanded or refined without doing extensive analysis or programming. For example, the payroll system's database must be complete enough so that when additional reports are needed to fulfill government regulations, the information is available and can be formatted to produce the required reports.

Methods must be built into the system to evaluate the effectiveness of the information used for nonroutine decision making. As part of the design of a system, provisions must be made for an ongoing audit of the system.

Procedures should be standardized. When standards are used, it is easier to modify systems and to train new employees to perform the tasks that make up the procedures. There will also be fewer discrepancies between the formal and informal organizational structure and procedures.

The determined objectives may not be able to be achieved due to constraints imposed by top management.

Constraints that Prevent the Achievement of Objectives

Typical constraints are money, equipment, materials, personnel, and time. One approach will be used during a feasibility study if management indicates that the sky is the limit. The project team will investigate all types of alternatives and then select the best one. The scope of the investigation is very different if management tells the project team that there are no funds for new equipment and no budget for additional personnel.

Time is another very serious constraint. Management might state that the new system must be up and running within a year. Time will need to be budgeted carefully. When time is a constraint, too little may be spent on the feasibility study and design phase. As a result, the system may not achieve its stated objectives.

In preparing the plan for the feasibility study both the objectives for the system and the constraints imposed by management must be considered.

Preparing a Plan

If you refer to Figure 6-22, on page 197, you will see where it was estimated that the feasibility study for the word processing system would take two weeks. The reason so little time was needed is that constraints were included limiting the scope of the project to two types of letters. In the initial investigation, three problems were identified: secretaries were doing clerical tasks and not utilizing their time efficiently; the cost per letter was too high; and it took too long to respond to letters. The analyst who did the initial investigation felt that word processing could be cost-justified. In reviewing the report, it was also apparent that there were a number of problems with the current system that could not be solved by a computerized word processing system.

Task Identification

The starting point in any planning process is to analyze what must be done. The analyst assigned to the project might review the report and the documentation and then formulate a list of tasks that need to be accomplished. The initial list might look something like this:

1. Study the workflow to determine whether problems exist outside the sales department that are relevant to the problems cited.
2. Interview the executive assistant who codes the letters and dictates the variable data. Find out how the letters are coded, why someone else cannot be authorized to sign the letters in her absence, and see whether some type of recording device can be used rather than dictating to a secretary.
3. Study the variable data used to determine whether there are typical responses that could be used to create more form paragraphs. This would decrease the amount of variable data that would need to be entered.
4. Interview the secretaries who type the letters to determine whether they have any suggestions. For example, what might be done to decrease the amount of interruptions that occur?
5. Determine the present standard of performance—how many letters does each secretary type in an hour?
6. Investigate the word processing software that is available as part of the office automation package on the mainframe. Also determine the cost, advantages, and disadvantages of obtaining dedicated word processing equipment or microcomputers.
7. Determine the cost of the present system.
8. Determine production standards for the new system—how many letters can be processed by a secretary in an hour?
9. Determine the time needed to complete the detailed systems work and to implement the system.
10. Estimate the cost of designing and implementing a word processing system.
11. Determine the number of workstations needed and where they should be located.
12. Determine the files needed and the approximate cost of creating and maintaining the files.
13. Write the final report and organize the documentation.
14. Prepare the necessary visual aids.
15. Make the formal presentation.

Scheduling Events

The next step is to prepare an ordered plan that indicates when each task or event will be accomplished. It is necessary to determine which steps can be done concurrently and which ones must be completed in a given sequence.

If you study the Gantt chart illustrated in Figure 7-4, you can see that when the events are put into some type of order and scheduled, all of the fact-gathering activities occur prior to the interview with the sales manager and the executive assistant. Since the major costs involved with the new system will be labor, cost estimates cannot be prepared until production standards are developed for the proposed system.

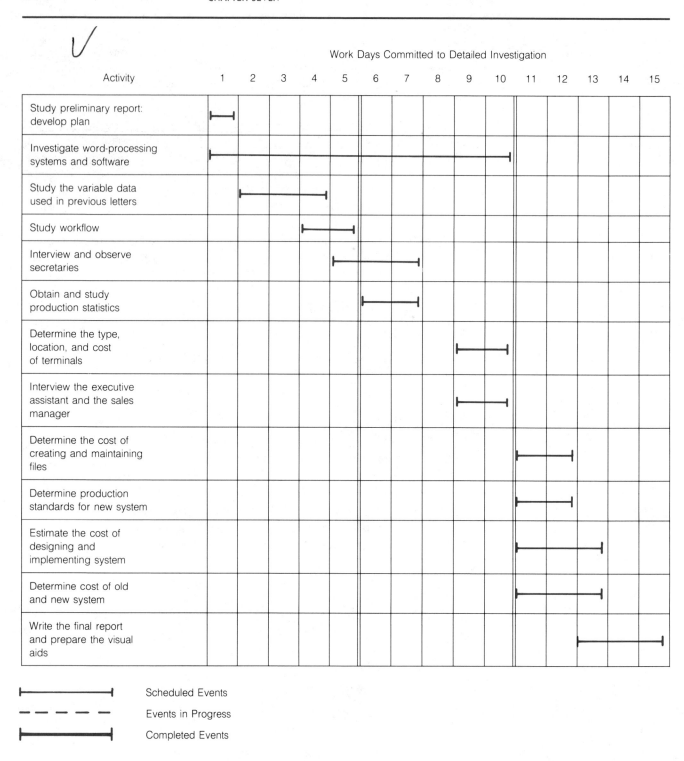

Work Days Committed to Detailed Investigation

Activity	1	2	3	4	5	6	7	8	9	10	11	12	13	14	15
Study preliminary report: develop plan	⊢⊣														
Investigate word-processing systems and software	⊢────────────────────⊣														
Study the variable data used in previous letters		⊢──────⊣													
Study workflow				⊢──⊣											
Interview and observe secretaries					⊢────⊣										
Obtain and study production statistics						⊢──⊣									
Determine the type, location, and cost of terminals									⊢──⊣						
Interview the executive assistant and the sales manager									⊢──⊣						
Determine the cost of creating and maintaining files											⊢──⊣				
Determine production standards for new system											⊢──⊣				
Estimate the cost of designing and implementing system											⊢────⊣				
Determine cost of old and new system											⊢──⊣				
Write the final report and prepare the visual aids													⊢────⊣		

⊢────────⊣   Scheduled Events

— — — — —   Events in Progress

⊢━━━━━━━━⊣   Completed Events

FIGURE 7.4   Gantt planning chart for the word processing detailed study

Few charts would break down the activities day by day as was done in Figure 7-4. The important concept to remember is that *planning must be done* and the project must be broken into small tasks than can be scheduled on a week-by-week or day-by-day basis.

The schedule should never be so tight that allowances are not made for contingencies that can, and will, occur. If the Gantt chart in our example were completed, showing the work in progress and then accomplished, the lines would tell the story. Some tasks or events might be completed ahead of schedule while others might be completed beyond the projected dates.

CONDUCTING A FEASIBILITY STUDY

During the feasibility study, data must be gathered and analyzed, and recommendations made regarding how the problems identified can be solved. In gathering data, documents can be studied, observations can be made, and it should be determined who performs each of the identified tasks. Data is also gathered by using questionnaires, visiting other installations, and contacting vendors.

After the data is gathered, it must be organized so that recommendations regarding the alternatives available can be presented. The formal presentation should include a review of the procedures used in gathering and analyzing data as well as suggested recommendations. The presentation should include the estimated cost of developing and maintaining the suggested solution.

Gathering Data

Some of the techniques used in the feasibility study phase are the same ones discussed in the previous chapter. The methods usually used to gather data are interviews, in-depth studies of existing documents, in-depth observations of workflow, questionnaires, visits to other installations, and presentations by vendors. For a number of reasons, no two feasibility studies will be conducted in exactly the same way. The problems identified during the investigation determine the action that must be taken. The constraints imposed by management also determine the scope of the detailed investigation. However, in any investigation, the basic questions illustrated in Figure 7-5 must be answered.

Basic Questions

WHAT?   —What tasks make up the procedures that are being studied?

WHY?   —Why are these tasks being done? Are they meaningful or done "because we always did it this way"?

WHERE? —Is there a possibility that the task could be decentralized and performed closer to where the data originates?

WHO?   —Who performs each of the tasks? Should someone else be assigned to some of the tasks?

WHEN?   —Is the task being done soon enough so that the resulting information is timely and relevant

HOW?   —Is there a better, cheaper, more effective way of performing the task?

FIGURE 7.5. Basic questions to ask of any system being studied

In planning the feasibility study, one of the first things that should be done is to determine how the data will be gathered. If interviews are to be used, a list should be made of who will be interviewed and the objectives of each interview. When the interviews are scheduled is sometimes critical. Figure 7-4 shows that the operational personnel (secretaries) are observed and interviewed before the executive assistant and sales manager. Since this is the second interview with management, the analyst is assuming that additional questions will be formulated based upon the interviews with the secretaries.

The Gantt chart also shows that documents and the overall workflow pertaining to the letters were studied before any interviews were conducted. The analyst should obtain as much background as possible about the topic being investigated before interviewing personnel who are directly involved with the system.

Interviews Interviews are a major source of data. During the previous phase of the study usually only management personnel are interviewed. During this phase of the study, the operational personnel who are doing the tasks will also be interviewed. Just as there is a difference between the formal organization and the informal organization, there often is a difference between how the manager thinks the job is being done and how it is actually being done.

Operational personnel often have good ideas that can be used in the design of the proposed new system. Although involving more people takes more time, a far better system will be developed if operational personnel are included in the investigation and design phase of the study. A list of the top ten reasons why systems are unsuccessful would show that the first reason is lack of communication with staff personnel during the investigation and design phase.

The same steps should be followed in conducting interviews as were previously outlined: determine who is to be interviewed and the specific objectives for each interview; make an appointment with the person to be interviewed (if the person is a staff member, contact the individual's supervisor); send a memo confirming the appointment; design a simple checklist that will assist you in conducting a successful interview; during the interview focus your attention on the individual being interviewed; and after the interview summarize the conclusions reached and send a memo to the individual interviewed. Figure 7-5 also illustrates the type of questions that should be asked during an interview.

In-Depth Study of Existing Documents A consultant hired to do a systems study would need more time to study documents than would the company's own analysts since the inhouse analysts are expected to remain current regarding changes in the organizational structure or in the organization's goals. Inhouse analysts and consultants usually start an in-depth study by securing the organizational chart, financial statements, policy manuals, and standards (procedural) manual.

An analyst needs the answers to six basic questions—who, what, why, where, when, and how—in order to determine what documents must be studied. Figure 7-6 lists the documents and reports analyzed for the word processing study.

Observations In virtually every company, there is a difference between the formal and informal organization. Furthermore, not all companies have up-to-date standards manuals and documentation. One of the major problems that exists within many companies is the lack of documentation and a current standards

Documents and Reports Studied for the Word Processing Study

Report	Purpose
Preliminary report and documentation from the first phase of the study	To obtain an understanding of the problem and find out what facts have already been determined.
External documents: articles	To provide external information regarding word processing.
Technical material from vendors	To describe hardware and software being used in word processing, and new advances in terminals.
Cost studies	To summarize cost figures regarding the use of word processing.
Industry standards	To determine the production rate for a secretary, typist, or data entry clerk.
Letters written	To ascertain if more standardized paragraphs can be used.
Financial records	To determine the cost of the proposed system by first determining the cost of materials, equipment, and labor.
Job classifications	To enable the analyst to decide who, other than a secretary, should be inputting the data.
Standards manual	To provide detailed information about specific tasks such as coding the letters received (to determine the standrdized paragraphs that can be used), typing the original letters, typing envelopes, and preparing the letters for mailing.

FIGURE 7.6. Documents and reports studied for the word processing system

manual. By making observations, the analyst can determine the accuracy of the standards manual and the difference that exists between the formal and the informal organization.

Some things can be learned only by observation. Remember that systems analysis involves forms design and retention, work simplification, records management, work measurement, ergonomics, and analysis and design of new procedures and systems. Observations are generally made to determine the workflow, the physical location and arrangement of workstations, who performs each task, and how each task is performed.

Workflow   In the case study regarding the sales-order system, it often took four or five days to process an order. One objective of a new system might to process a sales order in one day.

The analyst assigned to the feasibility study would probably observe the workflow. Once the order leaves the mail department, who gets the order? What does the person do to the order? How long does it stay on his or her desk? The analyst observes where the document goes, what tasks are completed at each workstation, how long it takes before the order goes on to the next workstation, how it is transported to the next workstation, and the distance it has to travel. Each task will be studied until finally the document ends up in the processed order file.

In studying the workflow, the analyst will determine whether workloads are unequal or whether some highly skilled employees are doing tasks that should be done by someone with a lower job classification. The analyst must also be concerned with the physical environment of the worker.

Systems Flowchart   An analyst should construct a systems flowchart that graphically shows the movement of documents and the tasks performed by each individual involved with the procedure. Figure 7-7 illustrates a systems flowchart that might have been constructed for the sales-order processing procedure.

If this were the actual flowchart for the Computer Products Incorporated sales-order procedure, an experienced analyst would see a number of things that need to be changed. For example, why should a manual credit check be necessary when a computer and a customer master file are available? Also why aren't all of the editing-type functions performed at one time so there is only one error report? You might also have observed that a check was not made to see whether the items ordered were available. Perhaps what is needed is a more comprehensive data entry and edit program.

However, the flowchart illustrated in Figure 7-7 meets the criteria given for a well-constructed flowchart and would be useful in explaining to management how the sales orders are processed and where improvements might be made. In preparing a presentation for management, an overhead transparency should be made of the present sales-order procedure.

Many benefits can be derived from a simple flowchart. The overall flowchart can be supported with subsidiary flowcharts that provide detailed information regarding what happens to incomplete orders that cannot be processed or to orders from customers who exceed their credit limit. Additional flowcharts can be prepared to illustrate what happens when the exception routines designated as B1 and C1 are executed. Also, the tasks performed by the sales department, designated as A4, should be detailed. Since a flowchart does not include a great amount of detail, it may be supported by a one- or two-page narrative.

Although a flowchart does not show time or distance, the analyst must be concerned with how long it takes each individual to perform his or her assigned tasks. One of the problems might be that the clerk who inspects the documents cannot keep up with the workload. The analyst might also elect to check the data control clerk's log to see how long the sales order remained in the data entry department.

Design and Location of Workstations   The working environment plays an important part in the productivity of the worker. An employee is usually distracted if his or her workstation is located in a traffic pattern. Key personnel who need to concentrate should not be located in the middle of a traffic pattern or near places where people tend to converse, such as at the copy machine or near the line printer.

One of the questions to be answered in the word processing study is where the workstations should be located that will be used for entering the customer's number, the paragraph codes, and the variable data. In determining the answer, the analyst would need to know which other employees the individuals entering the data would have to communicate with, what resources might be needed, and what impact (if any) the workstations would have on other workers. Old-style terminals were often very noisy and distracted people who were located in the same area. The letter-quality printers needed to print the letters might influence the productivity of the workers in nearby workstations.

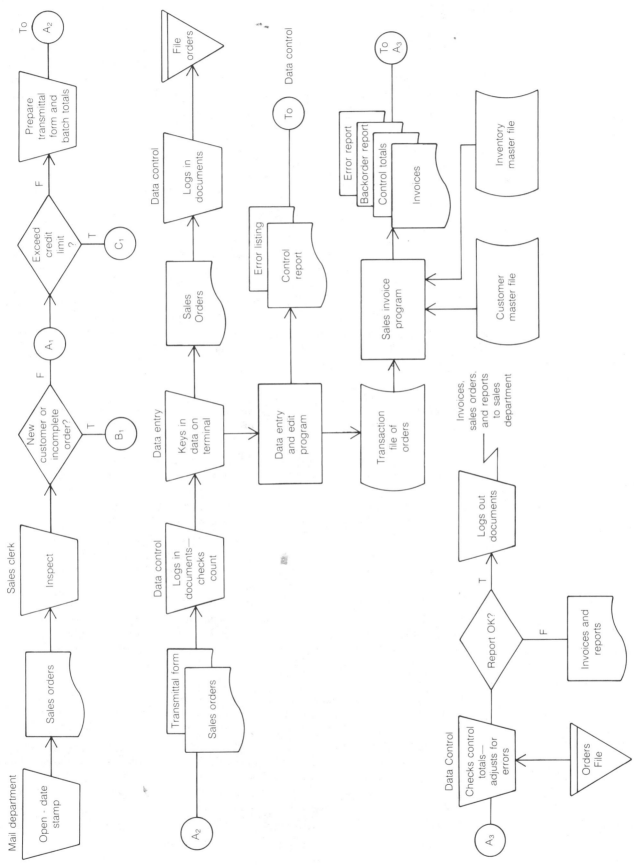

FIGURE 7.7. Systems flowchart illustrating how a sales order is processed

Who Performs Each Task?   The analyst can learn who is assigned to do various tasks by studying job descriptions and the standards manual. Since there may be a difference between the formal and informal organization, the analyst usually confirms by observation what he or she has found. As the analyst watches the employee perform a given task, one question always come to mind: "Is the right person performing the task?" A task is often performed by someone who is overqualified. This increases costs and may also cause the employee to become dissatisfied. Should someone classified as a secretary spend most of the day typing form letters, typing envelopes, and stuffing the envelopes?

Only by observation can an analyst determine how a task is performed. The analyst may be able to suggest changes that make the employee more productive. For example, a typist may be assigned the job typing data on a form that is approximately 4.5 inches long. Each form has only five lines of data. Almost as much time is spent positioning the forms in the typewriter as is spent typing the data on the form. Studies have shown that the use of continuous form paper (like the paper in line printers) would allow a typist to be 30 percent more productive. The savings in time and money would more than offset the additional cost of the forms.

In observing the executive assistant dictate to a secretary, the analyst saw that a great many interruptions occurred. While the executive assistant was answering questions or the phone, the secretary was totally unproductive. Could a Dictophone be used so that the secretary could be doing other tasks while the executive assistant recorded the variable data?

Use of a Questionnaire   Although not as effective as an interview, a questionnaire can also be used to obtain useful data. When a small amount of data is needed from a large number of people, a questionnaire may be an effective tool. The questionnaire should be easy to fill out and should contain only a few questions. Research shows that the response is poor when a long questionnaire is used and that the data obtained on the last few pages may not be very reliable.

The questionnaire should be in a checklist format. It should be appealing to the eye and the questions should be clearly stated. If a question is not clear usually there is no one to interpret what is intended. Once a questionnaire is designed it should be given to a test group that has the same background as the **population** for whom the questionnaire is intended. If the test group feels that parts of the questionnaire are unclear, those sections should be rewritten.

If a questionnaire is sent through the mail, a self-addressed, stamped envelope should be included. Sometimes there is some type of gimmick that goes along with the questionnaire such as a silver dollar. Some people feel guilty if they keep the dollar and do not answer the questions. Depending upon who receives the questionnaire and what control there is over the group, a 30 percent response is usually considered good. If there is some type of control over the group, such as a department head administering it to members of the department, it is possible to obtain a 100 percent response.

If only a sample population is used rather than sending the questionnaire to the entire population, the selection of the sample is very important. For example, suppose a household survey sampled one household in each square block of a city. If all the people surveyed had lived on a corner, the sample would not have been considered valid since usually certain types of people elect to live on corner lots. If the company doing the survey has a research and development department, there is usually someone in the department that can assist in selecting the sample and formulating the questionnaire.

Once the results are obtained, statistical tests can be used to determine the reliability of the data. Computer programs may be available on the system to analyze the data. If a statistical software package is not available on the system, it may be possible to have the results analyzed by contacting a college, a university, or a service bureau. If the data is already punched into cards, it will usually require only a minute or two of CPU time to perform the necessary calculations. This can save hours of manual work.

CPI Uses a Questionnaire   Assume that Computer Products Incorporated wishes to send out a questionnaire to former customers to find out why they are no longer sending in orders. Since a large number of people in different geographic areas is to be surveyed, a questionnaire may be an effective way of obtaining the data. The cover letter should be written very carefully in order not to irritate the person receiving the questionnaire.

The letters should be sent directly to the individual within the firm who is responsible for selecting the vendor with whom orders will be placed. Both the letter and the envelope should be addressed to the individual and not to the position. If someone who had been a good customer receives a letter addressed to "Purchasing Agent" rather than being individually addressed, it can be a source of irritation.

Assume that CPI will send out 1,000 questionnaires to a random sample of firms that had placed orders for at least two years but from whom orders had not been received during the last three years. Most of the orders received by CPI are placed through a sales representative who calls on the firm. Figure 7-8 and 7-9 illustrate the cover letter and questionnaire that was used. Top management made the decision to send out the questionnaire and to offer a 5 percent discount on the company's next order if the questionnaire was returned.

A different questionnaire could be sent to the sales representatives who called on the customers. Although the questions would need to be worded differently, the questionnaire could be designed to get the sales representative's opinion as to why the customer was no longer placing orders. The sales representative's answers could be compared with the customer's answers and some additional, and interesting, data might be obtained.

Questionnaires should be designed to gather specific data, easy to fill out, relatively short, and well-written. A questionnaire being sent to a large number of people should be tested on a small group before being sent to the total population selected for the sample.

Questionnaires that contain poorly stated or loaded questions may be more harmful than helpful. The answers to the questions may present a totally false picture of the problem.

Visits to Other Installations   When a computer system is being selected or a large comprehensive system is being designed, a visit to another installation may be worthwhile. The selection of the installation to visit is important. If possible, the installation to be visited should be determined by someone other than a vendor. A vendor would pick a company that is well-satisfied with the system rather than one that had experienced difficulty. To get a true picture of the system, more than one installation should be visited.

A competitor dealing in the same product line may not be receptive to a visit and that company's personnel would not be at liberty to discuss some of their software packages. A company with a different product line that does about the same volume of business may be a better choice. However, regardless

computer products inc.

1540 Christy Way
Bay City, Michigan 48710
517-682-9134

June 10, 1986

Mr. James Case, Manager
Case Office Supply Company
1410 Park Street
Yonkers, New York   02367

Dear Mr. Case,

An order has not been received from your firm for some time and
we are concerned about the possible reasons. If we have failed
you, we would like an opportunity to remedy the problem.

We appreciate your taking the time to complete the enclosed
questionnaire. As soon as we receive your response we will
send you a certificate that will entitle you to a 5 percent
discount on your next order. If possible, we would like to
have your questionnaire returned by July 1, 1986.

We have an exciting new product line and would like an
opportunity to show you some of our new products.

We hope your name will once again be on our preferred customers
list.

Sincerely,

Mathew Dwan, President
Computer Products Incorporated

Enclosures (2)

MD/lmd

FIGURE 7.8.   Cover letter for CPI questionnaire

computer products inc. CUSTOMER SURVEY

Please check the appropriate answer for each question. Be sure to sign the form so we may send you a certificate entitling you to a 5 percent discount on your next order.

☐ 1. If you no longer carry the line of products sold by CPI pleae check the box on the left. We would still like you to complete the questionnaire.

2. What were your experiences in dealing with CPI while you were on our preferred customers list?

   A. Orders (check all that apply)
   _____arrived in good condition.
   _____arrived in poor condition.
   _____were often delayed.
   _____were frequently backordered.
   _____were shipped on time.

   B. Adjustments to orders
   _____seldom had to be made.
   _____were often necessary because the wrong merchandise was shipped.
   _____were often necessary because the merchandise arrived in poor condition.

   C. The quality of CPI products was (as compared to competitor's products)
   _____excellent   _____average
   _____good   _____poor

   D. Prices were
   _____about the same as CPI's competitors.
   _____lower than CPI's competitors.
   _____higher than CPI's competitors.

   E. The sales representative who called on you (check all that apply)
   _____was well-informed regarding CPI's product line.
   _____was not helpful or prompt in answering requests for information.
   _____was not well-informed regarding CPI product line.
   _____was not helpful in supplying the information requested.
   _____made very few mistakes in placing orders.
   _____made occasional mistakes in placing orders.
   _____made numerous mistakes in placing orders.

3. If you wish to make additional comments, please write them on the back of this form.

4. Form submitted by: _____
                              Your name

   _____
                              Name of Your Firm

FIGURE 7.9. Questionnaire sent out by CPI

of the company chosen, sometimes it is difficult to get the CIS manager to give a true picture of problems experienced, since the initial decision to secure the system may have been a poor one. The manager may hesitate to admit a mistake was made.

Before visiting an installation, follow the same steps as were followed in making an inhouse interview. The initial contact should be made with the CIS manager and the reason for your visit should be stated. If you have selected the company because you wish to learn more about a particular software package that has been developed inhouse, an opportunity might develop to obtain the software at a cost considerably less than developing it on your own. Even if a number of modifications are necessary, the total cost may be less than developing the entire system.

If you are interested in finding out about the performance of the company's system—its hardware and software—you should study the firm's financial statements to determine how the company compares in size and in volume of business with yours. If the company has half the volume of business as your company, its computer system may be inadequate for your needs.

During your visit you should attempt to find out from company officials the amount of downtime they have experienced with their system and the reasons for its failure. What about software support? Does the vendor have personnel in the field who can answer questions? One owner of a small business system found that the manuals for his machine were incorrect. No one in the local sales office or the district sales office could supply answers to the owner's questions. He was forced to contact the owner of a similar system in Scotland and wait three weeks for his answers. If you were to visit his installation, do you feel you would get honest answers regarding software support?

Visits to other installations can be productive if care is taken in selecting the installations and if specific objectives for the visits are determined in advance. You may find the CIS manager being interviewed defensive and guarded or the person may be open and candid and provide you with a great deal of useful information.

Presentation by Vendors   Most vendors are willing to demonstrate new equipment and provide literature about equipment and software products. Often when new products are announced, regional seminars or workshops are held to demonstrate or explain the new product line.

Larger vendors, such as IBM, often present one-day conferences on such topics as ''Office of the Future,'' ''Communications Within the Office,'' or ''Telecomputing.'' The seminars are based on research and are not presented as a sales pitch for a given product.

Potential customers and users are often invited to attend various types of educational programs designed to present general concepts about new systems or software packages. In-depth training programs may also be offered in certain areas, such as data communications and distributed processing.

Working with a vendor will provide data pertaining to the vendor's product line. However, there are numerous publications, such as *Datamation* and *DataPro Report*, that compare various systems, software packages, and items of equipment. These reports are less biased than the information supplied by the vendor and will usually point out both the good and bad features of the hardware and software being discussed.

The analyst working on the word processing system might study two or three reports such as *Computerworld's Buyer's Guide Terminals & Peripherals* and obtain literature from several vendors. After reviewing the literature that compares the features of the various terminals and letter-quality printers, the analyst might ask three or four vendors to demonstrate their equipment. The vendors should provide in-depth material about their equipment, warranty, and service.

Selecting Software   Making a decision about the merits of software is more difficult. There are reports available that contrast software. For example, for microcomputers, there are a number of different database programs. Different microcomputer magazines have devoted issues to comparing different vendor products. In making a decision regarding software, the literature should be reviewed and then vendors invited to provide a demonstration. The potential buyer should:

- examine the available documentation and user's manuals;
- determine whether menus are used and whether there is a "help" menu for each major feature; and,
- try using the package.

When selecting software, a checklist of the features that *must* be included should be developed. Some word processing programs designed for microcomputers have far fewer features than do programs that are included as part of an office automation package or software that is designed for dedicated word processing systems. Some word processing software is very poorly documented and much more difficult to learn to use efficiently than are some of the more popular programs. The factors in the selection of software are:

- completeness.   Does it provide all of the required features?
- ease of use.   Is the software internally documented and are the manuals easy to use?
- reliability.   Has the software been completely tested so that valid results will be obtained and the programs will not abort?
- ease of learning.   How long will it take for employees to learn to use the software efficiently?

While vendors provide a great deal of good information, the analyst should keep in mind that the vendor's primary goal is to sell products. Vendors should be instructed not to make a sales presentation. Instead, they should be asked to make a factual presentation regarding their products.

CHECKPOINT

7. Why should the analyst assigned to the feasibility phase of the project study the documentation and the report of the initial investigation?

8. Can some of the routine decision-making processes be automated? Give an example other than the one given in the text.

9. Why is a project divided into small tasks that can be normally be completed in one week?

10. In planning the feasibility study, why might the analyst make a list of all of the tasks that must be completed?

11. What six basic questions should be asked about any procedure being studied?

12. Is there any difference in the type of personnel interviewed during initial investigation and during the feasibility study?

13. What type of documents were studied by the analyst assigned to the word processing project?

14. What four things can be determined by observation?

15. What are the two major advantages of developing a system flowchart?

16. What two things are not shown on a system flowchart that are of concern to an analyst?

17. Does it make any difference where workstations that have keyboards, VDTs, and printers are located in an office?

18. Should an analyst be concerned with the physical environment of the worker?

19. What is the best way to determine how a task is actually performed?

20. When should a questionnaire be used to gather data?

21. What basic rules should be followed in designing a questionnaire?

22. Why is the selection of a sample so important?

23. What can be gained by visiting another installation?

24. When listening to a presentation given by a vendor, what should the analyst keep in mind?

25. If you were investigating letter-quality printers to be used for word processing applications, what would you do?

26. Your company has recently obtained microcomputers and would like you to evaluate the spreadsheet programs that are available. Spreadsheet software displays data in rows and columns. The original data can be manipulated by inputting additional data and will answer "what if?" questions. How would you proceed to evaluate the software?

Organization of the Data

Most feasibility studies will require gathering data about the present system and about the alternatives available. One school system requested that a study be made to determine how their data could be processed more effectively. Since no one in the system had expertise in EDP, an outside consultant was hired. After studying the present system, the consultant investigated timesharing systems, service bureaus, and the cost of developing an inhouse system. Each possibility was costed out and evaluated. A decision was then made regarding which of the three alternatives should be pursued.

Figure 7-10 summarizes the major ways data is gathered. Although there are guidelines than can be used for gathering data, few guidelines exist that will help in analyzing data and in making the right decision. In some areas, such as the features to look for in a word processing program, a checklist can be developed that will be useful in making an evaluation. Since in many areas guidelines or checklists are not available, it is wise to have a project team analyze the data and make the decision regarding what will be recommended to the computer policy committee. Some of the decision-making sessions could end up in lively discussions since there may be differences of opinion regarding which alternative to pursue. In order to analyze the data and reach a decision, analysts need to be more than just good data processing technicians. Better decisions are usually made by individuals who are knowledgeable in EDP and also knowledgeable in the application area for which the hardware or software is to be used. The project team should be made up of individuals who have backgrounds in both EDP and the application area.

Gathering Data

- Review the available documents such as organization charts, procedural manuals, and documentation for the existing system.

- Interview managers, supervisors, and staff members who perform the tasks being studied.

- Interview people who have expertise in the area being studied who are not employed within the organization.

- Observe the dataflow and the way tasks are performed. Contrast the way the tasks are performed with the documentation.

- Develop and use questionnaires when a small amount of data is needed from a large number of people.

- Work with consultants. If no one within the organization has expertise in a given area, costly mistakes may be avoided.

- Study impartial reports on hardware and software products.

- Obtain literature and demonstrations from vendors.

- Develop a checklist that can be used to evaluate the features of various hardware or software products.

- Use the equipment or software. Determine the amount of internal documentation included to make the software "user friendly" and evaluate the manuals to see if they are easy to follow and written in nontechnical terms.

FIGURE 7.10. Summary of procedures to follow in gathering data

The recommendation regarding word processing may show that a substantial saving, as well as better customer relations, will result if a new system is designed for processing letters of complaint and requests for information. However, if the decision regarding CPI's sales system involves converting the system to an online transaction processing system, can the system be sold to top management on the basis that it will cost less money? Or will the system need to be sold on the basis that a realtime system could eliminate many of the problems that exist? By solving the problems that have been identified, fewer customers will be lost and the company's rate of growth (and profits) should accelerate. In regard to the sales system, the justification moves into intangible areas and requires more salesmanship.

The analyst presenting the report must be prepared to defend the position taken by the team and must make a good sales presentation. Often there are several good proposals presented—but resources for only one or two.

Estimating Cost

One of the most difficult parts of a feasibility study is to accurately estimate the cost of the remaining phases of the project, the cost of the hardware and software, the cost of implementing the system, and the ongoing operational costs.

The cost of the general and detailed design phases of the project must be based on the estimated time and materials that it will take to design the system. This would include the amount of time that analysts, support personnel, and consultants spend in determining the design for the alternative selected at the conclusion of the feasibility study.

At the end of the feasibility study, experienced analysts can project the cost for the next phase (general design) fairly well. The length of time needed to complete the design phase and the costs involved should be reasonably close (perhaps within 10 to 15 percent) to what the actual costs will be.

Until the detailed design phase is completed, it is not reasonable to expect that the projected costs for the systems development and for the implementation of the design will be 100 percent accurate. The same is true of the operational costs. In the past, the estimates made at the conclusion of the feasibility study for the design and implementation phases have been as much as 50 percent off. At this point, it is also impossible to predict with any degree of accuracy what the actual ongoing operational costs will be.

At the end of the next phase (general design), all projections involving costs will be reviewed. At that time, the cost of the detailed design and implementation phases can be estimated more accurately.

As each phase draws to a close, the estimated expenses should be compared with the actual expenses. By learning from past mistakes, analysts can develop skill in estimating costs more accurately. The revised projections for the next phase of a study should always be more accurate than the original estimates. It is also easier to estimate costs if software is purchased rather than developed inhouse.

Costs must include all factors and can be put into one of five general categories:

• Developmental	All costs associated with designing the system. These costs should be prorated over the life of the system.
• Hardware	The cost of the CPU, communication controllers, cables, and I/O devices. How costs are allocated depends on a number of factors.
• Software	Although the application software may be purchased or developed inhouse, when a new computer is obtained it may be necessary to lease the operating system as well as other major software packages. Cost allocation will depend on a number of factors.
• Implementation	One-time costs such as converting files to a new database, keying in additional data, training employees, and so forth. Usually the costs are prorated over the life of the system.
• Ongoing costs	Operational costs which include expenses for personnel, supplies, and overhead such as maintenance and power costs.

Standards must be developed regarding how each category of costs is to be treated. The cost of a new system compared to existing costs would be far more favorable if the developmental and conversion costs are not prorated but are used as part of the ongoing costs. Normally developmental costs are prorated over the estimated life of the system. Also the way hardware and software costs are treated has a direct impact on how the cost of a new system compares to the cost of the old system. If hardware is leased, the monthly charge is included as one of the ongoing costs. In order to have accurate comparisons, standards must be developed and *followed*.

PLANNING THE PRESENTATION Good presentations do not just happen—someone has to carefully plan them. The steps that should be followed in planning a presentation are illustrated in Figure 7-11 and explained in he following sections of the text.

> ✓
>
> Critical Steps in Preparing a Presentation
>
> - Define the audience.
>
> - Define the objectives of the presentation.
>
> - Organize the presentation.
>
> - Define all technical terms.
>
> - Prepare audiovisual materials.
>
> - Review the entire presentation.

FIGURE 7.11. Critical steps in preparing a presentation

Define the Audience A good presentation must be aimed at the specific audience for which it is intended. Usually an analyst knows exactly for whom the presentation is being prepared. This permits the analyst to select the right materials for the presentation and to develop the type of presentation that will be accepted by the group.

The presentation made to the computer policy committee, which is composed of middle-management personnel, might be very different from the presentation made to sales department personnel.

Define the Objectives The purpose, or objectives, of the presentation must be defined. Usually the objectives of the presentation made to middle management upon completion of the feasibility study are fairly consistent. The presentation should inform management of the current status of the present system and the problems that have been identified. The report should also describe the alternative solutions that have been investigated and the projected cost of each. The estimated cost for designing and implementing the alternative solutions should be accompanied by the recommendation of the project team or analyst.

Organizing the Presentation Unless the presentation is well-organized, the audience may miss the major points being presented. A rough outline should be made of the sequence in which the various topics will be presented. Once this is done, the analyst can check to see that there is a logical flow from one topic into another.

After the outline is developed, the analyst can determine what techniques should be used to present the various topics. The major points, conclusions, and recommendations should be supported by some type of visual aid.

A speaker for the Associates for Human Resource Development, in presenting a management seminar, indicated that the format of any presentation should include the following:

- Tell'em what you're gonna tell'em.
- Tell'em.
- Tell'em what you've told'em.

In other words, you should outline your presentation to your audience, make your presentation, and then summarize the material presented. During the same seminar, the spokesman also indicated that any presentation should be: simple and orderly, stress the objectives, have an outline, and follow a logical sequence.

Sometimes the project team must "sell" their recommendations to management. If the presentation is not well organized, the proposal is more likely to be rejected. Some individuals have a high percentage of their proposals accepted while other individuals seldom come up with a "winner." The difference could be in the organization of the material and the amount of enthusiasm shown by the presentor.

Define All Technical Terms

In both the verbal and written presentation, all technical terms must be defined. Nothing turns an audience of non-data-processing people off more quickly than a speaker who refers to abbreviations and acronyms such as POWER, SPOOL-ing, CICS, and MIS* without explaining their meaning. Usually a written report or outline is also given to the audience. The first or second page of the report should contain a glossary of the technical terms and acronyms. Other terms, such as *short-range* and *intermediate-range*, which could be interpreted differently by the audience than by the presenter, should also be defined.

Definition of terms is of particular importance in data processing since it is a dynamic field and new terms are being introduced each day. Data processing professionals sometimes attend highly technical meetings on new topics and come away confused because none of the terminology was explained.

At a conference attended by a large number of non-data-processing professionals one speaker stated that the "problem with data processing is that it has a vocabulary all its own." She then proceeded throughout her entire presentation to use technical terms and acronyms that were unknown to the audience.

Before making a presentation, review the material one more time and make certain that all terms that might be unknown to or misunderstood by the audience are defined in nontechnical terms. The use of undefined technical terms impresses no one—remember the guideline. Keep it simple.

Prepare Audiovisual Materials

Most medium- and large-size companies have support services that help in the preparation of audiovisual materials. If professional support is not available, there are still many types of audiovisual materials that can be prepared by the individual making the presentation. Most are inexpensive and do not take a great deal of time take. The audiovisual aids most commonly used are chalkboards, overhead projectors, flip charts, slides, films, and videotape. The computer can be used very effectively as a means of presenting visual materials and demonstrations of equipment can be given.

Chalkboards   The chalkboard is one of the oldest methods of presentation and probably one of the least effective. Material cannot be prepared in advance and

---

*POWER is an IBM software package added to some operating systems that provides for SPOOLing and multiprogramming; SPOOLing is the acronym for Simultaneous Peripheral Operations On-Line; **CICS** is an abbreviation for **Customer Information Control System** and refers to an IBM software package that can be incorporated into the user's operating system; and MIS is Management Information System.

the material placed on the chalkboard is often poorly written and difficult to read. Nonetheless, it is helpful to have a chalkboard available to illustrate the answer to a question that had not been anticipated. When using the board, the presentor must remember to write or print large enough so that everyone can read the material. Often the presentor fails to erase the board before adding new material and the whole effect becomes one of confusion.

Overhead Projectors   Overhead transparencies are one of the most effective methods of presentation. There are a number of different ways in which material can be prepared for an overhead projector. Many methods require very little time or skill. Any printed material can be duplicated and then put through a copy machine which will make a transparency. Original materials can be used to make a transparency unless they are written with a ball-point pen or blue ink. However, usually a Xerox copy makes a better transparency than the original printed material.

Typewriters are available with bulletin-board type that make oversized letters, which are easier to read on overhead materials than those produced by a regular typewriter. Kits of self-adhesive materials in different colors can be purchased for making bar graphs and certain types of charts.

Artwork can be used on the overhead materials to add interest. The artwork should be in good taste and add, rather than detract, from the presentation. The presentor who knows the type of people who will be in the audience will be in a position to determine what type of artwork will be received. A major point made in the presentation may have more impact if some type of graph or artwork is used to emphasize the point.

The presentor should bring some unused transparency masters and colored visual-aid pens to the presentation. The pens can be used to highlight transparency masters already prepared or to quickly construct a new master to answer an unexpected question.

In making overhead transparencies or other types of visuals, it is well to remember that each visual should stand on its own. Simple visuals are often more effective than ones that are complex. Often a graphic display emphasizes a point better than an explanation. The visual aids should be of a professional quality and in good taste. Care should be taken that cartoon-type graphics do not offend any one group of people.

Flip Charts   The material presented by using flip charts can be prepared in advance and can be made large enough so that the material is easy to read. The flip chart is usually placed on an easel and the presentor turns the pages as the presentation is in progress. Flip charts can be purchased in most art supply stores.

Flip charts are a tried-and-true method since nothing can go wrong with them during the presentation—there are no bulbs to blow or extension cords to worry about. Colored marking pens can be used in making the charts and to add facts while making the presentation.

Slides   Slides can be obtained from a number of different sources or can be taken by the person who will be making the presentation. If the presentation involves a recommendation to secure new equipment, slides illustrating how the equipment is used can be very helpful. The amount of detail that should be provided in the slide presentation depends upon the audience. Top management certainly would not be interested in how the terminal is used, but the sales personnel who would eventually be using it would be very interested.

If the company has a graphic or media department, there is usually someone in the department who will prepare slides from printed materials, illustrations, graphics, or pictures. With the photographic equipment available today, even an amateur can produce effective slides.

Slide/tape presentations make very good training materials since the presentation is self-contained and needs no explanation from a presentor. The back side of the tape contains the electronic signals that advance the slides.

Demonstration of Equipment   During the feasibility study, the specific hardware to be used may or may not be identified. As stated earlier, no two projects are identical and the scope of each phase depends upon its objectives. The recommendation for new hardware could be made either at the end of the feasibility study or at the end of the design phase of the project.

Usually top or middle management groups are not interested in seeing hardware demonstrations. An exception might be made if the investigation indicated that a particular piece of hardware, such as a new type of intelligent (programmable) terminal, could be used to meet the objectives identified for the modification of an old system or the development of a new system.

If the presentation is being made to operational personnel, a demonstration of any new equipment should be given. The demonstration to operational personnel would probably be given during the implementation phase of the project rather than at the end of the feasibility study.

If a demonstration is to be given, it should be informative and provide pertinent information regarding the operation of the equipment rather than being a sales pitch.

Use of the Computer   Often an analyst forgets that the computer can be used very effectively in making formal presentations. This is particularly true if a teleconference room is available that has large monitors. Some conference rooms have two or three large monitors that can be used simultaneously. A simple program is written that causes a series of charts, graphs, outlines, or graphics to be displayed on the monitors. For a small group, a microcomputer with a large colored monitor can be used effectively.

Today with some of the graphic systems available, such as the Aurora System, impressive graphic displays can be created with a minimum of effort. Some software that is available for either mainframes or microcomputers will cause whatever is displayed to be printed on the attached printer. Small, inexpensive printers which produce excellent colored illustrations are available that can be used with microcomputers or mainframes. The printed graphics can be used to enhance the written report.

Review the Presentation

Before making the presentation, review both the written material and the outline you will use for making the verbal presentation. Mark on the outline where you will use the visuals you have prepared. Check to see that all terms are defined and that the presentation flows smoothly. If there is time, set the materials aside for a day or two and then give them their final review. You may see things in a new light when you are no longer engrossed in the preparation of the materials. For example, you might find that the presentation is too busy. There can be an overkill in the use of visuals. The audience may be distracted by the movements of the presentor as he or she rushes from one visual to another. Keep the presentation simple.

MAKING THE PRESENTATION

The lists of hints shown in Figure 7-12 was initially compiled by the Associates for Human Resource Development. For presentation in the text, it has been reorganized. The list provides a useful summary of things that should be remembered when making a presentation. Most of the hints require no further explanation.

Hints For Making Successful Presentations

Know your subject.

Know your audience.

Prepare visuals. However, poor visuals are worse than no visuals!

Check equipment, sequence of visuals, flip charts, etc. beforehand.

Practice, if only in your mind.

Take control and keep control.

Images count.

Talk it, don't "speech" it.

Choose words to suit your listeners.

People hear what they want to hear.

Be sensitive to your audience reaction.

If you believe it, say it convincingly.

If you don't believe it, don't say it.

On controversial points, be tactful.

Repeat any actions committed to by individuals.

Don't read visuals.

Don't be sloppy (physically or in preparation).

Don't knock (criticize) others.

FIGURE 7.12. Helpful hints for successful presentations

The key to success is to know your material and your audience! Although in a technical area no one is expected to be able to answer all of the questions that might be asked, if misinformation is presented or graphic displays are inaccurate, the audience loses confidence in the presentor. At a recent presentation on a very technical topic, the presentor could not answer any questions that were asked. Although he was to have been an "expert in the field," he had a very limited background regarding the software product he was explaining. The audience soon lost confidence in the presentor and interest in the presentation.

The presentor should also take control and keep control. There may be someone in the audience who will want to take over, and this may distract other listeners. The individual's comments and questions may be entirely off the subject. When this occurs, a skilled presentor can tactfully regain control and suggest to the individual that they discuss the additional topics at some other time.

The hint to practice, if only in your mind, is a good point. The amount of practice needed varies with the individual. Some individuals do far better if the presentation is made from a well-developed outline but is not practiced. Such a presentation is delivered with more enthusiasm and the presentor is also more responsive to the audience.

The hint about checking equipment is very important. Arrive at the location where the presentation is to be made in time to check things out. Is there an extension cord for the overhead or the slide projector? Are there extra bulbs, and do you know how to change them? At a presentation made by a major vendor, the audience had to wait 15 minutes for the presentor to find an extra bulb for the overhead projector!

Most people have some last-minute feelings of apprehension when they are required to make a major presentation. They will be far less apprehensive if they have personally checked out all of the equipment that will be used in the presentation. Even an experienced presentor will be shaken if none of the visuals can be used because there is no overhead projector. Adequate preparation is the best tonic for apprehension.

CONCLUDING THE FEASIBILITY STUDY

When the project being used as an example was planned, three days were allocated to the preparation of the final report and visual aids. The following steps were taken to conclude the investigation.

I. The documentation was put in good order and an index was prepared. The documentation included the following items:
   A. the Gantt chart that illustrated the tasks to be performed
   B. overview of methods used to investigate the problem
   C. memos requesting interviews
   D. follow-up memos for all interviews
   E. completed interview forms
   F. summary of observations
      1. dataflow, including the systems flowchart
      2. performance of staff directly involved in the preparation of the letters
      3. recommendations regarding relocation of key workstations, files, and some equipment
   G. documents showing the cost of the proposed system and the present system:
      1. salary schedules and budget report
      2. calculations made by the analyst upon observations
      3. vendor proposals for the cost of terminals and letter-quality printers
      4. vendor proposals for the cost of dedicated word processing systems
      5. vendor proposals for the cost of microcomputers, letter-quality printers, and software
      6. figures showing the cost of materials for both the present and proposed systems
      7. calculations showing the estimated amount of CPU time

8. calculations showing the estimated cost of developing and implementing the system

9. articles showing production standards for secretaries and word processing technicians

II. The formal report (illustrated in Figures 7-13 through 7-15) was written.

III. Visual aids were prepared. Overhead transparencies were made of the report and all items included in the documentation. Although the analyst did not intend to use some of the visuals, they were included in case they were needed to answer questions from the audience.

IV. A formal presentation was made to the computer policy committee.

ADDITIONAL COMMENTS ON THE FINAL REPORT

The useful life of the system was estimated to be five years. The period of time estimated for the life of any systems depends upon a number of factors. In this case, the assumption was made that since CPI's computer was relatively new, no major changes were anticipated for five years. Since the project team elected to go ahead with the word processing software included in the office automation package, it is reasonable to assume that additional workstations can be added, when needed, to expand the system.

Notice that the cost for CPU time is charged back to the sales department. If the time is available and not being used, the cost per month is not really an additional cost to the company. No charge was shown for supplies or for the executive assistant's time. It is anticipated that the new system will decrease the time she spends on the letters. Since her job description will not change, and the savings in time is an unknown, the cost factor is considered as constant and is not included in the report. An assumption was also made that the cost of supplies will remain constant. The type of supplies will change but the cost will remain constant.

The figures could easily be manipulated to show either a larger or smaller savings over the estimated life of the system. Also, the savings in salary expenses are based on today's wages. The impact of inflation would tend to increase the savings over the next five years since the increase in the secretrial wages would be greater than the increase in the word processing technicians' wages. On the other hand, given current trends, the cost of the terminals might decrease.

In presenting cost figures, guidelines should be used so all estimates are developed in approximately the same way.

CHECKPOINT

27. What constraints might be imposed that would limit the scope of the feasibility study?

28. What must be done after all data is gathered? Is this harder or easier than gathering the data?

29. Why should the audience be defined prior to making a presentation?

30. Why should the objectives of the presentation be defined?

31. Why should an outline be developed for the presentation?

32. Why should technical terms be defined?

33. Why should visual aids be used?

34. How might a computer be used in making a formal presentation?

35. Why is the use of motion picture films limited?

36. Which is easier to use in making a presentation, a chalkboard or a transparency master?

computer products inc.   FEASIBILITY STUDY REPORT

PROJECT NUMBER:	1051
REPORT PREPARED BY:	Kay Walzcak
DATE:	August 10, 1986
	Page 1 of 3

NATURE OF THE INVESTIGATION

1. Study the dataflow and work assignments to determine what changes
   are needed in the office function that will be incorporated into
   the word processing system.

2. Determine if some of the information retrieval functions can be
   computerized.

3. Determine the cost of the present system.

4. Estimate the resources needed to develop a word processing system.

5. Determine the estimated cost of the proposed system.

MAJOR OBJECTIVES OF THE PROPOSED SYSTEM

1. Process requests for information and complaint letters on the same
   day as received.

2. Reduce the cost of handling the requests for information and complaints.

3. Enable secretaries to utilize their time more effectively.

4. Develop a more effective way of coding letters that are received.

5. Reduce the amount of variable data that needs to be keyed in by the word processing
   technicians.

6. Reduce the amount of time the executive assistant spends on coding letters,
   looking up material, and formulating the variable responses needed for some of the
   letters.

7. Automate some of the other office functions that are incorporated into the system.

8. Design a flexible system that can be modified to include other applications.

PROBLEMS IDENTIFIED WITH THE PRESENT SYSTEM

1. Responses to requests for information and complaint letters are
   sometimes delayed four or five days. Additional customer dissatisfaction is sometimes created by the delay.

2. High cost of each individual letter.

3. Inefficient use of the time of highly-qualified secretaries.

4. Workstations are not clustered properly.

FIGURE 7.13. The feasibility study report (page1)

computer products inc. FEASIBILITY STUDY REPORT

PROJECT NUMBER: 1051

                                                  Page 2 of 3

DOCUMENTATION AVAILABLE:

1. Summary of methods used to investigate the problem.

2. Exhibits include:

   a. Memos requesting interviews, interview forms, and follow-up memos.

   b. Summary of observations regarding workflow and the productivity of the personnel involved in processing the letters.

   c. Proposal, submitted by IBM, for a Displaywriter system.

   d. Proposal, submitted by Radio Shack, for a TRS-80 Model III system using SCRIPTSIT software.

   e. Proposal from Digital Equipment Company for programmable terminals.

   f. Proposal from Computers International Incorporated for a Daisywriter printer.

SUMMARY OF FINDINGS:

1. Procedural changes must be made that will increase the productivity of the executive assistant and the secretaries.

2. The project should serve as a pilot. After the word processing technicians are trained, other sales applications should be converted to word processing. Other departments might also be encouraged to use word processing and other features of the office automation software.

3. A Dictaphone should be used to record the variable data rather than having a secretary take shorthand. Due to the number of interruptions, the secretary is idle a large percentage of the time during the time that the executive assistant is dictating.

4. The executive assistant and the secretaries spend an excessive amount of time looking for data in files located on the opposite side of the office. A better information retrieval system is needed that would eliminate the need for secretaries to be away for their desks looking up the data needed to answer letters.

5. The physical location of the workstations places key people in the traffic patterns and away from other employees with whom they must work.

6. The word processing software included in the office automation package provides all of the functions required.

7. Two programmable terminals will be obtained. The terminals will be used for entering the required codes, entering variable data, and for some of the routine maintenance for records stored in the accounts receivable database.

FIGURE 7.14. The feasibility study report (page 2)

CᴾI computer products inc. FEASIBILITY STUDY

PROJECT NUMBER: 1051

Page 3 of 3

ESTIMATED COST OF DESIGNING AND IMPLEMENTING THE SYSTEM

Analyst	3/4 person/month	$3,000.00
Word processing technician	1 person/month	1,500.00
Secretary	1/2 person/month	1,250.00
Materials and supplies		300.00
Computer test runs and creating files		400.00
Training word processing technicians		1,000.00
Total Estimated Costs		$7,450.00
Cost of feasibility study		3,425.00
Total costs to prorate over 5 years		$10,875.00

COST OF PRESENT AND PROPOSED SYSTEM-PER MONTH

	PRESENT (actual)	PROPOSED (estimated)
Secretarial expense (3 employees)	6,750.00	
Word processing technicians (2 employees)		3,000.00
Workstation - intelligent terminal and printer		400.00
Charge-back for CPU and connect time		480.00
Development cost of system prorated over 5 years		181.25
	6,750.00	3,881.25

ESTIMATED SAVINGS:

Per month	2,868.75
Over life of system – 5 years	172,125.00

INTANGIBLE BENEFITS (Not reflected in cost):

1. Better service can be rendered to customers.
2. If fewer customers are dissatisfied, less time will be required to handle letters of complaints. In addition, sales could increase.
3. Since the executive assistant will spend less time in coding and dictating variable information, more time will be available for managerial functions.
4. The productivity of other workers may increase when the workstations are relocated. Changing the locations of some of the files and equipment may also decrease the time employees are away from their desks.

RECOMMENDATIONS:

It is recommended that the word processing software that is part of the mainframes office automation package be used. Files created using the word processing programs can be used for other applications. Also the existing database can be used to obtain the customers' names and addresses. The word processing technicians should also be trained to do some of required file maintenance for the accounts receivable database.

FIGURE 7.15. The feasibility study report (page 3)

37. Why is it a good idea to arrive early at the location where the presentation is to be made?
38. Why are all of the items listed under Concluding the Feasibility Study included as part of the documentation?
39. What types of costs need to be considered in determining the cost of the new system?
40. Why should standards and guidelines be developed for determining costs?

SUMMARY   The feasibility study does more than the name implies since

- data is gathered;
- data is analyzed;
- decisions are made;
- a comprehensive report is prepared that includes detailed documentation.

The analyst or team assigned to the project starts by studying the initial investigation report and documentation. After the analyst or team members understand the objectives of the study, a detailed plan is prepared. The project is divided into small tasks. Each task should have the potential for completion within a short period of time—often the guide used is a week or less.

Data can be obtained by conducting interviews, studying documents and reports, brainstorming selected topics, making observations, visiting other installations, listening to presentations made by vendors, working with consultants, and using questionnaires.

After all of the data is gathered it must be analyzed and interpreted in terms of the objectives of the study. Almost any study includes the primary objectives that were listed in Figure 7-3. The decisions made will be influenced by managerial constraints. Usually more than one alternative is explored. However, the project team should make a recommendation regarding which alternative should be selected.

Both a written and an oral presentation will be made in order to show the objectives of the study, what was done, the conclusions reached, and the recommendations of the committee. The documentation should be organized, indexed, and filed so it can be referenced easily. The report should be developed using the guidelines established by the company.

The presentation to the computer policy committee and to other interested groups should be well prepared. Effective audiovisual aids should be used to make the presentation clearer and more dynamic. The analyst must be at his or her best while making the presentation. A poor presentation for a good project may cause it to be shelved or dropped in favor of a project that has less potential but was presented in a more professional manner.

The computer policy committee or a similar group which hears the presentation and studies the written report will either: approve the recommendations as presented and schedule the next phase of the project; approve part of the proposal but ask for modifications to be made on some parts of the recommendations; shelve the whole project until some time in the future; or drop the entire project.

DISCUSSION QUESTIONS

1. James Blakelee, president of the New Homes Corporation, has indicated "users and management should not be involved in any phase of the investigation, design, and development of a new system. Analysts are hired to develop and implement systems." How would you respond to Blakelee's statement and convince him that management and users should be involved?

2. Blakelee also said " it is cheaper and faster for an analyst to work alone. Project teams tend to take too long to develop systems." What rationale would you use to convince Blakelee that when major projects are to be developed, a project team should be used?

3. The president of Shadowridge Development Corporation, Matt Langosh, would like to have "a state of the arts sales-order system developed." He also indicated that no new equipment could be obtained. The present computer system has a CPU, card reader, card punch, four tape drives, and three disk drives. What would you tell Langosh regarding the type of system than can be developed? What other constraints might have been imposed?

4. You are a newly hired analyst for Webster Department Stores. The sales manager would like a new system developed. You have been told by your EDP manager to start at once. The first thing that he would like you to do is to contact vendors who sell cash registers that also function as terminals. How do you feel you should begin the feasibility study? The initial investigation was conducted by another analyst who is no longer with the company.

5. The president of a very small construction company would like to automate some of the job-costing procedures. No one on the staff has worked with either microcomputers or minicomputers. He has asked you what you think should be done to automate the procedures. What suggestions would you make regarding available options?

6. What impact might the location of workstations and the design of the workstation have on productivity?

7. You would like to find out from all customers their opinions regarding a new product line. What might be the most effective way to obtain the information? If you were in charge of the project, how would you proceed?

8. In determining the cost of a proposed new payroll system, the cost of the new system was far less than the cost of the old system. In looking at the report, you observe that the only costs shown for the new system were the cost of salaries for the data entry operators and the rental of the necessary terminals. Based on the information presented in the chapter, what other costs should be included? If you were a consultant hired to evaluate the efficiency of the company's CIS department, what else might you question?

9. The company that you work for has a timesharing system. However, most of the variable data is keyed on to diskettes. The diskettes are then used as input into a converter that reads the data and stores it on magnetic tape. You have been asked to do a feasibility study to determine whether the jobs should be redesigned so that terminals could be used rather than diskette recorders. What factors should be considered? How would you proceed to gather data? Would you make any visits or contact any vendors? What in-house documentation would you study? In making your report to the committee, what type of visual aids would you use?

10. You just completed the feasibility study for the online payroll project. You are to make the verbal report for your project team. The report is scheduled for 10:00 A.M. Tuesday. How would you prepare for the presentation?

11. In the word processing example, what intangible benefits will be derived from the new system? Why is the assignment of the secretaries to a different project considered an intangible benefit?

12. In determining the cost of the new system, why was a period of five years used to prorate the cost of developing the system? How might the figures be presented so that an even larger saving would be shown? Could the figures be presented in such a way that the proposed system would be more costly than the old system?

Team or Individual Projects

1. In the outline that showed the steps taken to conclude the feasibility study, it was noted that vendor proposals were submitted for terminals, dedicated word processing systems, and microcomputer systems. You are to develop a checklist that might be used in evaluating different word processing systems. In developing the checklist:

   a. Obtain information regarding the features that are included in the word processing software for the IBM Displaywriter, Lanier EZ-1 Word Processor, or one of Digital Equipment Corporation's word processing units. Prepare a list of the major features that are included in the word processing software.

   b. Obtain manuals from two word processing programs that are designed to run on microcomputers. You may be able to obtain the manual from a friend that has the software, from a computer store that sells the software, or from your school. You may wish to evaluate PIE:Writer from Hayden Software Company, Wordstar from MicroPro, PC-Write from Quicksoft, or Volkswriter from Lifetree Software, Inc. Use the list prepared for evaluating a dedicated word processing system as a checklist and determine how many features are also included in the software programs designed for use with microcomputers.

   c. Prepare a report that includes:

      (1) A checklist showing the major features that are incorporated into each of the three programs you are evaluating.

      (2) The characteristics of the hardware needed to run the software. This would include the amount of memory, disk space, storage medium, and type of printer that can be used or is recommended.

      (3) The cost of the three systems. This should include both hardware and software.

      (4) A good visual aid that shows the major features of each of the three packages.

      (5) A visual aid that identifies the source of your information. You will find that many magazines such as *PC* or *PC World* feature articles that contrast word processing software.

      (6) The additional steps that you would take before making a decision regarding which system to obtain.

2. The sales manager was impressed with the results of the data obtained from the questionnaire sent to former customers. He would like to survey the sales representatives and determine why they feel so many former customers are no longer placing orders. Each sales representative will be sent the following information:

   a. A list of customers in his or her territory that have not placed orders in the last 18 months.

b. The average of the last five orders for each of the former customers on the list.

c. A copy of the cover letter and a summary of the results of the questionnaire sent to customers who have not placed orders.

Directions:

a. Prepare a cover letter that will be sent to each sales representative.

b. Prepare a questionnaire.

3. You have been asked to do a feasibility study regarding the possibility of replacing key-to-diskette recorders with terminals. At the present time, the company employs 12 full-time operators that either record data on the diskettes or verify what is recorded on the diskettes.

Directions:

a. List the tasks you will need to do to gather data. Whom would you interview? What observations would you make? What materials would you obtain and study? Would you contact any vendors? Prepare a Gantt chart that spreads the tasks to be performed over four weeks.

b. If possible, obtain an interview with a data entry supervisor, a computer information services manager, or an analyst who can provide you with information on the advantages and disadvantages of key-to-diskette recorders and terminals.

    (1) List the resources that you would use to obtain as much information as possible regarding the two systems before the interview.

    (2) Write a letter confirming your appointment.

    (3) Develop an interview form that has the questions you would ask during the interview.

c. Prepare a report of your findings. Use visual aids. Include in your report the nature of the investigation, the objectives of the new data entry system, the problems identified with the use of key-to-diskette recorders, the documentation available, a summary of findings, and your recommendation. If possible, include the cost of the two systems and the intangible benefits.

GLOSSARY OF WORDS AND PHRASES

CICS (Customer Information Control System) An IBM software package that supports online processing of data and a comprehensive database.

population Members of a common group or individuals that have certain characteristics in common. For example, one population on a campus is made up of all freshman students.

workflow The order in which various tasks are performed on data. When the workflow is studied, the analyst determines who performs each task, how it is performed, the distance documents travel between stations, and the length of time needed to complete each task.

STUDY GUIDE 7

Name _____

Class _____ Hour _____

A.   Indicate whether the following questions are true (T) or false (F) in the space provided. If false, indicate in the margin how the statement should be changed to make it true.

_____   1.   The major objectives of a feasibility study are to further define the problem and to determine the alternatives that can be used to solve the problem most effectively.

_____   2.   Management and users are not involved in a feasibility study.

_____   3.   The way the feasibility study is conducted and the alternatives studied have little impact on what will occur during the rest of the project.

_____   4.   All members of a project team should be from the CIS department.

_____   5.   Only small companies without CIS departments hire EDP consultants.

_____   6.   One of the advantages of hiring a consultant is that the individual hired has no preconceived ideas regarding the organization and its problems.

_____   7.   A small-shop specialist only needs to be knowledgeable about microcomputers, minicomputers, and available software.

_____   8.   One of the objectives for a new system may be eliminating the need for clerical employees by assigning additional tasks to more highly trained personnel.

_____   9.   If more decision points are added to a system, workflow and the production of reports will be accelerated.

_____   10.   When existing systems are redesigned, there is a tendency to increase the amount of paperwork and reports.

_____   11.   It is easier to train employees if work procedures are standardized.

_____   12.   During a feasibility study, more data is gathered by using questionnaires than by any other means.

_____   13.   In interviewing operational staff, a   difference may be found in their perception of a task and the perception management has of the task.

_____   14.   When an investigation is being made, the only questions that need to be answered are what tasks make up the system, why are the tasks being done, where should the tasks be done, when should the tasks be done, and how should the tasks be done?

_____   15.   In the word processing study, industrial standards were studied to determine the production rate for secretaries and word processing technicians.

_____   16.   The systems flowchart showing how sales orders are processed indicated the tasks performed but not who was assigned to perform each task.

_____   17.   The systems flowchart indicated that all of the sales orders were processed in a batch and on the same day they were received.

_____   18.   According to the systems flowchart, the program used for the data entry function edited the data as it was keyed in by the operator.

_____ 19. In making observations, one should always ask "Is the right person performing the task?"

_____ 20. Vendors only provide presentations on specific products.

_____ 21. When comparing the cost of the old and new system, only the ongoing operational costs should be used for the new system.

_____ 22. When making a presentation, you should only deliver the material and then in a summary recap the highlights of the presentation.

_____ 23. Presentations are usually better if the presentor does not develop an outline to follow.

_____ 24. Using a chalkboard is more efficient than using an overhead projector.

_____ 25. Each visual aid should be simple and should stand on its own.

_____ 26. Flip charts are effective, can be prepared in advance, and can be used without worrying "What might go wrong?"

_____ 27. Since it is difficult to make a videotape, this medium is seldom used.

_____ 28. In making a presentation to management regarding the proposed word processing system, the analyst should arrange for a demonstration of the programmable terminal and letter-quality printer.

_____ 29. Listed as an intangible benefit to which a dollar value is not attached is the additional time the executive assistant will have to perform other functions.

_____ 30. The word processing project was considered a "pilot project." Management hoped that the estimated cost benefits would encourage other managers to consider using word processing and some of the other programs available in the office automation software.

_____ 31. In the final report it was indicated that no procedural changes would be needed other than to have terminals used rather that typewriters.

_____ 32. In investigating word processing software, only programs that could be run on CPI's mainframe were evaluated.

B. Match the step listed below with statements regarding either the verbal or written report.

  a. define the audience   d. define technical terms
  b. define the objectives   e. prepare audiovisual aids
  c. organize the presentation   f. review the presentation

_____ 1. A glossary is provided at the beginning of the report.

_____ 2. May include practicing the presentation. Some people might "only do it in their mind" while others actually give the presentation.

_____ 3. Must be done if there is a good transition from one topic to another.

_____ 4. Must be done before the objectives for the presentation can be defined.

_____ 5. May require the help of support personnel who specialize in that particular area.

_____ 6. Includes making sure that the audiovisual aids are in the right order and indicated on the outline.

_____ 7. Must be done after the audience is defined and before the content of the presentation can be determined.

_____ 8. Often a presentation that has good material and visual aids is not well received because the person failed to _____.

____ 9. In order to make the presentation more interesting, it is wise to _____.

____ 10. Although it is important to _____, it is sometimes wise to avoid the use of technical terms when possible.

C. Match the objectives listed below with the statement. In some cases more than one answer is required.

a.	control by exception	e.	eliminate clerical work by highly-skilled employees
b.	automation of decision making	f.	minimum of paperwork
c.	usable output without reformatting	g.	information disseminated rapidly
d.	flexible system	h.	evaluation method built into the system

____ 1. Word processing technicians rather than secretaries would input the required codes and key in variable data.

____ 2. Sales representatives could use terminals to determine if the item was in stock or the customer's credit limit was exceeded.

____ 3. A list is printed each day of all items that are below the reorder point and need to be reordered.

____ 4. Additional information can be added to the database and additional reports made available without an extensive revision of the system.

____ 5. Each level of management received the type of report needed.

____ 6. In designing a new sales system, each manager was asked to justify the reports received and tell how they were used.

____ 7. When the sales-order data is entered into the new system by the sales representative, the order will be rejected if the customer has exceeded his credit limit.

____ 8. Word processing technicians rather than secretaries will make the necessary changes to the accounts receivable database.

____ 9. When the new sales system is implemented, more data will be displayed on the VDTs.

____ 10. When the new sales system is implemented, each sales representative will have a form that must be used whenever a complaint is received from a customer. CPI is very concerned that the problems identified in the old system are corrected.

D. Multiple choice. Record the letter of the correct answer in the space provided.

____ 1. The audiovisual aid that is probably used the most to present graphs and charts is
a. the slide projector.   c. the chalkboard.
b. the overhead projector.   d. the computer.

____ 2. Because of the cost involved, it is unlikely that _____ will be used unless the material can be obtained from a vendor or is used on a continuous basis to train employees.
a. overhead transparencies   c. slides
b. films   d. the computer

_____ 3. Is one of the cheapest, most effective visual aids, and one that can be prepared in advance. Color, cartoons, charts, and graphs can be presented.

    a. slides   c. overhead transparencies

    b. flip chart   d. videotape

_____ 4. In designing a new sales system, it might not be possible to develop an online, realtime system because

    a. management is not directly involved in the project.

    b. users are not directly involved in the project.

    c. constraints were imposed regarding budget and personnel that limit the type of system that can be developed.

    d. the sales representatives do not know how to use terminals.

_____ 5. Before a consultant is hired,

    a. his or background should be investigated.

    b. the tasks to be performed by the consultant should be defined.

    c. other employees within the organization should be informed as to why the consultant is being employed.

    all of the above must be done.

_____ 6. System flowcharts provide

    a. a graphic representation of the workflow.

    b. an indication of the length of time each task requires.

    c. information regarding the distance documents travel.

    d. all of the above.

_____ 7. Questionnaires should be used

    a. when a large amount of data is needed from a small number of people.

    b. when a small amount of data is needed from a large number of people.

    c. it is necessary to determine the difference between the documentation for a task and how the task is actually performed.

    d. in place of interviews.

_____ 8. The report for the word processing feasibility study included the nature of the investigation, the major objectives of the study, the problems identified with the present system, the documentation available, estimated cost and savings, benefits, recommendations, and

    a. the list of people interviewed.

    b. a system flowchart.

    c. a summary of findings.

    d. none of the above.

_____ 9. In order for people to be able to interpret the projected cost of the new system,

    a. only operational costs should be included.

    b. all developmental costs should be prorated over three years.

    c. guidelines and standards must be established showing how the project cost will be determined.

    d. only the cost of labor and supplies should be included.

_____ 10. The feasibility study

    a. requires less time and fewer personnel than the initial investigation.

b. requires less support from management than the initial investigation.
c. should be concluded with both a written and oral presentation.
d. is conducted in a more efficient and cost-effective manner when performed by an analysts rather than a project team.

Designing Systems

Texas Farm Bureau

The Texas Farm Bureau (TFB) in Waco, Texas is an independent voluntary organization of farmers and ranchers united for the purpose of analyzing their problems and formulating action to solve those problems. The members own, finance, control, and operate TFB. At the present time there are over 200 organized county Farm Bureaus in Texas which provide a variety of services to their members. According to the Farm Bureau's master plan for EDP each of the 200 local associations will be able to communicate directly with the centralized home office by using IBM XT personal computers.

The Farm Bureau was organized in the summer of 1920 and by years' end had a membership of 2,550. By the second year membership had grown to 10,841. Membership was in excess of 317,000 at the beginning of 1985. One of its major objectives is to secure economic benefits for its members. At the present time the TFB is composed of eight wholly owned subsidiaries: TFB Mutual Insurance Company, Texas Agricultural Marketing Developing, TFB Management Corporation, Texas Agricultural Services Company, TFB Building Corporation, TFB Investment Corporation, Texas Farm Bureau Rural Health Association, Inc., and TFB Underwriters Company. TFB has been instrumental in passing legislation beneficial to its members and keeps its members informed regarding new services and products.

One of the economic service programs, inaugurated in 1965 as the Texas Agricultural Service Company (TASCO), involves group purchase of tires, batteries, and some types of tillage equipment.

Located in Waco, Texas, the modern office building houses the executive director, division and department heads and the required support staff for the Texas Farm Bureau in Waco, Texas.

COMPUTER SERVICES AT TEXAS FARM BUREAU

Hardware

In regard to computer utilization, TFB has reached maturity. Management at TFB has always recognized the need for keeping its members informed and the need for relevant, timely information. In 1955 TFB's centralized data processing facility was established. By 1958 data was keypunched, verified, and processed by using two 403 Accounting Machines and a 604 Electronic Calculator. By the mid 1950s standards, TFB had a sophisticated system. In 1962 a second generation computer, an IBM 1401 with 8K of memory, was obtained. Although the 1401 was upgraded, a third-generation IBM 360 with 32K of memory was obtained in 1967. TFB later upgraded to an IBM 370 which was then replaced by an IBM 4341. As their EDP needs continued to grow, an IBM 4381 was added to the system.

Any of the peripherals can be utilized by either computer. Although either computer can be used for both batch, time-sharing, and network activities, the IBM 4341 with its DOS/VS operating system is used primarily for the development and testing of new systems. The IBM 4381, with its VM-1 operating system, is used for managing the network, batch jobs, interactive computing applications, and responding to queries from users of the network.

Software

Information accessed for either batch or online processing is stored in VSAM files. Data compression software developed inhouse condenses the data so that it can be processed faster with less storage space required. All of the major information processing systems have been developed inhouse. Although available software packages are investigated during a feasibility study, TFB has always felt it is more advantageous to involve the user directly in developing their software. A user from the functional area, often a supervisor, is usually assigned as a member of the project team.

According to our host, Shelby Tunmire, well-designed systems can be modified to take advantage of technological advancements and therefore can be used for a number of years.

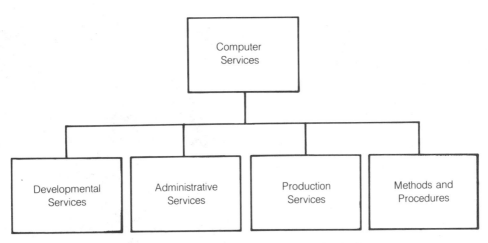

FIGURE IV-1 Computer Services at Texas Farm Bureau is divided into four functional areas.

COMPUTER SERVICES

Developmental Services

A project team which includes a user does the investigation, design, and implementation of the project. The team might include one or two programmer/analyst trainees who program from well-defined specifications, a lead analyst who is responsible for the project design and management, and a member from the user's department.

According to Shelby Tunmire, the programmer/analyst role requires those individuals to have a wide range of ability and knowledge. In doing the analytical work they must be able to interview the users to determine what is needed. Shelby feels programmer/analysts must be trained to ask the questions that will obtain the required information. Programmer/analysts learn to do this by receiving on-the-job training and working with their project leader. The team members must also become knowledgeable regarding the application, be able to convert the needs identified into program specifications, and be able

The computer operator monitors both the IBM 4341 and the IBM 4381. The hardware in the computer center includes two 2000 line per minute printers, 12 tape prices, and a substantial amount of online disk storage.

to communicate effectively. Team members must also be familiar with the control software, utilities, file structures, and interfaces that make it possible to build applications that can be incorporated into the organization's system.

Tools used by the project team include planning worksheets, data flow diagrams, hierarchy charts, and pseudocode. Standards for project planning, design, implementation, documentation, and the project audit have been developed inhouse. Ron Jones, the project leader for the TASCO project, indicated that it is difficult to remember when structured analysis, design, and programming have not been used. When some of the old programs are reviewed and modified, it is hard for him to believe that at one time those unstructured, disorganized programs were considered acceptable.

Walkthroughs are conducted to review the design of the system, the program specifications, the detailed logic plans, and the source code. The source code is written in standard ANSI COBOL. Once the application is completely tested and documented, it is released to production through Administrative Services.

Administrative Services

Administrative Services is responsible for putting jobs into production. This group is responsible for publishing the documentation, updating the documentation as applications are upgraded, providing security for the hardware, software, and online files. The department's security officer and the individuals who perform the administrative functions are members of this group.

Production Services

Once an application is considered operational, it is turned over to Production Services. Batch jobs, interactive computing, and network activities are supported 24 hours a day, five days a week. The computer center is staffed by 6 full-time and 2 part-time operators.

The Computer Network   The map constructed by Wilbert Curtis shows the communication network that will be established. Different colored pins are used so that at a glance Wilbert can tell which installations are operational, the ones currently being installed, and the ones to be installed in each of the final phases of the project.

Although only approximately one half of the XTs are in place, the impact of their utilization is already felt. According to Tunmire, the system has resulted in savings due to a decrease in the number of phone calls and in the number of printed reports. The IBM XT personal computers can be used as stand-alone systems or as terminals. Users can take advantage of the

A data flow diagram is used by Ron Jones, the project leader, to illustrate the second phase of the TASCO project to other members of the team

electronic mail facilities and make inquiries into the various data files.

Data Entry   Production services also include the centralized data entry department. The keypunches and verifiers that once dominated the data entry function have either been replaced by key-to-diskette recorders or terminals. As applications are streamlined, the key-to-diskette records will be replaced by terminals located within the user's area and operated by members of the user's department.

Methods and Procedures   Our host Shelby Tunmire, is Manager of the Methods and Procedures Department. Shelby started at TFB 17 years ago as a programmer trainee. He advanced to programmer, analyst, lead analyst, and to Manager of Development Services. At one time he also managed Production Services.

The mission statement and job description for his present position indicate he is responsible for the flow of information throughout the organization. In studying the procedures that make up the various information systems, Shelby must determine which procedures should be streamlined. In improving both manuals and computerized procedures, he will interface with management, users, and members of the other three groups that make up Computer Services. In order to effectively carry out his mission, Shelby must maintain a broad perspective of TFB and its goals and objectives of a more comprehensive management information system. Since standards and guidelines have been developed and are being followed, Shelby's role will be to make existing systems more efficient and cost effective and to further refine existing standards and guidelines.

Wilbert Curtis, the network supervisor, explains the communication network being established that will connect 200 IBM XTs located in the 200 county offices to TFB's system.

In his role as Manager of Methods and Procedures Shelby Tunmire interfaces with personnel from all other departments and studies the procedures used to input data, process data and utilize information.

A typical IBM XT workstation includes the computer with its builtin 10 MB hard disk and diskettes drives, a printer and a modem.

PCs WITHIN TFB

A decision was made to use the IBM XT for both standalone EDP functions and word processing and as an intelligent terminal to communicate with the organization's mainframes. The XTs are used by departments located in the Waco centralized office facility and in the remote county farm bureau offices.

A training facility with six workstations has been developed. New users receive instruction in the use of the hardware, the use of passwords, and in the software they will be using.

Under the leadership of Computer Services personnel, a user's group has been established. The purpose of this group is to share ideas in the use of the IBM PC and it's software.

SUMMARY

The Texas Farm Bureau's 30 year history of centralized data processing started with the utilization of electromechanical equipment and at the present time includes the use of two fourth-generation computers that are networked, along with IBM XTs, to form a comprehensive system.

As technical advancements are made, existing guidelines, standards, and procedures are reviewed. Within Computer Services, modern structured analysis and design tools are used by the programmer/analysts to investigate, design, and implement new applications.

In Section V we will see how one of TFB's newer applications was designed and how it functions as an operational system.

8 GENERAL DESIGN SPECIFICATIONS

Looking Ahead

After reading the text and completing the learning activities you will be able to:

- Determine the general guidelines that must be considered in preparing design specifications.
- Determine the major requirements of a computer system obtained to run transaction processing applications.
- Describe why it is important to develop modular systems.
- Identify who should be on a project team charged with the responsibility of developing general design specifications and tell why a project leader should be appointed.
- Identify the tasks that must be performed before a PERT chart can be developed.
- Explain how using a PERT chart to identify the critical path can determine the shortest period of time in which a project can be completed.
- Identify the differences that exist between the feasibility study report and the report for the general design phase of the project.
- List some of the reasons a request for a new mainframe or for minicomputers or microcomputers might be made.
- Determine what must be done prior to developing an RFB.
- List the items that might be included in the response to an RFB received from a vendor.
- Identify the reasons why any new computers should be able to communicate with the existing mainframe.
- Identify the advantages and disadvantages of downloading some of the EDP applications to microcomputers.
- Identify the standards that should be developed regarding the acquisition of microcomputers.
- Define and utilize the words and phrases listed in the end-of-chapter glossary.

INTRODUCTION

When the feasibility study report is submitted, various alternatives are presented and the members of the project team recommend the solution they feel is most advantageous to the organization. The computer policy committee studies the report and endorses one of the recommendations. If none of the alternatives seems desirable, the team is asked to gather additional data and to formulate a new proposal.

The proposal recommended by the project team may be rejected because the costs are higher than the amount budgeted for EDP. Or the department requesting the study may feel the proposed system is excellent but that there are insufficient funds to implement the recommendation.

If the team's recommendations are approved, the project is scheduled for the next phase. The feasibility study report establishes major parameters for the design phase. In this phase, the general or overall specifications for the system are developed. If the computer policy committee approves the alternative that proposes the inhouse development of a new batch sales-order system, a major parameter and constraint is imposed upon the team doing the design of the system.

The project team that completes the feasibility study will usually be assigned to design the system since the team members are fully aware of the shortcomings of the present system and of the objectives developed for the new system.

The team may be charged with the selection of a new computer system or may need to design a system that can be implemented on the existing computer. If distributed data processing concepts are endorsed by the firm, a mini- or a microcomputer might be acquired that would meet the specifications developed during the design phase.

There should be relatively few restrictions upon the team's activities. As the team members begin to design the system they may need to call upon additional resource people in order to explore new ideas. The analyst must not go into the design phase with the idea that the investigation is over and all that remains is to put the pieces together. There may be pieces that are missing—the exploration phase is not over.

Management and the user must be involved in this phase of the project. As the various pieces, determined by the specifications developed during this phase of the project, begin to fit together, the user must be consulted. Since the design must be approved by the user for whom the system is being designed, it is easier to gain final approval if the user is directly involved as each piece (or module) of the system is approved and fit into place.

When the general design is accepted, all major commitments have been made and there should be no major changes made during the detailed design phase. The user and mangement must be aware of this policy so that they realize the seriousness of the commitments made during this phase of the project. It is important that the design meets the specifications developed for the system and that the objectives for the system be achieved when the system becomes operational.

During the investigation phase, the old system's problems were identified and alternatives that could solve the problems were identified. What must be done was fairly well determined. The design phase determines how the recommended alternative will be implemented. During this phase, the framework for the system is developed and all major commitments are made.

At the end of this phase, a formal report will once again be presented. A very important part of the report will be the specifications for completing the detailed design and for implementing the system.

If the cost estimates developed during the investigation phase were fairly accurate and the analyst assigned to the design phase worked closely with management and the user, there is little chance that the project will not be approved and completed.

GUIDELINES USED IN DEVELOPING GENERAL DESIGN SPECIFICATIONS

Sometimes analysts feel that the system being developed is *their* system. This is untrue: the only reason for having analysts, programmers, operators, data entry operators, and a CIS department is to meet the needs of management and the end user. The general design stage should be oriented to the user, who must be directly involved and committed to the system being developed.

Generally it is the end user who must pay for the system out of the department's budget. It is also the department's personnel who must prepare the source documents for processing and who must live with the output from the system. The function of the CIS department is to provide the link between the generation of the input and the use of the resulting output.

How the data is processed is usually of no concern to the user or to management as long as the general guidelines identified in Figure 8-1 are followed. Each of the guidelines is important, and it would be difficult to rank each in order of importance. Some of these guidelines will be explained in the following sections of the text.

- Keep service to the user foremost in mind.

- Consider the impact of the new system on the total environment.

- Consider the systems and procedures with which the new system will interact.

- Provide for new applications and expansion.

- Develop systems that are independent of the organization.

- Eliminate the need for the informal organization.

- Automate routine, repetitive manual functions and determine the proper level of automation.

- Integrate systems but avoid undue complexity.

- Develop modular systems and programs.

- Develop clear, explicit specification that illustrate how the system will be viewed by the user.

- Build adequate controls into the system so that data is protected and processed accurately, and valid output is produced.

- Provide output in time to influence the transaction or similar transactions that occur in the immediate future.

- Include the correct amount of detail based upon the person for whom the report is designed.

- Provide the capabilities of obtaining ad hoc reports upon demand.

- Determine that the benefits derived from the system exceed the costs associated with the system.

FIGURE 8.1 Guidelines for developing general design specifications

Keep Service to the End User Foremost in Mind

An analyst should constantly strive to provide the best possible service and to anticipate the requests of users. For example, the credit manager might have requested a report that lists all customers with overdue accounts who have not made a payment within the last three months. A poor analyst would do no more that determine the input, output, and processing specifications and then turn the project over to a programmer. What would you do?

A good analyst would immediately see that a flexible program could be developed. Depending on what information was requested, the report could list: customers with overdue accounts; customers who had not used their accounts within a given period of time; or customers who have indicated dissatisfaction with CPI's, products or service. The list of options could be almost endless. A limiting factor could, however, be the amount of information that is maintained for each customer in the online database. The codes entered as variable data would determine the selection criteria used in determining what information and which customers would be listed on a report.

Good analysts attempt to determine the future needs of the user in order to build a more flexible system that will provide the most useful information. The cost of initially developing the system so that it can be expanded to include additional options must be contrasted with developing a system that meets the company's immediate needs plus the cost of modifying the system at some future date.

Anticipate the System's Effect Upon Its Environment

During the design phase, it is necessary to determine the impact the new system will have on its environment. The term *environment* is used in a broad sense. All changes that will be made in personnel, in the assignment of personnel, and in the physical surroundings of the workers must be considered. The changes must be anticipated during the general design phase so that the detailed design and the implementation of the project can be planned.

When the new system is operational, the physical surrounding must be ready and the personnel realigned. Changes affecting personnel should be planned well in advance so that any necessary training programs can be completed before the system is operational. The employees should also be kept informed of the changes that will occur.

The Impact of One System on Another

In determining the design for a new system, its relationship to other information and communication systems must be determined. For example, a new sales system cannot be designed without having a direct impact on some of the programs and procedures that make up the purchasing, inventory, and accounts receivable systems. For this reason, the project team may have members from the sales, accounts receivable, and inventory departments. The interrelationships that exist between systems become more important as systems become more integrated and share databases and other resources.

Provide for New Applications and Growth

Selecting The Computer System   In designing a system, the short-, intermediate-, and long-range needs regarding hardware and software must be considered. Figure 8-2 illustrates the hardware requirements for transaction processing. The computer system selected must be able to be expanded to meet future needs. If the computer selected initially operates at 90 percent capacity, it should probably be considered inadequate. As the company's volume of business expands and new applications are added, the computer system will be unable to cope with the additional data. If the system selected is one of a family, it is possible to upgrade to the next size computer. Usually only minor changes will need to be made in existing programs. If a microcomputer is being used, there may be no way to expand the system to meet the additional demands.

Requirements For Transaction Processing

- Sufficient CPU time to support additional applications.

- Ability to support a sufficient amount of online storage for existing and future applications.

- Ability to support enough terminals for the required online applications and queries.

- Availability of a query language.

- Good response time.

- Communication capabilities that enable the organization's computers to communicate with each other.

- Capability of printing the required documents and report.

FIGURE 8.2 Requirements of a computer system used for transaction processing applications

Volume of Data to Be Processed   When designing a system, the volume of data to be handled is an important factor. When requests for bids are sent out, one of the factors included is the size of the records and the number of records needed for the applications to be implemented on the computer system. The analyst, by studying past growth patterns and the economic trends for the industry, can project the future volume of data. The new system should be able to handle the volume of data projected for the next five to ten years.

Estimating CPU Time Needed   Before a new application can be designed for an existing computer system, the amount of CPU time needed should be estimated. The analyst must make certain that when the application becomes operational the required memory, CPU time, disk storage space, and terminals will be available. If there is not a sufficient amount of CPU time to process the data, the present system can be upgraded, a new system obtained, or some of the work can be **downloaded** to microcomputers.

Online Storage Space Available   Another limiting factor may be the amount of online disk storage, memory, or both that can be added to the system. When a computer is obtained, there is a maximum amount of online storage that can be configured with the computer system. However, when the hardware is initially selected, far less than the maximum amount of online storage is usually obtained. For a while it may seem that there is an almost endless amount of storage space available. Yet how quickly the space is used up! Unless guidelines are developed for the retention and deletion of information in online files, the existing space is often used up much faster than was anticipated.

Adding Terminals   A *port* is the entrance into a computer that may be used for the transmission of data to and from the computer. Usually only a given number of ports are available. An organization's present computer system might have a controller that supports 64 ports. However, it might be possible to add 32 more to the system. If more than a total of 96 ports are required, it may be necessary to obtain either more memory or a new computer. Another option might be to obtain microcomputers or a minicomputer system so that some of the work of the mainframe can be downloaded to the smaller systems.

Along with having the ability to add enough terminals to meet the needs of additional applications, users must have good **response time**. Response time is the length of time that it takes the computer to reply to a query or to process data. Sometimes adding additional **real memory** (memory within the CPU) or faster secondary storage devices will take care of the problem. During peak times, the users of one timesharing system had to wait almost two minutes for the computer to respond to a request for information or to process data. Figures were displayed at the bottom of the screen and the user was told the computer was working on the problem.

Although what is considered good response time differs depending on the application, many analysts feel that response time should be no more than one or two seconds. Research has indicated that programmer productivity increases when response time drops to .3 seconds.

The Computer's Ability to Communicate   When replacing an existing mainframe or obtaining a smaller computer system that will be used for distributed processing or for related activities such as word processing, the computer's ability to communicate with other computers within the organization is essential. The literature describing a computer designed for distributed processing will always indicate what computers it can communicate with. Hardware and software are available that make it possible to link most computers into a network. However, if the computers are not really compatible, it is more costly and may decrease the efficiency of the entire network.

Availability of Printers   Today because of the way systems are designed, there is less demand for printed reports. However, the trend today is to have several slow, letter-quality printers located throughout the organization and one or more high-speed line printers located within the computer center. It is important that there is a sufficient number of printers to keep up with the demand for printed reports. Workstations that include printers should be positioned so they are convenient to use for entering data and retrieving information.

Anticipating Problems   A problem regarding response time, lack of ports, or inadequate file space should be anticipated long before it occurs. A solution should be determined and a request made for the necessary additional hardware or software.

Make the System Independent of the Existing Organization

Although the organizational chart is studied to determine the information flow and the various areas of responsibility, the system should be designed so that it is independent of the existing organization. The system should be flexible. If the organization changes, for example, three existing departments are merged and a new department added, the reorganization should have very little impact on the information and communication systems. Each operational system should be reviewed and evaluated to determine what changes are needed to meet the requirements of the revised organization. If the systems have been well designed, only a few modifications should be necessary. It should not be necessary to redesign any of the major systems.

Avoid the Need for the Informal Organization

The informal organization is the way that the various procedures and tasks are performed. The formal organization is the one described in the procedures manual or other types of documentation. If the tasks to be performed have been properly analyzed and the new system carefully designed to meet the objectives, there should be no need for an informal organization. As new procedures are designed, they should be tested before they become part of the formal organization. If modifications must be made, the documentation for the procedure must be changed before the change is considered operational.

Modifications to systems and procedures are considered normal and are needed due to external or internal factors. If the documentation is updated to reflect the change and all individuals working with the system are aware of the changes, there should be little difference between the formal and the informal organization.

Automating Functions

Routine, Repetitive Functions Routine and repetitive functions that involve a substantial amount of time should be considered for automation. For instance, assume that every day a clerk in the sales department checks each incoming order to see whether it was submitted by a new customer or by a customer who exceeded his or her credit limit. The clerk must look up each customer on a master list in order to determine the customer's credit status.

By redesigning the sales system, both functions can be automated. The sales order form can be redesigned so that the sales representative submitting the order can check the appropriate place on the form to indicate the order is from a new customer. When the orders are opened, orders from new customers can be sent directly to the credit department. When an online sales-order system is developed, orders from new customers will still need to be submitted along with an application for credit to the credit department.

The program that edits the sales transaction information can be expanded to provide a credit check. Orders from customers who exceed their credit limit will be listed and will not be processed until the credit manager approves increasing their credit limits.

At one time, as orders were received, a clerk checked the item number and description of the product with a master listing. In an online system, this function is easily automated. As the item number is entered, the product description is displayed. The sales representative visually checks the description with the one listed on the sales order. If the two do not agree, the problem can be solved immediately and either the correct item number is entered or the description on the order corrected.

When manual functions are automated, each should meet the test of maximum service provided for the least cost. The method used must provide the most for the dollars invested. When manual functions are automated, customers and users must receive better service and more information rather than less service and information. When the statement is made "we had more information before we went to computers," the system probably was poorly designed.

Proper Level of Automation The guideline of establishing proper levels of automation is directly related to the ones that advise automating functions and integrating systems. What is the proper level of automation? Are there functions that should not be automated?

When computers were first used, management often had to be convinced that the equipment could perform various tasks more efficiently than individuals could. The computer was often considered little more than a status symbol, and management did not really think in terms of using computers to solve problems, to process data, and to provide a fast, reliable information retrieval system. There were all sorts of horror stories and articles published about colossal errors generated by the computer (usually the fault of the user, analyst, programmer, or data entry operator).

Today the computer is looked upon quite differently. Management, accountants, users, and the general population expect computers to solve all sorts of problems. However, the analyst designing a system must realize that there are some functions that should be done manually. For instance, an inventory control system could be designed to automatically generate purchase orders, a purchase register, and an exception report. The record for each item in the inventory file would need to contain the reorder point, the name of the vendor from whom the item is to be purchased, and a great deal of additional data. When the quantity on hand falls below the reorder point, data for the exception report would be stored in the appropriate file. The orders, purchase register, and exception report would be printed during the second shift. The order forms could be designed in such a way that after being printed they would be ready to be mailed. The orders would be untouched by human hands prior to their being sent on their way.

Is this the proper level of automation? For the inventory system to be effective the data in the master file would need to be updated to reflect the current demands for items in stock. The vendor's data would also need to be reviewed and updated to make changes in the price, the quantity to reorder, the shipping specifications, and the source of supply.

Management might feel that such a system is overautomated. Perhaps a better level of automation would be to print an expanded inventory exception report that would be reviewed by the purchasing agent. Items to be ordered as indicated on the report would be checked and the necessary adjustments made to the rest of the items. A terminal would be used to enter a minimal amount of data, usually only the item number, which would cause the desired purchase orders and the purchase register to be printed.

Here is another example of automation. A company sent out a questionnaire to 100 people, 30 of whom responded. The company does not have a program that will provide the statistical test that is desired on the data and it is unlikely that the required test will be used for future surveys. Management submitted a request to have a program written that will analyze the data and perform the statistical test. Is this the proper level of automation?

Since the program will not be used in the future and because there is little data, the computer policy committee should deny the request. It would be less expensive to use a calculator and to type the report. If the program is written, the programmer will make the identical computations with a calculator in order to test the program.

Integrate Systems but Avoid Complexity

Although systems should be integrated, the design of the individual programs that make up the system must be kept simple. As a general rule, systems can be integrated and share common online databases. For example, a common database could be used for payroll, personnel, and job-cost accounting functions.

The programs accessing the database should be internally documented, menu driven, and user friendly. To users, programmers, computer operators, and data entry operators, the system should seem simple and easy to understand. The complexity of the database software that manages and protects the integrated database should not be apparent to the individuals who use or maintain it (users, data entry operators, management, and programmers).

In order to avoid undue complexity, each program that accesses the database and makes up the system should:

- not depend on the internal processing of another program.
- accomplish only one major function.
- be divided into modules that accomplish only one task.
- provide a clear audit trail.
- be easy to maintain.

An example of a program that provides only one major function is a program that adds new employees to the payroll database. The section of the program that edits the data entered by the operator might be called the edit module—the task is editing the data. The list of the employees added to the database is part of the audit trail.

Programs that perform only one major function are easier to maintain than those that perform several major functions. At one time it was considered a challenge to write a program that performed several major functions and was so complex that other programmers had difficulty following the logic. Today such programs are rejected by CIS managers because they are often unreliable and difficult to maintain.

Develop Modular Systems

When a system is *modular* it can be designed, programmed, tested, and documented module by module. The system is first divided into procedures. Usually each procedure requires one or more computer programs. Each computer program is then divided into modules. Since each module is fairly independent of the others, a module can be removed from a program, or modified, without seriously affecting how the other modules will function.

Modular systems composed of modular programs are easier to design, implement, test, and maintain than nonmodular systems. Although a given program may be very large, when a modular approach is used, the program is divided into small, easy-to-manage modules that can be completed in a short period of time. The progress of the programmer can be charted, and management can see the progress being made.

Develop Clear and Explicit Specifications

During the general design phase, clear and explicit specifications must be developed. The specifications should not contain minute detail but should be complete enough so that the individuals who do the detailed design work know exactly what must be done. Macro or general specifications must be developed for the entire system, the procedures within the system, and the tasks that make up the procedures. The following indicates how the specifications for the system, procedures, and tasks vary:

System	Establish objectives for the total system.
Procedure	Describe what each component of the system will accomplish.
Task	Explicit specifications must be developed for both manual and computerized functions.

The specifications should indicate what computer programs are needed to process input, what each program will accomplish, and what controls are needed.

The analyst must also determine where and how data will be entered into the system. Equally important is how the data will be verified. Depending on where and how the data is entered, it could be key verified or it could be edited by the computer as it is entered into the computerized system. Most systems require the use of both **internal** and **external controls**. An internal control is an edit routine written by a programmer or provided by the operating system. An example would be a check to make certain that only numeric data is entered into a field designated as numeric. An external control is a procedure used to check the validity of the output. For example, batch totals of regular and overtime hours established by the payroll clerk are checked with the regular and overtime totals accumulated by the payroll edit program that determines the validity of payroll transaction data.

The specifications must also indicate the documentation that is required to meet the needs of management, the CIS department, and the end user. General specifications are also developed regarding how each procedure and the entire system is to be tested, audited, and maintained. Often after a system is considered operational, the maintenance is performed by programmers and/or analysts who specialize in modifying existing programs.

General guidelines should be established for documenting, testing, and evaluating procedures and systems. Specific guidelines that apply only to a particular system or procedure may also be added to the general guidelines. However, all members of the computer information services department should be familiar with the general guidelines and *must abide by them*.

CIS departments that are successful consider as essential:

- the development of clearly defined standards for all areas of systems analysts and design; and
- the requirement of following established guidelines and standards.

Standards and guidelines must be realistic. If audits indicate that the standards cannot be achieved, the standards must be revised so that they can be attained.

CHECKPOINT

1. What are the differences between the feasibility study phase and the general design phase of a project?
2. After the report for the detailed design phase is accepted, why might it be considered as a point of no return?
3. Should all of the investigation be ended when the design phase begins?
4. Are the users and management an integral part of the design phase?
5. Management has indicated that there can be no additional hardware or software obtained for the new sales-order system. What must the analyst determine about the present computer system before designing an online transaction system?
6. Refer to Figure 8-1 and for each situation described determine which of the guidelines to be considered in developing general design specifications was ignored.
   a. Specifications were provided for one large computer program that would provide all of the necessary processing of the sales data.
   b. When the specifications for the sales system were developed, the sales manager was not consulted.
   c. The specifications for the new sales system indicated that a clerk in the sales department would look up each customer's balance and credit limit to determine whether the credit limit had been exceeded.

    d. The analyst determined that if the sales system is implemented, there is sufficient capacity on the computer system being used as long as the volume of data is not increased or additional features are not added to the applications.

    e. The analyst in charge of the project has indicated that "since employees never follow the documentation, there is no need to develop documentation for the users."

    f. The analyst did not consult with the accounts receivable manager or the inventory manager.

    g. The new sales system is designed so that each time a customer exceeds his or her credit limit, the order is returned to the customer with a form letter that indicates the order cannot be processed. The credit manager is to receive a list of people to whom the form letter is sent.

    h. When the new sales system is implemented, three fewer data entry operators will be needed. Management has not been informed that fewer data entry operators will be needed.

    i. When the sales orders are processed, the availability of the item is not checked. An inventory shortage is not discovered until the shipping department determines there are not enough items on hand to fill the order.

PROJECT ORGANIZATION

The project team that completed the feasibility study will probably be assigned to the general design phase. Ideally, the team should be expanded to include individuals from other areas who will be directly affected by the new system. The sales-order system may have been investigated by two analysts and a member of the sales department. The representative from the sales department is probably from middle management and it is unlikely that he or she is totally committed to the project since other responsibilities will still need to be fulfilled. When the design phase begins, it would be wise to involve someone from inventory control and someone from accounts receivable.

The analysts will do most of the design work, but there must be ongoing communication between the analysts and the sales department, inventory department, and accounts receivable department. If members of these departments are directly involved in the project, there will probably be better communication and less resistance to change.

The Project Leader

Any time more than one person is involved in a project, there must be a leader. The leader must have authority to delegate assignments and must assume the responsibility for the project. Unless there is a recognized leader, the line of authority and responsibility is not established. Everyone even remotely involved with the project must know whom to contact in case there are questions.

If there is no leader, or if the leader does not have managerial skills, it is unlikely that the project will be completed according to schedule. A good leader must see that a workable plan is developed and then measure the progress of the project against the plan. It is important to determine which tasks can be completed simultaneously and which ones must be completed before other tasks can be started.

If a large team is involved in the design of the system, the CIS manager will appoint a leader and an assistant leader. The assistant must always be aware of the status of the project and must be authorized to make decisions in the absence of the leader. It is not wise to have co-leaders because no single line of authority is established.

Controlling the Design Phase

After studying the feasibility report, the team is ready to plan the design phase. The same principles that were followed in completing the feasibility study are once again followed:

- tasks are identified and broken into manageable units;
- tasks are ordered in the sequence in which they must be performed;
- the time needed to perform each task is estimated; and
- a planning chart is constructed.

The tasks that will be required to complete the detailed design must be determined *prior* to the completion of the planning chart since some tasks can be completed simultaneously.

Figure 8-3 illustrates the list of tasks that must be accomplished during the design phase of the word processing system. Since only one analyst is assigned to the project and the project is fairly simple, the PERT chart is less complex than the ones developed for larger systems. When a large, complex system is being developed, a PERT chart (or planning chart) for the entire system can be developed that only shows major activities and the sequence in which they must be completed. Each major activity will have its own planning chart. Figure 8-4 illustrates the PERT chart developed after the activities involved in the detailed design phase of the project were determined.

Team members can be individually assigned to specific tasks. Team sessions will be held to review each procedure and information will need to be shared. One analyst might be designing a procedure to produce a file used as input to a procedure being developed by another analyst. The second analyst will need to know the characteristics of the file designed by the first analyst and must know what information is stored in each record.

PERT Charts   PERT (Program Evaluation Review Technique) is a planning and analysis tool that illustrates the relationships between the tasks that must be performed. PERT is used to create a master plan for the control of complex projects. The technique was developed in the late 1950s by the United States Navy, Lockheed Aircraft Corporation, and the consulting firm of Booz, Allen, and Hamilton. PERT was used to estimate the shortest period of time required to complete the Polaris missile project. Since that time PERT has been widely used for both civilian and military projects.

Two key terms used in describing a PERT chart are explained below:

Activity   Application of time and resources to achieve an objective. It is measured in units of time, usually weeks, and is represented on the chart by an arrow. The arrows on a PERT chart are similar to the bars on a Gantt chart. However, the length of time is not proportionate to the length of the arrow.

Event   A point in time at which an activity begins or ends. It is represented on the chart by a circle.

Tasks Required to Complete the Design Phase of the Word Processing Project

1. Determine the overall objectives for the system.

2. Determine the procedures that make up the system.

3. Determine the specific objectives for each procedure.

4. Develop precise specifications for each procedure.

5. Review the specifications for each procedure with the members of the team and informally with management.

6. Test the proposed design.

7. Estimate the length of time needed to do the detailed design work.

8. Estimate the length of time needed to implement the system.

9. Determine the tentative schedule for the detailed design and implementation phases of the project.

10. Review, and revise if necessary, the cost estimates.

11. Review the office layout and prepare the new office layout.

12. Review the new office layout with the sales manager and the executive assistant.

13. Review, and revise if necessary, the personnel requirements and realignments.

14. Prepare the documentation and the final report for the design of the project.

15. Prepare the visual aids to be used for the presentation.

16. Make the presentation.

FIGURE 8.3 Activities to be accomplished during the design phase

Once events and activities have been determined a mathematical formula can be used to calculate the shortest period of time in which a project can be completed. First, the time needed to complete any activity must be calculated. This is done using the following formula.

$$\text{Time} = \frac{O + 4M + P}{6}$$

$$\frac{1 + 4(2) + 3}{6} = \frac{12}{6} = 2$$

O   The optimistic time estimate. How long it will take if everything goes well.

M   The most likely estimate. The normal time such activities take.

P   The pessimistic estimate. How long the activity should take under adverse conditions.

After the time for each activity is figured, the relationships between the various activities must be ascertained so that the critical path can be determined. The **critical path** is the sum of all expected activity times along the longest path leading to an event. The project will be finished on time providing each activity is completed on time. It may be completed on time if one activity takes longer to complete while two or three others take less time.

The PERT network shows how the tasks will be done by displaying the sequence in which activities must occur if specified events are to be reached.

Critical Path A PERT chart can be used to analyze any project that is well defined and broken into modules. Before a critical path can be determined, the way in which the modules are interconnected and the length of time needed to complete each event must be determined. Many of the calculations are based upon estimates (the optimistic, the most likely, and the pessimistic times). When used as a planning tool by individuals who have had experience in completing the tasks that make up a project, a PERT chart is very effective and the project will usually be completed according to schedule. Most texts on the principles of management provide detailed information regarding the development and use of PERT as a management tool.

Although the calculations have not been shown, double lines have been used in Figure 8-4 below between 1-3-4-11-12-14-16 to illustrate the critical path. If there is sufficient personnel to perform the events simultaneously, the project should take as long as the events along the critical path. All of the analysts will be involved in tasks 1 (determining the system's objectives, 11 (reviewing the procedures), 12 (testing the design), and 16 (completing the documentation and preparing the final report). In addition, the following individual assignments might have been made:

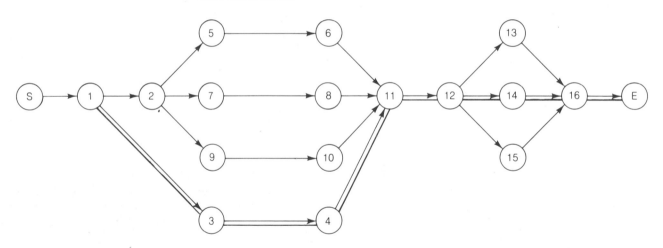

Event	Description
S	Start design phase
1	Determine system objectives
2	Determine required procedures
3	Prepare new office layout
4	Review new layout
5	Develop objectives for the creation, maintenance and security of files
6	Develop specifications for file procedures
7	Develop objectives for the procedure that creates the letters
8	Develop specifications for the letter procedure

Event	Description
9	Develop objectives for the address sticker procedure
10	Develop specifications for the sticker procedure
11	Review of all procedures
12	Test design
13	Determine schedule for next two phases
14	Review cost estimates
15	Review personnel estimates
16	Complete documentation and final report
E	End design phase

FIGURE 8.4 PERT chart for the word processing system

Analyst	Assigned to Complete
1	3,4
2	2,9,10,15
3	2,5,6,13
4	2,7,8,14

If only one analyst were assigned to a project, or if the CIS department had only one programmer/analyst, the critical path would be the sum of the times needed to complete each event.

Adding Time and Activities to a PERT   A PERT chart is easier to use and understand if it is labeled and the time taken to complete each task is shown. Figure 8-5 illustrates a detailed PERT chart that might have been developed for a program designed to display prompts telling the sales representatives what data is to be entered and to edit the data entered. The program specifications would have been developed during the general design phase of the project. The planning chart illustrated includes tasks that would be done during both the detailed design and implementation phases of a project.

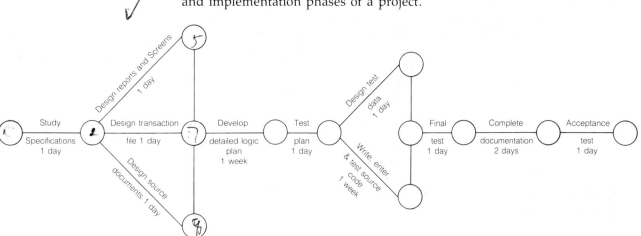

FIGURE 8.5   Adding time and activity to a PERT chart

Or you might assume the chart shown in Figure 8-5 was developed by a programmer/analyst who was given general design specifications and told to complete the detailed design, develop a detailed logic plan for the program, write the code, test the program, and complete the documentation. The program might be one added to an existing system, and a programmer trainee might be assigned to work with the programmer/analyst. Management would like the program operational as soon as possible and has asked how long it will take. Developing a PERT or Gantt chart will enable the programmer/analyst to answer the question.

As you look over the chart you might be unfamiliar with some of the terminology. A **detailed logic plan** is a step-by-step analysis of exactly how the computer will process the data. The logic plan should be modular. If the program is modular, the programmer/analyst and programmer trainee can both be writing, entering, and testing source code. Testing the design is referred to as a design **walkthrough**—programmers ''play computer'' and see what will occur as different records are entered and processed. Omissions or incorrect logic should be detected at this point. During the **acceptance test**, the users will follow the directions provided in the documentation, enter data, and work with the output. Each of these concepts—designing programs, testing the design, and acceptance testing, will be covered in greater detail in future chapters.

DESIGN PRINCIPLES   If **structured design** guidelines are followed, the procedure shown in Figure 8-6 will be followed. Although the major constraints such as the personnel, time, money, and computer system dedicated to the project must be considered, the analysts should be encouraged to be creative and to explore new methods. After each, an evaluation should be made to determine whether it meets the stated objectives. The user *should* also be consulted to see whether additional features are needed to meet the problems identified.

Structured Analysis and Design

Phase*		Task
FS	1.	Define the major goals and objectives for the system.
GD	2.	Divide system into subsystems and procedures.
GD	3.	Determine objectives and specifications for each procedure.
GD	4.	Divide procedures into small, manageable tasks.
GD	5.	Determine objectives and specifications for each task.
DD	6.	Develop detailed specifications for each program required. Files, screen, and reports are designed.
DD	7.	Divide each program into modules. Each module performs one major function.
DD	8.	Develop a detailed logic plan for each program.

* FS, Feasibility Study; GD, General Design; DD, Detailed Design.

FIGURE 8.6 Steps in the development of a structured system

In designing the individual procedures, the analyst should not be tied to the old system. The analyst must be creative and devise a method that is better than the old methods—both the computerized and manual methods must be more efficient and cost-effective. By studying the old system, its problems and shortcomings can be ascertained. If this is not done, the new system with its new methods may have the same old problems. In fact, if the proper level of automation has not been achieved, more problems might be created than solved by a poorly designed system.

Because each system is different, there is no cookbook solution that can be turned to listing "ten easy ways to design a computerized system." The analyst must gather data, analyze the data, and then based on observations, study of existing materials, brainstorming sessions, and working with the users come up with a solution that will solve the problem. The advantage of having a team assigned to the project is that collectively there is more experience to bring into the decision than if one individual were to design the solution. Also walkthroughs can be conducted so that each element of the design can be tested.

PROCEDURES NEEDED FOR THE WORD PROCESSING SYSTEM   The PERT chart, illustrated in Figure 8-4 identifies the tasks that must be performed in completing the word processing project. Figure 8-7 illustrates a **hierarchy chart** that might have been created to identify the four procedures that make up the system and the major tasks associated with each procedure. The hierarchy chart provides an overview of the entire system and includes both manual and computerized procedures.

Incoming letters are processed by writing the codes of the form paragraphs that can be used directly on the letter. If information not available in a file is required, the executive assistant will use a dictaphone and dictate any variable data that must be added to the form paragraphs to provide a response to the customer's letter. An alternative that might have been considered is to use voice input to create a file that could be reviewed and edited by the word processing technician and then incorporated into the letter responding to a complaint or request for information.

Before any letters are created, maintenance may be needed on the customer database. It may be necessary to add records for new customers, delete records for customers who no longer use their accounts, or make corrections in the information stored within a customer's record. The files that contain the form paragraphs may also need some maintenance. New text can be added and old text can be modified or changed. In performing the required maintenance, the database software may be used to maintain the customer database and the word processing software for maintaining the text files.

A word processing technician identifies the customer to whom the letter is to be sent, enters the codes for the form paragraphs that can be used, and keys in any variable data that is recorded on a voice-recording machine. A file is created that contains the complete letter (name, address, date, personalized salutation, and complimentary close). A separate file is also created that will contain the names and addresses that will be printed on **continuous-form** envelopes. Although the envelopes are printed in a separate procedure, they will be in the same sequence as the letters.

The final procedure is once again a manual one. Since continuous-form paper was used for printing the letters, the letters must be separated, the pin-feed strips along the edge of the paper removed, and the letters folded and inserted in the proper envelopes. A machine can be used to remove the pin-feed strips and to separate and stack the letters. Care must be taken that the proper letter is placed in the correct envelope.

If you were the analyst, you might have shown how it would be more efficient to use window envelopes. However, the sales manager objected to the use of window envelopes on the basis that printed envelopes are more personal and the extra cost of using continuous-form envelopes could be justified on the basis of the intangible benefits derived from a more personalized letter.

The hierarchy chart shows the operational procedures—what will occur on a day-to-day basis after the system becomes operational. During the implementation phase of the study, the text master file will need to be created. Since the existing customer database can be used for obtaining the required names and addresses, only routine maintenance will need to be done to make certain that all of the required addresses are in the file. Although the documentation for the operating system and the preparation of the material in the standards manual is not done during the design phase, the length of time needed to complete these tasks needs to be determined.

Also the selection of the employees to be trained for the newly-created word processing technician positions should be completed as soon as possible. The technicians must be trained prior to the system becoming operational. The personnel department should work with the analyst in determining the job descriptions, pay classifications, and amount of training needed.

Although it is not necessary to construct a hierarchy chart, it is necessary to divide the system into procedures. However, the chart provides a good graphic

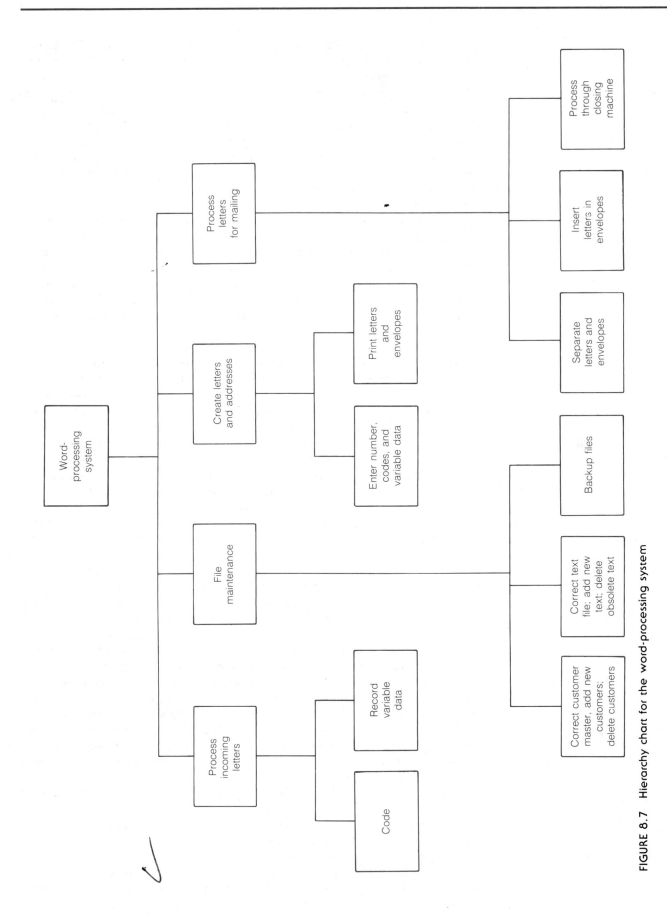

FIGURE 8.7 Hierarchy chart for the word-processing system

representation of the system. As the analyst begins to define the objectives and specifications for each task within a procedure, the chart might be revised to include additional tasks.

DESIGN REVIEW

Analysts, programmers, and students sometimes object to the number of steps and review activities involved in the analysis and design of a system or program. However, it has been statistically shown that if more time is spent on the analysis and design phases of a project, less time will be needed in the implementation phase. If omissions, duplications, logic errors, and other bugs are detected during the general design phase, corrections will be far less costly than if the errors are detected after the system is operational.

Unfortunately, some companies do not require the analysis and design phases of a project to be reviewed. Larger, more progressive companies with well-established departments insist upon reviews. In a walkthrough, the design is tested by members of the team. Each team member pretends that his or her data is following the flow outlined in the design. Bad data as well as valid data flows through the system. Each team member tries to determine whether valid results will be produced.

PREPARING THE GENERAL DESIGN PHASE REPORT

The format of the general design phase report will be similar to the one developed during the feasibility study. Figures CS8-1 through CS8-3 on pages 542 to 544 in Appendix illustrate a general design report that might have been developed for the CPI sales-order system.

The report should be prepared from the users' point of view. Enough information should be provided regarding the general design of the system to enable the user to determine whether or not the new design will solve the problems identified, will be relatively easy to implement, will be more cost-effective than the former method, and will provide the information needed to make the necessary management decisions.

Both a written description and a systems flowchart should be provided. The major objectives of the proposed system should be listed. Although the objectives may be similar to the ones listed on the feasibility study report, additional objectives may be included and the original objectives may be more specific.

A brief description should be provided for each of the procedures (automated and manual) identified on the hierarchy chart. The macro (or overall) design for each required program may also be given. The method used to present the macro logic plan should be consistent with the standards developed by the CIS department.

The projected time schedule for the completion of the project should be fairly accurate. Since the general design is completed, the cost figures should be more precise than the ones developed as part of the feasibility study report. Also included should be more precise costs for the day-to-day operation of the system.

Although the report is similar to the one illustrated for the feasibility study, investigation did continue throughout the entire general design phase. Also time and cost estimates that are more accurate and more detail should be provided.

The users should have been directly involved in the general design phase and should have approved each of the major items that will be included in the report. Unless users are in agreement with the report, it should not be presented to the computer policy committee. However, if users and management have been directly involved, the report will reflect their requirements and suggestions. When the report is presented, it should receive their full support.

The documentation should include the supportive materials used in making the decisions required to divide the system into manageable procedures and in developing the macro logic and specifications for each procedure. It might include memos, printed reports, articles, vendor's specifications for either hardware or software products, and a summary of phone calls made or received, as well as other items.

CHECKPOINT

7. Why is it a good idea to include someone from the sales, accounts receivable, and inventory departments on the project team responsible for designing the new sales system?

8. If the project team only has two members, why should one member still be appointed the leader?

9. What must be done before a PERT chart can be developed?

10. What must be done before the critical path can be calculated?

11. What is an activity and what is an event?

12. What are the basic concepts of structured design?

13. What is a hierarchy chart?

14. Why should the PERT chart and hierarchy chart be reviewed before developing specifications for each of the procedures?

15. The content of the report prepared at the end of feasibility study and the one prepared at the end of the general design phase are very similar. Summarize the major differences between the reports.

16. If good practices have been followed in the general design phase, why should the analysts feel their report will be approved?

SELECTING HARDWARE AND SOFTWARE

Selecting a new computer system (hardware and software), additional hardware for an existing system, or additional software is not an easy task. There are many options available and many good products on the market. The decision to obtain a computer may be based on any of the following reasons:

- Management of a small company with very few employees feels that acquiring a microcomputer system will solve some of the company's problems. Company data is currently being processed manually or by a service bureau. For example, a small construction company might be manually computing job estimates and maintaining all records for each construction job. These two applications alone could justify the acquisition of a microcomputer system.

- A medium-or large-size company finds it is running out of computer time and new applications cannot be added to their present mainframe. Or so many additional online systems have been developed that the response time for the users of a timesharing system is very slow. The company investigated adding more memory, file space, and terminals but found that a new system with greater capabilities could be obtained for not much more than the cost of their present system plus the cost of the additional components needed to make it adequate for their present needs.

- A large company with a centralized computer system and database decides that satellite systems should be placed in various departments so that some of the data processing functions can be distributed. For example, the sales department might want to maintain its own database, input the data to print the sales invoices, and generate some of its own reports. When this situation occurs, a major concern will be to make certain the new equipment can communicate with the existing computer in order to share resources. The sales department's condensed data must be added to the corporate database.

- A department within the company has a unique application that requires the use of a small computer system. It seems advisable to secure a small standalone system that can be dedicated to the application. For example, a large manufacturing company wants to develop a computerized maintenance system that can keep track of when a machine was repaired and who repaired the machine. In addition, the location of each repairperson and the assignment each is working on is to be available at all times. This particular application will not create data that is needed for other applications. Therefore, a computer dedicated to the maintenance application may be the best solution.

- When completing the feasibility study, the project team determined there was not enough CPU time on the present system to implement a new application. The project team must investigate the possibility of obtaining a new centralized computer system, upgrading the present computer system, or obtaining one or more small, standalone computer systems. The additional computers could be networked to the present system and share some of the mainframe's resources.

Steps in Selecting a Computer System

Because more alternatives are available, making a decision regarding either hardware or software is more difficult than it was in the past. A medium- or large-size company usually has been through the process and has established guidelines. A small company selecting its first computer may have no one within the organization who has been involved in the process. Management within such a company may find it very difficult to make a decision.

Often a feasibility study is made for the sole purpose of determining whether a new computer system should be obtained. A project team that includes professional CIS staff and representatives from the computer policy committee and from top management may be formulated to make the study. If the investigation indicates that there are applications which cannot be implemented on the present system or that additional computers are needed for distributed applications, further investigation must be made and the steps outlined below should be followed:

1. Determine the applications for which the computer will be used. It will make a difference if the computer is to be used by the art department to create graphics, by the engineering department for CAD/CAM, or by the payroll department for gathering and processing payroll data.

2. Determine the major requirements for each application. If the computer is to be used primarily within a department for batch processing, the specifications will be different than if an online transaction system is to be developed.

3. Determine the volume of data that will be processed for each application. A major concern is the number and size of the records that must be maintained in the online database. The number of transactions that will be entered on an hour-to-hour and peak-hour basis is also a major factor to consider.

4. Determine the software specifications. Each system should be studied and divided into procedures. The software required for each procedure should be determined.

5. Convert the requirements for each application, the volume of data to be processed and the software specifications, into hardware specifications. Different hardware will be needed for transaction processing of records originated throughout the world than will be needed to process the direct labor and material costs for a department. Forms are available that can be used to consolidate into a meaningful format all of the data that has been gathered regarding the application to be implemented.

6. Evaluate the data. There is sometimes a tendency for users to overestimate the requirements of their applications. For example, the sales department might feel that ten terminals are needed to input transaction data and maintain the database. Analysts who have studied the volume of data and the standards available regarding the amount of data that can be entered by an operator, might feel that five terminals will be more than adequate.

7. Work with vendors in obtaining additional information. At this point, it is an informal request for information and for specifications regarding some of their hardware and software.

8. Evaluate the data obtained from vendors and other outside sources such as *DataPro Report*.

9. Develop format **requests for bids**. The requests will list the required hardware and software specifications. The RFBs should be available to any vendor who wishes to submit a proposal. Some vendors will bid on the entire system—hardware and software—while others will only bid on selected hardware items or some of the software.

10. Evaluate the bids received from vendors. Again this is a very difficult task.

11. Ask for demonstrations or presentations from the vendors who seem to offer hardware or software that meet the specifications, have established reputations for quality products and service, and are within the cost figures that have tentatively been established for an entire computer system or for the hardware or software specified.

12. Enter into contract negotiations and select the system that offers the most for the dollars invested in the system. At one time, the bids received were the final price. However, because of the competitive nature of the computing industry, many companies are willing to enter into price negotiations.

Determining Need

An application should need a system rather than a system needing an application. Before a company considers what type of computer system to acquire, the reasons for obtaining the system should be identified. Specifications must be determined for the applications to be computerized in order to determine the type of system that is needed.

A few years ago a system might have been obtained and then the CIS department would determine what applications would be developed. Since management is more knowledgeable regarding applications well suited to computerization, this is no longer true. Many companies have far more requests for "computer power" than they can fill. However, some of the applications for which there are requests cannot be cost-justified.

Determining the Type of Computer System Required

Once the needs are identified, a decision should be made regarding the type of computer system required. There are many different options available. The options often include: obtaining one large centralized computer; developing a network of microcomputers or minicomputers; obtaining standalone microcomputers or minicomputers; and obtaining a large mainframe that acts as the host computer for numerous microcomputer or minicomputer systems.

The specifications for the system—hardware and software—cannot be developed until the needs and type of system required are determined. Although the specifications for the hardware and software should be developed separately, when the selection of the computer system occurs, the two must be considered as a unit.

Hardware Specifications

Precise hardware specifications should be developed so that all companies are submitting bids on computer systems that are comparable. A vendor might submit a bid on one type of computer system and also indicate some of the other options available. The RFB will usually specify the:

- amount of real memory needed;
- amount of auxiliary storage needed;
- lines per minute that the printer should be able to print; and
- the specific requirements for each I/O device specified.

If cards are to be used, the number of cards per minute should be specified for both the card reader and the card punch. When magnetic tape drives are to be considered, the CPI (characters per inch) that can be stored on the tape and the IPS (inches per second) that can be read should be stated. In addition to the speed of the printer, the type of printer that is desired should be specified. For each I/O device there are precise requirements that can be specified. When VDTs are required, the width of the lines required, the number of lines, and whether a monitor that displays text and graphics in color is required should be specified.

Some applications will also have unique requirements. Banks, for example, often list as a basic requirement a **MICR** (magnetic ink character recognition reader). Grocery stores would include a UPC (universal product code) scanner. An insurance company or magazine publisher that sends out a large number of renewal notices might want some type of scanner that can read **turnaround documents**

When the company begins working with a vendor, the vendor will want to know the anticipated volume of data to be processed. A large vendor may conduct its own needs analysis in order to recommend the most efficient configuration of hardware and software. A vendor does not want to recommend a larger system than is needed since the price may not be competitive. If too small a system is configured, the user may be unhappy because of poor response time and the time required to run some of the batch jobs.

The configuration selected may fall somewhere between the recommendations of the vendor and the specifications of the project team. Both recommendations may need to be tempered when the harsh realities of the budget constraints for the initial and ongoing support for the system are considered.

Software Specifications

The cost of the software can often exceed the cost of the hardware. Also there is a great deal of difference in the way vendors charge for control software. Prior to July 1, 1969, when IBM **unbundled** (separated the cost of services and software from the cost of hardware), the cost of a computer included the CPU and a basic operating system. Today some companies include the operating system with the price of the CPU, while others charge separately for the operating system. When writing the specifications for software, the required features must be specified.

Today most mainframes and minicomputers are **virtual systems (VS)**. Under a virtual system, a program is divided into segments (sometimes called *pages*) by the control software. Only the segment that contains the commands to be executed at any one point in time needs to reside in main memory. Segments of programs swap back and forth between real and virtual memory (VM is usually on magnetic disk memory). In a VM system, real memory is expanded to include the VM memory. Also most basic operating systems provide for multiprogramming—two or more jobs running concurrently—and for timesharing.

Features other than those that provide for multiprogramming and timesharing may not be included in the basic operating system. Since unbundling occurred, some companies charge an additional amount for utility programs, compilers, database software, and software that provides for networking computers and for telecommunications.

Although the security measures built into the system may be extremely difficult to evaluate, it is critical that both online data and programs be protected from accidental destruction and unauthorized use. One of the functions of an operating system is to protect files from being accessed by hackers and those who intend to commit a criminal act—obtaining information with the intent of selling it, changing programs so that funds or property can be obtained illegally, or stealing computer time.

When one college obtained a new computer system it took a couple of knowledgeable students only one day to "break the system." The students could get into any of the accounts and obtain any information they wished. For each way the students could get into unauthorized areas of real memory or files, the system programmer had to devise a way to prevent it from occurring. Within a matter of months, the defects in the operating system had been corrected.

When evaluating software, a checklist must be developed that indicates the features that are required. Although there are publications, such as the *DataPro Reports*, that evaluate operating system and other control software, it is harder to evaluate software than hardware. For each item on the checklist, it should be determined whether the item is included at no additional cost or must be purchased or leased.

Gathering Data   Once the specifications have been developed, data must be gathered about the computer systems that have the desired characteristics. Today there are a great many choices available. For example, one *DataPro* feature report compared 249 small business systems from 89 vendors. At the present time there are over 150 microcomputer vendors. In the medium- and large-size markets there are substantially fewer vendors. However, each vendor may have more than one computer system that meets the user's specifications. The geographical location of the company selecting the computer may influence the choice somewhat. In selecting a system, one of the guidelines used might be to make certain that a maintenance center is within a 30-mile radius.

Vendors will usually make formal presentations. They may suggest that the client attend conferences or workshops that highlight the system being considered. Visits should be made to other installations that have the same type of system being considered. Although it may not be feasible to do so, a typical mix of programs should be **benchmarked** on the computer systems being considered.

In a benchmark, a typical mix of programs is run on different computers and the results evaluated. The mix should include programs that have a large number of calculations as well as programs that require a great deal of I/O activity. The mix should require the use of tables, files, and sort/merge procedures. The same mix should be used on each system being considered and the results compared. Any difference in the length of time that the different systems took in processing a given amount of data should be noted.

When a benchmark is being run in an interactive mode, the computer being evaluated should have the same number of terminals in use as would be if the system were obtained by the potential buyer. Response time and the time needed to complete batch jobs may be entirely different if only one or two applications are being run than if 80 users are simultaneously accessing a system designed for a maximum of 96 online terminals.

**Information Received in
Response to RFBs**

In response to requests for bids for a complete system (CPU, I/O devices, and control software) a vendor will usually submit a report that includes:

- a summary
- cost sheets for the entire system
- a description of the hardware on which the bid is based
- a description of the control software needed
- a description of application software
- a conversion plan
- staff requirements
- educational programs for management, users, and selected members of the CIS staff
- maintenance agreement
- alternatives for financing the system
- terms and conditions of the contract
- optional bid items.

The optional bid items include either hardware or software that the vendor was not asked to bid on. However, in configuring the system, the vendor realizes that the additional hardware or software will be needed.

There is sometimes a great deal of difference in the maintenance contracts. Under some contracts only the period of time identified as the "normal working day"—from 8:00 A.M. to 5:00 P.M.—is covered. Other vendors have a longer period of time and sometimes include maintenance on Saturday or Sunday. Normally maintenance contracts are carried on mainframes and minicomputers. If the computer systems are leased, maintenance is included.

Maintenance contracts are normally not carried on microcomputers. Also unless an organization intends to purchase a large number of microcomputers, RFBs are not developed and made available to vendors.

Evaluating the Data

In analyzing data regarding the systems that meet the user's specifications, there are a number of points that must be considered. One very important issue is the reputation of the vendor. How do the users in the area feel about the service and support they receive from the vendor? Also when a call is made for service, how long does it take the vendor to respond to the call?

The performance of the vendor's hardware and software is also an important issue. Some companies keep records regarding the number of times the computer is "down" and a customer engineer had to be called to solve the problem. In some contracts, a penalty is imposed on the vendor if the system is down for more than a certin percentage of the time—sometimes the figure is as low as 1 percent.

One of the problems that can occur is when the hardware is from one vendor and a major control software package is from another vendor. Neither vendor wants to take responsibility for a problem that it is difficult to identify as being caused by defective hardware or software. When a call is made for service, how long does it take for the vendor to respond to the call?

In regard to service, some vendors maintain a phone diagnostic service. Within the mainframe is located a separate small computer that can communicate a problem over a regular telephone line to the vendor's diagnostic computer. Sometimes the precise problem cannot be identified and additional research must be done to solve the problem. As soon as the cause is determined and the solution found, the information is added to the vendor's diagnostic computer's database.

The computer systems that are being seriously considered should be compared on factors such as the amount of memory, type of memory, internal calculation speed, and methods used for storing data on auxiliary storage devices. Another important factor is the I/O devices available and how easily they can be added to the system. Two key issues are the maximum amount of real memory that can be added to the system and the maximum amount of online secondary storage that can be supported.

A potential buyer should also determine whether the computer being considered makes use of the newest technology available. In some cases, a model that is a little older but still well supported by the vendor might provide more computer power for the dollar than one that is considered on the "leading edge of technology."

Any special environmental conditions that are required must also be determined. These costs—such as additional wiring or air conditioning—must be considered when determining the cost of the computer system.

After all the data is evaluated, a decision must be made regarding what hardware and software is to be obtained. Microcomputers and minicomputers can often be purchased "off the shelf." The computer is available as soon as a contract is signed. When obtaining a large mainframe, the computer might not be available for several months after the order is placed.

Purchasing Versus Leasing

One of the innovative marketing ideas that IBM introduced was to permit users to lease rather than purchase equipment. It was often easier to sell **unit-record equipment** or a computer if the user did not have to invest capital. Few companies returned equipment, but if a company was unhappy with the equipment, it could be returned more easily under a leasing agreement than if the equipment had been purchased.

Medium- or large-size computer systems are usually leased rather than purchased. Since a relatively small cash outlay is needed, microcomputer or minicomputer systems are usually purchased.

The organizations's tax consultant should help make the decision regarding whether to lease or to buy a large computer system. Two of the major considerations are the tax advantages and disadvantages of both leasing and purchasing, and the present position in regard to the organization's cash flow and investments.

If a company intends to keep a computer for four or five years, it is usually less expensive to buy than to lease one. If the past is any indication of what may occur in the future, many companies will not keep a computer system that long. Because the computing power obtained for the dollars invested increases each year along with the computational speed of computers, justification can be shown for changing computer systems in order to obtain the advancements in technology. However, the cost of converting to a new computer must also be considered.

Third Party Versus Original Vendor

Sometimes equipment can be obtained more cheaply from a third party than from the original vendor. If obtaining a system from a third party is being considered, the project team must find out what guarantees, if any, are included. The project team must also determine what ongoing support will be available for the hardware and software.

If a company is willing to settle for less than the very latest technological advancements, some excellent buys may be available from a third party. The available systems may have been obtained by the third party because they were inadequate to meet the needs of the original user or because they are larger, more powerful systems than the original user needed.

CHECKPOINT

17. For what reasons should a study be conducted to determine whether a new system should be acquired?

18. Does the project team assigned to detemine the specifications for a new computer system and to assist in the selection of the computer system have the same type of membership as the team assigned to design a new sales-order system?

19. What is the first step that must be taken when the selection of a new computer system is begun?

20. What must be done before an RFB can be developed?

21. After the RFBs have been given to the vendors, what information might be included in a vendor's proposal in response to the RFB?

22. Do all vendors treat the cost of a basic operating system for a computer in the same manner?

23. What is a benchmark?

24. What additional factor should be considered when a timesharing system is being benchmarked?

25. If a company has more than one computer system, why should the systems be able to communicate with each other?

26. Why might a company elect to obtain a computer that has been on the market for five years rather than one that has just been announced?

27. Who besides the project team and the computer policy committee should be directly involved in the decision to buy or lease the computer system?

28. Why are computer systems sometimes obtained from a third party rather than the original vendor?

ESTABLISHING MICROCOMPUTER GUIDELINES

The use of microcomputers within a large organization has stirred up a great deal of controversy. Some reports indicate that the microcomputer has become a status symbol and often sits unused on an executive's desk. Other reports suggest that

The Invasion of Microcomputers

the use of microcomputers is more cost-effective than using minicomputers or mainframes. Also implied is that microcomputers are the cure-all for every problem!

While the software for microcomputers has been billed as being "user friendly," individuals with little or no background in working with computers have experienced varying degrees of difficulty in using the software. You may also have read that some companies have purchased as many as 1,000 computers in order to give each executive his or her personal computer. We have also been led to believe that the integration of microcomputers into the office environment has been totally disorganized. In the many reports that you may have read, what is truth and what is fiction? Where do microcomputers belong within the organization?

Results of Survey

A report commissioned by the National Association of Accountants and based on a survey of 240 individuals in 100 companies indicated that 81 percent of the companies had a formal policy regarding the purchase of microcomputers. Seventy-four percent had microcomputer training programs for employees, and 73 percent included microcomputers in the *integrated plan for the acquisition and use of computers*. Fifty-nine percent of the companies indicated that their computers were part of a network. The microcomputers either communicated with other microcomputers or with a centralized mainframe. The following indicates applications assigned to microcomputers:

Percentage	Application
80	budget forecasting
70	word processing
47	standalone accounting
36	mailing lists
28	cash management
23	timesharing

Advantages of Using Microcomputers

Microcomputers can be used to **download** some of the routine data processing functions. When this is done, mainframes can be used for more important jobs that require massive amounts of data, calculations, and decisions. Data can be transferred into the microcomputer's files, and processed, and printed output can be available immediately. Also using the microcomputer's ability to function as an intelligent terminal frees communication lines. Data is downloaded to a microcomputer. When all of the data is processed, the microcomputer transmits the updated data back to the centralized computer system.

Microcomputers are extremely portable. Although they require a reasonable amount of care such as being used in a "clean" environment—one with little static, dust, or dirt—special air conditioning and wiring is not needed. There is a great deal of good software available that can be used to keep track of budget information, project earnings, generate mailing lists, and so forth. The three most widely used programs provide for word processing, spreadsheets, and database management.

When a department obtains microcomputers, members of the department are often encouraged to develop additional applications. If a formal request had been made to computerize the application, it may have received such a low priority that the application would never be implemented. End users also learn more about computers and have a better understanding of what must be considered in designing new applications.

Disadvantages of Using Microcomputers

The microcomputer industry is going through what the computer industry experienced 30 years ago—a lack of standardization. There are around 150 different companies that manufacture microcomputers. Each of these companies may have several different models. Each model may use one of several operating systems. Often files created using one microcomputer cannot be processed on a different type of microcomputer. Also files created under one operating system cannot be used on a computer that has a different operating system. If you create a file on a double-sided disk, your friend cannot use the file on his single-sided drive.

Although the cost of a standalone microcomputer system (perhaps $3,000) seems minor, the cost must be multiplied by perhaps 100 times or more—the cost of obtaining systems for all executives and individuals who expressed a desire and a need. Also a not-so-hidden extra is the cost of the software. Software such as spreadsheet and word processing programs is often licensed for *a given computer*. The cost of basic software, such as the operating system and programs for word processing, spreadsheets, and database management, might be an additional $1,000 to $2,000 per computer. Maintenance is another factor. It is true that many individuals have had excellent service from their microcomputers, while others have had difficulty getting their computers operational. A maintenance contract usually runs about 1 percent of the cost price of the computer system per month.

Often a little knowledge is dangerous. Individuals without expertise are developing software that processes data. Are the results valid? Were the programs completely tested? Can the output produced be incorporated into the company's database?

Developing Standards

Standards or guidelines need to be developed in the following areas:

- *Justification.* What applications should be assigned to microcomputers and which ones should be assigned to larger computers? Who really needs his or her own microcomputer system?
- *Product selection.* Some companies have established rules that must be followed. Often only one or two types of microcomputers are approved.
- *Compatibility* Any microcomputer must be able to communicate with the mainframe and with the other microcomputers already obtained by the company.
- *Procurement.* Will the CIS department handle the acquisition of microcomputers in the same way as any other EDP equipment or will each department obtain its own computers?
- *Implementation.* When the computer arrives, will the CIS department be responsible for installing and checking out the system?
- *Training.* Will the CIS department be responsible for setting up inhouse training programs on use of the various software packages or will each department set up its own training programs?
- *Maintenance and supplies.* Will the routine maintenance of the equipment be done by CIS employees or will each department provide its own maintenance such as alignment and cleaning of the read/write heads?

- *Programming.* Will the department which obtains the microcomputers do its own program modification and programming? Often patches have to be put into an operating system to support a particular I/O device or an application program must be modified.
- *Control.* Who is responsible for seeing that established guidelines and procedures are followed? Is the security officer who is responsible for the safety of hardware, software, and information stored in files also responsible for the microcomputers?
- *General education.* The CIS manager is responsible for seeing that the CIS staff remains current regarding computer technology. Should the CIS manager also provide general education sessions on new microcomputer products and applications?

The guidelines established in each of the ten areas identified must provide for a conscious and planned integration of microcomputers into the organization. Some companies, under the computer information services concept, have established standards and a "help desk" for microcomputer users who wish to develop some of their own applications. As indicated by the survey and many other articles, guidelines and standards regarding the use of microcomputers have been and are being established. Each application must be evaluated to see whether it is more cost-effective and beneficial to use microcomputers or mainframes.

Just as testing and documentation of applications run on larger computers is a critical part of systems analysis and design, rigid testing and documentation standards must be followed in developing applications for microcomputer systems.

SUMMARY   After the feasibility study is reviewed, the tasks required to complete the general design phase of the project will be listed. The team must also determine the time needed to complete each task and which tasks can be performed concurrently. A PERT chart may be constructed that can be used to determine the least amount of time needed to complete the project.

After the general design phase has been accepted, all major commitments have been made. A commitment is made to a batch, online batch, or transaction processing system. A hierarchy chart may be prepared illustrating both the manual and computerized procedures that are required. The project team should also determine what databases, files, and reports are needed as well as where and how transactions are to be entered.

Although the detailed logic for each program is not determined, specifications and the macro logic for each program are developed. The report for the general design phase will include more detail and more clearly defined objectives than did the feasibility study report. The report for the detailed design phase will also include more precise estimates of the time, materials, personnel, and other costs needed to complete the study or to provide ongoing support for the system. The documentation submitted along with the report will indicate what additional data has been gathered, how the data was analyzed, and provide support for any additional recommendations.

No major changes should be made after the general design for the system is accepted. If the design is not acceptable, or if the projected costs are too high, the project team may be asked to modify the design.

If a substantial investment is to be made for new equipment or for new software, a special task force may be appointed to gather data and to make recommendations. The applications for which the hardware or software is to be used must be determined first. Next the volume of data to be processed and the specifications for the reports and records stored in the database or file is determined. After detailed specifications are determined, an RFB is prepared. After bids are submitted and reviewed and all available data is analyzed, a decision must be made regarding what hardware and software is to be obtained. Computers may be benchmarked to provide additional data regarding the capabilities of the systems. The company may elect to purchase or to lease the equipment either from a vendor or from a third party.

The thought that must be constantly kept in mind is that computer systems are selected *after* the applications for which they will be used are determined; applications should not be designed to conform to the characteristics of the computer system that was acquired *before* a needs survey was completed. Also although the analysts would like to have a new computer system, one of the constraints imposed on the design may be that no additional equipment can be obtained.

DISCUSSION QUESTIONS

1. Why should a system be divided into procedures and procedures into manageable tasks?

2. Refer to Figure 8-1 on page 257 and explain why each of the guidelines should be considered when developing specifications for a computerized system.

3. Refer back to Chapter 3 and review the information regarding the Gantt chart. What are the major differences between Gantt and PERT planning charts? If you were the CIS manager, which would you recommend that your employees use? Explain why you would have your employees use either the Gantt or PERT chart.

3. Why should no *major changes* be made in the design of the system after the general design report and recommendations are approved by the computer policy committee?

4. In controlling a project, of what value is determining the critical path?

5. What are the differences and similarities that exist between the feasibility study report and the general design phase report?

6. When the general design phase report was given for the sales order system, the sales manager indicated that "The design would not meet the objectives developed by the sales department." What would you do to determine why the sales manager did not support the recommendation? If you elected to call the project leader in for an informal discussion regarding the problem, what questions would you ask and what points would you be sure to include in the discussion?

7. The president of a small corporation stated "After we get our new computer, we will decide what applications should be computerized." If the president was a friend of yours, what advice would you give him? If he has not yet determined what type of computer system to purchase or lease, what would you suggest that he do?

8. What must be done before an RFB can be prepared and made available to vendors? What type of information would you expect to receive from vendors who submitted bids?

9. The president of your company has indicated the present computer system is not adequate. He further indicates that he will contact IBM to find out what type of computer is needed. In the past he has always relied on IBM recommendations and believes that is the only way to determine what type of computer system is needed. You feel that the present system never met the needs of the company. What suggestions would you make to the president?

10. A friend of yours has indicated that hardware is far more difficult to evaluate than software. Your friend also indicates there is no way to evaluate software—you just take what goes with the system! How would you respond to your friend's comments and what might you suggest that he investigate?

11. In one of your classes you overheard a student say that "The operating system, as well as educational classes and printed materials, are always included in the price of the computer." The student also stated that "it is far better to buy a computer than to lease one." How would you respond to the two statements?

12. Review the general design report for the sales-order system. Also review the information presented in Figure 8-1. After studying the report and the information presented in Figure 8-1, what guidelines seem to have been included in the general design of the sales-order system?

Team or Individual Projects

1. After reviewing the material in the text regarding microcomputers and the information presented in the Chapter 8 case study in Appendix I, prepare a report that might be included as an exhibit for the general design report which would include the following items:

   a. Rationale for providing each sales representative with a device that will function as an intelligent terminal. Indicate how each sales representative will use the terminal in order to obtain information, submit orders, in communicating with other sales representatives, or in doing some of the routine activities (submitting reports, keeping track of expenses, and preparing budgets).

   b. Rationale for using microcomputers that conform with the established guidelines as the intelligent terminals. Referring back to a, indicate for which applications the microcomputers would be used as a standalone system and for which functions it would be used as a terminal to communicate with the host computer and the centralized online databases.

   c. A recommendation regarding which microcomputer you would select. Include in your description of the microcomputer selected the amount of memory it has, the I/O devices that you think should be included, and the major software packages that you feel should be obtained.

   d. Prepare an outline of the presentation that you would make in regard to your recommendation to use microcomputers. You were told that some of the computer policy committee members favored the use of "dumb" (nonprogrammable) terminals. Indicate what visual aids you might use when presenting your portion of the report to the committee.

Directions:

In order to complete the assignment, you will need to research the microcomputer systems that are available and determine which ones can be used as part of a network. You will also need to determine what type of software is available for the computer you select and indicate how the software would be used by the sales representatives. In preparing your report you might wish to talk to people directly involved in EDP to determine their views regarding the use of microcomputers as intelligent terminals.

GLOSSARY OF WORDS AND PHRASES

acceptance test Programs and procedures are tested by the users before being accepted as operational. The users input the data and follow the documentation provided for each procedure.

activity When used in describing a PERT chart, an activity is the application of time and resources to achieve an objective.

benchmark Determining the efficiency of a computer when processing different types of data. Different types of jobs constituting a typical mix of programs are run on more than one computer system and the results are compared to determine which system produced the best result.

continuous forms Letterhead stationery, envelopes, or forms that are connected and have pin-feed holes that keep the forms aligned in a high-speed or letter-quality printer. The forms, letters, or envelopes are printed and then separated for distribution.

critical path The path leading to the objective that involves the greatest length of time. All of the expected activity times along the path are added and the sum is used to represent the shortest period of time in which a project can be completed.

detailed logic plan A well-defined course of action that can be followed by the computer to process data.

download Applications usually run on a mainframe are transferred, along with the required data, to a microcomputer system. The mainframe can then be assigned more complex tasks for which a microcomputer might not be able to be used.

event When used in describing a PERT chart, an event is the point in time at which an activity begins or ends.

external control The validity or security of the data is determined by the data control clerk or by an accountant. For example, the monthly payroll distribution report must balance with the combined total of the weekly payroll registers. An accountant does a spreadsheet and checks the totals.

hierarchy chart A chart which shows the relationship between the various procedures that make up a system.

internal control A check on the accuracy of the input or output that is performed by the program processing the data or by the operating system.

MICR An acronym for magnetic ink character recognition equipment. MICR readers are used offline to sort checks and online to read data. The characters to be read are written in magnetic ink on the documents.

real memory Memory located within a computer.

response time The length of time that it takes the computer to process data or a request for information entered from a terminal.

request for bids (RFB) Vendors are sent the specifications for a computer system, for software, or for a specific piece of equipment. Each vendor is asked to submit a bid. The bid describes the hardware or software, the maintenance agreements available, and the educational programs available for employees.

structured design A set of guidelines and techniques that assist an analyst in determining which procedures, interconnected in which ways, will best solve a stated problem. Structured design is also top-down—start with the system, divide it into procedures, and then divide the procedures into small, manageable tasks.

turnaround document A printed form or punched card produced as output that is later used as input. Typical examples are utility bills and insurance statements.

unit record equipment Electromechanical equipment controlled by a wired panel used to process data punched into the standard 80-column card.

unbundled The term used to indicate that the price of a computer was separated from the cost of maintenance, control software, utility programs, printed material, and educational programs.

virtual system (VS) Programs and data are divided into segments or pages. When a program is being run, only the pages containing the instructions being executed need to reside within real memory. Pages of data and instructions are moved back and forth between real memory and some type of fast-access secondary storage.

walkthrough A term used to denote the review of the design of a system, procedure, or program. In a program design walkthrough, the logic of the program is tested.

STUDY GUIDE 8

Name _____

Class _____ Hour _____

A. Indicate if the following statements are true (T) or false (F). If false, indicate in the margin how the statement should be changed to make it true.

_____ 1. During the general design phase, no further investigation is made.

_____ 2. Management and users should not be involved during the design phase.

_____ 3. After the general design report recommendations are accepted, there is still no commitment as to the type of system that will be developed.

_____ 4. During the general design phase of the project, "what" must be done is determined.

_____ 5. In the general design phase, specifications are only determined for automated procedures.

_____ 6. In the general design phase, detailed specifications are determined for each program.

_____ 7. Each procedure should accomplish a number of major tasks. This is what is meant by integration of systems.

_____ 8. The "proper level of automation" only means that there are some procedures that should not be computerized.

_____ 9. In developing the general design for the sales system, other systems such as inventory or accounts receivable should not be considered.

_____ 10. It is better to appoint co-leaders for the project team so that in the event one person is gone, the other person can assume responsibility for the project.

_____ 11. A hierarchy chart only illustrates the procedures that will be computerized.

_____ 12. Today there is a trend to spend less time on the design of a system and more time on the implementation phase.

_____ 13. In selecting a computer system, the first step is to determine what software is needed.

_____ 14. Before the computer selection committee develops an RFB, vendors should be invited to make presentations.

_____ 15. An activity is the point in time at which a task begins or ends.

_____ 16. A walkthrough is conducted to see whether there are errors or omissions in the design of a system or procedure.

_____ 17. The critical path leads from task to task and denotes the shortest pathway through the activities identified on a PERT chart.

_____ 18. PERT charts were first used as a planning device in determining how long it would take to implement a computerized system.

_____ 19. MICR readers can only be used to read data encoded in magnetic ink into a transaction file for processing.

_____ 20. If dial-up terminals are used, a company might have more terminals than ports.

_____ 21. In benchmarking a computer that will be used for distributed and transaction processing, the response time should be evaluated when only one terminal is in use.

_____ 22. In preparing an RFB it is important to determine how many records will need to be online and the size of the records.

_____ 23. The same hardware and software are needed for a system made up of all batch applications and for a system that has transaction processing as well as batch applications.

_____ 24. Most analysts feel that it is always better to lease a computer system than to buy one.

_____ 25. Most large companies have not established guidelines for the acquisition and use of microcomputers.

_____ 26. According to the survey taken to determine how large organizations were using microcomputers, the highest percentages were for maintaining mailing lists and for standalone accounting applications.

_____ 27. Standards developed for the acquistion and use of microcomputers usually specify that it must be possible to network any microcomputers obtained with the organization's mainframe.

_____ 28. Microcomputers can be used to download jobs from the mainframe.

_____ 29. If consideration is being given to obtaining a mainframe or minicomputer system from a third party, consideration must also be given to the guarantees and ongoing support that will be provided.

_____ 30. All vendors have "unbundled" and there is always a separate charge for an operating system.

B. Match the guideline ignored with the descriptions of what occurred during the general design phase of the project. Record the letter of the answer in the space provided. Some guidelines might be used more than once while others are not used.

    a. Clear, explicit specifications that illustrate how the system will be viewed by the user must be developed.
    b. Service to the user must be kept foremost in mind.
    c. Routine, repetitive manual functions should be automated.
    d. Undue complexity must be avoided.
    e. The proper level of automation must be achieved.
    f. The need for an "informal organization" should be eliminated.
    g. The impact of the system on its total environment must be considered.
    h. The system and procedures with which the new system will interact must be considered.
    i. Provision must be made for expansion and growth.
    j. The system must be independent of the existing organization.
    k. The systems and programs must be modular.

_____ 1. A company's payroll system is completely automated. Once the hours are entered, all the reports (including the paychecks) are generated automatically. Frequently people are paid the wrong amounts or paid when they did not work.

_____ 2. An analyst stays in his or her office and designs a "fantastic" payroll system. Although it saves the company money, the payroll manager does not have access to information that was used as part of the old audit trail.

_____ 3. Neither the inventory or accounts receivable managers were consulted when the new sales-order entry system was designed.

_____ 4. The credit check is still done manually.

_____ 5. One very complex program is used to create a file, add records to the file, delete records from the file, correct records, and generate three differenct exception reports.

_____ 6. After the sales-order system was designed and implemented, the sales manager asked who would be responsible for maintaining the files and for entering the orders received over the phone and in the mail into the system.

_____ 7. When the computer arrived it could not be installed because special wiring was needed to provide the required power.

_____ 8. Two months after a computer installed in the payroll department was operational, it was impossible to add more records to the database.

_____ 9. The high-speed printer for the payroll department was installed in the payroll accountant's office.

_____ 10. No documentation was provided for a new system. The analyst indicated people performing the tasks could determine how each task should be completed. The analyst claimed that people never use the documentation or follow the established guidelines.

C. Multiple choice. Record the letter of the correct answer in the space provided.

_____ 1. During the general design phase, specifications are developed
   a. only for the system.
   b. for the system and for the procedures.
   c. for the system, procedures, and programs.

_____ 2. During the general design phase, the factor that must be foremost in mind is making certain
   a. the new system is less costly than the old system.
   b. all procedures are computerized.
   c. the user gets all of the information needed and in a format that is meaningful.

_____ 3. Although systems should be integrated, the analyst
   a. must avoid undue complexity.
   b. must work only with the department that requested the study.
   c. should develop separate files or databases for each application.

_____ 4. When the program that processes the sales data was written, a check was made to see that only numeric data was entered for the quantity. This is an example of
   a. an external control.
   b. an audit trail.
   c. an internal control.
   d. none of the above.

_____ 5. A PERT chart is
   a. used to plan a project.
   b. used to control a project.
   c. used to plan and control a project.

_____ 6. A PERT chart can be constructed
   a. as soon as the tasks to be completed are identified.
   b. as soon as the time needed to complete each task is identified.

     c. after the tasks are identified, the time each takes is determined, and a decision is made regarding which tasks can be completed concurrently.

     d. None of the above answers is complete.

_____ 7. According to the survey, microcomputers are used within a large organization most frequently for

     a. timesharing applications.

     b. cash management and standalone accounting systems.

     c. budget forecasting and word processing.

_____ 8. Microcomputers can

     a. only be networked to other microcomputers.

     b. be networked to other microcomputers and to mainframes.

     c. only be networked if a mainframe is established as the host computer.

_____ 9. The feasibility study report shows what must be done while the general design report

     a. provides very specific details about what must be done.

     b. shows how the required tasks will be accomplished.

     c. gives specific and detailed specifications for the system, procedures, and programs that are needed.

_____ 10. The first thing that must be done in selecting a new computer system is to

     a. identify the amount of file storage that will be needed.

     b. determine what applications will be run on the system.

     c. ask vendors to submit bids.

_____ 11. One of the general guidelines that must be followed when developing specifications during the general design phase is

     a. that only exception reports should be printed.

     b. that all reports should be printed according to a predetermined schedule.

     c. all reports scheduled should be eliminated and all reports printed upon demand.

     d. The correct amount of detail must be included for the level of management that receives the report.

_____ 12. In the word processing application

     a. the database management software was used to maintain all files.

     b. the word processing software was used to make corrections and additions to the form-paragraph text.

     c. all procedures were computerized.

9 DETAILED DESIGN: INPUT AND OUTPUT

Looking Ahead

After reading the text and completing the learning activities you will be able to:

- Identify the factors that should have been determined prior to the detailed design phase.
- Identify the four major tasks that must be completed during the detailed design phase.
- Explain why an analyst starts designing a procedure by determining what output is needed.
- Determine what factors must be considered in determining the output needed, the medium to use, and the format of the report, display, or file.
- Identify the functions performed by a turnaround document.
- Explain how internal and external reports are designed.
- Identify the rules that are followed in designing screens.
- Determine what medium should be used when given a description of the output and how it will be used.
- Identify the reasons management is more concerned about the method used to input data than they were in the past.
- Identify the ways in which data keyed in by a terminal operator can be verified.
- Identify the techniques that should be used in designing source documents.
- Explain the relationship that should exist between the source document, display screen, and transaction record.
- Identify the type of information that might be included on a transmittal form and how the form might be used.
- Explain why and how codes are used in EDP.
- Define and utilize the words and phrases listed in the end-of-chapter glossary.

INTRODUCTION

The amount of work that must be done during the detailed design phase of the project depends upon what was accomplished during the general design phase. If a small project is being designed, the detailed work might have been done prior to the committee evaluation. If that was the case, the next step would be to implement the system.

As a result of the analyst working closely with the user, the feasibility study indicated *what* must be done. During the general design phase, the specifications developed indicate *how* the goals and objectives for the system would be accomplished. During the detailed design phase, the emphasis is placed on developing more detailed specifications as to how the computer system can be used to implement the design.

When large projects are being designed, the system specifications should be reviewed prior to committing the resources needed to do the detailed work. During the detailed design phase, the work is more technical in nature, and management and the users will not be as involved. However, management and users should be kept informed and continue to be involved in decisions that must be made regarding inputting data, processing the data, and outputting the results.

Figure 9-1 illustrates what should have been accomplished prior to the detailed design phase. During the detailed design phase of the project the following tasks will be completed:

The precise format of the output will be determined. Prior to this phase it might have been decided that a sales invoice containing specific information should be printed. Now the exact format of the report will be designed and approved by the user. Or it might have been decided that a screen should be created to display a customer's credit history. In this phase of the project, the screen will

---

Factors Determined Prior To The Detailed Design Phase

- The type of system to be developed—transaction or batch. In a transaction system, some programs may still be run in a batch mode.

- The hardware and software that will be committed to the system. A firm decision must be made regarding the acquisition of additional equipment.

- Resources that will be committed to the development of the system—materials, machines, methods, and personnel.

- Resources needed when the system becomes operational. The resources needed to maintain the system should also be determined.

- Accurate estimates of the time needed to complete the detailed design and implementation phases of the project.

- Specifications for the procedures and tasks needed to accomplish the stated goals and objectives.

- Overview of the computer programs needed.

- Transaction files, master files, and databases needed.

- Methods used to input transaction data.

- Displays, reports, and files created as output.

FIGURE 9.1 Summary of factors that must be determined prior to the detailed design phase of the project

be designed and submitted to the user for approval. Exact specifications and formats need to be determined for every type of output required.

Input formats will also be developed. Source documents should be designed to capture the data. Record layouts should be designed for card, diskette, disk, or tape input files. If terminals are to be used, the hardcopy needed to provide an audit trail and system security must be determined. The methods to be used for verifying variable data must also be decided.

Record layouts will be designed for databases, master files, and transaction files. Prior to this time the medium (disk, tape, or some other type of mass storage medium) should have been determined along with the major specifications. The major specifications would include how the file will be organized, and for each procedure, how the records will be accessed. Now, field by field, the records need to be designed.

Detailed logic plans for all programs must be developed. During the general design phase, the gross specifications for major programs were determined. Now very detailed logic plans must be developed and *tested* for all programs.

Although input and output, files, and procedures will be covered in three separate chapters, you should understand they are interrelated. The starting point is to determine the detailed design for the output—reports, displays, and files. *The output produced dictates the input and file requirements*. However, when the detailed logic plan is developed and control and audit procedures are added to programs, it may be necessary to add additional fields to the transaction files or databases. In a sense it is a cycle—determine output, decide what input is required, and create a detailed logic plan. When the logic plan is determined, additional output may be produced which requires more data to be entered or stored in the online databases.

INPUT/OUTPUT DEVICES AVAILABLE

Analysts must know which I/O devices are available on the computer systems they are using in designing and implementing applications. For each I/O device available on the system, analysts should have an in-depth understanding of the advantages and disadvantages of using the device for inputting, outputting, or storing data.

It has been estimated that approximately half of an analysts's time should be spent in some type of learning activity. In regard to I/O devices, analysts must be aware of the current state of the art regarding each device. At what point should voice input be considered rather than having the sales representative key in the customer's account number, item numbers, quantity ordered and other selected information? Or when should videodisk be considered as a replacement for printed catalogues?

Figures 9-2 and 9-3 illustrate the input media and devices used most frequently. The chart provides the name of the device, the amount of use, its relative speed, and a brief description. The codes used to indicate use and relative speed in Figures 9-2 and 9-3 are:

Use Codes	*Description*
L	Little or limited amount of use. L is assigned when the device is used for unique applications, such as MICR in the banking industry, or when use is still considered experimental.
P	Popular device used for a wide range of applications.
D	Declining use due to improvements in other media and devices.

Medium/Device	Use	Speed	Explanation
Card Reader	D	S	Used to read data, JCL, and source decks
Magnetic tape	D	VF	Should be thought of more as a storage and communication medium rather than as an input medium.
Hard disk	P	VF	Used to store transaction and master file data. Most widely used input medium for mainframes.
Videodisk	L	VF	Stored on videodisk are text, pictures, and graphics; the color reproductions are excellent. Used for educational materials, making presentations, catalogues, and advertising materials.
Diskettes	P	F	Most widely used input medium for microcomputers.
Scanners			
UPC	L	C	Widely used in the retail food industry; limited use for other applications.
OCR	P	S	Typewritten, handwritten, and printed turnaround documents can be read.
OMR	P	S	Widely used for test scoring, inventory, production control, and meter-reading applications.
MICR	L	S	Used by the banking industry to read and sort checks.
Light pens	L	C	Used in CAD/CAM applications for creating graphics and inputting changes for financial graphs and charts.
Microfilm	L	S	When computer-assisted retrieval is used, the computer is used to retrieve and display the data stored on microfilm.
Voice	L	F	Wide range of applications. However, in terms of total data entered for processing, voice input represents a small percentage of the total input.

FIGURE 9.2 Frequently used input media and devices (Page 1)

Relative Speed

VF	Very fast
F	Fast
S	Slow
C	The relative speed of the device depends on a number of variables and cannot be classified unless the conditions being considered are explained in detail.

The same codes are used for the output media and devices illustrated in Figure 9-4. If different individuals were asked to code the I/O media and devices, it is doubtful that all would agree on any one code for any one device. The codes refer only to a medium's use for input or output and reflect what seems to be the current trend.

Medium/Device	Use	Speed	Explanation
Terminals I/O	P	S	Terminals are used a great deal for keying in data that will be processed in an interactive mode. Since transmission into the system is based on the ability of the operator, it is slow and costly.
Cash register	P	C	Used for a wide range of applications.
Badge readers	P	C	Used as a replacement for time clocks; good applications have been developed for their use in payroll and job-cost accounting.
Sensors	P	VF	Used to provide continuous data into a computer for processing. Wide range of applications such as monitoring a patient or controlling production activities.
Input for microcomputers:			The following devices were initially developed for microcomputers but can also be used with minicomputers or mainframes. The use indicated is for microcomputers.
Mouse	P	C	Used to manipulate the cursor and to enter data.
Graphic tablet	P	C	Images created by using the graphic tablet are displayed on the VTD screen and can also be stored in a file. Widely used for creating graphic displays.
Joystick	P	C	Also called *paddles*. Used to produce graphics and also to input data into games and educational programs.
Robo stick	P	C	Used to manipulate the cursor and to create graphics.
Touch-sensitive screen	L	C	Data is entered by touching the screen. Although limited in use, it takes advantage of a natural instinct to point.

FIGURE 9.3 Frequently used input media and devices (Page 2)

Medium/Device	Use	Speed	Explanation
Card punch	D	S	Except for use as turnaround documents, very few applications require punched-card output.
Magnetic tape	P	VF	Tape output is often used to communicate with another computer or offline device, such as a printer or one that produces microfilm.
Hard disk	P	VF	Master files and databases are updated; new transaction files are created. One of the most widely used output media.
Diskettes	P	F	Microcomputers' most frequently used output media
Printers Line	P	S	Line printers are used for printing reports, checks, invoices, and documents.
Workstation	P	S	Many workstations consist of a terminal that has a keyboard, VTD, and character printer. The computer program controls what will be displayed on the VTD and printed by the printer.
Microfilm	L	C	COM can be produced as computer output or prepared in an offline operation from information recorded on magnetic tape.
Voice	S	C	Voice output is being used in an increasing number of applications. Its use may be overlooked when the system is designed.
VDT	P	C	Since database software and query languages have become more widely used, retrieval systems often use a VDT rather than printers to display data. Often the VDT provides the operator with messages and instructions generated by the operating system or with prompts programmed into application software.
Actuators	S	F	As sensors are used to collect and transmit data into the computer, the data is processed and the output is in the form of an instruction to an actuator. Actuators are used in process control, inspection of products, and in robotics.

FIGURE 9.4 Frequently used output media and devices

Although some of the major uses, advantages and disadvantages will be discussed for a few of the I/O devices, it is assumed you are familiar with most of the media and devices identified in the charts. If this is not the case, it is suggested that you spend some time reviewing the major characteristics of various I/O devices.

OUTPUT CONSIDERATIONS

The trend is to make wider use of voice output, displays on VDTs, and disk output, and to eliminate some of the hardcopy previously printed. Several major factors must be considered regarding output. The first step is to determine what output is really needed. External factors, such as legislation and competition, play

Output Needed

a vital role in determining what output must be produced. In regard to the electronic processing of data, the federal income tax guidelines are rather vague since they state only that there must be a clear audit trail.

The question of what output is needed should have been answered during the investigation and general design phase. Analysts learn to challenge the user regarding the output desired since some reports are never used. In the ongoing evaluation of a system, the use of various reports and other types of output should be challenged in order to eliminate unnecessary costs. One CIS manager found that by having many of the reports available upon request, rather than on a periodic basis, the volume of printed data was substantially reduced.

Any system must generate the information that is needed to effectively manage the business. Each level of management has different informational needs that must be considered. While the line officers may only be interested in a report that shows the five-year sales trend by type of product sold, the district sales managers may want detailed reports that show the volume of sales by sales representative, by type of product sold, and data regarding the customers who bought various types of products.

In the final analysis, the major function of any system is to produce relevant, timely information *when needed*.

Use of Output   How the output will be used determines the medium selected and also has a direct impact upon the detailed design phase. The questions that must be answered are: Will the output be visually referenced or will it be used as input for another procedure? Can some type of turnaround document be used? Is the information primarily for internal or external use? How timely must the data be? What output medium should be used?

The answers to the first four questions usually determine what output medium will be used. For example, if an external report is being prepared for the Internal Revenue Service it must either be printed or written on magnetic tape. If the major criterion is the "timeliness of the data," it may be displayed upon request on a VDT.

Visual Data   Data to be referenced visually can be printed, displayed on a VDT, recorded on microfilm, punched into cards, or stored in a file. Data stored in a file can usually be retrieved when needed and displayed on a VDT or printed.

The analyst must decide which medium offers the best solution in regard to the timeliness of the data and the cost of storing, retrieving, and displaying it. While printed reports offer certain advantages, they are often only valid for the precise moment they are printed. Although the report may be run daily, as time elapses it becomes obsolete since the current transactions for the day are not reflected in it. Perhaps the major advantage of the printed report is that it is still cheaper than providing additional online storage and VDTs for visually referencing data.

If an online sales system is implemented, the sales data is recorded directly into each customer's record as a transaction occurs. If the credit manager or a clerk wishes to know the status of a customer's account, the data is current.

In the past, printed reports were used almost exclusively when the data had to be referenced visually. Now the analyst should also consider the use of microfilm and online data files. There will probably always be a large number of printed reports that show the status of various accounts or activities at a precise moment of time—such as a schedule of accounts receivable, a payroll summary, or an inventory report. However, the volume of printed data in proportion to the volume of data processed is decreasing.

In the detailed design phase of the study, the format for visually referenced data is determined. The specifications from the design phase should provide the necessary guidelines.

Turnaround Documents   A printed statement in the form of a turnaround document, such as the one in Figure 9-5 is often prepared and sent to the customer. Customers detach a portion of their statements and return it with their remittances. Depending upon how it is printed, the returned portion can be machine-processed by either a mark reader or a scanner.

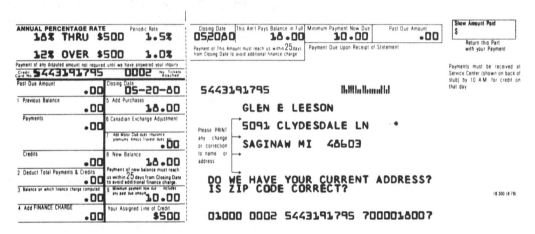

FIGURE 9.5   A typical turnaround document

Since the cost of preparation of input is constantly increasing while many of the costs associated with data processing are decreasing, thought should be given to the use of turnaround documents. Insurance companies, banks, and utility companies frequently use turnaround documents. With the cost of mark readers and scanners decreasing, other types of companies should also consider printing turnaround documents to be used in billing procedures.

If a turnaround document is to be used, the specifications have to be determined in the design phase. During the detailed portion of the study, the exact specifications and the layout of the document will be prepared.

Internal Versus External Reports   Certain factors must be considered in designing either internal or external reports. The primary concern is that the report contains all of the required information. It must also be easy to use and to handle. The report should be balanced on the form and appealing to the eye.

If multiple copies are needed, the analyst should consider the various alternatives. Usually no more than six-part form paper can be used. If more than six copies are printed at one time, some of the copies are difficult to read. If eight to ten copies are needed it may be possible to design the form so that **two-up forms** can be used.   Each of the two originals can have three or four carbon copies.

One company found that it was cheaper to print all single-part reports and to duplicate the required number of copies on a copier than to change the type of paper being used. The company's analyst determined that approximately five minutes of print time was lost each time the paper was changed. In addition, if only single-part forms are printed, an operator does not need to decollate the report, or separate the various copies. The reports prepared on a copier (a Xerox or some other type) are cleaner to work with and may be reduced to a size that is easier to store and handle.

Internal reports should be printed on **stock paper**. The headings are printed by the line printer as the report is being run. The analyst should design the report so that it contains all of the required information and is easy to read and use. Reports that are seldom used should probably be single-spaced while those that are referred to a great deal should be double-spaced. For the convenience of the user, headings should be printed on each page of the report. All reports should have the page number and date at the top of each page. If the report is printed more than once a day, the heading should also include the time.

External reports (see Figure 9-6) have traditionally been printed on preprinted forms such as the ones used for sales invoices or employees' paychecks. External reports should project a positive image for the company, be easy to handle, and contain all the required information. The quality of the paper should be high, and the artwork in good taste. Different colored paper can be used for the original and each of the copies to aid in the distribution of the report.

Most stationery companies carry a line of standardized forms that can be used for external reports. Such companies will print variable data, such as the customer's name, address, and phone number, on the forms. When standardized forms can be used, they are considerably less expensive than forms designed specifically for a company. The forms-company's representative should be consulted during this phase of the study. The representative's expertise will prove to be very helpful. The analyst and the user must determine the quality of the paper, the color of the paper, the color of the printing, and the number of copies needed.

If a report is designed so that it can be mailed without being inserted into an envelope, postal regulations require that it be a specified standard size. If the form is to be inserted into an envelope, the analyst must decide whether window envelopes or regular envelopes will be used. This must be determined prior to designing the form.

Since forms design is an important consideration, a separate section is included later in the chapter on the subject.

Timeliness of Data   If information must be up-to-the-minute, output should be available upon demand and reflect the current status of the account or item. When information is to be available upon demand, a transaction processing system should be developed and VDTs used to display the information.

Some companies now process sales over the telephone in a realtime environment. The online databases (customer and inventory) are updated as each sale is processed. If the item is unavailable, the customer can be told when it will be available or informed what items must be substituted for the desired item.

If the customer has exceeded his or her credit limit, the call can be referred to the credit manager. Credit managers must be able to display on their VDTs a complete and current history. By reviewing the information, they can make a decision regarding the transaction. The system would be ineffective if the information displayed is not timely and does not include all current transactions.

CHECKPOINT   1. In regard to a printed report or transaction file, what factors are determined during the general design phase and what factors are determined during the detailed design phase?

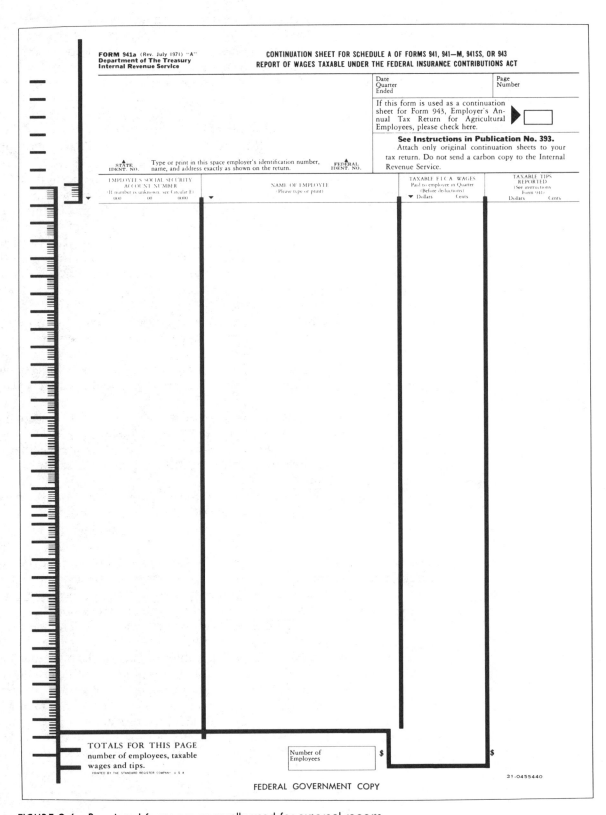

FIGURE 9.6   Preprinted forms are generally used for external reports

2. Is the following statement true? "Determining the format of the output is the first step in the detailed design of a procedure. Once this is done, there will be no change in the format or content of the output."

3. What five major factors relative to output must be determined for each program?

4. What difference does it make if a printed report is going to be used internally or externally?

5. An analyst is designing a sales system and indicates to the user that each morning the status of each customer's account and the status of each inventory item will be printed. The reports will be used by the personnel who will take the orders over the telephone. How would you react to this aspect of the system design?

6. What are two-up forms? How might they be used?

7. Why does the intended use of the output make a difference in the design of a system?

8. What is a turnaround document?

CONCLUSIONS REGARDING OUTPUT DEVICES

Study the output media and output device chart illustrated in Figure 9-4. If you are not familiar with any of the devices, or disagree with the use and relative speed codes, do some additional research regarding the medium or device. You may wish to read articles, obtain material from vendors, or consult with professional programmers or analysts.

Although the dynamics of EDP include improvements in existing I/O devices and the development of new ones, at the present time the following trends are occurring. Punched cards, magnetic tape, magnetic disk, and printed output were once considered the traditional output devices. Now terminals have been added to the list of output devices found in most medium- and large-size installations. Excluding standalone microcomputer systems, well over 80 percent of all minicomputers and mainframes are timesharing systems and terminals are widely used for displaying information.

While the volume of printing continues to increase, it now represents a small percentage of the total output. VDTs are used for displaying selected data and eliminate the need for some of the reports that once were printed. Many types of printers are available. The applications for which the printer will be used determines the type of printer that should be obtained.

The use of punched-card output is decreasing. Magnetic tape is being used for specialized applications, such as backing up files. The use of magnetic disk output and output displayed on VDTs is increasing.

Cassette tape and floppy disks were first used with minicomputer systems. Today cassette tape is seldom used with minicomputers. It is, however, a popular input, output, and storage medium for microcomputers. On larger computer systems, diskettes were once considered as a replacement for punched card input and output. Decreases in the cost of hard-disk storage and terminals have cause applications that once used diskettes to be redesigned. Although both hard disks and diskettes are available for microcomputer systems, diskettes continue to be very popular. One of the major concerns regarding diskettes is the lack of standardization in the size of the diskette and in the manner in which data is formatted.

COM (computer output to microfilm) and voice output might still be considered as untraditional since they are only used for certain types of applications and are not always available in what might be considered a typical data processing installation. When COM is used rather than printing large volumes of printed data, it is less expensive and less space is needed to store the microfilm. Information stored on the microfilm can be retrieved faster than information stored in a printed report. COM provides low-cost storage for historical data. Figure 9-7 provides a brief summary of the typical uses for each major output medium.

Medium	Typical Application
Punched cards	Used as turnaround documents in cost accounting, billing, and inventory control applications.
Printed reports	Required for paychecks, billing statements, financial statements, invoices, letters, reports, and documentation.
Magnetic disk (hard)	Storing transaction files, databases, operating systems, application programs, documentation, and text files.
Magnetic diskettes	Storing transaction files, databases, programs, operating systems for microcomputers, documentation, and text files.
Magnetic tape (reel)	Storing backup files, historical data, reports, and transporting data between computer systems.
Magnetic tape (cassette)	Storing files and programs used on some microcomputer systems.
VDT	Displaying data as it is entered, processed, or retrieved from online databases. Displaying messages from the computer to the operator.
COM	Storing price lists, catalogues, inventory listings, and documents.

FIGURE 9.7 Typical uses for output media

CHECKPOINT

9. You are an analyst working for a large organization that has the following output devices: punched cards, magnetic tape drives, magnetic disk drives, terminals with VDTs, COM, and voice output. Which output device would you recommend for each of the following situations?
   a. Sales invoices are to be prepared.
   b. A new database for an online application is to be created.
   c. The accounts receivable database is to be backed up.
   d. Every week a 500-page price list is prepared in duplicate for the 200 branch stores.
   e. Sales representatives who call on customers make inquiries into the status of customers' accounts from the offices of the customers.
   f. Turnaround documents are to be prepared. When a portion of the document is returned by the customer it will be read by a scanner.

10. You are a self-employed electrician and recently obtained a microcomputer system. You want to be able to prepare bids, bill customers, and do routine accounting applications. What I/O devices would you obtain for your system?

11. Today would a company selecting I/O devices for use with a microcomputer or minicomputer system elect to use punched cards or diskettes?

12. What should an analyst know regarding the output devices that are available?

PRINTED REPORTS

External Reports

In designing external or internal reports, the analyst or programmer must work with specifications such as those illustrated in Figure 9-8. The specification could have been prepared and submitted as part of the final report for the general design phase of the project. Since the specifications are detailed, some analysts would feel they should be developed during the detailed design phase of the project.

In studying the sales-order system documentation form for the accounts receivable statement, you can see why the analyst must consider for each program the output requirements, the input requirements, and the processing that will be required. The specifications form indicates that the information to be printed comes from three sources: control information entered from a terminal by the operator who initiates the program; the accounts receivable database; and calculations.

In preparing the print layout forms, the analyst or programmer uses the specifications form to determine what must be printed. Next the analyst designs the form. In designing the form the analyst must:

- Make certain the form provides all of the required information.
- Create a form that is easy to handle. The size of the envelope that will be used, as well as how the form will be processed after it is printed, has a direct impact on its design.
- Make certain the forms represent the organization favorably. The quality and color of the paper as well as how the data is arranged should be considered. Depending upon the form being designed, an attractive logo or some type of artwork may be used to enhance the appearance of an external document.
- Work with a representative of the company selected to print the statements. Often stock forms are available that can be customized. The company's sales representative may have helpful suggestions and oversized forms that can be used in creating the design.

The form is designed on a print layout form like the one illustrated in Figure 9-9. Each square represents one print location. The four points already listed must be considered. In creating the design there are very few rules. The rules that must be followed are: a heavy line is used to show the actual size of the form; the information to be printed on the form by the forms company is printed (company name, due date, and so forth); the location where the variable data will be printed by the program is shown with Xs; and dashes are used to show where a form is to be perforated. Figure 9-9 illustrates how the Accounts Receivable Statement might be designed using the data provided on the specification form illustrated in Figure 9-8 on page 306.

Paychecks and sales invoices may be prenumbered. When numbers in sequence are to be printed on documents, the size and location of the field is shown by using red Xs. Numbers are used to provide control over the forms. In the case

computer products inc. REPORT SPECIFICATIONS

PURPOSE	PROJECT NUMBER	1051
Specifications for the Accounts Receivable Statements and the Control Total Report	PREPARED BY:	Aguilar
	DATE:	July 14, 1986

GENERAL SPECIFICATION FOR THE ACCOUNTS RECEIVABLE STATEMENTS:

1. Statements will be printed on three-part preprinted forms.

2. A portion of the statement received by customers will be returned along with their checks.

3. The statement will be mailed in a window envelope.

4. The preprinted forms should have the following data:
   a. Company logo, name, and address.
   b. Phone number used when additional information is required.
   c. Captions for payment due date, new balance, account number, closing date, old balance, total payments received, total charges made, and finance charges added.
   d. Column headings for month and day of purchase, reference number, description of transactions, charges, and payments/credits.

5. Printed on the back of the statement will be detailed information regarding whom to contact for adjustments, how to make inquiries regarding the statement, and how finance charges are calculated.

VARIABLE DATA TO BE PRINTED BY THE LINE PRINTER:

Field	N/A	Length	Source and Comments
Firm Name	A	25	AR database
Address 1	A	25	AR database
Address 2	A	25	AR database
Address 3	A	25	AR database
Zip code	N	5	AR database
Payment due date	A	8	Operator's terminal
New balance	N	8	Calculated
Account number	A	10	AR database
Transaction-month and day	N	4	AR database
Reference number	N	6	AR database
Description	A	30	Generated from transaction code
Previous balance	N	8	AR database
Total payments	N	8	Calculated
Total credits	N	8	Calculated
Total charges	N	8	Calculated
Finance charge	N	8	Calculated
Closing date	A	8	Operator's terminal

*N = numeric A = alphabetic

FIGURE 9.8 Specifications for the accounts receivable statement

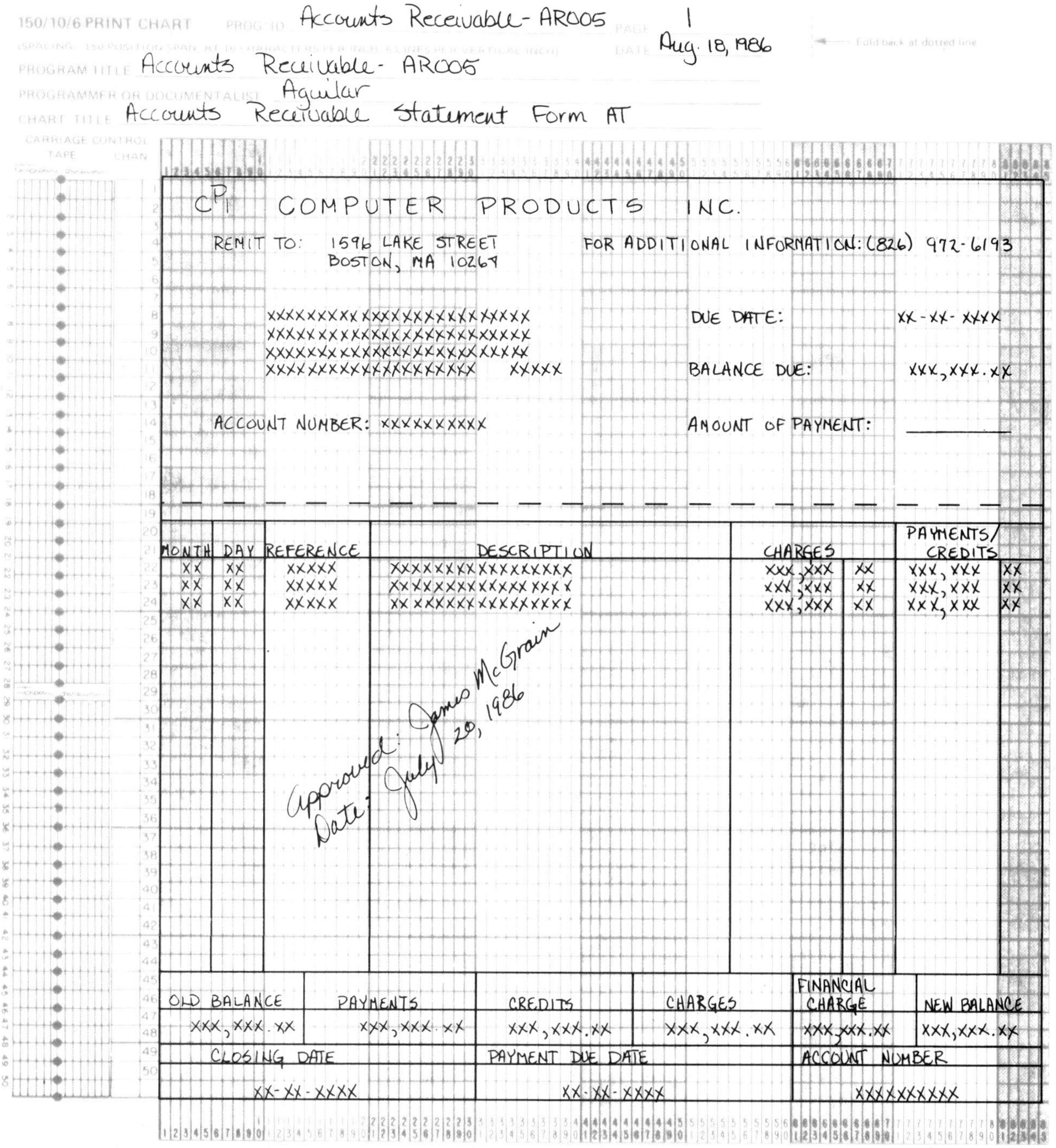

FIGURE 9.9 Layout for the accounts receivable statement

of paychecks or sales invoices, all numbers must be accounted for. Either a check was written to a particular individual or it was voided. A security officer would insist that all checks be numbered and that the checks must be stored in a vault.

After the intial design is completed, the form is discussed with the user. In this case it would probably be discussed first with the project team and then with the accounts receivable manager. Both the analyst and the manager should check to see that all of the required information is printed on the form. If the manager approves the design of the form, the approval should be indicated on the form *in writing* The accounts receivable manager should be asked to write "Approved by," the date the design was approved, and then sign his or her name and title. At some time in the future, the manager cannot say that the form is unsatisfactory because an opportunity to review it was not provided.

The analyst must work with the user in selecting the quality and color of the paper, the style and size of the typeface to be used, the width of the lines used to separate the various sections of the form, and the color of the ink to be used. Shading can be used to highlight certain areas of the form such as the new balance and the due date.

Before the forms are printed, the forms company will submit a proof to the analyst. The proof will show the exact size of the form and the size and style of the typeface to be used. In order to determine exactly how the report will look, the report can be run on stock paper. A transparency made from the proof can be laid over the report to determine whether any adjustments need to be made on the proof.

Internal Reports

Internal reports are also designed on print layout forms. Layout forms often have 132 print positions since this is considered the standard number of horizontal print positions available on a line printer. Figure 9-11 was prepared by using the specifications provided in Figure 9-10. When designing internal reports, the constant data (such as ACCOUNTS RECEIVABLE CONTROL TOTAL REPORT) is printed on the print layout form. The Xs represent the variable data that will be printed.

After an internal report is designed, it should be inspected and approved by the user.

DISTRIBUTION OF REPORTS

The documentation for programs should indicate which control totals, such as the ones illustrated on Figure 9-11, should be checked before the report is distributed. The *run sheet* (directions for the computer operator) should also indicate what, if anything, should be done before the report is considered ready for distribution. It may be necessary to decollate or "burst" the report.

Decollators

A **decollator,** or separator, is used to separate multiple-part forms and to remove the carbon paper. Decollators come in models that can separate from two- to eight-part forms in one pass. If a four-part separator is used for four-part forms, the carbon paper is removed and wrapped around the carbon rewind spindles. The original report and the three carbon copies will be stacked in their own stackers. Optional features for the decollator include marginal slitters for trimming the pinhole edges of forms and a center slitter for separating two-up forms.

The operator loads the forms and aligns the pin-feed holes with the pins from the pin belt. Once the forms are manually aligned, the operator adjusts the speed control to be the most effective speed for the job. The operator is also responsible for removing the carbon paper from the rewind spindles, removing the forms from the stackers, and emptying the trim container.

Burster

A **burster** is used to separate continuous-form reports. It should be capable of handling a wide range of paper weights and form sizes. Many bursters today are capable of slitting the sides of the forms, bursting, stacking, and imprinting the forms.

computer products inc.

PURPOSE   PROJECT NUMBER: 1051

Specifications for the Accounts Receivable   PREPARED BY: Aguilar
Statements and the Control Total Report   DATA: July 15, 1986

GENERAL SPECIFICATIONS FOR THE CONTROL TOTAL REPORT:

    1. The report is generated as the Accounts Receivable Statements are
       printed.

    2. Two-part 8 1/2" by 11" stock paper is to be used for the
       report.

    3. A suitable heading containing the date should be printed at the
       top of the report.

    4. Control totals are to be printed for the following fields:

       a. Previous balance

       b. Charges for period

       c. Credit granted

       e. Finance charges added

       f. Current unpaid balance

    5. The report will be verified by the accounting department. The
       report cannot be released to the sales department or the customers'
       statements mailed until the cause of any discrepancy is determined.

    VARIABLE DATA TO BE PRINTED ON THE REPORT:

            Field   N/A   Length   Source and Comments
        Previous balance total   N   11   Calculated
        Total of all charges   N   11   Calculated
        Total of all payments   N   11   Calculated
        Total of all payments   N   11   Calculated
        Total finance charges
          added to balances   N   11   Calculated

FIGURE 9.10 Specifications for the accounts receivable control total report

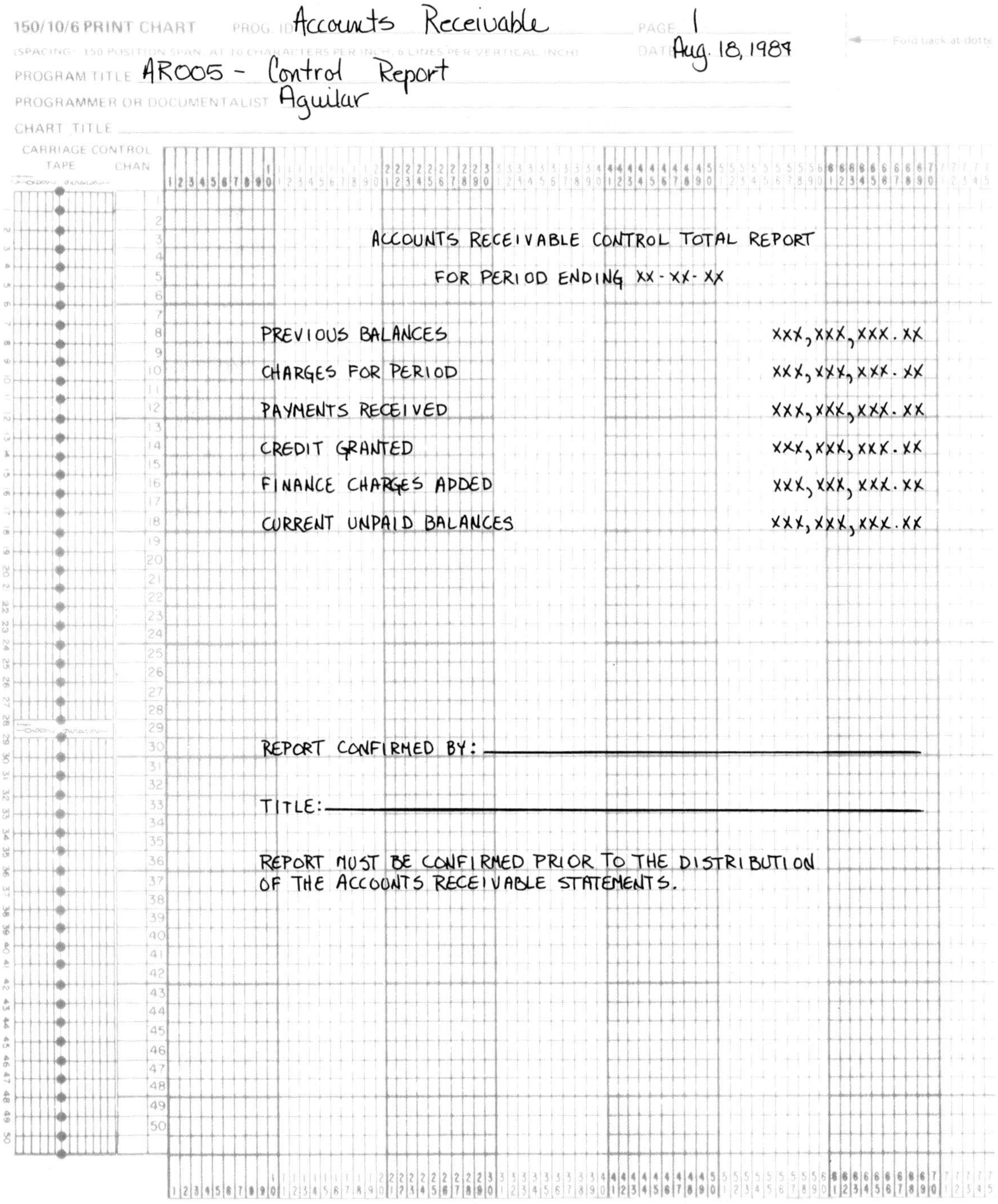

150/10/6 PRINT CHART   PROG. ID *Accounts Receivable*   PAGE *1*
(SPACING: 150 POSITION SPAN AT 10 CHARACTERS PER INCH, 6 LINES PER VERTICAL INCH)   DATE *Aug. 18, 1989*
PROGRAM TITLE *AR005 – Control Report*
PROGRAMMER OR DOCUMENTALIST *Aguilar*
CHART TITLE

```
              ACCOUNTS RECEIVABLE CONTROL TOTAL REPORT
                     FOR PERIOD ENDING XX-XX-XX

        PREVIOUS BALANCES                      XXX,XXX,XXX.XX

        CHARGES FOR PERIOD                     XXX,XXX,XXX.XX

        PAYMENTS RECEIVED                      XXX,XXX,XXX.XX

        CREDIT GRANTED                         XXX,XXX,XXX.XX

        FINANCE CHARGES ADDED                  XXX,XXX,XXX.XX

        CURRENT UNPAID BALANCES                XXX,XXX,XXX.XX




        REPORT CONFIRMED BY: _____

        TITLE: _____

        REPORT MUST BE CONFIRMED PRIOR TO THE DISTRIBUTION
        OF THE ACCOUNTS RECEIVABLE STATEMENTS.
```

FIGURE 9.11   Control report for the accounts receivable statement procedure

Some bursters have an optional feature that slits the two-up forms apart and then merges the two parts. Another optional feature folds and glues forms as they are being slit, burst, and stacked. Firms that send out advertising fliers would be interested in this feature. Models equipped with the imprinting feature have an impregnated ink roll and rubber patches that can contain a small amount of data such as signatures, return address information, and advertising notices.

Output Retention   Usually a printed report has a very short useful life since the information quickly becomes obsolete. Based upon company policy, government regulations, and the wishes of the user, the analyst must determine how long a report should be retained. Reports that might be referenced upon occasion for a long period of time can be converted to microfilm.

If a report or source document is destroyed prematurely, reconstructing the information can be very expensive. Sometimes companies destroy documents that are needed for a tax investigation and it becomes very expensive to reconstruct the audit trail to the satisfaction of the tax auditors.

When a printed report is no longer needed, it should be destroyed according to the record retention policy. Depending upon the nature of the report, the policy might state that it is to be shredded and then recycled or that it is to be burned. Highly confidential reports may need to be destroyed in the presence of an officer of the firm.

There must also be retention schedules for output punched into cards or records stored on disk or magnetic tape. It is very important that the individual in charge of the library containing the magnetic disk packs and tapes knows how long each file is to be retained. The programmer or operator who sets up the job control language for the execution of the job must also know how long the files are to be protected so that the internal file label will also protect the data.

CHECKPOINT

13. When designing an external report or statement, what four factors should the analyst keep in mind while creating the design?
14. What are the rules that must be followed in designing an external report?
15. Why should the analyst have the user sign the report layout form and indicate that the form is acceptable?
16. On an external report, the information that is printed on the form layout is there for what purpose?
17. On an internal report, how will the information lettered on the print layout form get printed?
18. What are some of the factors that must be considered when determining how long a printed report is to be retained? If a report is seldom used but must be kept as part of the audit trail, what might occur two or three weeks after the report is printed?
19. What is a decollator?
20. What is a burster?
21. You overheard a newly hired computer operator say to his friend "I never use job documentation. As soon as a report is printed, I send it to the department charged for the CPU and print time." If you were the CIS manager, how would you respond to his remark?

DISPLAY SCREENS

Software is available that make it easy to lay out VDT screens. The programmer keys in the required format on the screen, gives the format a name and then stores the format in a file. Whenever the format is to be used as a display in a program, it can be called into the program by using its name. A programmer can also create display screens within a program.

There are a few basic rules that must be followed in displaying information on a screen. The important points to remember are:

- Clear the entire screen between formats. There is usually a "clear screen" command that can be used.

- Format the output so that it is easy to read. For example, don't clutter up the screen with unnecessary information, always display directions or error messages in the same place on the screen, and leave space between items so that the information is easy to read.
- Prevent **scrolling**. Unless delays are coded into programs, information is displayed on a VDT faster than most people can read. One screen of information should be displayed at a time. When the operator is ready, a specified key is depressed and a new screen of information is displayed.
- Don't overuse color. Often monitors that display information in color are used for terminals or for microcomputers. Carefully controlled use of color can make the information more understandable; uncontrolled use of color adds confusion.
- Be consistent. For example, all instructions may be displayed at the bottom of the screen.
- Develop and use simple conventions such as having an operator enter a *1* or a *y* for a positive response to a question or a statement.
- Make certain all directions are clearly stated.
- Test all screens. Have someone totally unfamiliar with the program load the program and enter the required data *without using the help option.*

Figure 9-12 illustrates two screens that might be used for the maintenance subsystem menus. The first screen asks the operator to determine which database is to be updated. The second screen asks the operator whether new records are to be added, records are to be deleted, or changes are to be made in existing records. Both screens provide a ''help'' option and a way of returning to either the main menu or the menu that was displayed prior to the screen.

Figure 9-13 illustrates a screen that might be used to tell the operator what must be entered to create a record for a new customer. Since more data must be entered than can be recorded conveniently on one screen, only a portion of the data required is entered on any one screen. As illustrated in 9-13, the prompt tells the operator what data is to be entered. As each character of data is keyed in, the / is replaced with the data. When the ''enter'' key is used, the remaining slashes are replaced with blanks. After all of the data has been entered for one screen, the operator is asked to visually check the data. By keying in the number of an incorrect field, the invalid data is removed and the operator can key in the correct data. In addition to visually verifying the data, the programmer should provide edit routines which also check the validity of data entered from the keyboard.

Figure 9-14 illustrates three screens that might be used in a sales-order entry system. The first screen asks the operator to enter the customer's account number. As soon as the number is entered, the customer's name and address is displayed. At the bottom of the screen is the instruction to enter a *y* if the correct record was retrieved. Any response other than a *y* will cause the screen to clear and the following message to appear:

```
ACCOUNT NUMBER = = = = = = = = = = = = = = = = = = = = = >//////////
```

The operator is given an opportunity to enter the correct account number. If the operator entered a *y*, screen 2 appears and the operator enters the item number.

MAINTENANCE MENU

   1. ACCOUNTS RECEIVABLE
   2. INVENTORY
   3. SALES
   4. HELP
   5. RETURN TO SALES-ORDER MENU

AFTER YOU MAKE YOUR SELECTION, DEPRESS THE ENTER KEY. = = = = = = = = = >

ACCOUNTS RECEIVABLE MENU

   1. ADD RECORDS TO THE DATABASE
   2. DELETE RECORDS FROM THE DATABASE
   3. CHANGE DATA STORED IN EXISTING RECORDS
   4. HELP
   5. RETURN TO THE MAINTENANCE MENU

AFTER YOU MAKE YOUR SELECTION, DEPRESS THE ENTER KEY. = = = = = = = = = >

FIGURE 9.12 Typical menu screens

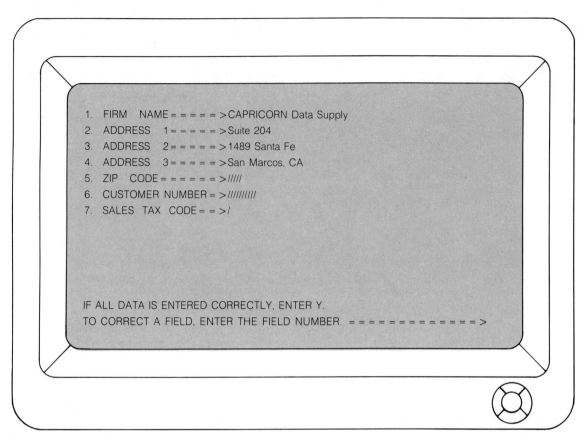

```
1.  FIRM    NAME = = = = = >CAPRICORN Data Supply
2.  ADDRESS    1 = = = = = >Suite 204
3.  ADDRESS    2 = = = = = >1489 Santa Fe
4.  ADDRESS    3 = = = = = >San Marcos, CA
5.  ZIP   CODE = = = = = = >/////
6.  CUSTOMER NUMBER = >//////////
7.  SALES  TAX  CODE = = >/
```

```
IF ALL DATA IS ENTERED CORRECTLY, ENTER Y.
TO CORRECT A FIELD, ENTER THE FIELD NUMBER. = = = = = = = = = = = = = >
```

FIGURE 9.13   One of the screens used to create a record for the accounts receivable database

The inventory record is retrieved and the description of the item is displayed. If the description is incorrect, the operator enters any response other than a *y*. The following would then appear on the screen.

```
ITEM NUMBER = = = = = = = = = = = = = = = = = = = = = > //////
```

If the correct item number had been entered, screen 3 is displayed. After the operator enters the quantity either the message ITEM IS AVAILABLE is displayed with the information that follows, or the following will appear on the bottom of the screen.

```
ITEM IS NOT IN STOCK.

ENTER 1 TO BACK ORDER THE ITEM.

ENTER 2 IF A SUBSTITUTION CAN BE MADE.

PLEASE ENTER 1 OR 2 = = = = = = = = = = = = = = = = = = = = >
```

Note that two conventions have been used in displaying information on the screens: instructions are always displayed at the bottom of the screen, and the operator always enters a *y* for a positive reply.

Some formatted screens look very much like a form that your might fill out with a pen or pencil. Once all of the data is entered, the operator is asked to review all of the fields. Until the enter key is used, the operator can position the cursor

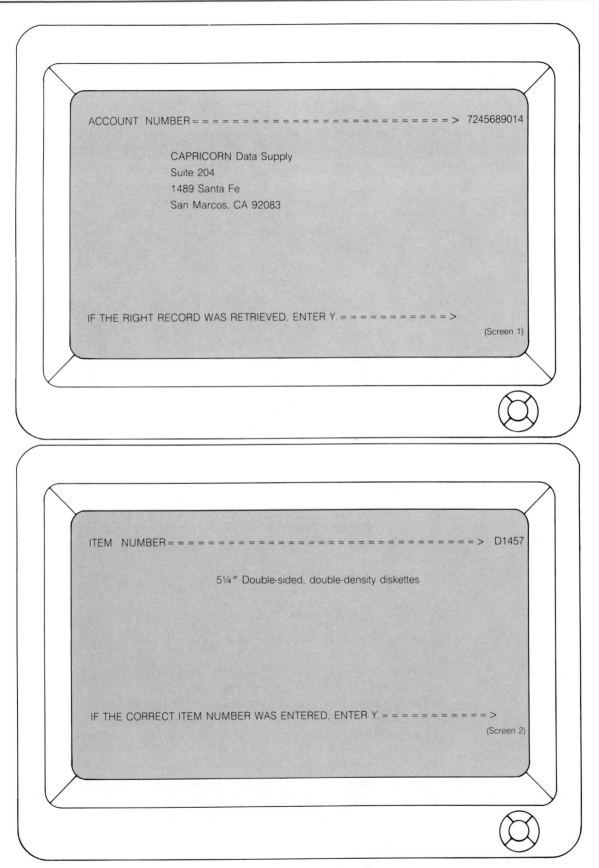

FIGURE 9.14 Screens for the sales order system

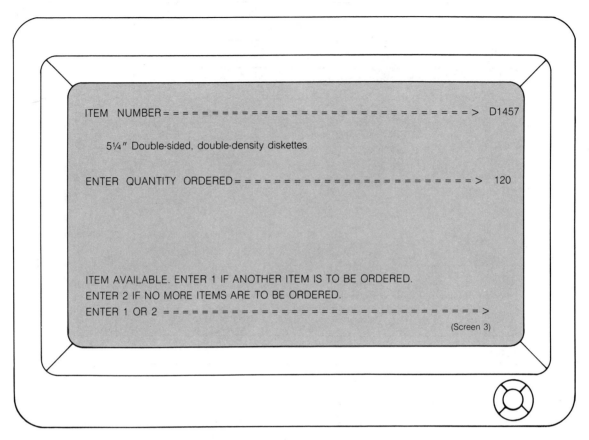

FIGURE 9.14   *(continued)*

and make any changes that are required. Once the enter key is depressed, the data entered by the operator enters the system as input.

Some analyst might prefer to have a message such as the following displayed:

```
IF THE RIGHT RECORD WAS RETRIEVED, ENTER Y.

IF AN INCORRECT RECORD WAS RETRIEVED, ENTER N.

ENTER Y OR N = = = = = = = = = = = = = = = = = = = = = = = = = = = = = > /
```

However, both the external documentation for the program and the internal documentation displayed when the operator enters an *H* explains that any response other than a *y* will provide another opportunity to enter data.

Forms similar to print layout forms are available for designing screens. However, most analysts prefer to do the design directly on the screen. While most screens can display 80 characters horizontally and 25 vertical lines, some screens display fewer or more characters per line and fewer or more lines.

Terminals are sometimes available that can be used to obtain information similar to what might be obtained from a receptionist. For those types of applications, adding graphics to the display can make the information visually more attractive and may also make it clearer. The possibilities regarding the use of VDTs for displaying various types of information are only limited by the imagination and resourcefulness of the analyst.

22. What are the basic rules that should be followed in designing screens?

23. What two conventions were used in the sales-order screens?

24. How should screen designs be tested?

25. Is software available that can be used in designing screens?

MAJOR CONCERNS REGARDING INPUT

Analysts often fail to look at the importance of the input function since it once seemed to be one of the more routine aspects of the data processing cycle. Also, in the early days of EDP, the only input medium was punched cards.

While the computational costs and the cost of storing data online are decreasing, the costs of preparing some forms of input have increased. If the input method requires an operator to key the data, the cost has increased due to the increasing cost of labor. Depending upon the application, the cost of preparing input may be anywhere from 30 to 50 percent of the operational data processing cost. Because of the increase in cost and the large number of options available, analysts are more concerned about the way data is entered into the system.

Figure 9-15 illustrates the ten major concerns regarding how data should be captured and entered into the system.

Ten Major Concerns Regarding Input

- What input is needed?

- How and where is the input created?

- How should the source documents be designed?

- What format should be used for the input records?

- What medium should be used for recording the input?

- Can a turnaround document be used?

- What provisions are made for establishing a clear audit trail?

- What program controls should be built into the programs that process the input?

- Should the data entry function be centralized or distributed?

- How can the data entry operator verify the data?

FIGURE 9.15 Ten major concerns regarding input that must be considered by the analyst

Sources of Input

Input to be processed generally comes from four major sources: transaction files; master files and databases; control information; and data entered from terminals.

Transaction files primarily contain variable data that is used for a limited time. The files are stored on media such as punched cards, magnetic tape, diskettes, or hard disk. A payroll transaction file that contains the data regarding who is to be paid and the regular and overtime for the hourly employees is used for one pay period. The processed data becomes part of the information stored in the payroll master file or database.

Master files and databases contain constant and updated information. The payroll master file or database may be used over the entire life of the payroll system. Although the same file or database is used, the records stored in the file are updated, obsolete records are deleted, and records for new employees are added. A volatile database such as the accounts receivable database for an online sales system changes second by second.

Control information is often needed. For example, for each pay period a record of information that contains fields such as the dates for the paychecks and payroll register, the pay period of the month, and the beginning check number is stored on disk. Or the required fields of information can be keyed in by the operator when each job is initiated. Each pay period, the control information changes. Control information is entered once per program.

As more jobs are redesigned as online, realtime applications, more data is entered directly into the system by using terminals. Data entered from terminals can be used to create transaction files, update master fields and databases, and as input for programs designed as interactive or transaction processing applications.

A typical job usually requires a limited amount of transaction data since the majority of the data is stored in a master file or database. The following section of this chapter will deal primarily with providing the answers to the ten major concerns regarding how transaction data should be captured and entered into the system. Chapter 10 will cover files and databases while Chapter 11 will discuss procedures and programs.

Input Needed

The input needed for any program is determined by the output desired. The analyst must ask the following questions: What information is already in the master file or database? What constant data is required that can be entered from some type of control record? What information must be supplied by using some type of transaction file? What data should be stored in and accessed from **tables**? What information can be calculated by the program?

Any time the use of a transaction file is being considered, or the data is to be entered from a terminal, the analyst must challenge each field to determine whether the data is already in a master file or might be included in a table. The analyst must be concerned that all of the data required to produce that output is entered into the program in the most efficient and cost-effective manner.

How and Where Data Is Generated

How the data is generated, and where it is generated, has a direct impact on a number of other questions. The analyst must determine whether transaction processing is feasible and whether a source document is needed. If it is not feasible to develop a transaction processing system, the analyst must determine whether distributed or centralized data entry would be more effective.

In a cost-accounting system, much of the data is generated when material is put into production. The analyst should attempt to provide a reliable means of entering data directly into the system from the factory. Data collection devices or special terminals can be used to enter some of the data. If data collection devices are designed for factory use, the keys should be larger and the lettering on displays should also be larger so it is easier to read messages or confirm data being entered.

In a retail sales system, wands or some other type of scanner may be used to read price tickets. When charge sales are made, special readers are available that make it possible to use the data stored on the customer's charge card. Whenever possible, the manual keying of data should be eliminated. In a retail sales application, the only variable data that a clerk might need to key in is the quantity of a given item that is purchased.

Format of the Source Document

Traditionally, data was first recorded on a source document. The source document was either prepared internally or received from a vendor or some other external source. Documents submitted by an external source were usually visually audited to make certain that all the required data was on the form. Some documents were **conditioned** by a clerk, who date-stamped and numbered the documents and added codes, such as the customer's number or a transaction code.

Analysts had little control over the format of externally prepared source documents. However, the source documents that were prepared internally were controllable by the analyst.

Designing the Source Document

The formats for the input records and the source documents should be determined simultaneously. Since commitment to the type of system has already been decided, there are specific considerations that will influence both the design of the source document and the format of the input.

The source document can be designed as soon as it is determined what data is needed and where and how it is to be entered into the system. The analyst should work with the data entry supervisor and the supervisor of the personnel who will record the data on the documents. The design of the documents should permit the personnel recording the data to do so as easily and rapidly as possible. Check boxes can be used, which reduce the time needed to fill out documents and minimize recording errors. For ease in transcribing data into a machine-processable form, the locations for each field within the record should be specified on the document.

The input record should be designed so that the flow of data on it is the same as on the source document. This decreases the time needed to record data and also reduces errors. The document should also be designed so that data is recorded from top to bottom and from left to right.

When a data recorder (keypunch), key-to-disk recorder, or key-to-tape recorder is used, the following considerations influence both the source document design and the input record format:

1. Fields that can be duplicated should be grouped together and placed in the leftmost positions on the record.
2. Fields that must be manually keyed should be grouped together to the right of the fields to be duplicated.
3. If possible, all numeric fields should be grouped together and all alphabetic fields should be grouped together.
4. Fields that are to be skipped should be placed in the rightmost positions on the input record.
5. When multiple input records are to be used in the same job, the fields common to all records should be in the same relative locations.

Input Format

The input format must be designed concurrently with the source document. The factors that must be considered are: record length, field size, use of codes, and the relationship of the source document to the input record.

The size of the record may be determined by the medium used. For example, some diskette records are limited to either 80 or 128 characters. Card records are limited to either 80 or 96 characters. When working with hard disk or tape records, the size is determined by the programmer.

The analyst must understand the characteristics of the data entering the system and determine the field size that should be used. For example, if each customer is assigned a number and the firm now has 9,945 customers, the analyst should allow five positions for the field. If five positions are not reserved for the field, as soon as 54 more records are added, the field will not be large enough to handle the account number. Field sizes are usually determined by studying historical data, projecting future needs, and providing for growth.

Well-designed source documents and input formats help to reduce errors and increase the productivity of operators. If the analyst feels that the data entry personnel are not as productive as they should be, the low productivity may be the result of: poorly designed source documents, inadequate instructions (documentation), lack of reference materials, poor quality of the data recorded on the source documents, or unskilled operators.

Input Medium

An input method that requires a minimal amount of data conversion should be selected. If punched-card recorders (keypunches), diskette recorders, key-to-key tape recorders, or terminals are used, the data is usually recorded on a source document and then transferred to the machine-processable medium (cards, diskettes, tape, disk, or directly into a transaction file.) In some cases, there is no other reasonable solution.

Verification of Data

When punched cards, diskettes, or magnetic tape (key-to-tape) are used, the data is generally rekeyed during the verification processing. Although rekeying the information usually detects recording errors, it is time-consuming and expensive. It takes just as long to verify the data as it did to initially record the data.

When key-to-disk recorders or terminals are used, other methods can be employed to decrease the length of time needed to record and verify the data. Because of the various options available for verifying the data, two terminal or key-to-disk operators can enter and verify as much data as three card-punch, diskette, or key-to-tape operators.

Since key-to-disk units are often controlled by a microcomputer, the same verification methods can be used as those used for verifying data entered on a terminal. A VDT is available, which is used to display the data entered. The operator can visually verify and confirm each field as it is entered. Online files can also be used to assist the operator in confirming the validity of certain fields. For example, an abbreviated version of the customer master file is online during the entering of sales-order data. As the customer's number is keyed in, his or her name is displayed on the VDT. The operator can visually confirm the validity of the customer's number. This method not only catches errors made in keying in the data, but also detects errors made in recording data on the source document.

Other Input Devices

Other available input devices, such as scanners, mark readers, and voice input, make it unnecessary to key the data into some type of machine-processable medium. There are some excellent applications that use scanners or mark readers to scan turnaround documents or price tickets so that labor-consuming conversion of data to a machine-processable form can be avoided.

Providing an Audit Trail

When devices such as the diskette recorder are used, an identifying number from the source document is keyed into the input record. In recording cash receipts, the receipt number and the date are both entered in the input record, which establishes an audit trail. Printouts of the transactions can be used as a visual reference and also further define the audit trail.

When terminals and realtime systems are designed, there must still be a clearly defined audit trail. There may be a temptation to eliminate the required audit trail in order to save time and money. This will usually prove to be false economy since any errors that get into the system will be extremely difficult to find. Also, in the event of a tax audit or an external audit, it will be very difficult to confirm entries recorded into the system. In some cases, there would be no alternate method of handling transactions in the event of a system failure.

When a customer of the First Savings and Loan of Saginaw, Michigan makes a deposit to his or her account, the customer's number is keyed in by the teller. Displayed on the terminal are the customer's name and balance. Although the name would provide enough visual confirmation, there is a one-in-a-million chance that a wrong number could be keyed in that would retrieve a record for a customer with the same name. By displaying the customer's balance as well as his or her name, the odds are increased against entering a wrong number and retrieving another customer's record with the same name and balance.

Although the data goes directly into the system as the transaction is being completed, a deposit slip is manually prepared, which helps to provide an audit trail. It also provides additional data that can be referenced visually if needed. If the system goes down, the bank operates as usual and records the transactions manually. As soon as the system is up, the transactions are entered into the system. The analyst who designed the system was very aware of:

- the need to verify all data;
- the need to provide a well-defined audit trail; and
- the need to provide a means of recording transactions in the event that the system was not available.

VERIFICATION METHODS

Several methods of verifying input have already been illustrated. If incorrect data enters the system, it is usually very costly to make the necessary corrections. Also, how expensive would it be to have your operator record a quantity of 100 rather than 10 for a shipment of sports cars? The shipping charges for sending the cars to the customer and then of having them returned would be only one of the costs. While the 90 extra cars were in transit they would not be available to other customers (which could result in a loss of sales) or could be damaged.

As many methods as necessary should be used to verify data entering the system as input. Some of the methods used are:

- *Key verification.* A second operator rekeys the data already recorded. This method is used for verifying data recorded in punched cards or on diskettes and magnetic tape.
- *Visual verification.* This sometimes proves to be the least effective method since the same operator both records and verifies the data. The operator may have a built-in assumption that "since I recorded the data it must be correct."
- *Visually displaying an identifying characteristic.* When using a terminal, a part number is entered. Displayed in the VDT is the description of the part, which is then visually confirmed by the operator.

- *Use of self-checking numbers.* An optional feature is available on a punched-card data recorder or a diskette recorder that will verify self-checking numbers. The computer can also be programmed to reject numbers that have been transposed or have one or more wrong digits. Check digits and self-checking number routines can be effectively used for numbers in a series, such as student numbers, account numbers, part numbers, or invoice numbers.

- *Batch or control totals.* The amount written on the cancelled check is encoded on the check. The encoding process uses a magnetic ink to record the amount on the check. A batch total slip is made out that contains the number of checks in the batch and the total of all the individual amounts within the batch. A relatively small number of checks, typically 20 to 25, should make up a batch. As the checks are entered into the system, the computer keeps a record count and a total of the amounts within each batch. If there is a difference between the amount entered on the batch total slip and the amount accumulated by the computer, this difference is printed on the report. The report also lists each check number and the amount of the check as well as the batch total. Someone must take the printed report and the batch of checks and determine where the error was made. Either one of the checks was encoded incorrectly or the batch total slip was wrong. In the first case, where a check was encoded incorrectly, a correcting entry will need to be made.

- *Hash totals.* Sometimes numbers are added to produce a meaningless total called a hash total. For example, totaling the quantity of all items purchased. When the records are entered and processed, the hash total is compared to the original total. If the two totals agree, it is an indication that all quantities were entered correctly and all records were processed.

- *Checking between a range of numbers.* The numbers on the orders being processed on a given day should fall between, say, 4999 (the last number from the previous day) and 6001 (the next order number that will be on all of the orders processed by the next day). If the order number recorded on the input record does not fall within that range, an error message will be generated.

- *Reasonableness test.* Based upon past history, some input can be checked to see if it is reasonable. For example, because of longstanding company policy, it is unlikely that any employee will have more than 20 hours of overtime. If more than 20 hours of overtime are recorded in an employee's current transaction record, an error message will be generated as the data is being edited.

- *Verification of codes.* The pay and fringe benefits are calculated for employees based upon their payroll status. Assuming that the valid status code must be either an H (hourly), S (salaried), T (trainee), or a P (part-time), an error message would be generated if the code used was not an H, S, T, or P.

- *Verification of data type.* Some input fields should contain only numeric data while others should contain only alphabetic data. The fields can be edited to make certain that only the right type of data is recorded in each field.

- *Verification that certain combinations of data exist.* For example, all students may be coded with either a W or a V. The V denotes a non-workstudy student while the W indicates that the student is on work-study. The only valid account numbers for a workstudy student are 2155 and 2156. Any other account number for a W-coded student in invalid.

- *Sequence check.* If the numbers in the source documents are serial and the documents are in order, the input records will also be in numerical sequence. A check can be made by the program to determine whether the records are in either ascending or descending sequence.

DISTRIBUTED DATA ENTRY

When the data entry function for the sales-order processing procedure is centralized, the following problems may be cited: incomplete documents, invalid product numbers, and incomplete codes. If the data entry function is centralized, the operator has no way of determining what additional data should be recorded on the document. The same is true when invalid product numbers are recorded. The operator is not familiar enough with the product numbers to realize when the numbers and the descriptions do not match. The data entry department is separated from the source of the information needed to determine the validity of the data recorded on the documents. If the source document does not contain an explanation of the acceptable codes, the operator has no way of knowing when an invalid entry is made.

The operators entering and verifying the data are generally well trained and highly skilled in the use of the equipment. In some cases, better-designed documents could increase productivity and prevent some of the errors. In most cases, the operator is not expected to determine the validity of the data recorded on the documents. Unless there is an obvious omission of data or an invalid code, the operator keys exactly what is recorded on the document.

A solution may be to decentralize the data entry function. This can be done even if additional methods of recording data, such as punched-card data recorders and diskette recorders, are used. The disadvantage of distributed data entry is that data entry is no longer the primary job of the person entering the data. This broadening of responsibility can result in less productivity and less effective use of the hardware.

One advantage of decentralizing the data entry function is that it places complete responsibility for the validity of the data on the department requesting the services of the computer information services since the data is recorded in the user department. If a document is incomplete, the operator has an opportunity to determine what data should have been recorded. As a member of the department, the operator is more aware of the important numbers and other codes. Therefore, fewer errors are likely to occur. Less transcribing of data from one form to another is needed since the operator understands what should be recorded on the document.

The analyst must decide whether the total cost is less when equipment is used inefficiently by a member of the user's department who understands the data or when the equipment is used by a skilled operator who is not a specialist in handling the department's data. As part of the total costs, the intangible cost of delaying processing because of incomplete or invalid data must be considered. When information is entered by using a terminal located in the user's department, as the operator enters data, error messages can be displayed. The operator has the necessary resources to look up information and make whatever corrections are required. Fewer records will be rejected and users will receive better service.

The trend today is to decentralize both the data entry and the EDP functions. Therefore, users are more accountable. In one school system, a great many errors occurred in the payroll system. The data entry function was subsequently distributed and each payroll clerk was made responsible for the validity of the input. When the data entry function was distributed, almost all input errors were eliminated.

CHECKPOINT

26. Why are analysts so concerned with how, when, and where data enters a computerized system?
27. What are the four major sources of input?
28. What factors influence the amount of variable data that must be keyed in, read from a transaction file, or entered by using voice input, scanners, or some other input device?
29. Why should the source document, VDT screen, and the input record be designed at the same time?
30. What four factors should be considered when designing an input format?
31. If a data entry operator is unproductive, what are some of the factors that should be investigated?
32. Why should as many methods as possible be used to verify data entering the system?

TRENDS REGARDING DATA ENTRY DEVICES

You might wish to refer back to Figures 9-2 and 9-3 which summarize some of the characteristics of the leading input devices. Some of the trends that are occurring regarding data entry devices and media are summarized in the following paragraphs.

Most companies with large volumes of data are using some input medium other than punched cards. In the past several years there have been no major improvements in punched-card recorders. Most of the manufacturers who until recently manufactured punched-card recorders are no longer producing the equipment. The data recorders available on the open market are generally reconditioned equipment.

Since the equipment was first manufactured, a number of improvements have been made in key-to-diskette recorders. One of the most recent improvements is to increase the density of the data stored on the diskettes. As density increased, the time needed to retrieve records decreased. Microcomputers and a few of the minicomputer systems will continue to rely on diskettes as an input, output, and storage medium.

Most companies that used key-to-tape recorders for high-volume jobs are now changing, or have changed, to terminals or key-to-disk recorders. Although key-to-tape systems are still available, it is doubtful that additional improvements will be made in this type of equipment.

Most independent key-to-disk systems are controlled by a microcomputer or a minicomputer. The systems are used to record data at the points where the transactions occur. After all of the data is recorded, the key-to-disk system is put in a record mode and the condensed and edited data is transmitted into a centralized database.

An increasing number of terminals are now being used at remote locations to enter data directly into the systems. Terminals come in a wide variety of shapes, forms, and sizes—some so small they can be held in the palm of a hand and others as large as a small desk. The terminal configurations may make use of one or more of the following devices: a keyboard, a VDT, a ten-key pad used to record numeric data, a printer, a mark reader, or a scanner.

Many companies are replacing "dumb" terminals with intelligent (programmable) terminals. Intelligent terminals have their own microprocessors and often have a storage device such as a cassette tape recorder, bubble memory, a chip-memory device, or some other type of storage medium. After all of the data is keyed in, edited, and stored, the data is transmitted in a **burst mode** to a centralized, host computer.

Optical readers are probably best suited for applications that require a limited amount of input prepared by a large number of people. Each individual needs to record very little data on the input medium; therefore, the likelihood of making errors is greatly reduced. Many new optical readers that read typewritten and handwritten documents are available at a low enough cost that they can be considered for both minicomputer and mainframe systems. Some are also advertised for use with microcomputer systems.

A touch-tone telephone is also a very inexpensive terminal that can be used for entering numeric data into a system. As our society becomes more involved in electronic transfer of funds, computerized news services, and shop-at-home services, wider use will be made of the ordinary telephone as an input device.

Voice input systems are available for microcomputers, minicomputers, and mainframes. As further advances are made, there will be more business applications that utilize voice input. At the present time, there are sales-order entry systems that make it possible for sales representatives to verbally enter sales data into the centralized system. The major problem is still dialect. The computer does not understand dialects or accents that it has not been trained to recognize. Some of the interesting uses of voice input involve the use of robots. Robots in both factories and health-care facilities can respond to verbal instructions.

DESIGNING SOURCE DOCUMENTS

During the design phase it was determined: who should enter the data; where the data would enter the system; what data would be entered; when the data would be entered; and how the data would be entered. The documentation from the detailed design phase included the specification for the source documents, screens, and input records that need to be designed.

External Source Documents

Figures 9-16 and 9-17 illustrate the specifications developed for the Computer Products Incorporated sales-order form. Although most of the orders are placed by sales representatives in a transaction processing mode, some customers use CPI's form and submit their orders through the mail. Other customers use their own order forms, while others phone in orders. When customers phone in orders, the information is written on a CPI order form. Figure 9-18 illustrates the sales-order form. Since the order form is supplied to customers, it can be referred to as an external source document.

The analyst should make certain that external documents represent the company well. The documents should be in good taste, attractive, easy to complete, and should provide the required information.

Although only 10 percent of the orders are either mailed in by the customers or phoned in, a procedure had to be developed for handling these orders. When the sales department receives the documents they are checked for completeness. If data is missing and cannot be supplied by the sales-order clerk, either the sales representative or customer is called. The completed sales orders are divided into two stacks. One stack contains the orders from new customers and the second the orders from established customers. The orders from new customers are stamped at the top of the form with the date received, logged in the new customer log, and sent to the credit manager.

computer products inc. REPORT SPECIFICATIONS

PURPOSE   PROJECT NUMBER   1051

Specifications for the sales invoice and
the control total report   PREPARED BY: Aguilar

                                          DATE   July 20, 1986

                                                              Page 1 of 2

GENERAL SPECIFICATIONS:

1. The customer or clerk who receives the order by phone should be able
   to identify the data to be entered as easily as possible. Boxes or
   a series of blanks should be used for recording critical data such
   as the customer's number.

2. The format displayed on the screen and the format of the transaction
   record should match the flow of information on the sales order.

3. A form number will be preprinted on the document to serve as a
   reference number when a customer makes an inquiry regarding an order.

4. In processing the sales orders, the date and a sequential order number
   will be stamped on the document. A place should be provided for this
   information which will be used in establishing an audit trail.

5. Four-part forms should be printed. The four copies will be used as
   follows:

   Copies 1 and 2: Retained by the customer. When referring to the order,
                    the customer will use the preprinted order number and
                    the date recorded on the order.
   Copies 3 and 4: Submitted to CIS for processing. Copy 3 will be
                    attached to the sales invoice and filed by customer
                    number. Copy 4 will be filed by preprinted order
                    number.

6. When orders are placed by phone, copies 1 and 2 are retained and
   filed by the person taking the order. Copies 3 and 4 are processed
   in the same manner as the orders mailed in by customers.

REQUIRED DATA:

Field	A/N	Length	Description
Preprinted order number	N	6	Used in forms control.
Date	N	8	Used in referencing the order.
Customer number	N	10	Used to retrieve the customer's record from the database.

FIGURE 9.16 Requirements for the sales order source document (Page 1)

$C^P I$ **computer products inc.** REPORT SPECIFICATIONS

PURPOSE
  Specifications for the sales invoice and
  the control total report

PROJECT NUMBER   1051

PREPARED BY: Aguilar

DATE   July 20, 1986

Page 2 of 2

Field	AN/N	Length	Description
Customer name and address	AN	U*	Provides visual identification.
Shipping code	A	4	Provides routing information.
Sales representative's number	N	3	To be used in determining commissions and in sales analysis.
Sales representative's name	AN	U	Provides visual identification.
Items ordered	AN	5	Used to retrieve the required record from the inventory database.
Quantity ordered	N	5	Used in calculating cost and decreasing the quantity on hand.
Selling price	N	U	Used as a visual reference. The price stored in the database will be used for making calculations.
Extension of price times quantity	N	U	Visual reference only. Extension will be calculated.
Sales tax	N	U	Visual reference. Will be calculated on basis of sales tax code recorded in the database.
Discount	N	U	Visual reference. Will be calculated by the computer using the discount code.

*Items identified with a ''U'' are for visual reference only and are not
entered into the transaction record.

VERIFICATION OF INPUT

  Visual verification will be used to determine whether the correct customer
record and inventory records were retrieved. All numeric data will be
edited. Codes will be verified for validity. A quantity total and document
count will be established.

FIGURE 9.17 Requirements for the sales order source document (Page 2)

☐ NEW CUSTOMER

☐ ESTABLISHED CUSTOMER

ORDER NUMBER: 000001

C^PI computer products inc.

1596 Lake Street
Boston, MA 10264
(826) 972-6193

SOLD TO:

DATE:

ACCOUNT NUMBER: | | | | | | | | | | |

SHIPPING CODE: | | | | |

SOLD BY:

NUMBER: | | | |

ITEM NUMBER	QUANTITY	DESCRIPTION	UNIT PRICE	EXTENSION

Date: XX/XX/XX	INVOICE TOTAL
Time: XXXXX	SALES TAX PERCENTAGE
	MERCHANDISE AND TAX TOTAL
ORDER ID: XXXXXX	DISCOUNT
	NET PAYABLE

FIGURE 9.18 Sales order form

The sales orders from established customers are stamped at the bottom of the form with the date, time, and order identification number. This information will become part of the input record and will help provide an audit trail. After the orders from the new customers have been processed by the credit department, the orders are returned to the sales department, logged in, stamped, and become part of the batch to be processed. Also included in the batch are orders received over the phone.

A batch total is taken on quantity. The fourth copy of the order is filed by the preprinted order number, the third copy is sent with the transmittal form (Figure 9-19) to the clerk who will key in the data. Figure 9-20 illustrates the record layout form used by the programmer to define the fields in the transaction record that will be created. The transaction records are stored in a file and after the data has been entered, the orders will be processed in a batch. Batch processing is used so that the invoices printed on the high-speed printer will be printed in a batch and the control totals can be compared to the totals on the transmittal form.

When the account number is keyed in, the customer's record is retrieved from the database and the operator checks to make certain the correct record was retrieved. The program then checks to make certain the customer's credit limit has not been exceeded. When the item number is keyed in, the product description is displayed.

The sales orders are referred to as external source documents since many of the orders originate outside of CPI. Each copy of the four-part form is a different color which helps in identifying how each part should be handled. An analyst must make certain the form is easy to fill out, includes codes whenever possible, and represents the firm well. Documents should be fairly uniform in size so that they are easy to handle, file, and store. The form illustrated is 8½ by 11 inches.

Other External Source Documents Other external source documents are submitted by vendors and creditors. Typical examples are sales invoices submitted by other vendors to CPI, utility bills, tax notices, and statements for services provided. Often a date stamp similar to the one illustrated in Figure 9-21 is used.

The date the document is received is stamped on the document along with the other information. The person who authorizes the payment usually fills in the vendor number and the account number to be debited. A serial number could also be stamped on the document. The flow of information on the source document will usually not match the order in which the data is keyed into the system since each document is different. However, there are some common procedures that are used in handling these types of documents.

Procedures Used in Processing External Documents The documents are usually sent to the department responsible for **conditioning**. The date is stamped on the document as a reference to when the document was received. The document is checked for completeness, account numbers and codes may be added, and questionable documents are referred to the person designated to investigate the problems. If the due date is not on the document, it must be added. Most firms do not want to pay bills until the actual due date.

After the documents are coded and conditioned, they are sent with a transmittal form to the person responsible for determining the accuracy of the data entry function. The documents are logged in and sent to the person responsible for entering the data. Every person who handles the documents should keep an accurate record of how many documents were received, when they were received,

$C^P I$ computer products inc. SALES-ORDER TRANSMITTAL FORM

NUMBER OF DOCUMENTS: _____

DOCUMENT NUMBERS: FROM_____ TO_____

DATE:_____   TIME_____

TOTAL QUANTITY OF ALL ITEMS ORDERED: _____

- -

TOTALS PRINTED ON THE COMPUTER REPORT:

ITEMS BACKORDERED: _____

INVALID ITEMS: _____

ORDERED BY CUSTOMERS WHO
EXCEEDED THEIR CREDIT LIMIT: _____

TOTAL OF ALL THREE ITEMS: _____

The top portion of the transmittal form is completed by the sales order clerk and attached to the source documents. The bottom portion is completed by the data control clerk. If the two totals, the one taken by the sales order clerk and the total printed on the computer report, do not agree, the cause of the problem must be determined.

If the two totals do not agree, each transaction record must be reviewed and the discrepancy found. Invoices cannot be printed until the two totals are reconciliated.

Form NO: CIS 089
08/15/86

FIGURE 9.19 Sales order transmittal form

RECORD LAYOUT AND TEST DATA

RECORD NO.
Project Number 1081

RECORD DESCRIPTOR
Sales Transaction Record

Invoice date	First order number	Last order number	Order date	Control Data — The four fields of data are keyed in by the terminal operator as soon as the job is initiated. The order numbers are used as a range check.
N	N	N	N	

Account number	Ship-ping code	Sold by	Item number	Quantity	Order ref. number	
N	A	N	AN	N	N	

FIGURE 9.20 Control and transaction record layouts

Date:	XX/XX/XX
Vendor Acct. No.	_____
Acct. No. to Debit	_____
Authorized by:	_____
Due Date:	_____
Reference:	XXXXXX

FIGURE 9.21 Typical date stamp used on external source documents

and when they were returned to the sender. *Everyone who has any contact with the documents is responsible for the safety and integrity of the data. All data—processed or unprocessed—is confidential.*

Internal Source Documents

Documents are needed to supply data to the CIS department or to the data entry operator within a department. Documents are needed regarding additions, deletions, changes, or corrections to be made to databases. *Information stored in a database should not be changed without written authorization.* This is necessary in order to prevent the misuse of information and to ensure its integrity. Usually a form supplies the written authorization and is considered as part of a well-defined audit trail.

Internal documents are usually designed and printed within the organization. Figure 9-22 illustrates the form used to make changes in the accounts receivable database. Each internal form may be printed on a different color paper for ease of identification. All forms must include a form number and the date the form was designed.

Boxes that contain the required codes make filling out the form easier and faster, and there is less chance of making errors. The codes, such as credit and business classification codes, are explained on the form. This also reduces the possibility of errors. Because of the limited amount of information required on the form, it will probably be filled out in longhand rather than being typed. Blanks showing the number of characters that can be entered are used to encourage people to print rather than to write the information. Using the blanks (one space for each character that can be entered) also shows where abbreviations must be used.

When a terminal is used to make the changes or to add new customers to the file, a formatted screen should be designed that looks very much like the source document. The operator fills in the blanks (a / may be used to indicate the number of characters that can be entered) in much the same way the form would be filled in manually. A source document is still needed since it provides a way of collecting the required data and establishes the authorization for the change made in a database.

CHECKPOINT
33. What is the difference between a programmable terminal and a ''dumb'' terminal?
34. What factors should be considered in designing external source documents?
35. On source documents, why are lines or squares used for recording information?

Field size
1

C^PI **computer products inc.** CUSTOMER DATABASE CHANGE FORM

Field size
1

[N] New customer. Fill in all sections of the form.

[D] Delete customer. Fill in customer's account number and name.

[C] Change data in existing record. Fill in the customer's account number and name. Enter the new data for the fields that must be changed.

10
— — — — — — — — — —
Account Number

25
— —
Name of Customer

25
— —
Address 1

25
— —
Address 2

25
— —
Address 3

5
— — — — —
Zip Code

1
Credit Code:
[1] Credit limit 5,000.00   [3] Credit limit 50,000.00
[2] Credit limit 10,000.00   [4] Unlimited credit

1
Business classification code:
[1] Wholesaler   [3] Retailer
[2] Jobber   [4] Consumer

1
Discount code:
[1] 5 Percent   [3] 10 Percent
[2] 7 Percent   [4] 12 Percent
                          [5] 15 Percent

Date record was added, changed, or deleted: _____

Authorized by: _____

Form No: CIS 077
08/12/86

FIGURE 9.22 Accounts Receivable database change form

36. Why are the codes identified on the source document?

37. What relationship should exist between the source document, display screen, and the design of the transaction record?

38. What information is recorded on the transmittal form? Why might the two totals for items ordered disagree? Why should the error(s) be found before the invoices are printed?

39. Why were the sales orders prenumbered if an order number was stamped on the bottom of the form?

40. Why are the first order number and the last order number entered as part of the control information?

CODES

During the punched-card era, when cards were used both for the transaction and master files, it was necessary to condense the data as much as possible to avoid using an excessive number of cards. Also, keypunching and verifying long names and product descriptions were both time-consuming and expensive.

As direct files were supported and written on disk, a code (or key) was needed to randomly retrieve records from the files. As more data is entered from terminals and more queries are made into online databases, the use of codes becomes increasingly important.

When files or databases are designed, it is important to determine the relationships that exist between the fields of data that make up the records within the files. Analysts and users must determine in what ways they may wish to access the data. It must be possible to extract from the database only the information needed, based upon selection criteria (codes) entered into the system.

Codes are used primarily to:

- condense data;
- classify and identify data;
- retrieve or select data; and
- enable one or more courses of action to occur based upon the value stored in the code field.

Codes should be easy to use. Whenever possible, *mnemonic* codes should be used. A mnemonic (or memory aid) code is easier to remember and will result in fewer errors. Also, when mnemonic codes are used, the individuals who audit the documents or reports may be able to identify errors more readily. For example, the following identification number for a performer might contain the following codes:

F P N C 1578   12 82

F = Female   P = Performer   N = Nonvocalist   C = Clarinet player
1578 = Sequential file number   12 85 = Month and year last placed

An individual code should be used for each attribute. Assume that a one-character field was used to code the following information:

F — Single females
G — Married females
M — Single males
N — Married males

It would have been much better to use a one-character field to indicate if the individual was married or single and a second one-character field to indicate if the person was male or female. Better use of mnemonics can be made and the program used to abstract the data from the file or to do one of two things based upon the sex or marital status of the individual will be easier to understand.

Types of Codes	**Simple Sequence Codes** A simple sequence code has no relation to the characteristics of the data. It is possible to code an unlimited number of items with the least number of digits. However, little information can be obtained from the code. Student number is often a simple sequence code. If the numbers are issued in sequence, it is possible to determine in what semester the students first registered. Very little else can be determined from the code.

Block Sequence Codes   In a block sequence code, blocks of numbers are reserved for different classifications. Traditional accounting charts are excellent examples of block sequence codes. Often the 100 series is reserved for current assets while the 200 series is reserved for fixed assets. By looking at the account number, the accountant can readily determine the major classification of either the debit or the credit portion of the entry.

Group Classification Codes   The Zip codes used by the United States Postal Service can be broken into the major, intermediate, and minor classifications as follows:

4	86	02
Section of country	Region within section	Subdivision of region

Significant Digit Codes   Significant digit codes are often used for inventory items that require clerical reference. The codes must be able to be related to the article that is being retrieved and must be easy to change and to extend. For example, the item number might be:

$$1 \ 01 \ 015 \ 01$$

This tells anyone who is familiar with the code that the article should be:

$$1 = \text{round} \quad 01 = \text{iron} \quad 015 = 15 \text{ inches long} \quad 01 = \text{cast}$$

Decimal Codes   Libraries generally make use of decimal coding although it is equally well suited to coding correspondence and other similar material that requires classification on the basis of subject matter. Unlimited expansion is possible. Decimal coding is not as well suited to machine data processing since the code numbers are of varying lengths and tend to be somewhat longer than other types of codes.

Choice of Coding Methods	Usually product codes and codes used to identify various types of customers are already established within a company. When a new system is being designed, the coding methods should be reviewed. If a new code is to be developed, the code designer should consider: the type of material to be coded; the use to be made of the coded data; clerical efficiency and economy; and the acceptability of the code to the personnel who will be working with it.

If codes are chosen that will confuse individuals who must use them, there will be a higher error rate. Often codes are explained on the source document, which helps to reduce errors. Within the computer, it is possible to break the code into its components to check that each component is valid.

CHECKPOINT

41. Why are codes used so extensively in EDP?
42. Zip codes are examples of what type of coding system?
43. Libraries usually make use of what type of coding system?
44. What is a mnemonic code?
45. In the student database, two digits are used to identify the high school from which you graduated. If you graduated from Riverside High School and a 44 is recorded in your record, is the 44 considered a code?

SUMMARY

In the detailed design phase, the formats of the output and input are determined. Each program must be designed so that the input is efficiently processed and valid output is produced. Although the output is designed first, the specifications from the design phase should show the source of data needed for the output and specify what medium is to be used. The analyst must know whether the data is read in from a file, entered from a terminal, or calculated. However, it is not until the detailed design phase that the precise format of the output is determined.

When the layout of the output is reviewed by the project team and the user, omissions may be found that require additional input or calculations. Changes in one part of the data processing cycle—input, processing, or output—usually require changes in the other two parts of the cycle.

Although there must be enough printed output to provide a clear audit trail and to provide evidence that the input was processed accurately, the necessity for any printed report must be established.

Analysts, users, and management are all concerned about the high cost of preparing and entering data into the system. Today there are many options to choose from regarding where data is recorded, when it is recorded, who records the data, and how it is recorded.

The analyst and user must weigh the advantages and disadvantages of the various input media available. Regardless of the method used, the data must be verified in as many ways as possible and a well-defined audit trail must be established. Valid output can only be produced when error-free input is processed.

Well-designed source documents and screen formats help to eliminate errors. When the data is to be handwritten on forms, blanks or squares should be provided for recording the data. Also codes should be identified and checklists should be provided whenever possible.

The analyst must provide for the safety and integrity of the data as it moves from one location to another. When online systems are used, provision must be made for alternate methods of handling input in the event of a system failure.

Although a formal presentation is made at the end of the detailed design phase, the designs for source documents, input records, and output should be approved because the users were directly involved in determining what was needed and how it should be formatted. The documentation for the detailed design phase will include record layout forms, print layout forms, screen designs, and source documents.

DISCUSSION QUESTIONS

1. In regard to output, what decisions are made in the feasibility study and what decisions are made in the detailed design phase of the project?

2. A programmer/analyst used the output specifications for the paycheck program and designed a new paycheck. During the implementation phase of the project, it was discovered that a check number was not printed on the check. When the accountant saw the check, he was concerned that a number had not been printed and the payroll department manager was angry because not enough information was printed on the check stub. If you were the CIS manager investigating the problem, what might you expect to find? Why was the accountant concerned that the checks were not numbered? As the CIS manager, how would you handle the situation? There are 99,957 preprinted paychecks in stock.

3. Sally Mitchell, a newly appointed programmer/analyst, was asked to convert the existing payroll system to an interactive program. Sally was to determine which programs should be run in an interactive mode and make the necessary conversions. Sally made the decision that all programs that printed reports would be run in a batch mode and all of the same reports should continue to be printed. Do you agree with Sally's decisions? What else do you feel Sally should have done?

4. If no constraints were imposed upon the design of the sales/accounts receivable/inventory system, how would you recommend that each of the situations described be handled? Remember that each sales representative has an intelligent terminal. Give the rationale for your decision.

   a. Under the old method, cash received on account was batched and processed the day after it was received. The company receives mail four or five times a day. Under the new system, should the same method be used?

   b. Before an order can be placed, the customer's balance must be checked to see whether it exceeds the customer's credit limit. How can this be done in an interactive mode? What should be added to Screen 1 which is illustrated in Figure 9-14?

   c. Sales representatives call on customers and must be aware of all price changes. Under the old system, price changes were mailed out once a week. When a price change occurs, the catalogue price is obsolete until a new catalogue is printed. If sales orders are processed in an interactive mode, how might the price changes be handled? Is more than one solution available?

   d. In the design of the sales/accounts receivable/inventory system is there any place where a turnaround document might be used? Keep in mind that scanners that can read typeface are now relatively inexpensive.

   e. Which of the following reports would you design as internal reports on stock paper and which would require the use of preprinted forms?
      (1) sales invoices
      (2) daily control report listing all orders placed
      (3) daily report listing all items that had to be placed on backorder
      (4) accounts receivable statements
      (5) daily exception report listing items that were below the reorder point and those items within 10 percent of the reorder point
      (6) daily exception report of all customers who were within 10 percent of their credit limit
      (7) credit memo for credit granted to a customer for returned or defective merchandise

   f. Which of the above reports might be considered as part of an audit trail?

5. Why is it important that the documentation include the retention time for all files and reports?

6. A new member of the payroll department was told that the payroll registers that were over four years old no longer needed to be retained in the semi-active file. The employee disposed of them by throwing them in a wastebasket. The department manager found the registers and was very upset. What should have been done? If you were the payroll manager, how would you handle the situation?

7. Why are analysts and management more concerned about the methods used for entering data into the system than they were in the past?

8. You have been asked to design a form used to add records to, delete records from, or change the records stored in the inventory database. What data must you have before you can design the form? What guidelines or standards would you follow in designing the form?

9. Discuss both the advantages and disadvantages of distributed data entry. You are the payroll manager and would like to have one of your employees maintain the payroll database. Give the rationale that you would use to convince the CIS manager that your request should be implemented.

10. What is the major disadvantage in using card data recorders (keypunches), key-to-tape, key-to-diskettes, or some types of terminals for entering data into a computerized system?

11. An analyst that you just hired has the responsibility of designing the procedure to be used for inputting the data concerning the weekly payroll into the system. The analyst has the data entry operator key in the employee's number, name, regular and overtime hours, salary, number of exemptions, prior earnings, and amount previously withheld for social security. While all of the fields are visually verified by the operator, additional verification methods are not used. The data keyed in is recorded in a transaction file. After all of the data is entered, the transaction file is used as input to the payroll register program. You have called in the analyst to discuss the procedure. How would you handle the situation? What would you tell the analyst? What might you suggest the analyst do to learn more about inputting data, files, and databases?

12. An analyst that you supervise designed a new student database. The college is rather small and most of the students come from 1 of 25 local high schools. At the present time, there are 7,894 student records. A portion of the record layout to be used in creating the database is illustrated below:

Field size	Contents
25	Name of high school graduated from
25	Name of the family doctor
25	First name of student
25	Last name of student
6	Either ''male'' or ''female'' is entered
4	Student number
30	Name of father
30	Name of mother
30	Curriculum student is enrolled in, such as Business Administration.

If the rest of the file is similar to what is illustrated, what problems regarding content and field size would you point out? Why would this be of concern to a CIS manager or the department manager whose budget is charged for the cost of developing and maintaining the system?

Team or Individual Projects

1. Prepare the report specifications for the check that will be used by Computer Products Incorporated to pay their accounts payable and design a voucher check. The top portion of the voucher check is deposited and the bottom portion containing the listing of invoices or statements being paid by the check is retained. Three-part forms are needed since one copy is sent to the customer, one copy is sent to the accounts payable department, and one copy is sent to accounting.

   Before designing the check, you may wish to look at some examples of voucher checks. You may have received one through the mail or examples may be shown in an accounting text.

   In the accounts payable database is stored the name and address of the company to whom the amount is owned. For each invoice or statement, there is an identifying number, code used to print a description, the amount to be paid, and the date on which the amount must be paid. According to CPI policy, checks are printed two days prior to the due date. The operator enters the dates for which checks are to be printed and each account is searched to see whether there are payments to be made. The program that prints the check is run Monday through Friday.

   Directions:
   a. List the report specifications on a form similar to the one illustrated on pages 326-327.
   b. Use a print layout form and design the voucher check.

2. Prepare a print layout form for the internal report that lists to whom the check was paid, the vendor's account number, check number, and amount of the check. An appropriate heading is to be included, each page is to be numbered, and the total amount paid is to be printed after all checks are listed.

   Directions:
   a. List the report specifications on a form similar to the one illustrated on pages 326-327. The program that prints the voucher checks generates the data to be printed on the report. After the voucher checks are printed, stock paper is put in the printer and the report is printed.
   b. Use a print layout form and design the accounts payable check voucher report.

3. The records stored in the inventory database contain the following information:

Field	Size	Field Type
Item number	5	AN
Date item first stocked	6	N
Description	25	AN
Items on hand	6	N
Items flagged for delivery	6	N
Cost price	7,2	N
Selling price	7,2	N

Reorder point	6	N
Reorder quantity	6	N
Vendor codes (up to 4)	4 each	AN

When the size is shown as 7,2 it indicates that the field is seven digits long with two places beyond the assumed decimal place. The field might have been shown as XXXXX.XX.

Under field type, N is used for numeric data, AN for alphanumeric data, and A for alphabetic data.

The inventory file is somewhat incomplete but has enough information for the required activities.

Directions:

a. Prepare a source document that would be used to record the required information to add, delete, or change records.
b. Design the screen (or screens) that would be used as the data is entered by the terminal operator.
c. Indicate how each field of information entered would be verified.
d. Based on CPI's product line, what type of code might be used in determining the item number? You may need to refer back to the first part of the case study to answer the question.

GLOSSARY OF WORDS AND PHRASES

burst mode Data is transmitted or printed without interruption.

burster A device used to separate the various pages of a report. Bursters can also slit the pin-feed hole margins from forms.

COM (computer output to microfilm) the method used for producing microfilm from computer output.

Conditioning documents Forms are checked for completeness, codes are added, and an attempt is made to identify any incorrect data.

decollator A device used to remove the carbon paper and separate the various copies of a report. The process is called decollating the report.

OCR (optical character recognition) Human-readable data is converted into computer input by using light reflected from either handwritten or printed characters.

OMR (optical mark recognition) Marks are converted into computer input. The placement of the mark determines how the data is interpreted for processing. OMR equipment is often used for inputting data recorded on test scoring sheets and on survey forms.

scrolling Information is displayed on a VDT without interruptions. If a large amount of data is displayed, lines scroll off the top of the screen as new lines of data are displayed on the bottom of the screen.

stock paper In a given company, several types of paper may be referred to as stock paper. Common sizes, such as 11 by 13¾ and 8½ by 11 inches are found in most installations. Usually one-part, two-part, three-part, and so forth stock paper is available. When stock paper is used, all printing is added by the company's line printer or letter-quality printer rather than being preprinted on the form.

tables Areas reserved within real or auxiliary memory that contains data that can be accessed either randomly or sequentially. Using tables decreases coding. Tables are also called arrays.

two-up forms A format permitting identical forms to be printed side by side. For example, if eight copies of the W-2 are needed, two originals each with three carbon copies can be printed side by side on two-up forms.

UPC (universal product code) A code developed for the supermarket industry for identifying products and manufacturers on-product tags. The 10-digit bar code can be read by optical scanning devices.

STUDY GUIDE 9

Name _____

Class _____ Hour _____

A.   Indicate whether the following statements are true (T) or false (F) in the space provided. If false, indicate in the margin how the statement should be changed to make it true.

_____   1.   The hardware and software to be committed to the project are determined in the detailed design phase.

_____   2.   Report specifications are developed during the general design phase of the project.

_____   3.   The programs needed to process data and a detailed logic plan for each are determined during the general design phase.

_____   4.   When doing the detailed design work necessary for a program, the input formats are usually determined before the output formats.

_____   5.   When a detailed logic plan is developed, it may be necessary to revise the input and output formats.

_____   6.   The organization for which you work as an analyst has a card reader, magnetic tape and disk drives, terminals, and a high-speed printer. Since your company does not use scanners, videodisks, or voice input and output, it is not necessary for you to learn about these devices.

_____   7.   Although card readers are considered very fast input devices, the use of cards for either input or output is decreasing.

_____   8.   Magnetic tape is often used for backing up files and for storing history files.

_____   9.   Pictures, graphics, and text can be stored on videodisk.

_____ 10.   OCR (optical character recognition) equipment can only read marks placed in designated areas.

_____ 11.   Light pens are used to input data into programs that create graphics or in CAD/CAM applications.

_____ 12.   Today if a company phases out the use of punched cards, it is almost certain that diskettes will be used in place of the cards.

_____ 13.   Sensors can be used to input data into a system. A typical application is the use of sensors to monitor patients.

_____ 14.   Graphic tablets can only be used with minicomputers.

_____ 15.   A mouse is used to enter large amounts of data.

_____ 16.   Magnetic tape is the most widely used input medium for minicomputers and mainframes.

_____ 17.   Diskettes are used exclusively for microcomputer systems.

_____ 18.   Since fewer reports are being printed, in the future it is unlikely that printers will be considered as part of a basic computer system.

_____ 19.   Scanners are often used to read turnaround documents.

_____ 20.   There are many types of scanners available. Usually they fall into one of the following categories: UPC, OCR, OMR, or MICR.

_____ 21.   All reports should be printed on continuous-form stock paper.

_____ 22.   In designing sales invoices, the analyst should consider how the document will be handled for mailing and must also consider postal regulations.

_____ 23. In some applications, punched cards are used as turnaround documents.

_____ 24. The use of magnetic disk is limited to storing transaction files and databases.

_____ 25. COM (computer output to microfilm) is used to store price lists and catalogue information.

_____ 26. Red Xs are used to indicate variable information to be printed on a preprinted form.

_____ 27. The information printed on the form design for an internal report represents constant information that will be printed along with the variable information.

_____ 28. After a form is designed, written approval should be obtained from the user.

_____ 29. After an external report is designed and approved, the company that will print the forms submits a proof form for final approval.

_____ 30. One of the rules that is generally followed in displaying screens of information is to clear the entire screen before displaying new information.

_____ 31. Each time a screen is to be used, it must be designed. There is no way to store screens used in more than one program.

_____ 32. Analysts within the organization should be able to develop their own conventions that will be used in designing screens.

_____ 33. When data is to be entered, the only prompt that is displayed is the name of the field, such as ACCOUNT NUMBER.

_____ 34. When the operator who enters the data has an opportunity to visually check the data, there is no need for the programmer to include input validation routines in the program.

_____ 35. Terminals are sometimes available that can be used by individuals who wish to request information similar to what might be obtained by asking a receptionist.

_____ 36. The four major sources of input are transaction files, master files and databases, control information, and data entered from some type of terminal.

_____ 37. Transaction files are used over a long period of time while a master file may only be used for the current fiscal period.

_____ 38. Control information is data that is entered once at the beginning of a program such as the beginning and ending sales-order number.

_____ 39. When the sales-order data is keyed in, it is not necessary to have the accounts receivable database online.

_____ 40. Traditionally, data was first recorded on a source document and then keyed into some type of machine-processable record by an operator.

_____ 41. The size of the record stored in a transaction file could be determined by the medium being used.

_____ 42. At the present time your college has 895 employees. Therefore, the field size to be used for the employee number should be three digits in length.

_____ 43. If data is recorded in a punched card or on a diskette and then verified by a second operator, it takes only half as long to verify the data as it took to record the data.

_____ 44. The data recorded on the sale order is keyed in by a terminal operator and stored in a transaction file. A report listing the records keyed in by the operator might be considered as part of the audit trail.

_____ 45. A code is used in developing item numbers. The numbers are not in any particular sequence and contain letters and digits. A self-checking number routine could be written using a mathematical formula to determine whether there is a transposition or the number has wrong digits or letters.

_____ 46. As a general rule, whenever a field that contains all digits is entered, an edit routine should be used that makes certain no letters or special characters were recorded in the field.

_____ 47. Because professional data entry operators are more productive than employees in the sales department (who perform other tasks as well as data entry), the trend is to centralize data entry.

_____ 48. Optical readers are well suited to applications that require large amounts of data be recorded on a document and then entered into a computerized system.

_____ 49. The number preprinted on the top of the sales order is used to establish an audit trail.

_____ 50. When a date stamp like the one illustrated in Figure 9-21 is used, the data that must be recorded regarding the transaction is clearly identified and in a uniform format.

B.   Multiple choice. Place the letter of the correct answer in the space provided.

_____ 1. A device used to separate various pages of a report.
   a. decollator   b. burster   c. OCR

_____ 2. Used to read a code printed on a container or label such as one that might appear on a box of corn flakes.
   a. OCR   b. UPC   c. OMR   d. MICR

_____ 3. In designing a system, one of the critical issues is
   a. the quality of paper used to print internal reports.
   b. the color of the paper used to print external reports.
   c. the timeliness of the data.

_____ 4. The trend today is to
   a. print more reports.
   b. print only exception reports.
   c. display more data on VDTs.

_____ 5. Terminals, or VDTs, are considered a standard output device since
   a. about half of today's minicomputers or mainframes support terminals.
   b. 60 percent of today's minicomputers or mainframes support terminals.
   c. over 80 percent of today's minicomputers or mainframes support terminals.

_____ 6. In designing screens it is important that
   a. conventions are established and followed.
   b. instructions are clearly stated.
   c. an opportunity is provided to correct data that is entered incorrectly by the operator.
   d. the factors listed under a, b, and c are followed.

_____ 7. The output specification form
   a. only lists specifications for the report or display.
   b. lists the specifications and the variable data.
   c. lists the specifications, variable data, and the source of the data.

_____ 8. External reports should be well designed, contain all the required information, and
   a. always be printed in a batch mode.
   b. be printed on a letter-quality printer.
   c. should present a positive image of the organization.

_____ 9. Information stored in a database should
   a. only be updated once a week.
   b. not be changed without written authorization.
   c. require very little maintenance since most databases are not very volatile.

_____ 10. A typical chart of accounts used in accounting is an example of a
   a. simple sequence code.
   b. group classification code.
   c. block sequence code.

_____ 11. Codes are used primarily to
   a. make it difficult for someone to obtain information.
   b. condense data.
   c. make it easier to obtain information.

_____ 12. Using a reference number, such as the order number, on the sales order, transaction record, and sales invoice
   a. is unnecessary if the customer's number is on all three items.
   b. requires additional keystrokes and is considered costly.
   c. helps to establish a clear audit trail.

_____ 13. Your project leader indicated that "If data is visually verified while being entered, no other verification of the data is necessary." Other analysts would probably disagree because
   a. all data should be key verified.
   b. all fields should have a self-checking number.
   c. as many ways as possible should be used to verify the correctness of any data that enters the system.

_____ 14. The contents of the overtime hours field could be edited.
   a. by using self-checking numbers.
   b. by checking between a range of numbers.
   c. by using a reasonableness test.

_____ 15. "Human-readable data is converted into computer input by using light reflected from either handwritten or printed characters" describes
   a. OCR.   b. ORM.   c. MICR.   d. UPC.

10 DETAILED DESIGN: FILES AND DATABASES

Looking Ahead

After reading the text and completing the learning activities you will be able to:

- Describe the differences between databases and files.
- Identify when a file rather than a database should be used.
- Identify the applications for which magnetic tape should be used and define the major terms associated with magnetic tape.
- Define the major terms associated with the use of magnetic disks.
- Identify the major differences between ISAM, VSAM, direct, and sequential files.
- Describe the differences between the primary and alternate keys and explain how each is used to retrieve records from a VSAM file.
- Identify the major advantages provided when optical-disk memory systems are used rather than magnetic-disk memory systems.
- Explain how after the first invoice for a customer is processed in a hierarchial database, the record containing the second invoice for the customer can be randomly retrieved.
- Identify the major characteristics of a relational-model database.
- Explain how each of the three types of database languages is used.
- Explain the type of data stored within a data dictionary.
- Define and utilize the words and phrases listed in the end-of-chapter glossary.

INTRODUCTION

Files and databases are the backbone of any data processing system. The computer industry changed completely once it became possible to store massive amounts of data online. The first breakthrough came with the use of magnetic tape. However, records stored on tape can only be accessed sequentially. Therefore, when IBM introduced the 305 RAMAC (RAMdom ACcess), data processing made a quantum leap into the future. Magnetically coated aluminum disks stored volumes of data on concentric tracks. In a fraction of a second a read/write mechanism found the correct magnetic spot and retrieved data from any point on the disk. The RAMAC's 50 spinning disks contained 5 million characters of information. By today's standards, RAMAC would rate very poorly. However, it marked the beginning of a new era in file processing.

Regardless of their function, all files and databases should be considered a collection of records. However, you will note as you read the material regarding databases, different terms are used to describe a database than are used to describe files. The records within the files or databases can be organized in different ways and serve different purposes. Files and databases can be stored on different media such as magnetic tape, magnetic disk, or **optical-disk memory systems.**

An analyst must be familiar with the technical advancements that are being made in storage media and in the software that supports files and databases. In designing a system, the analyst must determine when the data should be part of an online database and when a file should be used. Figure 10-1 summarizes some of the differences that exist between files and databases. Often management or users use the term database when they are referring to files stored on a medium such as magnetic tape or disk.

If a database becomes too large, a great deal of computer time (called overhead) may be required to manage the database. Often within an organization there is a database administrator who is responsible for the integrity and management of the databases. Data stored in both files and databases must be updated so that it is relevant and timely. Procedures must also be developed and followed that provide for the security of the data.

USES OF FILES AND DATABASES

Files and databases are used for the following functions:

Databases. Sometimes described as a collection of interrelated files, databases generally serve a wide range of applications. Databases are generally online at all times since they are usually used to support interactive transaction processing systems. Records are updated as transactions occur. Additional maintenance is required to make corrections or changes in records, add new records, or delete records from the database.

Master files. The records contain constant and updated information used as input to a number of different programs. If the file is processed in an update mode, a record can be read, selected fields changed, and the record rewritten back into the original location in an updated form.

Transaction files. The records contain data that is used for a relatively short period of time. Transaction files are created by keying in data from source documents or by recording data that enters the system while transactions are being processed. Often the file is considered as a "backup" for the transactions that entered the system from remote locations and is also part of the audit trail.

Temporary files. One program may produce three or four different output files that contain information used as input to other programs which produce additional reports. The temporary files may be sorted prior to being used as input into another program. The programs that use the files may perform additional calculations on the data, print reports, and then delete the temporary files.

Files/Database

Files	Databases
1. Supported by programming language and operating system.	1. Requires an additional software package that interfaces with the operating system.
2. Organized as a sequential, direct, or indexed file.	2. May be hierarchical, network, or relational model.
3. Separate files for each system or application.	3. Provides data for multiple applications.
4. New programs are written whenever different data stored in the file is to be accessed. Programmers write the programs.	4. User-friendly query languages used to access and format data. Interactive query languages can be used by nonprofessionals.
5. Less information can be obtained from a given amount of data.	5. More information can be obtained from a given amount of information.
6. Depending upon the organization, only one key may be available to randomly access records.	6. Usually a record stored in the file can be accessed using one of many fields that contain unique data.
7. Redundancy of data in separate files.	7. Redundancy of data eliminated.
8. Less integrity of data since each separate file might not be updated.	8. Greater integrity of data as only one source needs to be updated.
9. Program/data dependency.	9. Program/data independency.
10. Data dictionary not supported.	10. Usually supports a data dictionary.
11. Data management is more difficult.	11. Database manager responsible for the creation and maintenance of the database.
12. More labor intensive which results in higher programming and maintenance costs.	12. Less labor intensive which results in lower costs.
13. Data is less vulnerable as it is often only online while being used.	13. Since databases are often online at all times, the data is more vulnerable.
14. Less "overhead" needed and operating costs are less.	14. More "overhead" needed and operating costs are greater.

FIGURE 10.1   Comparison of files and databases

Communication between programs. When a limited amount of memory is available, it is often necessary to divide a large program into three or four segments. The data processed by the first segment is transmitted by means of a file to the second program.

Communication between computer systems. Because it is standardized, magnetic tape is often used as the communication link between computer systems.

Queues and libraries. Most operating systems are dependent upon having a large amount of online disk storage for the various components of the operating system. Queues (temporary files) are also needed for jobs that are to be executed and for information to be printed or punched into cards.

Backup for both master and transaction files. The importance of file backups cannot be stressed too much. While most large organizations with mainframes have established backup procedures, many microcomputer users do not back up files. If a diskette is damaged and cannot be used, it may take hours of time and effort to recreate the files. If a microcomputer has only one disk drive, a copy of a diskette can still be made.

In general design phase of the project, the various files or databases that are needed are defined. As the output requirements are identified, the data needed to produce the output is determined. Usually the medium that the file will be stored on is also determined in the general design phase. During the detailed design phase, the analyst must decide the size of each field and the type of data that will be stored in the field. If the system being created is closely related to an existing system, the database administrator may approve the addition of new fields to an existing database.

LOGICAL AND PHYSICAL RECORDS

The blocking factor used in storing records on physical devices such as magnetic tape and disk is determined by the programmer or the analyst. The larger the blocking factor, the faster records can be processed. For example, in one system if 80-byte tape records are blocked 100 records to a **physical record**, it takes less than a minute to copy the tape onto a backup tape. However, if the records are unblocked, the job takes more than four minutes. A physical record contains the amount of data that will be read from or written on the medium being used when one ''read'' or ''write'' command is executed.

Also in using magnetic tape, when a large blocking factor is used, more data can be stored on the a reel of tape than when a smaller blocking factor is used. As indicated in Figure 10-2, 2.5 times as much data can be recorded in two inches of tape when five **logical records** are recorded in one block. A logical record contains data about one person or subject.

FIGURE 10.2 Blocked and unblocked physical records stored on magnetic tape

PROCESSING DATA STORED ON MAGNETIC TAPE OR DISK

Figure 10-3 illustrates the concept of tape or disk processing. If a file stored on tape or disk contains 240-byte records blocked ten records to a physical record it would take in excess of 4,800 bytes of real memory to create or process the records stored on the tape. Usually two buffers are used for each input and output file. Each time an application program is stored in the computer's memory for execution, the appropriate number of buffers is allocated for the storing of information read from or written on magnetic tape or disk. The amount of space reserved for each input and output buffer depends upon the size of the physical record being read or written.

When a file is used for input, a physical record is read into the buffer during the execution of one ''read'' command. The operating system's software takes care of unblocking the records. The programmer is concerned with processing logical records. While the records from the first buffer are being processed, the operating system takes care of filling in second buffer. After the records in the first buffer are processed, provision is made to process the records from the second buffer. The first buffer is filled with new records while the records in the

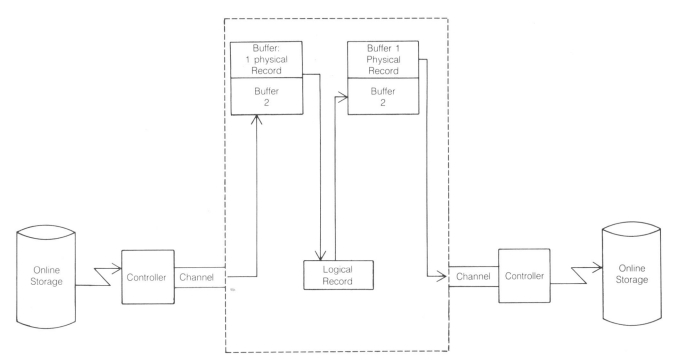

FIGURE 10.3 Physical records are read into and from memory. Operating system software unblocks the records so that a logical record can be processed. The logical records are then reblocked to form physical records

second buffer are being processed. Each buffer holds the number of records identified as the blocking factor. In addition, there must be room within the CPU's memory for the commands which manage the blocking and unblocking of records.

While separate buffers are maintained for both input and output operations, when records stored on magnetic disk are being updated, the same channel, controller, and disk drive are used for the read and write commands. The controller acts as a "traffic cop" and makes certain the incoming blocks of data do not get in the way of the data being written out to disk. If data is read from tape and written on disk, separate channels, controllers, and storage media would be used. Channels are used to transfer the data to and from the CPU and the controllers. The channels for large mainframe computers are often located below the floor; microcomputer channels look like pieces of brightly colored ribbon and are usually visable. The controller may be built into a disk or tape drive or may be a free-standing unit that is located adjacent to the disk or tape drives. A microcomputer controller is usually a circuit board that is located within the CPU housing.

Some operating systems do not support the blocking or unblocking of records. When this occurs, the records should still be blocked in order to more effectively use the storage space on the disk or the tape. The programmer will have to supply the instructions needed to block or unblock the records. Also some computer systems have a fixed record size of perhaps 512 bytes. If small records are being processed, it may be up to the programmer to see that more than one record is recorded in the 512-byte block.

When selecting a system and determining the language that will be used, the magnetic tape and disk file-handling facilities supported by the system should be investigated.

TYPES OF STORAGE MEDIA

Records that make up a file can be stored in punched cards or on diskettes, hard disk, cassette tape, magnetic tape (reel), a **MSS system**, or an optical-disk system.

The use of punched cards for files is almost nonexistent. Cards are bulky, must be stored in a humidity- and temperature-controlled environment, and can store a limited amount of data.

Diskettes are widely used for storing files and databases used with microcomputer systems. The major problem is using diskettes is that each computer system formats data differently and standards are not available regarding how data must be stored and organized on a diskette. Although one computer company has indicated that diskettes created on different computer systems can be read by their computer, as a general rule, a diskette written under one operating system cannot be read by a computer that has a different operating system. Diskettes are very inexpensive, can be mailed, and the amount of data stored on 8-, 5 ¼-, or 3-inch diskettes is increasing. Accessing data stored on a diskette is slower than when hard disks are used.

Cassette tapes, similar to the ones used on voice recorders, were first used on terminals and minicomputer systems. When microcomputers were first made available as "personal computers" cassette tapes were used for storing both files and programs. Shortly after personal computers were available, most companies developed low-cost diskette drives that replaced cassette tape recorders. Larger versions of cassette tapes are used with some minicomputer systems for tape backup. The tape is approximately the size of a videotape and can store a large amount of data.

Mass Storage Systems

Mass storage systems designed by IBM and Control Data Corporation combine the low cost of tape with the random access capabilities of disk. Control Data Corporation's MSS stores data on 770-inch strips of magnetic tape that are stored in cartridges. Each cartridge is two inches in diameter and four inches in length. The tape stored in the cartridge can hold 50.4 million bytes of data. It is difficult to comprehend the magnitude of MSS systems. One of the larger IBM systems can store up to 472 billion bytes of data—equivalent to the amount of data stored on 4,720 disk packs (3330s) or on a tape library containing 47,300 reels of tape.

To the programmer, using data stored on an MSS system is the same as using data stored on magnetic disk. The software which manages the MSS system selects the cartridge that is needed and takes care of transferring the files to magnetic disk for processing. After the job is finished, the updated data is transferred back to magnetic tape, the tape returned to the cartridge, and the cartridge stored in its proper location—all of which occurs without the help of an operator. Although the transfer of data is fast, MSS is primarily considered as a replacement for conventional magnetic tape. The process is too slow to be considered for databases that must be used in interactive systems.

Only organizations that have large mainframes can cost-justify the use of MSS systems. Also, few improvements have been made in the systems since they were first announced.

Magnetic Tape

Although there are many unique applications for magnetic tape such as the MSS system described, when the term magnetic tape is used, most people are referring to the standardized reel of magnetic tape. The major disadvantage is using magnetic tape is that records can only be accessed sequentially.

The major advantages of using magnetic tape are that it is an inexpensive method of storing large quantities of data and it is standardized. If either the standard **ASCII** or **EBCDIC** code is used, the tape can be input for a large number of different types of computers and other peripherals such as offline printers, print subsystems, and COM systems. Often reports such as those submitted to the Internal Revenue Service or to a credit union are on magnetic tape.

Some of the newer magnetic tape subsystems have controllers that take the responsibility of sequentially for the right record, reading the record, and transferring the data into the computer for processing. This frees the host computer for other, more meaningful, tasks. Some drives also support the updating of tape records. This means that a record can be read, selected fields changed, and the record rewritten *without creating a new tape file.*

Storing Data on Tape   The amount of data that can be stored on tape and the rate at which the data can be transferred from tape into the CPU are determined by the **CPI (characters per inch)**, the blocking factor, the **IPS (inches per second)**, and the type of tape drive used. Some of the newer tape drives provide for **streaming tape.** Streaming means that the data is written onto the tape on the fly—without starting or stopping between blocks of data.

CPI is also referred to as **BPI (bytes per inch).** The term is used to indicate the number of characters that can be stored on one inch of magnetic tape. If the EBCDIC code is used for storing data on nine-channel tape, the data can be packed and two digits can be stored in each location. The most common CPIs are 800, 1,600, 3,200, and 6,250. When IBM announced their new 3480, tape-drive density was increased to 38,000 CPI. The drive also uses 4- by 5-inch cartridges rather 10 ½-inch reels. The transfer rate between the drive and the mainframe was also increased to 3MB (megabytes) per second. A megabyte is a million bytes of data.

Usually only very large installations run 6,250 (or higher) CPI tape. The most common density used is 1,600 CPI. Often when reports are to be submitted on magnetic tape, the specifications often indicate that 1,600 BPI, nine-channel tape, with data recorded using the EBCDIC code must be submitted. In obtaining tape drives, a premium is paid for increased speed and density.

Another factor that determines the amount of data that can be stored on a reel or cartridge of tape is the size of the **IRGs (interrecord gaps).** Some tape systems require only .3 inches for IRGs while other systems require .75 inches or more for the IRGs. The IRGs (also called IBGs or interblock gaps) provide for the deceleration and acceleration of the drive mechanism during the read/write operations.

While the size of the IRGs is determined by the hardware being used, the analyst or programmer determines the blocking factor for a file. The analyst must consider what programs will use the tape being created and the amount of memory that can be allocated to the file.

Protection of Magnetic Tape Files   Usually the operating system has a number of features that provide for the integrity and safety of data stored on tape. The software usually performs a **parity check** on data written into a file or read from a file into the CPU. Label checking is provided, which prevents the data on the tape from being overwritten inadvertently and also checks that the right file is being processed.

After a block of data is written on tape, a parity check is performed to make certain the data is valid. If an error is detected, the tape is backed up and the record rewritten. If necessary, the system stays in the loop (writing and checking) ten or fifteen times. If the record written on the tape is still invalid, a message will be transmitted to the operator indicating that there is a hardware failure on the tape drive. If the problem was created by a fleck of dust, when the tape is backed up and the record rewritten the problem is usually corrected. If the job cancels, either the tape is defective or there is a hardware problem. If the operator has failed to clean the read/write heads this could cause the job to cancel.

The *header label*, illustrated in Figure 10-4, is written on the tape from information supplied by the programmer. A *trailer label* is placed at the end of the file. The labels usually contain the date the tape was created, the expiration date (how long it will be protected), and the name of the file. Whenever a file is created, commands from the system check whether the expiration date on the tape in use is past. If it is an unexpired file, the operator will be notified. It is impossible to write over data that is protected by an unexpired date without knowing it has been done.

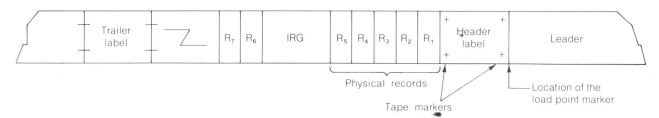

FIGURE 10.4   Header labels help to protect the data stored on magnetic tape

When the tape is mounted as input, the system checks to see that the right reel of tape is being used. The label on the tape is checked with information supplied by the programmer. Usually the information is supplied to the system by using job control language. In some cases the data used to check the validity of the file being used is internal to the program that uses the file.

Tape is also protected against overwrite by a file protection ring. A plastic ring is inserted into the tape reel whenever the tape is used for an output file. Thus the expression "no ring no write" is often used. Once the file is created, the ring is removed. If a tape without a ring is mounted on a drive that is to produce an output tape, the operator is notified.

The CIS manager should also see that additional provisions are made for the protection of files (either tape or disk). Clearly defined procedures must be developed for:

- externally labeling the files so they can be identified;
- storing the files in a location where there is limited access—only authorized personnel should be in the area and have access to the files;
- storing the files so that they are protected against damage from fire, water, or other catastrophes; and
- backing up files. The backups should be stored in a separate location.

When selecting a system, the project team should determine what provisions are made for checking the validity of data. If the system does not support a parity check, the programmer should provide additional routines in the program that will check the validity of data being read in from a storage medium or being written as output.

CHECKPOINT

1. What development in the data processing industry marked the beginning of a new era in data processing? Why was such a totally different data processing environment created?
2. What factors concerning the use of files and databases are determined during the general design phase? What remains to be done during the detailed design phase?

3. What seems to be the major problem concerning the use of diskettes?

4. When programmers use data stored on an MSS system, are they concerned with how the data is organized on the magnetic tape that is stored within a cartridge? Is a computer operator responsible for transferring the data to magnetic disk for processing?

5. What is the major disadvantage in using magnetic tape?

6. What is the major advantage in using magnetic tape?

7. When magnetic tape is being used, who determines the blocking factor? Why should tape be blocked rather than reading, processing, and writing one record of tape at a time?

8. What are the differences between the EBCDIC and ASCII codes?

9. In processing disk or tape records, how are buffers used?

10. What is a parity check?

11. Why are header labels written on tape?

12. For what type of application is magnetic tape used? Would it be appropriate to store a database on magnetic tape?

Magnetic Disk

Disks drives can be obtained that hold a single disk or several disks that are referred to as a pack or spindle. When selecting a computer system, very important considerations are the type of disk drives supported by the computer system and the amount of software support available for file and/or database handling. The team selecting a new computer system should be concerned about the following issues.

Capacity. Analysts are concerned about the amount of data that can be stored on one disk pack or drive and the number of drives that can be attached to the computer system. Often one of the limiting factors regarding the use of microcomputers is the amount of online data that can be stored on disks.

Organization and access methods. The organization method determines how data will be stored within a file or database. The access method refers to how the data can be obtained from the file or database. The organization method used determines how data can be accessed and how long it will take to randomly or sequentially retrieve records. Figure 10-5 shows the relationship between the organization of the file and the way records can be accessed. Some of the older computer systems that are still being used make it difficult to work with files; newer computer systems have improved file-handling capabilities.

Access time. Access time is the length of time needed to find a record stored on disk, read the record, and transfer the data into the CPU for processing. Access time is made up of seek time, rotational delay time, and transfer time. *Seek time* is the length of time needed to position the read/write heads so that they can read or write a physical block of records. *Rotational delay time* is the time needed for the disk to revolve so that the required records are accessible to the read/write heads. *Transfer time* is the length of time needed to transmit data from the disk into the CPU. If you read the ads or brochures available for various disk drives, you will see there is a great deal of difference in the time needed to access records depending upon the type of disk drive being used.

Relative cost of storing data. Ads and articles often emphasize the cost of storing a megabyte of data on different storage media. An article in *High Technology* stated that storing data on magnetic disks costs from $25 to $140 per megabyte. The wide range in cost is an indication of why an analyst should be concerned about the megabyte cost of storing data on disks.

Organization	Access	
	Sequential	Random
Sequential (tape & some disk)	Yes	Not possible
Indexed sequential (ISAM & VSAM key-sequenced)	Yes	By key
Direct	Usually not possible	Generation of address or use of relative record number

FIGURE 10.5 Relationship between organization and access methods

Special features. Some disk drives have removable packs while others have nonremovable packs. Other drives use **data modules** or **Winchester packs.** A data module has a sealed unit that contains both the read/write mechanism(s) and the disks. The same read/write head that recorded the data will read the data stored on the disk. More costly drives may have separate read/write heads for each track. These drives are called **head-per-track** or **fixed-head drives.** Since the read/write mechanism need not move in and out, data can be retrieved faster than when conventional disk drives are used.

Cylinders and Tracks   A disk is divided into physical areas called **cylinders.** Visualize a long-playing record. A recording surface, similar to the groove on a long-playing record, is called a **track.** If a disk pack has six disks with a total of ten usable surfaces, a cylinder of data includes records stored on all ten recording surfaces. If a pack has six disks, usually the top of the first disk and the bottom of the sixth disk are not used. Since this is the case, only five access arms with read/write heads that read the bottom of one disk and the top of a second disk are needed.

An analyst should know how much data can be stored on a track (one recording surface for a cylinder) and how much data can be stored in the entire cylinder. When establishing a new file or database, the programmer or analyst must either tell the computer how many cylinders are needed or how many records will be stored in the file or database. If the analyst must tell the computer how many cylinders are needed, guides are available from the manufacturer. Figure 10-6 illustrates a partial table available for the IBM 3380 DASD (direct access storage device).

By referring to the partial table for the IBM 3380 DASD, you can see that the smaller the records, the larger the blocking factor. Once the size of the record is determined, the table should be used to determine the blocking factor. For example, if 480 byte records with keys are being stored, the complete guide indicates that 39 records can be stored on one track. If 480 byte records without keys are being stored, a blocking factor of 48 can be used. The larger the blocking factor, the faster the records stored in the file can be processed.

Records with keys are indexed records and can be accessed either sequentially or randomly. Records without keys could be either sequential or direct files. Records stored in direct files do not have indexes but have a different type of addressing scheme that makes it possible to retrieve records randomly.

Bytes Per Record Block				
Record Blocks	Without Keys		With Keys*	
Per Track	Minimum	Maximum	Minimum	Maximum
1	23,477	47,476	23,241	47,420**
2	15,477	23,476	15,241	23,240
3	11,477	15,476	11,241	15,240
4	9,077	11,476	8,841	11,240
	. . .			
48	469	500	233	264
49	437	468	201	232
	. . .			

* A key is a field that can be used to randomly retrieve a record from a file. In a indexed file, the key is used to search an index which tells the cylinder and tract where the record is located.

** The maximum block size supported by IBM-supplied access methods is 32,760 bytes.

FIGURE 10.6 Partial table for the IBM 3380 DASD

Since the partial table shown in Figure 10-6 is for the IBM 3380, you might be interested in a few more facts about the 3380 direct access storage device. Each storage drive in the 3380 configuration contains a sealed head-disk assembly. each sealed assembly has two **actuators**. An actuator consists of the arm mechanism that moves in and out to access the tracks where data is stored in the read/write mechanisms. Simultaneously each actuator may be transferring data along different paths. There are 15 tracks per cylinder, 885 cylinders can be accessed per actuator, and each track can hold 47,476 bytes of data. If you do some simple mathematics you will find that there are:

712,140 bytes per cylinder
630,243,900 bytes per actuator
1,260,487,800 bytes per spindle (or per disk drive)

The term spindle is used to denote the number of disks that can be accessed in one sealed-in unit. When referring to nonremovable disks, the term spindle rather than pack is used.

Record and File Size   When allocating space, additional cylinders should be provided for adding records to the file. Also, when determining the size of the record, some space should be left for adding more fields. When designing an application, it is difficult to project what things might occur that will require adding more fields to a record. For example, assume that the payroll system only accumulated data on a calendar year. Management decides later they want totals maintained on a fiscal basis as well as on a calendar basis. New fields will be needed to store the additional data. However, if an excessive amount of space is left in records when a file is initially created, it will take longer to process the data stored in the file. Since the blocking factor will be smaller, more seek commands will be required to process the data stored in the file.

No single guide exists to tell the analyst how much extra space to leave in the record or the file. Some analysts feel that from 10 to 15 percent more space than is needed should be initially allocated to a file and to the size of the record. The analyst must be knowledgeable about the disk subsystem being used and must use good judgement. One factor to be considered is the volatility of the file—how many records are normally added or deleted during a given period of time. If the record size must be changed because of adding data to the file, all of the programs that process the file may need to be changed and recompiled. However, if the records can be defined dynamically at run time, the programs will not need to be recompiled. Again it is a question of the type of support that is provided by the operating system.

Reference manuals for the computer system being used often recommend that files be reorganized in order to provide greater efficiency in accessing records. The operating system may keep track of which records are accessed most frequently and store those records where they can be accessed faster than records that are not used as often. Files are generally reorganized during a period of time (such as third shift) when there is not a high demand upon the system.

The programmer or analyst must also provide the label information that will be used in accessing files. When a file is requested, the operating system software will make certain that the correct data is being accessed. If a new file is being created, the operating system software will make certain that data protected by an unexpired date will not be overwritten.

Dynamic Allocation   When timesharing systems or large databases are used, the allocation of the space on either removable or nonremovable disks may be controlled by the operating system's software. This is called **dynamic disk allocation.** The operating system is told the size of the file (number of records and the size of each record) so that space will be reserved for the file or database. Analysts, programmers, and operators are not concerned about where the data will be stored. In some systems, the records are clustered into groups and can be accessed with one read operation. However, the clusters of records may be recorded anywhere on the disks. The system uses **pointers** to keep track of where the records are stored so they may be retrieved either randomly or sequentially. A pointer is either an **address** (cylinder, track, and location of the record on the track) or another indication of location. The pointer can contain the address or can use some type of mathematical formula to calculate the address.

Optical-Disk Memory Systems

Lasers are used to record data on optical-disks—a mark for a binary 1, no mark for a binary 0. As many of 10 billion bytes of data can currently be stored on a single 14-inch disk. Some systems hold as many as 500 disks. However, the storage capacity of the individual disks continues to increase. The storage capacity, combined with random access and fast retrieval might eventually make obsolete magnetic tape and micrographic products used for archival applications.

At the present time, all optical-disk products are nonerasable read-only or read/write disks. The read-only feature is beneficial in storing historical data that should not be changed. The read-only disks might also be used for publishing. Texts of magazines, newspapers, and books could be placed on optical disks by publishers. The texts could be transmitted electronically to subscribers' terminals or sent physically. Software might also be distributed on read-only optical disks.

Hundreds of megabytes could be stored on low-cost 5¼-inch optical disks. The cost might be about the same a today's 5¼-inch Winchester magnetic disk drive that holds 20 megabytes of data. Drives using optical disks as small as 2 inches, some of them erasable, are expected within a very short period of time (perhaps before you read this material).

The first high-capacity system introduced was capable of storing four **gigabytes** (4,000 megabytes) of data on one side of a 14-inch nonerasable disk. This is the same amount of data as could be stored on:

> 40 reels of magnetic tape
> 50,000 microfiche
> 2 million double-spaced lines of text
> 64,000 pictures

The single-quantity price per drive is about $150,000 which brings the cost to about 8 cents per megabyte, while magnetic tape storage costs approximately 15 cents. Magnetic disk storage for a megabyte of data is in the $25 to $140 range.

Still in an experimental stage, the goal of the optical-disk industry is to market low-cost erasable optical disks. While Japan is concentrating on medium-capacity drives, American firms are striving to produce small, low-cost drives that might replace the 5 ¼-inch magnetic drives. The small drives might be able to hold up to 550 megabytes of user data and be affordable for microcomputer users. By 1990 we may see as much as 91 percent of the sales in the small disk market being optical disks.

Files and databases would be organized and accessed in the same manner as when they are recorded on magnetic disk. Because the data is more densely recorded on optical disks, the access time would be less.

File Organization

In selecting a computer system and a programming language, one of the first questions that should be asked is what methods are supported for the organization of files. At the present time the BASIC language used on some microcomputer systems does not support indexed files. Working with direct files that can be accessed randomly is far more difficult than when mainframes with better file-handling methods are used.

The four most popular forms of file organization are sequential, **ISAM (indexed-sequential access method), VSAM (virtual storage access method),** and **direct.** Both ISAM and VSAM are terms associated with IBM operating systems and file organization. However, other companies support files that have a different name but are organized and accessed in a similar manner.

Sequential files.

Records stored in sequentially organized files can only be accessed sequentially. When a record must be changed, the file is searched until the record is found. The record is then updated and rewritten back in its former location in the file. Usually records stored in sequential disk files can be processed faster than records stored in indexed files and accessed sequentially.

If the analyst knows there will be very little, if any, file maintenance required and that the file will always be accessed sequentially, the file should be organized sequentially. Since indexes are not required, less space will be needed to store the records. Also if an index is not consulted or some type of hashing scheme used to calculate an address, records can be processed much faster. James Martin, a noted authority on files and databases, has indicated that if more than 15 percent of the records in a file are normally processed, the file should be organized as a sequential file. Some analysts might use a figure as high as 50 percent in determining whether to organize a file sequentially.

When the program that creates the file is written, the methods used to organize the file, the record size, and the blocking factor are determined. Since records stored in transaction files, temporary files, and files used to communicate data between computer systems or programs are usually processed sequentially, those types of files should be organized sequentially.

ISAM Files   When a file is organized as an indexed-sequential file, each record has a key that is used to retrieve it. The key might be the student's number, the employee's social security number, or the item number. When record is to be randomly retrieved, the key is part of the data recorded in the transaction file. A search is made of the index (stored on a separate cylinder) to determine where the record is stored.

Some systems support the use of multiple keys for the same file. For example, the student's number might be the primary key while the student's city code or curriculum might be a secondary key.

Although random processing of an indexed-sequential file is slower than when direct files are used, indexed-sequential files are very easy to work with and offer a great deal of flexibility. If records are deleted or added to the file, the file must be reorganized so all of the records are still arranged sequentially by the value of the key. Neither the programmer nor the analyst is normally concerned about the location of the records stored within the file.

Allocating Space   In allocating space for the file, the analyst must provide for the **index area**, the **prime area** where most of the records are stored, and the **overflow area** where records are stored on a temporary basis. For example, a given track contains the following student records:

10   20   30   40   50

If the blocking factor is five and the entry track is filled, when a record for student 25 is added it will be inserted between 20 and 30. The record for student 50 will be forced into the overflow area. After the record is added, the track will contain the following records:

10   20   25   30   40

The record for student 50 is in the overflow area. It can still be accessed either sequentially or randomly since the track index indicated that it would be on the track. If it is not located on the track, a search is made of the overflow area. The system's software takes care of the search and it is of little or no concern to the programmer or analyst.

The amount of space (the number of tracks per cylinder) reserved for overflow depends upon: the size of the file; the average number of new records added before the file is reorganized; and how often the file will be reorganized. When some languages are used, the number of overflow tracks per cylinder is fixed. Other languages, such as PL/I and RPG, permit the programmer to specify the number of overflow tracks per cylinder. It is also possible to reserve one or two more cylinders for additional overflow. The record forced into overflow will be in the same address where it had originally been recorded unless the overflow area is full. If this occurs, it will be recorded in the space allocated for additional overflow.

Reorganization of ISAM Files   If the system software does not provide for automatic reorganization of the file so that the records stored in overflow can be repositioned in their proper locations, the analyst must provide instructions regarding how often the file is to be reorganized. Some files should be reorganized on a daily or weekly basis. Others can be reorganized less frequently. How often the file is reorganized depends upon the size of the file, the number of tracks per cylinder allocated for overflow, and the volatility of the file.

When the operating system does not provide for the reorganization of an ISAM file, two very simple programs are needed. The first program reads the records stored in the ISAM file sequentially and creates a new sequential file. The new file can be used as a backup file for the ISAM file. A second program reads the records stored in the sequential backup file and creates a new ISAM file. The reorganized ISAM file is generally rewritten over the old version of the ISAM file. Figure 10-7 illustrates the two-step process. One program could have been written to perform two functions. However, in keeping with the structured concept that each program should perform one major function, two programs should be written. Provisions must be taken so that program 1 is always executed prior to running program 2.

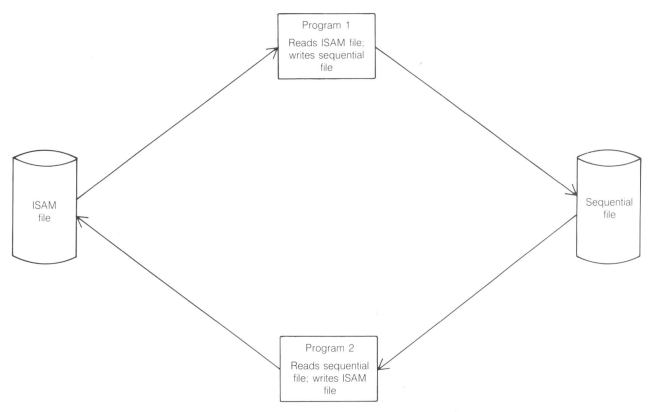

FIGURE 10.7   Reorganizing and ISAM file

After the two programs are run, all the records in the overflow area are correctly positioned in the prime area. There are no records in the overflow area. Two copies of the file exist— the ISAM file and the sequential backup file. For additional security, the sequential file may be copied to magnetic tape and stored in an offsite vault.

When the second program is run, records flagged for deletion can be omitted from the recreated file. It is also possible to roll certain totals back to zero and to make other minor modifications in the records stored in the file. In documenting the procedure, the analyst must make it very clear that the first program must be executed prior to the second program. The programmer can write the second program in such a manner that it will not execute unless the first program comes to a normal job ending.

If the overflow area is full when an effort is made to add a new record to the file, the file is closed and a message is transmitted to the operator that there is no room in the file.

Use of ISAM File   Depending upon the operating system being used, there may be some restrictions placed upon the use of indexed sequential files. For example, it may be impossible to sort the records stored in the file. In this case, the sequential version of the file is sorted.

Indexed sequential files are often used for master files. Although accessing records randomly is slower than when the file is organized as a direct file, the increased flexibility in the use of the file may offset this disadvantage.

VSAM Files   Virtual Storage Access Method is an access method that is supplied with the operating systems used on large IBM computers. Records stored in VSAM files can be accessed from 1½ to 3 times faster than records stored in ISAM files. VSAM provides three types of file organization:

- **Key sequenced.** This type of organization is the most popular and is the type usually being referenced when an analyst or a programmer refers to a VSAM file. The records are in order according to a unique key field and can be accessed by using the key field. Although records can be accessed faster and less file maintenance is required, it has many of the characteristics of an ISAM file. However, most programmers would elect to use VSAM files.
- **Entry-sequenced.** Records are stored in the order they are entered without regard to content.
- **Relative record files.** Records are stored and retrieved by referencing the relative number of the record within the file. For example, to obtain the 105th record in the file, 105 is used as the key.

Most VSAM files are organized as key-sequenced data sets because this provides the opportunity to access records both randomly and sequentially. The rest of the discussion regarding VSAM files refers to this type of file organization.

Control Intervals   In a VSAM file, the records are organized in **control intervals**. At the bottom of each control interval is additional space. There are also some control intervals that are entirely free of data. Records are added in their proper position within an interval. Figure 10-8 illustrates the concept of control intervals and how the file is reorganized. When all the extra space is used in a control interval and another record needs to be added, a split automatically occurs and a new control interval is created. One of the advantages often mentioned for VSAM is that the files are automatically reorganized. However, most operation supervisors feel that if the files are reorganized by an operator periodically records can be accessed more efficiently.

Accessing Records   When VSAM files are created, a primary key (such as student number) that is unique to each record is used. The primary key can be used to randomly retrieve records. It is also possible to define an alternate key which is not unique to one record. In a payroll master file, the employee's social security number might be the primary key and an employee number defined as an alternate key. A separate record, with the employee number, might be maintained for each job that the employee worked on. Because of the way the alternate keys

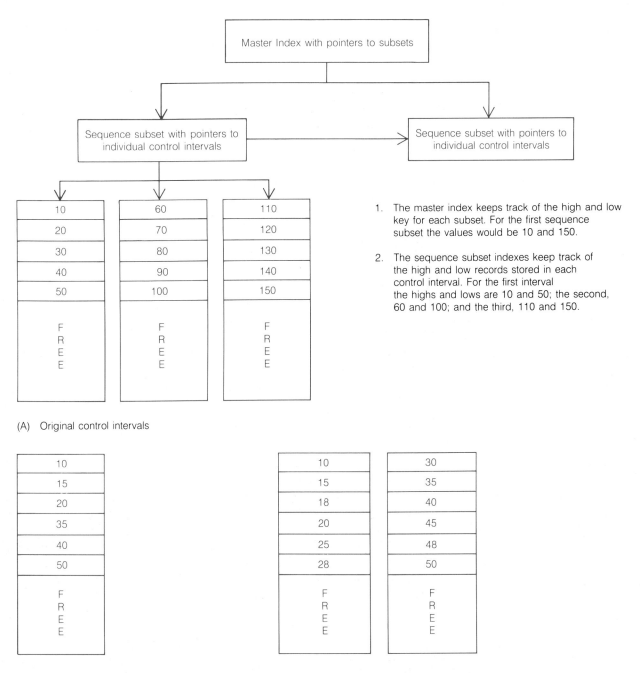

1. The master index keeps track of the high and low key for each subset. For the first sequence subset the values would be 10 and 150.

2. The sequence subset indexes keep track of the high and low records stored in each control interval. For the first interval the highs and lows are 10 and 50; the second, 60 and 100; and the third, 110 and 150.

(A)   Original control intervals

(B)   Records are added and there is less free space. Note how records 15 and 35 were added in the correct location.

(C)   As additional records are added, the free space is used up and a split occurs. The sequence index would now show that the first interval has records 10 through 28; the second records 30 through 50.

FIGURE 10.8 As additional records are added to a VSAM file, the records are inserted into their proper location. Eventually a split will occur.

can be used, a VSAM file has some of the characteristics of a database. For example, the primary key might be an account number. Once access is gained to the first record for an account, an alternate key can be used to access the rest of the records within the group.

When an ISAM file is utilized as input, the programmer must indicate whether the records are to be accessed randomly or sequentially. When a VSAM file is used, the programmer can indicate that the first record of a group is to be accessed randomly and the remaining records within the group are to be accessed sequentially. Figure 10-9 illustrates how inventory records having primary and alternate keys as well as **variable length records** might be organized. The main record, accessed by record number, contains most of the information regarding the inventory item. The secondary records for the group contain the detailed information regarding the shipments received.

Delete Code	Item No	Record number	Product description	Quan. on hand	Reorder point	Reord. quan	Cost price	Selling price	Vendor Codes				
									A	B	C	D	
	B0900001	00010	5¼" Diskettes	250	50	100	24|00	38|50					

Primary record — one per item.

Delete code	Item No	Order number	Date	Quan	Cost
	B0900001	1	09/01	50	23|00

Delete code	Item No	Order number	Data	Quan	Cost
	B0900001	2	09/10	10	24|50

Delete code	Item No	Order number	Data	Quan	Cost
	B090001	3	10/15	100	24|00

Secondary records — one per shipment.

FIGURE 10.9 Variable length records with a record and alternate key

In the inventory illustration, the record number is unique to each master record and is used to randomly retrieve the master record from the file. If the file is a VSAM file, all of the records identified by the same item number that are within the group (record number 10) can be read and processed. The secondary key is item number and it is not unique. There are a group of records with the same item number. In the example, each shipment of merchandise has a record that contains the detailed information pertaining to the transaction. The information pertaining to individual shipments must be retained if the company uses either the **FIFO (first-in first-out)** or **LIFO (last-in first-out)** inventory method. If an order was received for 60 items, the first shipment would be completely used up and a delete code would be placed in the first field of the record. The quantity for order number 2 would be reduced from 100 to 90. The quantity in the master record would be reduced to 190 items and a new cost price would be calculated. The cost of the diskettes sold would be 50 @ $23.00 and 10 @ $24.50. This is the information that must be used in calculating the cost of goods sold and in determining the company's gross profit margin. As new shipments are received, additional records are inserted after the one illustrated as order number 3. The records flagged for deletion would be eliminated when the file is reorganized.

Keys Used for a Student Master File   Your student master file (or database) may be organized in a manner similar to the inventory example. Your master record contains all of the general information such as your name and address, high school code, birth date, social security number, and so forth. The primary key might be your social security number. The alternate key is your student number and will be used to identify all of the records that contain information regarding the courses you have taken. Recorded in the shorter record for each course might be the course number, hours of credit, instructor code, semester taken, and grade. While you might have taken 10 courses and have 10 additional records, the next student might have taken 50 courses and have 50 additional records. Only the actual space needed to record the data pertaining to the records is used.

Deleting Records When using VSAM files, records can be deleted without an operator physically reorganizing the file. Access to records stored within the file is faster and the diagnostics regarding error conditions that might have developed are much better than when ISAM files are being used.

Training Required Most companies that have large systems are reorganizing their ISAM files as VSAM files. When a company decides to convert from ISAM to VSAM files, programmers and analysts are often asked to attend an in-service training program that provides from six to eight hours of instruction.

Direct Files Data stored in direct files can be accessed randomly faster than records stored in a key-sequenced file. Direct files have no index and require less storage space than key-sequenced files. Either the location of the record within the file is used as part of the transaction records or a mathematical formula of some type is used to determine the address of the record. Depending upon the system being used, the addressed can be the cylinder, track, and record on the track or the physical location of the record within the file. Or the record may be retrieved by using a record's relative number.

When direct files are used, a greater burden is placed on the programmer or analyst than when key-sequenced files are used. Online files used in telecommunication or transaction processing applications are often organized as direct files. Master files that require a great deal of maintenance should be organized as direct files or as key-sequenced files so that records stored within the files can be accessed randomly.

FILE ACCESS

Records stored within files are accessed either sequentially or randomly. Once the file is created, whenever it is utilized the programmer must indicate how the records are to be accessed. Although records stored in a sequential file can only be accessed sequentially, it is possible (depending upon the language being used) to access only the records that fall between different ranges or limits.

In some cases it is not easy to determine how records stored within a file should be accessed. If a large percentage of the records are to be processed and the transaction file is in sequence by the key used for the key-sequenced file, it is usually faster to access the file sequentially. The time saved in accessing the master file records has to be evaluated against the time that it takes to sort the transaction file into the desired sequence. If a small percentage of the records are to be accessed, the records should probably be accessed randomly.

Volume Table of Contents (VTOC)

A VTOC, or volume table of contents, is maintained by the operating system on one of the disk's cylinders. The VTOC is used by the operator in determining where on the pack a file is located. If for any reason the operator misplaces the job control language (JCL) that indicates where a file is located, a **utility program** which prints the contents of the file can be run. For each file, information pertaining to its location on the pack, the creation date, expiration date, and the label information is maintained. If a file has an index, the location of the index, prime area, and area reserved for additional overflow are all indicated in the VTOC.

If an operator wishes to know whether there are any unexpired files on a pack, the VTOC can be consulted. Usually a pack history is also maintained that tells what files are on the pack, where they are located, when they were created, and how long they are to be protected. One of the functions of the librarian is to maintain the history for the disk packs and tape files.

Alternate Tracks

When data is written on disk, a parity check is performed to make certain the data is stored correctly. If it is not, the system will try until it has been unsuccessful a given number of times or a successful write operation is achieved. If the data cannot be written on the track, the system usually provides for it to be written on an alternate track. When this occurs, a message is transmitted to the operator.

If a number of bad tracks are detected on a pack, it is usually designated as a "work pack" and retained for storing small files that are used in testing programs.

Checkpoint

13. What five general factors should be determined when considering the acquisition of a disk subsystem?
14. Why should an analyst know the number of bytes of data that can be stored on a single track?
15. What factors make up the access time needed to retrieve a record of data and bring it into the CPU for processing?
16. If an analyst knows the number of bytes (or characters) in a record, how can the blocking factor be determined? What is considered to be the ideal blocking factor for records stored on disk?
17. What is an actuator?
18. What is meant by dynamic disk allocation?
19. What are the four most popular file organization methods?
20. Why are VSAM files preferred over ISAM files?
21. Why should an ISAM file be reorganized?
22. What happens when a record is added to a VSAM file and the control interval is already filled?
23. What advantage is there in being able to define both a primary and an alternate key?
24. Do both the primary and alternate keys have to be unique values such as an individual's social security number or employee number?
25. How is data recorded on an optical disk?
26. How does the amount of data stored on a single 14-inch optical disk compare with the amount of data stored on a reel of tape?
27. Why might 5 ¼-inch optical disk drives replace the present Winchester hard-disk drives?

DATABASES In the early 1970s, 30 percent of the computer systems in use supported interactive computing and online terminals. Currently, more than 80 percent of the systems (excluding microcomputers) support interactive computing. Some microcomputers are *multitasking* (can execute more than one program at a time) and are also used for interactive computing.

A database is a collection of interrelated data stored together with controlled redundancy which efficiently serves one or more applications. The **data items** (the smallest units in the database) are stored so they are independent of the programs that use the data. A common and controlled approach is used to add new data, to modify data, and to retrieve items within the database. If there are multiple databases entirely separate in structure, the system is said to have a collection of databases. Supporting most interactive computing systems is a comprehensive database. With the increased use of databases there are also trends to have:

Increased accountability. Often there is a database administrator who is responsible for determining what goes into the database and for making certain the data stored in the database is relevant to the needs of the organization. The data must also be protected from unauthorized users.

More centralized databases. The databases are shared by application programs and also employed by the end users when making queries. This results in less duplication of data and lower maintenance cost.

More distributed terminals. Although the control of the database and its contents is centralized, the data entry and information retrieval functions are decentralized.

More complex programming support software. The database software and operating system that provided the interfaces between the application programmers and the database are becoming more complex. However, for the end users, the systems are user friendly and for the programmer, the layers of complex software make it easier to write application programs.

Sophisticated database software for small business systems. At one time only large mainframe users could afford database software. Now excellent software is available for microcomputer systems. DBMS software is one of the three major programs obtained by many microcomputer users.

Types of Databases **Tree or Hierarchy Structure** One type of database is based upon what is referred to as a tree or hierarchy structure. Figure 10-10 illustrates a simple tree and the type of records that are needed to support the tree. In the illustration there are three customers that have one or more unpaid invoices.

The content of the customer's record would include data that seldom changes such as the customer number, address, city, state, zip code, sales representative's number, credit limit code, and so forth. In a tree structure, the customer record must also contain the **relative record number** of where the data pertaining to the first invoice is stored. The relative record number establishes the pointer that is used to retrieve the other records that belong to the customer. When the customer record is retrieved, the link field identifies where the first invoice record is stored. After the first record is read and processed, the link field in the invoice record 1 points to the second invoice record, and so forth. When a zero is found in the link field, all of the invoices (or payments or credits) for the customer have been processed.

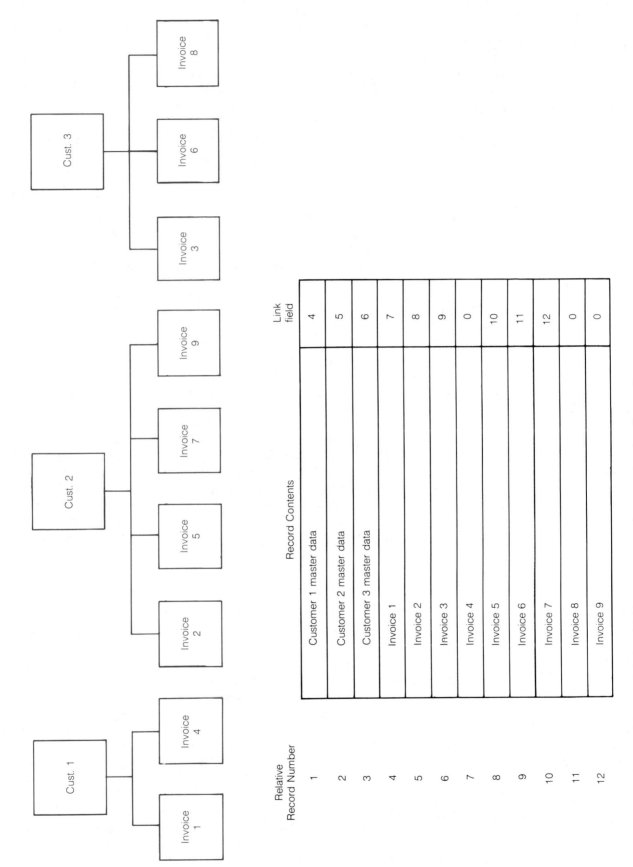

FIGURE 10.10 Tree structure with link fields that serve as pointers to the detail records

Relative Record Number	Record Contents	Link field
1	Customer 1 master data	4
2	Customer 2 master data	5
3	Customer 3 master data	6
4	Invoice 1	7
5	Invoice 2	8
6	Invoice 3	9
7	Invoice 4	0
8	Invoice 5	10
9	Invoice 6	11
10	Invoice 7	12
11	Invoice 8	0
12	Invoice 9	0

When an invoice record is to be deleted, the pointer that previously pointed to that particular record is changed. Since a pointer no longer provides the connecting link needed to access the record, the invoice record will no longer be processed. At this point the record is logically, if not physically, deleted. Records that are logically deleted can be "collected" and removed from the list. The records may be added to a history file. Additional records can also be added to the lists. The system adds records by referring to a pointer that identifies the address of the first unused record.

Much of the discussion regarding files organized as ISAM, VSAM, or direct files is still relative to the discussion of databases. The analyst must allocate enough space so there are unused areas for the insertion of records. Maintenance programs must also be executed that will remove obsolete records from the database and release the space for the addition of new records. The databases must be backed up on a regular basis and history files must be maintained as part of the audit trail.

The ability to process a VSAM file both randomly and sequentially in the same program gives VSAM files the same general characteristics as simple linked lists. However, hierarchial structures of databases can become far more complex than those illustrated.

Relational Databases   **A relational model** must be visualized as a two-dimensional table. Unlike the hierarchial or tree database, there are no pointers. The relationships that exist must be determined before they can be used in answering queries. However, since pointers are not used, there will be redundancy in the fields used to establish relationships. The relational database is easy for the end user to use but the processing required to retrieve the desired information may be extremely complex.

Each record occupies one row and each column contains a data item. In Figure 10-11 below there are two relational databases illustrated. The first is named CUSTOMER and the second INVOICES. Above each column is the name of the field. When a database is described, it must be given a distinctive name and each field within the records is also given a name. Note that CNUMBER is used in both databases. A request for data might be made as follows:

CNUMBER	CNAME	CADDRESS	CCITY	CSTATE	CZIP	CBALANCE
10	George Supply Company	1813 Longview	Denver	CO	23567	50000.00
20	Computer Supplies	1845 State Street	Chicago	IL	15678	10000.00
30	Capricorn Data Supply	1513 E. Vista Way	San Marcos	CA	92067	3000.00

CUSTOMER

CNUMBER	DATE	INV.NO	AMOUNT	SCODE
10	091686	1	40000.	A234
20	091686	2	5000.	B234
10	101986	3	10000.	A234
30	111386	4	3000.	C423
20	121886	5	5000.	B234

INVOICES

FIGURE 10.11   Two relational databases

```
SELECT  CNAME, BALANCE
FROM    CUSTOMER
WHERE   CBALANCE > 40000
```

Listed on the VDT would be George Supply Company, 50,000.00. Customer number 10 is the only one that has a balance greater than 40,000. A request could be made to retrieve information regarding invoice number 5 by writing a request as follows:

```
SELECT  CNAME, CADDRESS
FROM    CUSTOMER
WHERE   CNUMBER IS IN
            SELECT DATE,SCODE,AMOUNT,DATE
            FROM INVOICES
            WHERE  INV.NO = 5
```

In this example the customer's name and address will be retrieved from the CUSTOMER database and the details of the transaction will be retrieved from the INVOICES database. A third database that contains detailed information regarding the shipping codes (SCODE field) could also be developed. The keywords used to retrieve data are SELECT, FROM, and WHERE.

Databases can be accessed by using a **query language** similar to the one illustrated. Once the selection criterion is given (such as BALANCE > 40000), the information requested in the SELECT statement will be listed or printed for all customers that meet the qualifications. Most of today's database software permit the user to specify a report heading, column headings, and the totals desired. Column totals can generally be accumulated. However, only a limited number of mathematical operations can be performed.

Usually the person requesting the data can indicate whether the records obtained from the databases are to be sorted into any particular sequence. If more information had been shown for the CUSTOMER records, you could see it would be possible to list all customers whose balance exceeds their credit limit or all customers who are within 10 percent of their credit limit. Based on the data specified and the selection criterion, a sequential file is created that contains the information. The file may then be sorted in the sequence specified and the report printed.

Once a database is created, the records can be used as input into a program that is written by an application programmer. If a number of mathematical calculations are required, it may be necessary to write a program rather than to extract the information by using the query language.

Database Languages

Depending upon the database being used, there are generally three types of languages that are supported by the database software. The first is called a **data description language (DDL)** and is used to define the logical structure of the data. Often the programmer or user merely answer questions such as:

```
COMPUTER                              KEYED IN BY OPERATOR
NAME OF DATABASE:        CUSTOMER
NUMBER OF RECORDS:       20000
RECORD KEY:              CNUMBER
```

```
FIELD SIZE:            6
DATA TYPE:             N
FIELD:                 CNAME
FIELD SIZE:            26
DATA TYPE:             AN
```

For each field within the record, the programmer must indicate the name of the field, the size of the field, and the type of data stored within the field.

The second type of language is a **data manipulation language (DML)** that provides commands for storing, retrieving, updating, adding, and deleting records from a database. Usually all the necessary I/O commands are available as part of the DML subset. The DML commands may be used in the procedure division of a COBOL program. There also is a query version available that can be used for sorting records and writing reports in a nondatabase environment.

The third type of language that is available is a query language. A query language is a nonprogrammable and nonprocedural language that can easily be employed by either users or management. A query language may also be used for accessing and storing small amounts of data in a complex database. Although one example was already given, another illustration using a slightly different type of query language will also be used.

Suppose someone asks you, ''How many of your analysts who now live in Boston can speak German?'' The person then indicates he would like to know their names, addresses, and telephone numbers. You would go to the terminal and sign on. After the system identifies you as a qualified user, you indicate you wish to make a query into the database by using an interactive query language. You might key in a statement similar to the one that follows:

```
FROM THE PERSONNEL DATABASE PLEASE LIST THE NAME, STREET,
CITY, STATE AND PHONE NUMBER OF SYSTEMS ANALYSTS WHO LIVE
IN BOSTON AND SPEAK GERMAN
```

Will it work? Yes, if all the key words have been defined for the system. Key words fall into different categories:

1. *Database name.* PERSONNEL DATABASE is the name of the database that is to be searched.
2. *Data items.* NAME, STREET, CITY, STATE and PHONE NUMBER are some of the data items that have been identified as being contained in the databases.
3. *Connectors.* AND and OR are connectors used to indicate compound conditions that must exist in identifying the data to be retrieved.
4. *Commands.* LIST, SORT, COUNT, and TOTAL are commands that are usually available.
5. *Special meaning words.* The software package may initially have small vocabulary of words that can be used to provide certain functions. Additional words and phrases can be defined by the user and added to the system vocabulary. LIVE, SPEAK, and OF would need to have been defined for the system.

In defining the terms that can be used to provide various functions, the user must make certain the terms are understandable and easy to work with. Other words, called *null words,* can be added to make the statements more natural. These words will be ignored by the system when it translates the user's request into some type of object code.

Usually the query language supports the use of relational operators (equal to, not equal to, less than, greater than or equal to, and greater than or equal to), numeric constants, and the use of nonnumeric literals or constants.

Another way to write the same request would have been to enter:

```
PERSONNEL
JOBCODE EQ ANALYSTS SKILCODE EQ ''GERMAN'' LOCATION
   EQ ''BOSTON''
LIST NAME, STREET, CITY, STATE, PHONENUMBER
```

Other types of query languages carry on a dialogue with the user. For example, the following dialogue might have taken place:

```
Terminal display:   WHAT DATABASE DO YOU WISH TO USE?
User:               PERSONNEL
Terminal:           ON WHAT FIELDS DO YOU WISH TO SEARCH?
User:               JOBCODE EQ ANALYSTS
Terminal:           ARE THERE OTHER QUALIFICATIONS?
User:               SKILCODE = ''GERMAN''
Terminal:           OTHER QUALIFICATIONS?
User:               LOCATION EQ ''BOSTON''
Terminal:           OTHER QUALIFICATIONS?
User:               NO
Terminal:           WHAT DATA DO YOU WANT ON YOUR REPORT?
User:               NAME, STREET, CITY, STATE, PHONENUMBER
Terminal:           SHOULD THE DATA BE SORTED IN A GIVEN
                    SEQUENCE?
User:               NO
Terminal:           WHAT HEADINGS DO YOU WANT ON YOUR RE-
                    PORT?
User:               ''ANALYSTS LIVING IN BOSTON WHO SPEAK
                    GERMAN''
Terminal:           ANOTHER HEADING?
```

The dialogue continues until the system knows the number of lines per page, how to center and space the report, and whether the system's date or a date from the terminal is to be used.

Many different types of query languages are in use today. The analyst working with the user must identify the frequently used programs that are needed to maintain the database and to provide the required reports. These programs will be developed in the usual manner since they will execute faster and more efficiently than when the user enters a natural language or carries on a conversion with the system regarding information needs.

With very little training, secretaries or users can learn to use the query language that generates those on-the-spot reports that are often needed by management. Prior to the development of databases and query languages, much of the information stored in the files was not available to the user *when needed*. Today the situation is much improved. Recently, for example, a Delta College counselor needed to know the names, addresses and phone numbers of all female students between the ages of 21 and 30 who lived in Midland, Michigan. The list was to contain the names of students currently enrolled as well as the names of students who had been enrolled in the past five years. An operator indicated that Delta College's student database was to be used to generate a report. The operator carried on the required conversation with the computer. In less than ten minutes from the time the request had been received, a nicely printed report was available. A database that contained information on more than 80,000 students had been interrogated.

Database Software

When a database is not used and data is to be obtained from the file, an interrupt in the application program occurs and the operating system assumes the responsibility for obtaining the data and making it available for the program.

When a database is used, additional software, referred to in Figure 10-12 as the database management system, provides the interface between the request for data from the database and the operating system. The additional software takes care of a great number of functions.

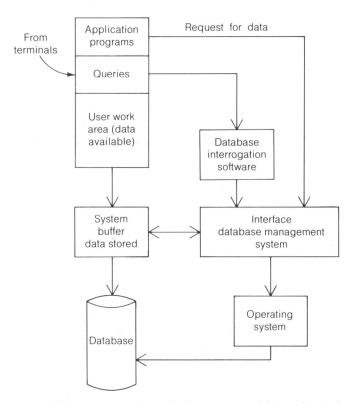

FIGURE 10.12 Additional interface software is needed when a database is used

When a query is made by a user, additional interface software is needed. The **database interrogation software** permits the user to make a request for data in an interactive mode. Simple English commands, using a few key words, are used to retrieve the desired information.

Databases are used for a wide variety of purposes and differ in the way they are organized and function. Databases can be constructed for scheduled operations and on-demand operations. For scheduled operations, the data to be used is planned in detail when the system is designed. Files are updated and reports are generated according to schedule. This type of processing is also called transaction-oriented processing.

On-demand operations occur when requests are made by terminal operators or by end users. Analysts must decide whether the on-demand operations will also be considered as realtime applications. In realtime applications, the data is updated as the transaction or request for information is processed.

The way a database is organized depends upon how it will be used. The designer must work with the users to determine what relationships exist and what type of information will be extracted by using a query language. Programs will be designed, written, tested, and documented for what are considered normal inquiries. The query language will be used for spontaneous, or unscheduled, queries made by the user.

Features of a Comprehensive Database System

A comprehensive database should provide for:

1. Accessing the database for scheduled operations such as the ones executed by many of the typical applications programs.
2. Handling ad hoc requests for information from users.
3. Non-data processing personnel using the system. A convenient English, nonprogrammer-oriented language must be provided.
4. Flexibility. Programs and databases are constantly changing. It must be possible to change programs without changing the database, and it must be possible to change the database without changing the programs.
5. Shareability. Different applications programs must independently be able to share an integrated database.
6. Relatability. Relationships must exist among the different logical records that make it possible to chain from one to another. In a database, logical records make up the physical database.
7. Integrity. With all the users sharing the same data with all the relationships involved, it is impossible for each user to be responsible for the items stored in the database. The integrity of the data must be maintained by providing consistency checks and input/output editing. An audit trail and some type of backup system must also be provided.
8. Elimination of redundancy.
9. Access flexibility. The data stored in the database must be accessible in a number of different ways. It must be possible to access the data on the basis of a number of different access keys and logical qualifications rather than on the basis of a single key.
10. Security of the data. Certain items may require high levels of authorization to access. Data items must be fully protected from unauthorized intrusions, either accidental or malicious. The database administrator must have the ability to assign and control access privileges of users.
11. Realtime accessibility of the items stored in the database.
12. Rapid response time. The time that it takes to retrieve a record, process the data, and rewrite the updated items is of vital concern in a realtime environment.
13. Tuning the database system to make it more efficient. It must be possible to do this without changing the applications programs.
14. Data migration. Items must be relocated so the ones referenced the most can be accessed quickly and conveniently. In some systems, the data migration is handled automatically by the database software.
15. Handling of both low and high volumes of data.
16. Batching repetitive and related I/O requests.

17. High performance and efficiency. Each software package that must be implemented to support the database objectives is an additional burden on the system's design, organization, implementation, and performance. The attractiveness and ease of use of a feature for the user is highly dependent on the performance and cost involved.
18. Compatibility with the previous systems.
19. Interfacing with future systems.

DATA DICTIONARIES

A **data dictionary** is a centralized file that contains data about data. It is a database about the databases managed by the dictionary software. Data dictionaries have become popular due to the increasing size and complexity of data and information systems. **Data dictionary software** includes a: definition language for defining entries to the dictionary; manipulation language for inserting, modifying, and deleting entries from the dictionary; means for validating the inputs to the dictionary; and a means to prepare reports pertaining to the database.

There is no commonly accepted data dictionary standard. Some vendors have developed dictionaries that can be used without database software. However, with the complexity of some databases, it would be almost impossible to manage them with the use of a dictionary. Most dictionaries contain information about and maintain the relationships between entries in the following categories:

1. field or data items
2. file or record type
3. database organization
4. transactions that use the items
5. source documents where the items originate
6. reports that use the items
7. programs that use the items
8. systems that use the items
9. users who can obtain information regarding the different data items

Reports on any or all of the entries in any given category can be printed. The data dictionary software provides and manages a database about databases and all the factors that relate to their creation, utilization, and security.

If a data dictionary is not available, a form similar to the one illustrated in Figure 10-13 should be developed. A separate form is made out for each master file or database. In Figure 10-13 both the customer master record layout and the variable length record used to record transactions are illustrated. The information at the top of the form gives the name of the file, the primary and secondary keys, file organization, date created, and record size. The retention data is used to protect the records stored in the file from being overwritten as new information is stored on a disk.

All of the programs that use the ARMASTER file are listed at the top of the forms and the fields that make up the records are listed along the left side. The Xs indicate in which programs each field of data is used. The form also lists the location of each field, the length of the field, and the type of data stored in each field. The data types such as AN (alphanumeric), NP (numeric packed), Z (unpacked numeric), and B (binary) are indicated for each field. Both analysts and programmers should be aware of the various ways data can be represented within the computer and on various storage media.

By referring to the form, an analyst or maintenance programmer can see which programs use each of the data fields. If a change is made that influences one or more fields in one program, it may also affect all of the other programs that utilize the data.

FILE NAME: ARMASTER
ORGANIZATION: VSAM
RECORD SIZE: 200

PRIMARY KEY: CNUMBER
SECONDARY KEY: SNUMBER
CREATED: Sept 1986
RETENTION: Dec 1999

Location	Field Length	Type	Data	Sales transaction	Cash received	Credit/returns	Aged trial balance	Inactive customer	Mailing list	Customer activity	Acct. Rec. statements	Monthly update	Annual update	New customer	Change	Correct
1	1	AN	DELETE-Deletion code	X	X	X	X	X	X	X	X	X	X	X	X	X
2-7	6	AN	SNUMBER-Secondary Key	X	X	X	X	X			X	X	X	X	X	X
8-17	10	AN	CNUMBER-Primary Key	X	X	X		X	X		X			X	X	X
18-42	25	AN	NAME-Customer Name	X	X		X	X	X	X	X			X	X	
43-67	25	AN	ADDRESS1	X					X	X	X			X	X	
68-92	25	AN	ADDRESS2	X					X	X	X			X	X	
93-117	25	AN	ADDRESS3	X					X	X	X			X	X	
118-120	5	NP	ZIPCODE	X					X	X	X			X	X	
121-121	1	AN	CCODE-Credit code	X		X				X				X	X	
122-122	1	AN	BCLASS-Business class	X										X	X	
123-126	6	NP	FDATE-First order data							X			X	X		X
127-131	9	NP	BALANCE-Beg Balance							X	X			X		X
132-135	6	NP	LDATE-Last order date	X				X		X				X	X	X
136-140	9	NP	APURCHASES-Yearly aver.					X		X				X	X	
			Year-to-date totals													
141-145	9	NP	YPURCHASES	X						X				X	X	X
146-150	9	NP	YCASH-Cash paid		X					X				X	X	X
151-155	9	NP	YCREDIT-Returns or cr.			X				X				X	X	X
156-160	9	NP	YDISCOUNTS-DISCOUNTS	X						X				X	X	X
161-165	9	NP	YINTEREST-Interest ch.							X				X	X	X
166-167	3	NP	SREP-Sales rep	X				X	X					X	X	
168-169	3	NP	DISTRICT-Dist. code	X				X	X					X	X	
170-200		NP	Unused area											X		
VARIABLE RECORD																
1	1	AN	Deletion Code	X	X	X	X	X	X	X	X	X	X	X	X	X
2-7	6	AN	SNUMBER	X	X	X	X			X		X		X		X
8-8	1	AN	TRANSI-Transaction type	X	X	X	X				X	X		X		X
9-12	6	NP	TDATE-Transaction date	X	X	X	X				X	X		X		X
13-16	6	NP	TREFNO-Reference number	X	X	X	X				X	X		X		X
17-19	3	AN	TSHIPCODE-Ship. code	X	X	X	X				X	X		X		X
20-24	9	NP	TAMOUNT-Invoice total	X	X	X	X				X	X		X		X

FIGURE 10.13   File utilization chart

Included as part of the documentation is also a record layout form similar to the one illustrated in Figure 10-14 on page 376. Although some of the information on the utilization chart is duplicated, a record layout form provides a better visual image of the relationship of the fields.

Checkpoint

28. What are five current trends regarding the use of databases?
29. What is a data item?
30. What is a relative record number?
31. Why is the relative record number of the record containing transaction data stored in the customer's master file record?
32. What is the significance of a zero stored in the link field of a transaction record?
33. How should you visualize the records and fields stored in a relational database?
34. In working with the database illustrated in Figure 10-11, what type of words are SELECT, FROM, and WHERE?
35. If a query language is used, is it possible to print report headings and totals?
36. Are the records stored within a database sorted?
37. What are the difference between a DDL language and a DML language?
38. How would you support or disagree with the statement that "a DML language is designed to be used by a programmer or analyst"?
39. An analyst indicated to you that "query languages are designed primarily for analysts." How would you either defend that statement or prepare a rebuttal to the statement?
40. Indicate whether the following statements were included in the list of features that make up a comprehensive database. Record a T if the statement is a feature found in most databases and an F if the statement does not represent a feature found in a database.
    a. Information within a database can only be accessed by using the student's identification number.
    b. A database should be designed in such a way that it can only interface with existing systems.
    c. When an application program is changed, the database must also be changed.
    d. Each database is used by one application program.
    e. If the database is finetuned to make it more efficient, all of the programs that use the database will need to be changed.
    f. Anyone who has access to the records stored in a typical database can obtain all of the information stored in the database.
    g. Records stored in a database can only be processed in a batch mode.
    h. Items that are used most frequently should be positioned where they can be accessed in the least amount of time. This is called data migration.
    i. It must be possible to chain from one record to another. The relationships may need to be defined by the programmer or the person making a query into the database.
41. What information is contained in a data dictionary?
42. A new analyst was hired by a long-established company. The analyst was amazed to find out that the company did not have a data dictionary. The

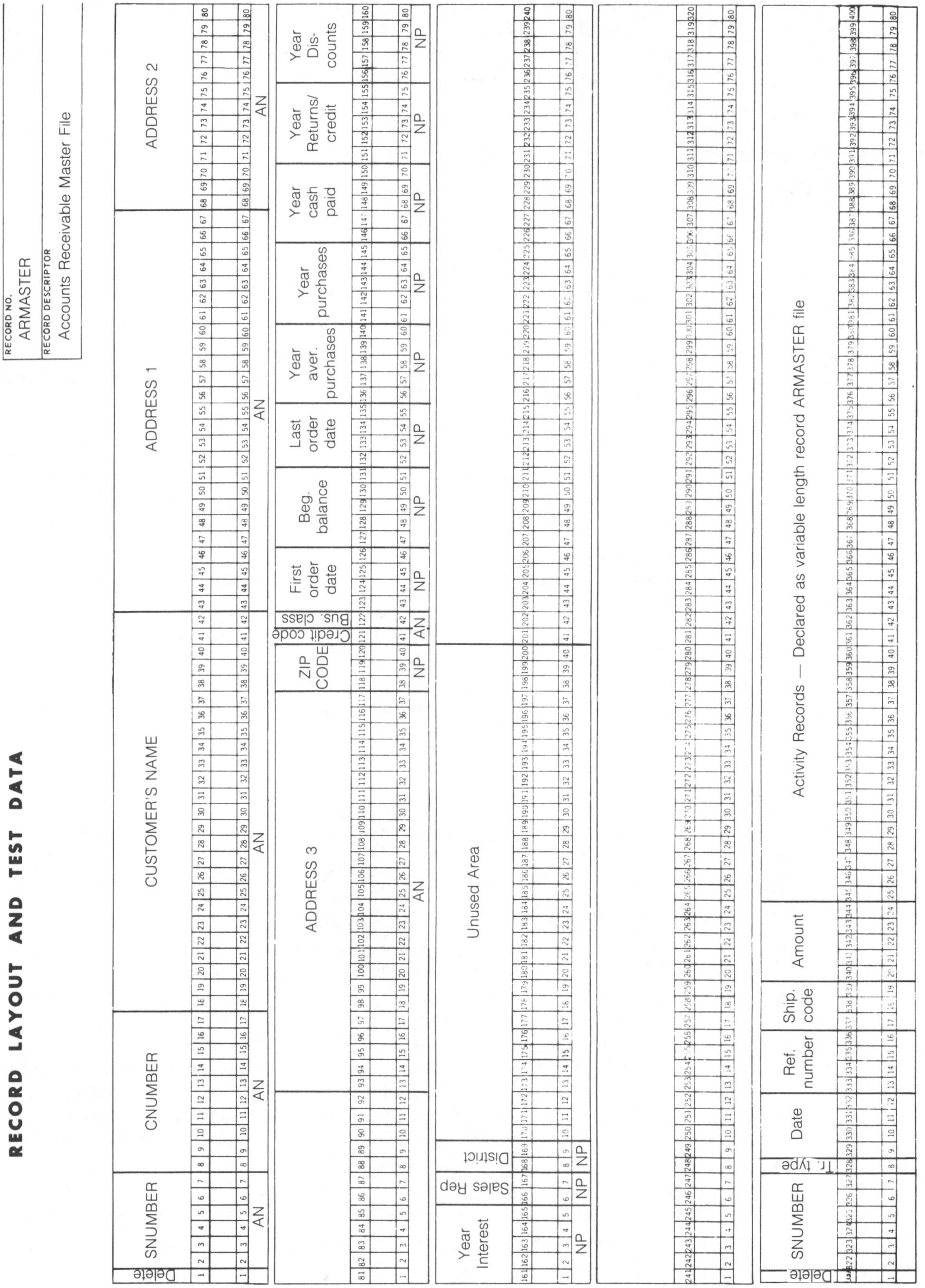

RECORD LAYOUT AND TEST DATA

FIGURE 10.14 Record Layout for ARMASTER file

CIS manager indicated that one would not be obtained. What action would you recommend that the analyst take?

43.   What information is shown on a record layout form?

SUMMARY

In the general design phase, the analyst determines whether a file or database is to be used and the medium on which it is to be stored. In the detailed design phase, precise specifications for each field of data within the file or database are determined. The size of each field and the type of data to be stored within the field must be determined.

At the present time magnetic tape is still widely used to back up files, transport data between systems, and for submitting reports. Historical data can be stored more cheaply on magnetic tape than on magnetic disk. Data that is used infrequently may be stored on magnetic tape and then transferred to magnetic disk for processing. The major disadvantage is using magnetic tape is that records can only be accessed sequentially.

Magnetic disk is the medium used most frequently by organizations that have minicomputers or mainframes for storing files and databases; diskettes are widely used by microcomputer users. Depending upon the operating system and language being used, files stored on disks can be organized as sequential, ISAM, VSAM, or direct files. VSAM records used in a program can be accessed both randomly and sequentially. The primary record is accessed randomly; secondary records are then accessed sequentially.

Databases should be used for storing data that is used in a wide range of applications. When using a database, there are usually a data description language, a data manipulation language, and a query language available. A database administrator should be responsible for the integrity and safety of the data stored within the database. Data dictionary software is used to keep track of the data stored within the organization's databases and files.

For each field of data stored in a file or database, an analyst must determine how the data is to be stored. The choices available vary with the operating system and the programming language being used.

Files and databases are two of the most important assets of an organization. Their contents must be current and accurate. Programs must be available to retrieve information and format it in a manner that is acceptable to each level of management.

Discussion Questions

1.   As a recently hired analyst, you are unfamiliar with the magnetic tape drives that are used by your computer system. What information would you want to find out about the magnetic tape drives? Should there be standards or guidelines developed regarding when magnetic tape files should be used? If standards are not available, what standards would you suggest?

2.   As an analyst, what information should you be aware of regarding cylinders and tracks? When using a particular disk drive, how would you find out what blocking factor should be used for an ISAM, sequential, or direct file? When a hashing formula is used to determine an address or when an index is used, what information makes up the address used to randomly retrieve a record stored on a disk?

3.   If ISAM, and VSAM file organizations were supported, which would you elect to use for the inventory master file? Give the rationale for your answer.

4. What considerations make up the total access time needed to retrieve a record? Will the access time be more or less if head-per-track rather than moving head drives are used?

5. If you obtained a microcomputer system and you could obtain a 20MB Winchester drive for three times the cost of a drive that reads data from a 5¼-inch diskette, which would you select for your second disk drive? The microcomputer you have selected has one built-in 5¼-inch diskette drive. Give the rationale for your decision.

6. A company has the policy that all files that are not used at least once a week are to be stored on magnetic tape. When the file is to be processed, the records are transferred from tape to hard disk. Why would a company have such a policy? What are the advantages and disadvantages of the policy to store semi-inactive files on magnetic tape?

7. Why is MSS not considered as a replacement for magnetic disk drives? In your opinion, why have the companies that manufacture MSS tape systems not developed new, smaller models?

8. The CIS manager for your installation has indicated that he is not interested in obtaining database software. One of the major problems that is often mentioned is almost all of the programming effort is in maintaining existing systems. Very few new systems or programs are developed. Managers complain they do not have the information they need *when* it is needed or in the format that is needed. What rationale would you present to your manager in support of obtaining database software?

9. Why should either a manual or an online database dictionary be maintained?

10. Why should master files and databases be backed up? Why are magnetic tape backups sometimes stored in an offsite location?

Team or Individual Projects

1. Consult with someone on your campus who is familiar with the contents of the student database and find out what data is stored in the database for each student and how the data is used. Assume that an actual database cannot be used but sequential, direct, ISAM, and VSAM files are supported.

Directions:

Answer the following questions and do the activities that are listed.

a. What type of file organization would you recommend?

b. Do a file layout that indicates the sequence of the fields, the size of each field, and the type of data that would be stored in the field. Use a record layout form similar to the one used in the text.

c. What key or keys would you recommend?

d. In each situation, indicate whether you would access the file sequentially or randomly. The records are in student number sequence. You also need to indicate whether you would use the master file or a sequential copy of the file.

(1) The file is to be backed up on disk and on magnetic tape.

(2) A change program is available that makes it possible to change fields where data such as the student's telephone number, address, or curriculum code is stored. At any one time less than 1 percent of the records are changed.

(3) Grade reports are printed for all students.

(4) A report listing students in sequence by curriculum code is to be printed.

(5) Mailing labels are to be printed for all students who have a grade point average of 3.5 or higher.

(6) A program is available that makes it possible to change a student's grade. This program must be run if a faculty member recorded the wrong grade or if an incomplete is made up or changed to a failing grade.

e. List the advantages of creating a database rather than a master file. Be specific and cite actual examples of the way a query language might be used by someone who has the authority to access the information stored in the database.

2. Consult either advertising material or articles printed in some of the current publications and prepare a report on one of the database packages available for a microcomputer. In your report include:

a. The type of computer system needed to support the database. Be sure to include the amount of real memory and file space needed. Also indicate the type of disk drives that must be used and the operating systems that the software will run under.

b. The major features that are included in the software.

c. Information regarding how the files created using the database software can be used.

d. A description of the type of file organization (hierarchical or relational), the number of records that can be stored in the database, the number of keys that a record may contain, and the maximum size of the records that make up the database.

e. The features that are incorporated in the software to protect the records stored in the database.

GLOSSARY OF WORDS AND PHRASES

actuator An actuator consists of the access arm and read/write mechanisms that read or write data stored on disks. If there is more than one actuator on a drive, several read/write operations can be made simultaneously. The data travels thorough different pathways to reach the CPU or to be stored on disk.

address A location in real memory or on disk. The address for data stored on disk usually contains the cylinder, track, and record location on the track.

ASCII (American Standard Code for Information Interchange) A seven-bit code that is used to represent up to 128 unique numbers, symbols, and letters. Used more frequently on minicomputers and microcomputers than on mainframes.

blocking factor The number of logical records within a physical record. The blocking factor for a file is determined by the analyst.

BPI (bytes per inch) The number of bytes of data that can be stored on one inch of magnetic tape.

CPI (characters per inch) See BPI.

control interval A term that describes a storage area on a disk used to store records that are part of a VSAM file. The system software keeps track of the records stored in each control interval.

cylinder A portion of a disk that can be accessed with one position of the access arm. A track is that portion of a cylinder that can be accessed by one read/write head.

data dictionary software Programs that maintain a file of data about the individual fields (data items) that make up records. By interrogating the dictionary, an analyst or programmer can determine the size of a field, what type of data is stored in the field, and the programs that access the data stored in the field.

data item The smallest unit in the database. Data items are also called fields.

data module A sealed unit that contains the access arms, read/write mechanism, and the disks. Also called Winchester packs.

database interrogation software Software that permits individuals with very little knowledge of computers or databases to obtain information from the database.

data description language (DDL) Used for defining the logical structure of the data.

data manipulation language (DML) Provides commands for storing, retrieving, updating, adding, and deleting data items from a database.

direct organization   An organization and access method that does not require the use of indexes. A hashing scheme is used to calculate the address of where data is stored on a disk. This makes it possible to randomly retrieve records from a file without first consulting an index to determine where the records are stored.

dynamic disk allocation   The process by which the system's software takes care of allocating the space needed for a file.

EBCDIC (Expanded Binary Coded Decimal Interchange Code)   An eight-bit binary code that may represent up to 256 unique letters, symbols, or numbers. A ninth bit is usually provided for use in a parity check. EBCDIC is used more frequently with mainframes than is the ASCII code.

entry sequenced   Records are stored in a VSAM file in the order they are initially recorded.

FIFO (first-in first-out)   An inventory method where the first items purchased are assumed to be the first items sold.

fixed-head drive   A disk drive that has read/write heads positioned above each track. Data can be accessed faster since no arm movement is required.

gigabyte   One thousand megabytes; one billion bytes of data.

head-per-track drives   See fixed-head drives.

index area   The area where the indexes containing the addresses of the records are stored.

IPS (inches per second)   Used to express the speed with which magnetic tape can be read.

IRG (interrecord gaps)   The spaces required between the physical records stored on tape which allow for the deceleration and acceleration of the tape during a read or write operation.

ISAM (indexed-sequential access method)   Records stored on magnetic disk can be accessed randomly or sequentially. For each program, the access method is determined. When records are accessed randomly, an index is checked to determine the cylinder and track on which the record is stored.

key-sequenced   When used to describe the organization of a VSAM file, records are stored in sequence by a key such as student or part number. As new records are added to the file, the records are inserted into their proper location by key sequence.

LIFO (last-in, first-out)   An inventory method in which the assumption is made that the last item purchased is the first item sold.

logical record   One record of information as defined by the programmer. A logical record contains information about one person or object. The contents of a logical record are processed by the computer program.

MSS (mass storage system)   A system used to store massive amounts of data. Before the records are processed, the information is transferred from the strip of magnetic tape onto magnetic disk.

optical-disk memory system   Laser technology is used to write and read information stored in disks. More data can be stored on a 14-inch optical-memory disk than can be stored on a magnetic disk of the same size. Because the data is stored more densely, information can be retrieved faster.

overflow area   A temporary location in an ISAM file that is used for storing records. When the file is reorganized, the records stored in the overflow area are inserted into their proper location in the prime area.

parity check   A check made by the operating system software to determine whether the correct manner of bits is activated to represent a given character. Depending upon the system, in one location either an odd or even number of bits must always be activated. When an invalid count is detected, the system software usually provides commands for retrying the operation that produced the invalid check.

physical record   One or more logical records read or written as a group. The programmer or analyst determines the blocking factor at the time the file is created.

pointer   A device used to determine the storage location of the next record to be retrieved. A link field may contain the address (pointer) of the next record within a group that should be processed. In some databases pointers are used to ''chain'' files together.

prime area   The area within an indexed-sequential file where data records are stored.

query language   The user communicates with the computer in English and is guided through the process of writing a program that abstracts from the database.

relational model   An organizational method use to create a database. The database is visualized as a two-dimensional table: rows contain records and the columns contain the individual fields. The relationships between the items stored within the database are determined when information is retrieved from the database.

relative record number   The number of a record stored within a file. If John's record is the fifth one in the file, the relative record number is 5. When using a direct file, the relative record number may be used rather than a key field to randomly retrieve a record.

streaming tape drives   Data is read from or written on tape without pausing each time a block of records is read or written. A continuous stream of data is recorded or read from tape.

track   See cylinder.

utility program   Programs often supplied by the vendor that perform common tasks such as copying or sorting files. Some utility programs are developed inhouse and then used by all members of the CIS staff.

variable length records When records are to be stored within a file, the programmer can indicate that different record lengths are to be used. By using variable length records, space can be saved on the storage medium being used.

Winchester pack See data module.

VSAM (virtual storage access method) A file organization method that makes it possible to use both primarily and secondary keys. Within a program, records can be accessed randomly and sequentially.

STUDY GUIDE 10

Name _____

Class _____ Hour _____

A. Indicate whether the following statements are true (T) or false (F) in the space provided. If it is false, indicate in the margin how the statement should be changed to make it true.

_____ 1. The development of RAMAC making it possible to retrieve records stored on disk sequentially marked a new era in EDP.

_____ 2. If online databases are available, all data to be stored on magnetic disks should be recorded in an online database.

_____ 3. When a file is used, a program must be written to extract information from the file.

_____ 4. A file can be organized as a relational model.

_____ 5. If a database becomes too large, a great deal of overhead is required to manage the database.

_____ 6. Managing a database and using the information stored within a database is more labor intensive than when files are used.

_____ 7. Databases are generally online at all times.

_____ 8. Transaction files contain records that are normally used for a short period of time.

_____ 9. When a file or database is updated, a record is read, the data stored in one or more fields changed, and the record is rewritten in the location from which it was retrieved.

_____ 10. When data stored on an MSS is used, the programmer must write the commands needed to transfer the data from magnetic tape to magnetic disk.

_____ 11. When streaming tape drives are used, data is written onto the tape without stopping between blocks of data.

_____ 12. The highest density used to record data on magnetic tape is 6250 BPI.

_____ 13. When magnetic tape or disk is used as a storage medium for a file, the blocking factor is usually determined by the operating system.

_____ 14. When processing data stored in tape or disk, the computer processes a physical record.

_____ 15. Buffers are temporary storage areas within real memory and usually differ in size and number depending upon the program being run.

_____ 16. A parity check is only made when data is stored within real memory.

_____ 17. In selecting magnetic disk drives, the project team should compare the capacity, organization methods supported, access time, and cost per megabyte of data.

_____ 18. When creating a new file, only enough space should be allocated for the records initially recorded in the file.

_____ 19. If the total amount of space needed for the fields identified for the records is 197 bytes, no more than 200 bytes should be reserved for each record.

_____ 20. In a relational database model, pointers are not used.

_____ 21. SELECT, FROM, and WHERE are keywords used to retrieve information from a relational database.

_____ 22. When a query language is used to retrieve information from a database, in any one query only one database can be used.

_____ 23. A data manipulation language is used to tell the computer the significant facts about the database—such as the size of the file, size of the records, and fields stored within the records.

_____ 24. When a query language is used, AND and OR are called connectors.

_____ 25. Some query languages permit the user to define his or her own special meanings for certain words.

_____ 26. One of the disadvantages of using a database is that additional software is required that must serve as the interface between the user and the operating system.

_____ 27. It is possible to fine tune a database without changing the application programs.

_____ 28. Data dictionaries can only be used to maintain data about a database.

_____ 29. Most large organizations only have one very large database that stores data for all of their online applications.

_____ 30. A record layout should be completed for each file or database and included as documentation.

B. Multiple choice. Record the letter of the correct answer in the space provided.

_____ 1. A file created by keying in data from a source document is usually referred to as a
    a. database.   c. transaction file.
    b. master file.   d. temporary file.

_____ 2. Queues are used for storing
    a. transactions.   b. jobs to be executed.   c. backup files.

_____ 3. A recording surface on disk is referred to as a
    a. cylinder.   b. actuator.   c. track.

_____ 4. Another name for a data module is
    a. actuator.   b. Winchester pack.   c. head-per-track drive.

_____ 5. One billion bytes of data is called a
    a. megabyte.   b. gigabyte.   c. billabyte.

_____ 6. In a tree or hierarchial structure, records within a group are retrieved by the system software referring to
    a. a key stored in an index.
    b. a relative record number.
    c. an alternate key.

_____ 7. When a zero is recorded in the link field,
    a. the record will not be accessed.
    b. it is the first record of a group of transaction records for a customer.
    c. it is the last record of a group of transaction records for a customer.

_____ 8. Can be used to determine in which programs and systems a field of data is used.
    a. database   b. dictionary.   c. sequential file.

_____ 9. EBCDIC is
  a. an eight-bit code.
  b. widely used to store data with the real memory of a mainframe.
  c. a code that is sometimes used to store data on magnetic tape or disk.
  d. all of the above.

C. Matching. Indicate which of the media is described, or should be used, in the statements provided.

a. cassette tape   f. MSS
b. Winchester drives   g. head-per-track drives
c. magnetic reel tape   h. moving-head drives
d. 5 ¼-inch diskettes   i. optical-disk systems
e. hard disk

_____ 1. Because it is standardized, it is used to transport data between large computer systems.

_____ 2. Used as a substitute for magnetic tape. Data is transferred to magnetic disk for processing.

_____ 3. Each track has its own read/write mechanism.

_____ 4. The most widely used medium for storing databases and online files.

_____ 5. May be used for submitting reports to the Internal Revenue Service or to a credit union.

_____ 6. Data is stored more densely than when reel tape or magnetic disks are used.

_____ 7. At the present time, its use is limited to personnel computers and to storing files that can be retrieved sequentially.

_____ 8. Has a sealed-in unit that contains the access arm and the read/write mechanism.

_____ 9. The type of disk drive that usually requires the least amount of time to access records randomly.

_____ 10. Is still considered experimental but may eventually be used as a replacement for hard disk.

_____ 11. At the present time, is considered a ''standard'' storage medium for microcomputer users who wish to retrieve records randomly.

_____ 12. At the present time it is used more widely than any other medium for storing information recorded in online databases.

_____ 13. Would only be used by a large mainframe system for storing files that need not be online all of the time.

_____ 14. Was the first medium used for storing programs and files used by microcomputers.

_____ 15. Are often used for distributing software to microcomputer users. Operating systems and software packages such as word processing and database programs are stored on this medium and then sold.

D. Match the description or statement regarding magnetic tape with the correct term.

a. IRG   c. IPS   e. logical record
b. CPI   d. header label   f. physical record

_____ 1. The speed with which tape can be read.

____ 2.   Read by one read operation and stored in an input buffer.

____ 3.   The space between physical records.

____ 4.   The density used to record data on magnetic tape.

____ 5.   Processed by the computer.

____ 6.   Used by the operating system to make certain the correct reel of tape has been mounted by the operator.

E.   Match the file organization described with the terms listed. In some cases a blank is left to show where the word would be inserted in the statement.

       a. VSAM   b. ISAM   c. direct   d. sequential

____ 1.   Has an overflow area used to record the data that was "bumped off" the track when a new record was added.

____ 2.   The organizational method that is usually used for backup files.

____ 3.   Supports both a primary and alternate key.

____ 4.   According to James Martin, _____ organization should be used for files that have an activity ratio greater than 15 percent.

____ 5.   Must be reorganized or the overflow area will become filled.

____ 6.   Most _____ files are created as a key-sequenced data set rather than as an entry-sequenced data set.

____ 7.   _____ files do not require the use of an index. However, records can still be accessed randomly.

____ 8.   Records are stored in control intervals.

11

DETAILED DESIGN: PROGRAMS AND PROCEDURES

Looking Ahead

After reading the text and completing the learning activities you will be able to:

- List the factors that must be considered in segmenting a system into procedures.
- Identify the four major considerations that must be satisfied when design standards are developed.
- Determine what procedures should be implemented in order to provide for the safety and integrity of information stored in files or databases.
- Describe the type of programs that must be available to provide the required file or database maintenance.
- Identify the questions that must be answered regarding each procedure within a system.
- Determine when utility programs should be used.
- Explain why tables are used to store data.
- Identify the items that should be included in the specifications for a problem.
- List the major reasons a detailed design report should be prepared.
- Identify the major items that should be included in the detailed design report.
- Define and utilize the words and phrases listed in the end-of-chapter glossary.

INTRODUCTION

Although the output required may be determined first, it is not necessary to design the files and determine the input formats prior to deciding what procedures are needed to create and maintain the system. Experienced analysts find that determining the required output, files, input, and procedures is an ongoing process during the entire detailed design phase. Although the output and input specifications are approved by the user, as the various phases of the detailed design phase are completed, revisions are required. For example, it might be possible to combine two reports into one or, based upon the information available, expand a report to include additional information.

The analyst must determine what procedures are necessary to:

- provide the required reports;
- maintain the accuracy and currency of the databases and files;
- provide a clear audit trail;
- provide the necessary controls that ensure the validity of the output; and
- ensure the security and integrity of data entering the system or stored in files.

After all phases of the detailed design phase of the project are completed, a formal report should be prepared. Many of the items included as part of the documentation are approved as they are developed. One of the most important items in the report is the detailed plan for implementing the new system or procedure. The projections detailing the resources that will be needed at specific times in the implementation phase of the project must be accurate. Although the user may ask for minor modifications, it would be very unusual for the project to be dropped after the detailed design phase is completed.

SEGMENTING THE SYSTEM INTO PROCEDURES

How a system is segmented into procedures depends upon a number of factors. However, if structured design concepts are followed, the ~~system will be divided into subsystems and each subsystem into procedures. Each procedure should have one major function~~.

When realtime systems were first developed, there was a tendency to develop a single program that edited the data, produced a number of reports, and updated several files. When this was done, providing internal controls, external controls, and a clear audit trail was very difficult.

Today there is a trend, when either batch or transaction processing systems are developed, to divide the system into small procedures. Each procedure provides one major function.

FACTORS TO CONSIDER IN SEGMENTING A SYSTEM

Characteristics of the Hardware Being Used

How much memory is available and how many files can be online at one time must be considered. File space and the allocation of memory are very real concerns for the analyst working with a small business system. When a small business system is used, more programs and intermediate files may be required. Figure 11-1 illustrates the complexity of the online, realtime sales invoicing procedure. The sales data is entered by the sales representative. In the illustration, the four output files are organized as sequential files. The transaction file is created to function as a backup in case something happens to one of the master files or databases. The data stored in the transaction file may be transferred to magnetic tape and retained for the current fiscal period.

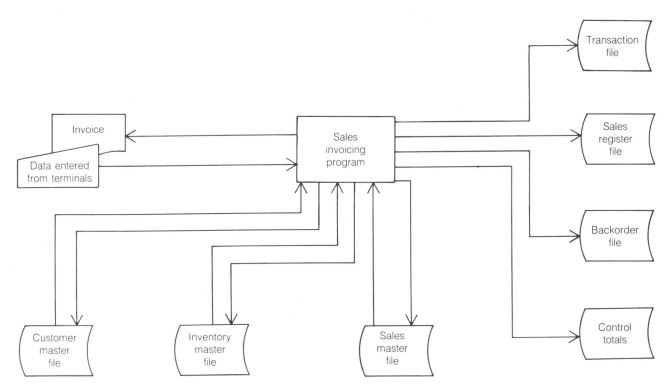

FIGURE 11.1   An online transaction processing system

The sales register file is used to print a register that is part of the audit trail and also provides a summary of the sales submitted each day by the sales representatives. If the customer calls about an order, the report may be used to determine the details of the transaction. The records could be sorted into sales representative sequence and the data used to update a file containing sales information regarding the performance of each sales representative.

If a microcomputer is used, it may be necessary to enter the data and create a transaction file. During this process, an incomplete customer and inventory file could be online. The records in the customer file might only contain the customer's number, name, and unused credit. The inventory records might have the item number, description, and quantity on hand. With this much information online, the operator could check to see that the right customer number was entered and that there was enough unused credit so that the order could be processed. When the item number was entered, the product description would be displayed for confirmation. The quantity entered would be subtracted from the quantity on hand to determine whether there was enough on hand to fill the order.

If there is not enough file space available to have all three master files online as the data is entered from terminals or from a microcomputer keyboard, a totally different design will need to be created. A transaction file will need to be created that contains the relative data regarding each transaction. Each of the master files will need to be accessed individually to determine the status of the customers' accounts, and the status of each inventory item, and to record the sales information in the sales master file. The segmentation into procedures will be totally different than when all three of the master files can be online simultaneously.

Characteristics of the Software Being Used

Timesharing applications that are transaction processing systems will be segmented differently than batch applications. If the system software does not provide for storing report data as files that can be queued by the operator when the printer is available, the analyst includes specifications for creating a separate file for each type of report that is to be printed. A separate program will then be written that reads the data stored in the file and formats the report.

Philosophy of the Analyst

Some analysts hesitate to develop an integrated program such as the one needed for the sales invoicing program illustrated in Figure 11-1. Figure 11-2 illustrates how two different analysts might produce a payroll register and update the payroll master file. In the integrated approach, the edited time card records are read, calculations made, the payroll register printed, a transaction file is produced, and the payroll master file updated. The transaction file will be used to generate additional reports and to print the paychecks.

In the second approach, the first program prints the payroll register and writes the transaction file. An audit is done of the payroll register to make certain that the right people are paid the correct amount. Until the payroll department indicates the register is correct, the payroll master file will not be updated or the reports printed from the transaction file. If errors are found, corrections can be made to the edited time file and the payroll register program rerun.

If an integrated approach is used, when errors are detected, the master file must be recreated by using the backup file. After this is done, the payroll register/update program can be rerun. In the final analysis and in terms of total time required, the more conservative approach might require less computer time and produce better results.

Controls to Be Built into the System

In the edit program that is used to produce the edited time file, the payroll manager might have insisted that the input be edited as follows:

1. Numeric fields are checked to see that no alphanumeric data is entered.
2. The department number is greater than 0 and less than 21. There are twenty departments ranging from 1 to 20.
3. Regular hours cannot exceed 35. Company policy states that regular pay is based on 7 hours per day for a total of 5 days per week.
4. Overtime cannot exceed 20 hours. No more than 4 hours of overtime can be worked by an employee in any one day for a maximum of 20 hours per week.
5. Salaried workers do not have a transaction record unless it contains a no-pay code or a part-pay code. If a part-pay code is recorded, the amount field in the transaction record must contain an amount greater than 0. Unless a part-pay or no-pay record is entered, salaried workers automatically receive their normal pay.

The following controls could also have been built into the edit program:

1. The total number of employees paid in each category is printed and compared to the total established by the payroll department.
2. The total regular hours and overtime hours for hourly employees is accumulated and compared with the total established by the payroll department.

If no error messages are printed and the batch totals agree with those accumulated by the payroll department, the file created is used as input for the payroll register program. During the payroll edit program, the same totals are accumulated. The payroll register is audited to determine whether the correct

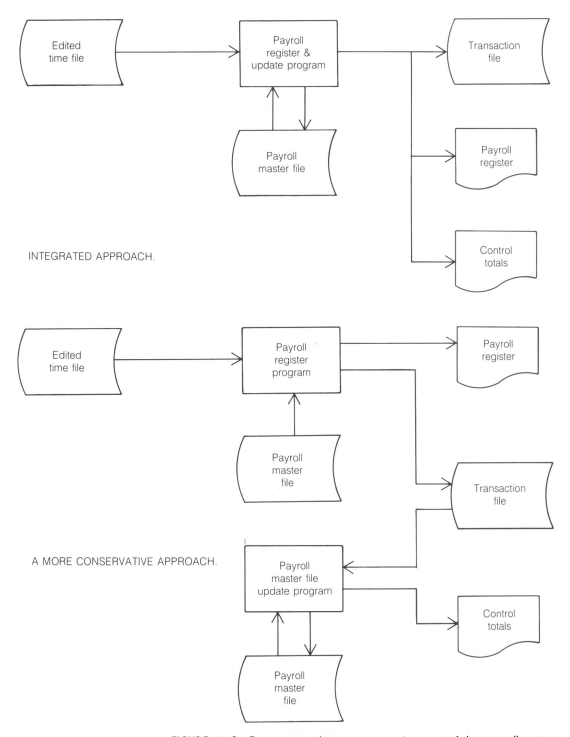

FIGURE 11.2 Two approaches to segmenting part of the payroll system

people are being paid in each department and for the correct hours. Salaried employees are checked to see that each employee is paid the correct amount.

When program specifications are developed, the analyst must look at each item of data that will be used in the calculations and determine whether there is any way to check its validity. For example, when a new payroll record is being added to the file, the pay classification code can be checked to see whether it is greater than 0 and less than 11. The pay level code can be checked to see that it is greater than 0 and less than 7. When new records are added to the file, a confirmation report should be printed and then *checked* by the payroll clerk.

The analyst must also consider whether any totals that have already been established can be used to help determine the validity of output. When the payroll register is printed, the gross pay, deductions, and net pay fields are all totaled. When the paychecks are printed, the same totals can be accumulated and compared with the ones printed on the payroll register. When the quarterly federal withholding tax report is printed, the total of the federal tax and FICA withheld must be checked. A total is taken of the federal tax and FICA totals printed on each register and compared with those printed on the quarterly report. If corrections were made, the individual totals may have to be adjusted by the amount of the corrections.

Audit Trail Required

A well-defined audit trail should be established for each transaction that enters new data into the system. When a pay raise is authorized for an employee, a change form signed by the employee's supervisor is filled out. The change is entered as part of the batch and a confirmation report is printed. The form can be checked against the confirmation report to see that the new amount is correctly recorded in the employee's record. For each change, the confirmation report would list the employee name, employee number, field changed, old value stored in the field, and the new value.

Sometimes manual procedures are used to provide part of the audit trail. When deposits are made in a bank account, the deposit slip filled out by the customer and visually verified by the teller serves as a backup for the transaction and also as part of the audit trail. If a teller's cash doesn't balance at the end of the shift, each source document (deposit and withdrawal slips) can be compared to a printout to determine the source of the problem. A transaction file of all entries into the system is created as part of the audit trail.

DESIGN STANDARDS

There should be standards that indicate the approach to be used in segmenting a system into procedures. The factors discussed earlier in the text (such as avoiding undue complexity and designing systems with the proper level of automation) must be considered when standards are developed.

Standards should be flexible enough to permit the analyst to design a system that is unique and fulfill the specific needs of the company. The analyst must know the characteristics of the hardware and control software, the basic philosophies of the company, and internal and external factors that will affect a system. The user must be involved in the general and detailed design phases. Standards must provide for:

- *the integrity of the data.* Regardless of how sophisticated a system is, analysts have failed in their assignment if invalid data is processed.
- *maximum utilization of the system.* A prime consideration is often the amount of throughput. The productivity of the computer as well as that of the individuals who interface with the system is a concern of management.
- *ease of maintenance and modification.* With an increasing percentage of the CIS budget spent on modifying existing systems, management is concerned about the time and resources needed to maintain and change systems.

- *user-friendly systems.* CIS personnel, users, and all levels of management must interface with computerized systems. Programs must be internally documented in easy-to-understand, nontechnical language.

Breaking systems into small, logical procedures that perform specific functions usually results in systems being easier to understand, work with, and manage. The computer may also be more productive than when highly complex, integrated systems are designed that perform functions before any check is made on the validity of the output. In a highly integrated system there may not be a well-defined audit trail and it may be more difficult to build in controls.

Transaction processing systems will be more complex and should only be designed and implemented in installations that have competent and experienced analysts and programmers.

Throughout the text some of the source documents, input layouts, report layouts, files, and procedures required to create and maintain a sales/inventory/accounts receivable system have been discussed. The system is far more limited than one actually designed for a medium to large company. The system was purposely limited in scope in order to keep the discussion as simple as possible. In presenting the material, the basic concepts of systems analysis and design are considered more important than the characteristics of the system being designed.

FACTORS THAT MUST BE DETERMINED FOR EACH PROCEDURE

Regardless of the type of hardware or software being used, for each procedure or program the analyst must determine the factors listed in Figure 11-3. Although program specifications may have been included in the general design report for the project, the specifications may be modified or expanded during the detailed design phase.

In the detailed design phase, the emphasis is on the development of precise specifications for input, output, files, and databases. As each procedure is designed, the analyst must question:

- ''Is the proper level of automation achieved?''
- ''Is the user getting relevant, timely information, with the proper amount of detail?''
- ''Where, when, and how will data enter the system for processing?''

PROCEDURES REQUIRED FOR ALL SYSTEMS

Regardless of the system developed, procedures must be established that provide for the security of files, file maintenance, and the execution of programs according to a predetermined schedule.

File Security

The security of files and the data stored within each file is the responsibility of the analyst. Files should be protected in the following ways:

Physical backup. Files should be backed up according to the directions provided in the documentation for the system. Usually one backup file is onsite and a second is maintained in an offsite location. The copy kept in the offsite location must be current.

Offline storage. When a file is not required to be online, the disk or tape should be stored in an area that has limited-access and is as disasterproof as possible. Often a large, fireproof vault is used. Only authorized personnel should be able to obtain the files. Immediately after use, files should be returned to their proper storage location.

Major Questions That Must be Answered
Regarding Each Procedure Within a System

1. Should the computerized portion of the procedure be designed as a batch or transaction processing application?

2. What output is needed? When is the output needed? What medium should be used for the output?

3. Should a scheduled report be printed or should the report be printed on demand? Can more satisfactory results be achieved by making queries into the system rather than by printing a formal report?

4. What internal or external documents are needed for recording data and establishing an audit trail?

5. What data already exists in files or databases that can be used as input? Are intermediate files needed?

6. What additional input is needed? What medium should be used and where should the data be entered into the system?

7. What techniques can be used to validate the input?

8. Can tables be used to increase the efficiency of the program?

9. What control information (such as the pay period of the month or the month of the year) must be entered by the operator? How can the control information be validated?

10. What functions must be performed by the data control section, operations manager, or EDP auditor?

11. What tasks need to be completed before reports can be distributed?

12. What documentation is required for CIS personnel, for the end user, and for management?

13. What interrelationships exist between the procedure being developed and other procedures within the system?

14. What programming maintenance is needed to ensure the currency, safety, and integrity of the data? (Some programs may need to be changed to reflect a new policy, a new law, or a change in an environmental condition.)

FIGURE 11.3 Questions that must be answered during the detailed design phase of the project regarding each procedure

Retention protection.   Files stored on magnetic tape and disk should be protected by using an unexpired **retention date.** If the label has an expired data, files can be accidentally destroyed.

Authorized access.   When online files are used, an account number and password system should be used to make certain only authorized personnel have access to the information.

Access protection.   When a record is in use, a "lock" should be placed on it so that another user cannot obtain the record until the first user releases it. Usually the operating system software or the programming language being used has software routines that make it impossible for more than one person to use a record at a time.

Authorized changes only.   Changes, other than normal updating, made to records stored in files must be properly documented and authorized. In this category are changes in salary, credit limit, reorder points, and so forth.

Review changes made to programs. Changes in programs must be reviewed, tested, and documented before being considered operational. When a change is made in one program, the interrelationships that exist between the program to be modified and other programs within the system must be studied. Unless this is done, programs that use the database or files used by the modified program may abort, destroy some of the records, or produce invalid results.

Establish policies. Policies concerning the handling, storing, and backing up of files should be established and printed in the CIS department's standards manual. For each database, master file, and transaction file the backup and other security provisions must be documented and the procedures followed by operations personnel. A security officer or EDP auditor should be responsible for ensuring that the documented security provisions are followed.

File Maintenance

Often programs are divided into three major classifications—**maintenance programs**, query programs, and report programs. Assume that the accounts receivable master file is already created and contains the fields illustrated in Figure 10-14. The file utilization chart illustrated in Figure 10-13 is included as part of the documentation for the system. Maintenance programs are required to add new customers to the file, delete inactive customers from the file, make changes to data stored within the file, and correct mistakes. The same type of program is required for making corrections that resulted from mistakes and making normal changes. However, the source of the data may differ and in keeping with the policy "Avoid unnecessary complexity," two programs may be used rather than one. A program must also be available that will create a copy of the file that can be used as backup.

In the past, one program might have been written that contained options for all of the required maintenance. Today a menu approach would be used. The menu displays the programs that are available. A help option would be included on the menu that would provide documentation regarding each program.

The change program makes what are considered normal changes to data stored within a file. Most master files and databases contain volatile data. Although the data is referred to as constant and updated information, changes occur on a day-to-day basis. Unless the changes are made, the integrity of the output is affected.

Regardless of how many controls are built into a system and how edit routines are used to validate data, errors do occur. A program must be available that can make the required corrections. If an error gets into the system, corrections may need to be made to more than one file or database. If a report is already printed, a correction in red should be written by hand along with a notation of why the data was adjusted.

When maintenance programs are run, the required change must be recorded on a source document, authorized, and a verification report printed.

Scheduling Programs

Because of the need for a clearly defined audit trail and a verification report of the changes made to the data stored within a file, maintenance programs are run on a scheduled basis in a batch mode. For example, all changes to the payroll database would be made just prior to running the weekly, report-producing payroll programs. Changes to the accounts receivable and inventory master files or databases would need to be made on a daily basis prior to the execution to the transaction processing programs.

In the final documentation for the system, a schedule of when the programs should be run is included. Often jobs must be executed in a particular sequence. For example, in the payroll system, the maintenance programs would be run and the file backed up prior to the edit program. Each time a program is run that updates the master file or database, the file should be backed up again.

Although some report-producing programs may be run during the 8:00 A.M. to 5:00 P.M. time period, the report may be queued to a file and printed during the second or third shift. If only one line printer is available, print jobs that occupy the printer for more than 20 or 30 minutes should be executed during other than ''prime time.''

In determining when jobs should be scheduled, the analyst must work closely with the operations manager. During the feasibility study, it should have been determined that enough CPU time, file space, and print time was available to support the proposed system.

DEVELOPING PROGRAMS

Report Programs

Many reports are produced by using data that is already accumulated in a database, master file, or transaction file. The major source of input is the data already available when needed. For example, let's assume a new payroll system is designed. When it is time to print the W-2 forms to send to each employee, someone might discover that the year-to-date FICA paid by each employee has not been accumulated. Or in a sales management system, it might be found that the total dollar value of this year's sales has not been accumulated so that it can be compared to the sales made in previous years. If data has to be accumulated after the fact, it is expensive and time-consuming.

When a new system is being designed, each report should be challenged. Questions such as ''How is the report used?'' ''Should the report be printed on a scheduled or demand basis?'' and ''Can the user obtain more timely information by using a query language than by waiting for a report to be printed?'' must be answered.

Using Utility Programs

It is not always necessary to write a separate program to perform each required task. There are usually a number of utility programs available that can be used to work with files. Analysts, programmers, and operators must know what utilities are available and how each is used. Utility programs are usually available that can: copy the file to another medium; reblock the new output file; change the record size of a new output file; display the contents of selected records on a printer or display terminal; retrieve a given record from a file; change fields within a selected record; and reorganize or condense the file.

Often reports must be printed according to a predetermined sequence. Before the report can be printed, the records stored in the file must often be sorted. A utility sort program can be used. If only a small portion of the records in the file will be used, three procedures are required. Figure 11-4 illustrates the three procedures. If a language such as **COBOL** is used, one program can be written that will perform all three functions. Other languages may not support internal sorts. When this is the case, the programmer will write the program that extracts the data. A utility sort program is used to sort the records and a report-producing program is written by the programmer. When three programs are required, the documentation may indicate that one job with three tasks is to be executed.

Using Tables

Tables can be used effectively to reduce the time needed to obtain input, reduce the size of a master file or transaction record, and write more efficient programs. When tables are used, loops can be established within a program that fills a table or utilizes the data stored in tables. In addition, less code is required to describe tables than is required to establish individual fields that can store the same amount

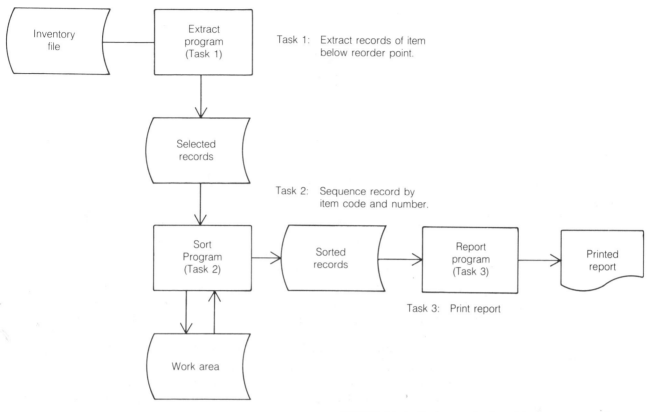

FIGURE 11.4 Tasks required to print a sequenced report

of data as a table. A table should be thought of as a group of mailboxes for an apartment complex. Your mail may be stored in row 4, column 8. Unless you know the location of your mailbox, it is obvious that you will not be able to retrieve your mail.

Tables can be part of an input record stored on cards, magnetic tape, diskettes, or disks. Other tables can be established within the memory of the computer, filled by reading data in from a file, and then used to extract the data that is needed. To illustrate the effective use of a two-dimension table, let's assume that hourly employees are paid on the basis of one of ten job classifications. Within each pay classification, there are six pay levels. The pay level is determined by seniority and job performance. Figure 11-5 illustrates how the pay table should be visualized. The row represents the pay classification and the columns show the pay levels within each classification.

Subscripts In each employee's master record is the pay classification and pay level, requiring three bytes. When an employee's gross pay is to be calculated, the values stored in the fields are used by the computer to calculate the unique address within the table where the rate of pay is stored. If an employee's pay classification is 6 and pay level is 4, 9.50 will be retrieved from the table. In the illustration the two numbers (6 and 4) are called **subscripts.** Whenever individual elements of a table are to be accessed, one or more subscripts must be used. One subscript would be used for a one-dimensional table; two subscripts (one for row and one for column) for **two-dimensional tables.** Subscripts must be whole, positive numbers within the range of the table. Therefore in the illustration the row subscript must be an integer from one to ten and the column subscripts an integer from one to six.

Pay Classification (Rows)	Pay Level (Columns) 1	2	3	4	5	6
1	6.00	6.12	6.25	6.50	7.00	7.50
2	6.50	6.75	7.00	7.25	7.75	8.50
3	7.00	7.25	7.50	7.75	8.00	8.75
4	7.50	7.75	8.00	8.25	8.75	9.50
5	8.00	8.25	8.50	8.75	9.25	10.00
6	8.25	8.50	9.00	9.50	10.00	10.50
7	8.75	9.00	9.25	9.75	10.25	11.00
8	9.00	9.25	9.50	10.00	10.50	11.50
9	9.25	9.50	9.75	10.25	11.00	12.00
10	9.50	9.75	10.00	10.50	11.50	12.50

FIGURE 11.5 Table containing pay rates

Advantage in Using Tables There are several advantages in using a table. If all of the pay rates are increased, the file with 60 fields is changed. This is much easier than changing each individual employee's pay rate. Also some space is saved. Often pay rates include data four places beyond the decimal. When this is the case, each rate requires six digits. If the pay classification and pay level are recorded in the employees' payroll records, less space is needed. Also the pay classification codes may be used to obtain job titles or descriptions from another table.

For another example of how a table might be used, let's assume that the sales representative's name is to be printed on the sales invoice. Each customer's record contains the sales representative's number. Since the data is packed, only two bytes are needed. The sales representative's names are loaded into a single-dimension table from a file. When a new sales representative is hired to replace someone who has left the company, only the file that contains the sales representative's names needs to be changed.

Often the easiest way to accumulate data is to read some type of transaction file, accumulate the data in a table stored within the memory of the computer, and then after all data is read, perform the required calculations. When the results of opinion polls or questionnaires are evaluated, tables must be used to accumulate the data. After all data is accumulated, the results are evaluated and a report is printed.

Tables are considered as one of the major sources of data. Analysts and programmers must know where tables can be used effectively.

CHECKPOINT

1. The major activities accomplished during the detailed design phase of the study are the output design, input layouts, file layouts, and segmentation of the system into procedures. Is each activity distinct and separate from the other activities? Could the analyst start by determining the procedures and programs that will be needed rather than determining what output is required?

2. "When transaction processing systems are developed, there is far less data-handling and there is less likelihood of errors than when batch systems are used. Therefore, fewer controls need to be built into the system." Are these two statements true or false?

3. With larger, more powerful computers and more databases available, there is a tendency to integrate systems. Should complexity be used to evaluate the effectiveness of the system?

4. What five major factors must be considered when determining how to segment a system into procedures?

5. Design standards must provided for what four major concerns that must be considered when a system is designed?

6. How can information stored in files be protected?

7. What types of programs must be available to provide the required maintenance for a file or database? Should the programs that provide the maintenance be run in a batch or transaction-processing mode?

8. Who determines when operational programs will be run?

9. What is a utility program?

10. Why are tables used for storing data?

11. If the pay rate table was called PRATES, the following calculation might be made in a payroll program:

$$\text{GROSS} = \text{PRATES (CLASS, LEVEL)} * \text{HOURS}$$

In the illustration what are the CLASS and LEVEL fields called?

THE SALES/ACCOUNTS RECEIVABLE/INVENTORY SYSTEM

At the end of the initial design phase of the sales/accounts receivable/inventory system, a report was submitted that included: an overview of the system; the objectives of the proposed system; a hierarchy chart illustrating each of the three subsystems; and a detailed explanation of each subsystem. The detailed explanation included an overview, objectives, and general specifications for the procedures that make up each subsystem.

During the detailed design phase the project team must:

- refine the specifications for each procedure and develop detailed specifications for each program.
- design report and screen formats. Print layout forms and screen formats should be developed and approved.
- design the source documents.
- determine the input formats needed.
- determine the contents and organization of master files and databases.

Often during the detailed design phase, additional procedures will be added to the system. Although structured design guidelines indicate that each procedure or program should accomplish only one major task, there are other factors that must be considered in dividing a system into segments.

Program Specifications

The degree to which the program specifications are defined during the detailed design phase depends upon the CIS department's standards. The experience and ability of the programmers assigned to the project will also determine the amount of detail included in the specifications. If the specifications are highly detailed, programmers feel the challenge is removed from their job and that only routine coding remains. If the specifications are not detailed enough, an inexperienced programmer may not provide sufficient internal programming checks or establish a clear audit trail.

A form similar to the one illustrated in Figures 11-6 and 11-7 might be used to provide the necessary program specifications. The programming specifications are for the program that processed the orders taken over the the phone and mailed in by customers. The report specifications for the sales invoice and control total report are illustrated in Figures 9-16 and 9-17 on pages 326 and 327. The sales-order form, transmittal form, and sales transaction record are illustrated in Figures, 9-18, 9-19, and 9-20 on pages 328, 330, and 331. You may wish to refer back to the previous illustrations and description of the sales-order procedure. Reviewing the materials mentioned will give you a better idea of what the program that processes the sales-order data must accomplish.

The Program Specification Form could be expanded to include columns to show who designed each of the supporting documents, such as the sales-order form, transmittal form, sales invoices, and control report. Some of the detailed work, such as designing the reports and source documents, could have been done by a programmer. The specifications for the validation of input and and the output controls required could have been determined by an analyst.

Under the heading of Special Considerations, any internal or external policies that influence the way the data must be processed should be noted. The following items might be included under Special Consideration: all accounts that are within 10 percent of their established credit limit must be flagged; salaried personnel are not eligible for overtime; or only students with a GPA above 3.8 are to be included on the report. The programming specifications do not tell *how* the data is to be processed but do tell *what* must be done.

In reviewing the program specification form for the sales invoicing program you will note that some of the program controls used to validate the input and to confirm the validity of the output were also included in the program that created the transaction file. When the control totals were compared after the transaction file was created, it may have been necessary to correct some of the records stored in the transaction file. The operator making the correction could have keyed in wrong data. Whenever an opportunity is available to edit data, it is wise to do so. Provisions should be made for the "it just couldn't happen" occurrences because occasionally errors that should never have occurred do. In a card-oriented payroll procedure, the keypunch operator keyed in 400 hours rather than 40 hours. The verifier operator made the identical mistake. However, when the card was used as input to the edit program, the error was detected. A statement was included in the program to check to see whether the regular hours were greater than 40. Most CIS managers, analysts, and programmers can tell you stories about errors that never should have happened.

PRESENTATION OF THE DETAILED REPORT TO MANAGEMENT

The detailed design report serves primarily as another checkpoint. Is the project on schedule and will the informational needs of management be fulfilled?

The analysts and programmers involved in the detailed design will usually continue working on the project and it will move from the detailed design phase into the implementation phase with very little interruption. If the analysts have worked with the user, and if the user has approved the various reports and source documents, there is very little danger that any of their work will be rejected.

The formal report becomes the plan of action for the future and also serves as a control over the project. Management is interested in knowing how the actual cost of developing the detailed design for the system compares with the projected cost. Were modifications required in the design that will be more costly or less costly than the original estimates? Was the detailed design completed in the time allocated?

The report also serves as a method of communication. Management and members of the computer policy committee are made aware of how the new system is being developed and of the benefits that will be derived from using the system.

computer products inc.   PROGRAM SPECIFICATION FORM

SYSTEM	Sales/Inventory/Accounts Receivable	PROGRAMMER:	KAY WALZCAK
PROGRAM	Batch sales-order processing	DATE:	September 5, 1986
ANALYST	Martha Aguilar	PAGE	1 of 2

1. ORIGIN OF DATA

   Sales order forms submitted by customers and orders called in by customers.

2. SOURCE DOCUMENTS

   Sales-order form. When the order is taken over the phone, the order is prepared
   by the order taker.

   Transmittal form containing the number of documents, document numbers, and date
   of the transactions.

3. INPUT

   A. Control record keyed in by the operator. The fields required are invoice date,
       first order number, last order number, and order date. The record was created
       by the operator who keyed in the data recorded on the sales orders.

   B. Transaction file. When the data was recorded in the transaction file, numeric
       fields were edited to determine that only digits were recorded in the field.
       When the account number was keyed in, the customer's record was retrieved and
       their name displayed on the VDT. The operator visually confirmed the customer's
       name displayed with the one on the sales order. The sales representative's
       number is compared to see whether it is within the allowable range. After the
       item number is entered, the item description displayed is visually compared
       with the one on the sales order. After the quantity is keyed in, it is
       visually compared with the one on the order. The order reference number keyed
       in is checked to see whether it is one of the numbers within the range
       established by using the first and last order numbers.

   C. Customer master record. The customers' name and address will be used in
       printing the sales invoices.

   D. Inventory master record. The description and selling price are obtained for
       each item sold.

   E. Sales representative sequential file. The sales representative's names are
       stored in a one-dimensional table. The representative's number from the
       transaction file will be used as the subscript to retrieve the sales
       representative's name from the table.

FIGURE 11.6 Program specification form   (Page 1)

computer products inc. PROGRAM SPECIFICATION FORM

SYSTEM	Sales/Inventory/Accounts Receivable	PROGRAMMER:	KAY WALZCAK
PROGRAM	Batch sales-order processing	DATE:	September 5, 1986
ANALYST	Martha Aguilar	PAGE	2 of 2

4. PROGRAM CONTROLS NEEDED TO VALIDATE THE INPUT:

   A. The transaction records must be sequence checked. The records should be in sequence by order reference number.
   B. The order reference numbers must fall within the given range.
   C. The sales representative's number must be within the acceptable range.
   D. All numeric fields must be validated to make certain they contain only numeric data.

5. OUTPUT CONTROLS

   A. A count of invoices printed must be determined. The total must agree with the one shown on the order transmittal form.
   B. The total of the quantity field printed at the end of the job must agree with the adjusted total shown on the transmittal form.

6. OUTPUT REQUIRED:

   A. Invoice records are added to the customer database. Each record contains the customer number, date, reference number, amount, and shipping code.
   B. The inventory master file record and the individual lot-purchased records are updated by decreasing the amount of the sale.
   C. A file containing items that must be placed on backorder.
   D. Sales invoices.
   E. Control report.

7. DISTRIBUTION OF THE REPORTS:

   A. Prior to releasing the reports, the data control clerk must confirm the accuracy of the control total report.
   B. After confirming the control totals, all four copies of the sales invoices, the sales orders, and the transmittal form will be transmitted to the sales department.

8. SPECIAL CONSIDERATIONS:

   A. Unless the entire quantity ordered can be shipped, the item will be backordered.
   B. When an item is placed on backorder, the statement ''Item Backordered'' will be printed on the invoice in the area normally reserved for the unit price and extension. A separate program is run to generate letters that inform the customer when the items will be shipped.

FIGURE 11.7   Program specification form   (Page 2)

If any last-minute changes are to be made in the system, they must be made at this point—before implementation begins. It is essential that careful planning be done for the implementation phase of the project. Obviously, some events must occur in a given order. For example, until the computer arrives, the programs cannot be tested. Unless the site is ready, the hardware cannot be installed.

Successful implementation of a system depends upon good project management. Planning for the implementation phase is more critical than for any other part of the project since more resources will be committed and the planning process should start in the design phase. In terms of total time, more time is spent on the implementation phase of most large projects than on any other phase. If management would like the system operational in the shortest period of time, a PERT chart should be constructed. The analyst must determine the sequence in which the various tasks must be completed, the length of time each task will take to complete, and which, if any, of the tasks can be performed simultaneously. After this is done, the chart can be constructed, and how soon the system can become operational can be determined.

One of the key factors is to have sufficient personnel available, when needed, to perform the required functions. During the conversion phase, it may be necessary to perform some tasks under both the old and new method. This will require additional personnel. However, if the implementation is well planned, personnel should be available when needed. It may be necessary to budget for additional professional personnel who can be obtained from an agency that specializes in placing people on temporary assignments.

Report Outline   A written report as well as an oral report should be prepared. The written report must be supported by the necessary documentation. The documentation will consist of the source documents, file specifications, report layouts, and program specifications.

Most of the items included in the documentation have already been approved by the user who will be working with the preparation of the input or the results of processing the output. The items included in the documentation also become the programmer's working documents as the project moves from the detailed design phase into the implementation phase.

There should be a suggested format for the formal report in the standards manual. Although the precise format of the report would vary depending upon the scope of the project, the following points should be covered in both the written and oral presentation.

I. Overview of the system.
   A. Functions.
   B. Benefits.
II. Hierarchy charts of the system.
   A. Overview of the entire system showing the various subsystems that will be developed.
   B. Separate charts for each subsystem which illustrate the procedures that will be developed to process the data and to produce the various reports.
III. Brief overview of each of the subsystems and its procedures.
   A. Functions performed.
   B. Controls built into the system.

IV. Plan for implementing the system. A PERT or Gantt chart should be developed to provide an overview of how the entire system will be developed. Along with the chart should be a brief explanation regarding the resources needed for each of the following items:

A. Site preparation. What needs to be done and when the work must be completed should be documented. If a large amount of work is needed, the architect's detailed drawings should be appended to the report.

B. Hardware. The CPU or the required peripherals and other offline equipment must be available for:

1. testing individual procedures and for testing the entire system; and

2. training personnel who will work with the system.

In the event that the computer is not delivered according to schedule, a penalty clause should be enacted and an alternate test site should be selected.

C. System software. If the control software (components of the operating system) is to be secured from vendors other than the computer manufacturer it must be scheduled to arrive in time to adequately test it before beginning to implement the new system.

D. Additional supplies. The following items need to be ordered and scheduled to arrive prior to being needed:

1. Source documents.
2. Preprinted forms for external reports.
3. Internal forms.
4. Preprinted cards.

E. Programming. A schedule must be determined for when the various programs that make up the system need to be designed, coded, tested, debugged, and documented.

F. File conversion. Files generally must be converted prior to testing the system. Additional personnel may be needed in order to convert the files.

G. Training of the personnel who will be working with the system. A schedule must be set up to accommodate the personnel who need additional training.

H. Conversion plan. The type of conversions will depend upon the type of system being developed. Some systems can be phased in while others must be implemented all at one time.

I. Evaluation. How the system will be evaluated in order to determine whether it meets its stated objectives.

J. Documentation. Date when all the documentation must be completed. The final documentation must adhere to the standards established by the CIS department and must be completed prior to the date the system is considered operational.

V. Actual costs of the detailed design phase compared with the estimated costs.

A. The actual cost of developing the system should be compared with the projected costs. There should be very little difference in the two sets of figures.

B. The actual hours involved in the detailed or actual design phase should be compared with the projected hours.

VI. Implementation costs.
   A. The costs projected during the general design phase should be reviewed.
   B. Although the figures may need to be revised, the costs projected at this time should be within 5 or 10 percent of the actual costs involved in implementing the system.

VII. Ongoing operational costs.

VIII. Supportive documentation.

Purpose of the Oral Presentation

The oral presentation is made to the computer policy committee and should be prepared in advance. Although by this time it is not necessary to "sell" the system to the users and management, the report should be well organized, and the presentor must be able to answer questions that arise during the presentation.

Overhead transparencies or slides should be made of the charts or graphs that will be referenced. Whenever statistics are presented, such as the projected costs compared to the actual costs, slides or transparencies should also be used.

Since the written report is distributed to the members of the committee and to other individuals who have been directly involved in the development of the project, the objectives of the oral presentation are to: provide two-way communication between the project team and the individuals who will be affected by the system; provide a progress report to determine whether the project is on schedule; and allow input into the detailed design by individuals who will be directly or indirectly affected by the system.

Although it is not anticipated that additional features or changes will be suggested, the project team should leave the door open for additional input from management and the end users. Additional controls or enhancements to the system could still be included. Changes to the proposed system will be cheaper and easier to make prior to the system being implemented rather than after the system is operational.

CHECKPOINT

12. What might occur if the programming specifications developed during the detailed design phase are "too detailed"?

13. Why must the programmer be aware of the policy requiring that an item be backordered if the entire amount is not available for shipment?

14. What major items should be included in the programming specifications?

15. Since the end users have been directly involved in the detailed design phase of the project, of what value is the detailed design report?

16. What is meant by the statement that "the formal report becomes the plan for action"?

17. When the formal report is given, is it anticipated that a number of changes will be suggested?

18. Why should a planning tool such as a PERT or Gantt chart be used?

19. What practical purpose will the documentation serve?

20. Why should the actual costs and time of developing the detailed design be compared with the projected costs?

SUMMARY The four major tasks involved in the detailed design phase of the project must be considered a looping process. Let's assume that the system was first divided into procedures. Next the output and file requirements were determined and each was designed. If management wants an additional report, the hierarchy chart for the system will need to be changed, the new report designed, and additional fields may need to be added to the database. If end users and management have been directly involved in the detailed design phase, when the report is given, there should be no major changes recommended. However, the project team must be receptive to any suggestions that might be made.

There are often well-established guidelines for the design of reports, documents, files, and databases. Since each system is different and because more systems are developed as transaction-processing rather than batch systems, precise guidelines for dividing a system into procedures are usually not available.

In designing programs, programmers and analysts must take advantage of the resources available. This might include use of utility programs, tables, and query languages.

The final report for the detailed design phase becomes the plan of action for the implementation of the system. Since more resources are committed to the implementation phase than any other phase, efficient project planning and management is essential.

DISCUSSION QUESTIONS

1. A friend of yours made the following statements: "During the detailed design phase you determine what procedures are needed. After the procedures are determined you decide what output, input, and files are needed. Once the procedures are established, changes should not be made." How would you respond to your friend's statements?

2. Explain why a system might need to be segmented differently if a microcomputer was used rather than a large mainframe. Use as an example the sales transaction program for processing sales transactions entered by the sales representatives.

3. Could a small retail establishment that had a microcomputer system still process sales in a transaction-processing mode? Could the availability of the item and the customer's credit still be checked?

4. If you were the analyst designing the payroll system, would you develop one program that would process the transaction records, update the payroll master file, and the payroll register or would you develop two programs as illustrated under the caption "a more conservative approach" in Figure 11-2? Provide the rationale for your answer.

5. What relationships exist between the "major questions that must be answered regarding each procedure within a system" and the specifications that are developed for a program?

6. You are a CIS manager and one of the programmers within your development made the statement "I never use utility programs; I write all of my own programs." As the manager, how would you react to the programmer's statement and what might you suggest that the programmer do in regard to the utility programs available for your computer?

7. One of the analysts on the project team to which you are assigned has indicated that there "is no value in preparing a report for the detailed design phase of the project." Use the outline provided for the detailed design report

and indicate how each of the items identified might be used. Also indicate why the reports should be prepared before the implementation phase of the project begins.

8. Your CIS manager has indicated that once the detailed design is completed, no further changes or additions can be made to a system. During the presentation of the report, an additional report is requested by the user. As an analyst, you feel that the request is valid and that all of the data required could be available in the database. Your manager has indicated that "after the system is implemented he might consider printing the additional report." What rationale would you use in trying to convince your CIS manager that the changes required to provide the additional report should be made now rather than after the system is operational?

Team or Individual Projects

1. Assume that you are designing the system for a company that has a very small microcomputer system with a very limited amount of file space. Since the operating system provides for multi-tasking, two terminals are available and it is possible to input sales data while another program is running. Diskettes are used both for storing the master files and for the transaction files. Unfortunately the system was obtained some time ago and only 10 MBs of storage are available on the hard disk drive. Hard disk is used for storing some of the components of the operating system and files that are being processed. When a file is required, a utility program is used to copy the file stored on a diskette onto hard disk for processing.

Directions:

a. Redesign the sales invoicing program that was illustrated in Figure 11-1 so that only one major function is performed by each program. Sales transactions will be processed as they occur and the sales invoices printed on one of the two letter-quality printers that are part of the computer system. Most of the sales are made on charge accounts rather than as cash sales. When a cash sale is made, an invoice is printed and marked "paid in cash." When the sales register is printed, both the cash and sales charge transactions will be itemized.

b. For each procedure that requires a program, draw the system flowchart that shows the input, program that processes the data, and the output created.

c. Indicate what type of controls will be included in the sales invoicing program that will help to make certain that the data keyed in by the operator is correct. In most cases, the customer is in the store, makes one or more purchases, and waits for the invoice to be printed.

d. Assuming that the store employs four sales persons and two employees who perform routine office functions, who would you recommend key in the sales data? Should each sales representative key in his or her own sales information or should one of the office employees be assigned the task? The office is in the front of the store in a modern open area.

e. Develop a schedule of the sequence in which each job will be run. You are to assume that the sales invoicing program will run continuously during the period of time that sales are being made.

f. Assume that a new hard disk could be obtained that can store 100 MBs of data. The cost of the drive is $3,000. What factors would you present to management along with your request to obtain the new disk drive? Is there any way that you might show that it would be cost-effective to obtain the new drive?

2. The following list of programs has been identified as being required to maintain the VSAM inventory master file, back up the file, update the records, and produce the following reports. A brief description of each program that affects the inventory master file is provided. The programs are not listed in any particular order.

a. Add new items to the master file.

b. Print a report of items that are below the reorder point.

c. Flag obsolete records for deletion. The records will be retained by item number in an inactive history file. The record cannot be deleted unless the quantity on hand is zero or unless management has indicated that the value of the inventory is to be written off as a loss.

d. Correct records that contain data that was entered incorrectly. An incorrect order number, quantity, or price could have been entered.

e. Copy the file to a sequential disk and to tape a backup file.

f. Print a report listing out-of-stock items.

g. Add new shipments of merchandise to the inventory file. Each shipment will be recorded in a separate record and the quantity onhand in the master record will be updated.

h. Print the inventory analysis report described in Chapter 10 of the case study.

i. Change records for what might be considered normal changes. The reorder point, reorder quantity, or vendor code might be changed.

j. Print a report of items that have changed in price due to the addition of new shipments.

k. Although not part of this subsystem, when the sales invoicing programs are run, the inventory records are updated. The quantity on hand is reduced by the number of items sold.

The majority of all sales are made between 8:00 A.M. and 8:00 P.M. (local time). Although CPI permits some people to work flexible schedules, most employees work either from 7:00 A.M. to 4:00 P.M. or from 8:00 A.M. to 5:00 P.M.. In the CIS department, however, one or more data entry and computer operators are assigned to each of three shifts.

Directions:

a. Review the information presented in the text and in the case study regarding the inventory.

b. Draw a chart that indicates how frequently each program should be run (daily, weekly, monthly, or on demand). Also include a recommendation for when the report should be run: during "prime time," second, or third shift, and in what particular sequence. Classify each program according to the following:

      M   = Maintenance
      U   = Update program
      F   = Full report (all items listed)
      E   = Exception report

Assuming the inventory records are in sequence by item number, indicate whether the file should be sorted prior to printing any of the reports. For each program, indicate whether the records would be retrieved randomly or sequentially.

c. Assuming that the inventory records were stored in a database, list the type of information that the inventory manager might wish to obtain by using a query language.

d. Although each program should provide one major function, adding new items to a file and printing a report would still be considered one major function. Review the programs and indicate whether you feel any should be combined or divided into two programs. After you have made your decisions, draw systems flowcharts for each program showing the input, name of the program being executed, and output.

GLOSSARY OF WORDS AND PHRASES

COBOL (COmmon Business Oriented Language)   COBOL is used more frequently than any other language to program business applications designed to run on mainframes.

maintenance programs   Software which adds, deleted, or changes records stored in a file. Unless files are maintained, the information becomes obsolete and relevant, timely reports cannot be produced.

retention date   The last day that information stored in a disk or tape file will be protected from accidental overwrite. Once the date is past, the information stored in the file is no longer protected.

subscript   A positive integer used to identify a field of data stored in a table.

two-dimensional   An area in storage that is visualized as having rows and columns. To access data stored in a two-dimensional table two subscripts are needed. The first subscript identifies the row and the second identifies the column.

STUDY GUIDE 11

Name _____

Class _____ Hour _____

A. Indicate whether the following statements are true (T) or false (F). If false, indicate how the statement should be changed to make it true.

_____ 1. During the detailed design phase there are no ongoing investigations and decisions made during the general design phase should not be changed or modified.

_____ 2. After the formal report for the detailed design phase is approved, all of the documentation presented is approved by the users of the system.

_____ 3. Since their involvement tends to slow down the development of the system, users should not be involved in the detailed design phase.

_____ 4. Today the trend is toward developing fewer and more complex programs that perform several major functions.

_____ 5. One of the most significant factors to be considered in designing a system is the amount of online file space available.

_____ 6. In the online sales invoicing program, the sales register file is organized as a VSAM file. The file is used to print the sales register report.

_____ 7. Sometimes a version of a master file might be online that contains only two or three needed fields of data. One example might be a condensed version of the customer master file that would contain only the customer's number, name, and balance.

_____ 8. When program specifications are developed, the analyst must look at each data item and determine whether there is any way to check its validity.

_____ 9. Any standards that are developed regarding how systems are designed must provide: for the integrity of the data; maximum utilization of the system, ease of maintaining and modifying programs; and a user-friendly system.

_____ 10. The major factors regarding each procedure within a system might be considered a checklist that should be completed before program specifications are developed.

_____ 11. If the label information provided for a disk or tape file has an expired retention date, the records can be accidently destroyed.

_____ 12. Since the changes to the payroll master file are keyed in by a payroll clerk, written authorization for changes should not be required and forms are not needed.

_____ 13. In determining when jobs should be run, it is only necessary to determine whether a job should be run each day, once a week, or once a month.

_____ 14. When a file is to be sorted, it is always necessary to write three separate programs. The first program extracts the records needed from the master file. The second program sorts the files stored in the file created by the extract program. The third program prints the reports from the records stored in the file that has been sorted.

_____ 15. When tables are used, the amount of code needed to describe and process data may be reduced.

_____ 16. In obtaining a field of data stored in a table, subscripts must be used.

_____ 17. If a two-dimensional table is created, it is still only necessary to use one subscript to access any item of data stored within the table.

_____ 18. Unless the program specifications developed during the detailed design phase are specific and detailed, programmers tend to lose interest in the program.

_____ 19. Any policy that directly affects how the program will process the data should be listed under the topic Special Considerations.

_____ 20. The report given at the end of the detailed design phase provides information, is an evaluation of the detailed design phase, and provides an implementation plan.

_____ 21. More resources are generally used in designing than in implementing a system.

_____ 22. During the detailed design report, recommendations for changes cannot be made by users.

_____ 23. Since each member of the computer policy committee has a copy of the detailed design report, the use of visual aids such as slides or transparencies in presenting it is unnecessary.

_____ 24. The detailed design report only includes the actual cost of developing the detailed design.

_____ 25. How employees are to be trained is not a consideration in implementing a system.

B. Indicate how each of the following programs might be classified. The classifications are:

      a. maintenance programs   c. report programs
      b. query programs

_____ 1. New records are added to an existing file. A confirmation report is printed.

_____ 2. The inventory manager wants to know exactly how much of product X is onhand at this precise moment.

_____ 3. A listing of all items that are below the reorder point is printed.

_____ 4. Changes are made to master file records to correct data that had previously been entered incorrectly. A confirmation report is printed.

_____ 5. At the end of the year, all of the year-to-date totals in the payroll master file records are set to zero.

C. Multiple choice. Record the letter of the correct answer in the space provided.

_____ 1. Program specifications are developed primarily for the benefit of the
      a. user. b. analyst. c. programmer.

_____ 2. One of the most critical items included in the detailed design report is
      a. the comparison of actual costs with budgeted costs.
      b. the print layout forms that have been approved.
      c. the detailed plan for implementing the system.
      d. the design of the source documents that will be needed.

____ 3. Provides an excellent overview of how the system is segmented into subsystems and procedures.

    a. PERT chart b. Gantt chart c. hierarchy chart

____ 4. The three major purposes of the oral report covering the general design phase are to provide two-way communication between the team and individuals who will be affected by the system, give a progress report, and

    a. go over the program specification forms.

    b. review the subsystems and procedures that will be developed.

    c. allow additional input into the detailed design.

    d. none of the above.

____ 5. General file or database maintenance programs are run

    a. once a week during prime time.

    b. once a month during the second or third shift.

    c. daily during prime time.

    d. daily during a time when the computer is not being used for transaction processing.

____ 6. Both the analyst and the operations manager feel that a particular program should be run during prime time. However, in order not to utilize the printer for such a long period of time

    a. the program should be run at night.

    b. the information to be printed can be stored in a print file or queue and then be queued to the printer during the second or third shift.

    c. the program can be divided into smaller programs so that smaller reports will be printed.

____ 7. When designing a microcomputer system, one of the most limiting factors may be

    a. the size of the computer's memory.

    b. the fact that only one rather slow printer is available.

    c. the limited amount of online storage space available.

____ 8. Data keyed in by the operator using a terminal is

    a. only visually verified.

    b. only edited by the program.

    c. often visually verified and edited by the program.

    d. always key verified by a second operator.

____ 9. One of the key issues in designing a system is to

    a. always print a report for every program that is run.

    b. develop a separate program to update a file rather than updating records where data is entered into the system.

    c. provide a clear audit trail.

____ 10. Sometimes manual procedures are used as well as automated methods in order to

    a. provide an alternate system that could be used in case the computer is down or there is a power failure.

    b. provide a clearer audit trail.

    c. provide a means of checking for an error that might have occurred when data was entered.

    d. provide for a, b, and c.

D. Match the factor to be considered with the description of the situation or problem:

a. Characteristics of the hardware being used
b. Characteristics of the software being used
c. Philosophy of the analyst
d. Controls to be built into the system
e. Audit trail to be provided

_____ 1. A guideline that each program should provide one major function is followed.

_____ 2. Only single-batch programs can be developed.

_____ 3. When a microcomputer is used, it might not be possible for all of the files to be online at the same time. Therefore, the sales transaction program will have to be segmented into several programs.

_____ 4. The payroll manager would like to determine that the right number of people will be paid for the correct number of hours before the payroll register is printed.

_____ 5. Interactive, timesharing applications cannot be developed. Therefore the sales data is keyed in, verified, and then edited by the computer before the sales invoicing program is run.

_____ 6. The inventory manager would like to see a printout of the changes to be made to the inventory file *before* the records are actually updated. Therefore a transaction file is created that is later used to update the inventory master file records.

_____ 7. Since the transaction is keyed in by the teller, it is not necessary to have a deposit slip filled out manually.

_____ 8. Before new items are added to the online inventory file, a printout of the items to be added is printed from the transaction file.

_____ 9. The analyst designing the new accounts payable system has combined the programs that formerly added new records, deleted records, and changed data in existing records into one program. It is felt that this is more efficient.

_____ 10. Although multiple users can use the system and more than one program can be executed, teleprocessing is not fully supported and sales information cannot be transmitted directly into the system by the sales representatives.

SECTION V

Implementation and Evaluation

Texas Agricultural Service Company

PHASE I

TASCO (Texas Agricultural Service Company) is organized to purchase tires, batteries, tillage equipment in large quantities. Shipments from the manufacturers of the products are stored in a warehouse located in Waco, Texas. The orders received from distributors are processed and shipped to the retail distributors. Distributors located near Waco often pick up and deliver their own orders.

The Old System

Under the old system, sales orders were sent to Waco for processing. Data was keyed in from the source documents by data entry operators. After all orders had been keyed in and verified, the data was processed in a batch mode and the sales invoices were printed. Until the order was processed it could not be determined if there was enough merchandise on hand to fill the order. Usually it took a minimum of two days to process an order. If a distributor wanted to pick up an order on Tuesday, the distributor was required to call in the order on Monday. However, if the merchandise was not available this fact could not be determined until Tuesday when an attempt was made to process the order.

Objectives for the Order Entry System

One of the major objectives was to provide for immediate processing of orders. It must be possible for a distributor to arrive in Waco, verbally order the merchandise, and have the merchandise loaded within a matter of minutes. Another objective was to be able to update and adjust the inventory in

Programmer at a terminal in his office. Using terminals that are hard-wired to the IBM 4341, programmer/analysts can enter source code and test programs.

a transaction-processing mode. Immediate feedback must be available regarding shortages.

It was determined that a menu approach should be used that would be user-friendly. Besides processing orders, printing the necessary documents, and updating the inventory, when a query regarding an order is made it must be possible to display the order on a formatted screen.

PROJECT DESIGN

It was estimated that the first phase of the TASCO project would take six months to complete. During the initial investigation and the feasibility study phase, the old system was studied. The project team, under the direction of Ron Jones, had to become familiar with how orders were placed, processed, and filled. The supervisors responsible for the warehouse operation were interviewed to determine the problems with the old system and what suggestions they had for the new system. The inventory and accounts receivable systems were also studied since

one of the objectives was to update the inventory and receivable records affected by each transaction.

During the design phase a hierarchy chart was prepared that illustrated the subsystems that would make up Phase I of the TASCO project. The order-entry subsystem was designed to process phone orders rather than orders submitted on sales order forms. This would help to eliminate delay and provide the distributors with immediate feedback.

Project Implementation

Figure V-1 illustrates an example of a planning chart (project estimating worksheet) that was prepared for the implementation phase of the TASCO project. By looking at the chart you can see the estimated man hours needed to design, code, and test the programs. The chart also shows the estimated machine hours needed to develop and test the programs and the estimated cost of the implementation phase. Each task item is given a name and number. The chart indicates the prerequisite for each task, the estimated time needed to complete each task, the target

DATE: _____11-17-84_____

PROJECT NUMBER: __A3273__

TOTAL MAN HOURS __417__

TOTAL MACHINE HOURS __23__

TOTAL COST __$ 4,300.00__

PROJECT NAME: __TASCO Printers in Warehouse__

PROJECT COORD. __DS__   SUPV. __RJ__

EST. START DATE: __11-26-84__
EST. TARGET DATE: __2-01-85__

TASK ITEM	TASK STEP	PRE-REQ STEP	POST-REQ STEP	EST. MAN HOURS	EST. MACH.HOURS	TARGET START DATE	TARGET COMP'L DATE
Inventory							
Design Menu map - TINVSIO	15			4		11-30	11-30
Code Menu map - TINVSIO	16	15		6		12-10	12-11
Design Basic info map - TIN2SIO	17	15		4		11-30	11-30
Code Basic info map - TIN2SIO	18	17		6		12-12	12-13
Design price map - TIN3SIO	19	15		3		11-30	11-30
Code price map - TIN3SIO	20	19		3		12-10	12-10
Design counts map - TIN4SIO	21	15		3		12-03	12-03
Code counts map - TIN4SIO	22	21		3		12-10	12-10
Design mass price chg map - TINSSIO	23	15		3		12-03	12-03
Code mass price chg map - TINSSIO	24	23		3		12-11	12-11
Design inq map - TIN6SIO	25	15		3		12-03	12-03
Code inq map - TIN6SIO	26	25		4		12-11	12-12
Design process selection CICSTINV	27	15	28	1		12-04	12-04
Code process selection CICSTINV	28	27	29	3		12-14	12-14
Test process selection CICSTINV	29	28		2	1	12-17	12-17
Design process basic info CICSTIN2	30	17		4		12-04	12-04
Code process basic info CICSTIN2	31	30	32	6		12-12	12-13
Test process basic infor CICSTIN2	32	31, 29		6	1	12-17	12-17
Design process price info CICSTIN3	33	19		4		12-04	12-04
Code process price info CICSTIN3	34	33		6		12-12	12-13
Test process price info CICSTIN3	35	34, 29		6	1	12-18	12-18

Figure V-1   A worksheet is used for the tasks involved in implementing the project

start date for each task, and the target completion date. Note how many of the tasks are started and completed according to schedule.

Programming The design, coding, and testing of the programs that make up the first phase of the TASCO project required an estimated 417 man hours. In addition, documentation for operators in Production Services and for the data entry operators had to be completed. User manuals were also completed which provided written procedures for the utilization of the system. Although the project team was responsible for the project from the initial investigation to the final phase of its implementation, users were directly involved in the creation of the User's Manuals.

Before the system was turned over to Production Services, the new system was run in parallel with the old system for one week. During this time the system was operated under the direct control of Developmental Services and the users were not yet directly involved with its operation.

THE OPERATIONAL SYSTEM

Entering the Data

Although Sandra Kay Pieper is keying in data from source documents, other clerks within the department key in data received from distributors who phone in their orders. As the item number is received and keyed in, the corresponding record is randomly retrieved from the VSAM inventory file. If the item is not available, the distributor is so informed. If the item is available, the TASCO clerk repeats the item number and reads the description of the item displayed on the VDT. If the wrong

While Lillian Bancale removes an invoice from the printer, Mary Olson signs onto the system in order to make a query regarding a customer's order.

Sandra Kay Pieper, one of the TASCO clerks enters data into the system by using a terminal. Each workstation has a keyboard, and a VDT.

Tom Green calls out the name of the individual who will assemble the order and the items included in the order.

Checker Andy Holmes checks each invoice to see that the correct number of tires have been assembled.

record was retrieved, the distributor will inform the clerk so that the correct record can be retrieved. As soon as the quantity is keyed in, the quantity sold is subtracted from the quantity on hand.

Only the customer's number, item number of the product ordered, and the quantity ordered is keyed in. The customer's name and address, product description, and product cost are retrieved from the online files. When the operator enters the code required to indicate that the order is completed, the computer makes the necessary calculations to complete the order and the customer's record is updated for the amount of the sale.

Printing the Invoices

As soon as the TASCO clerk enters the order, the information to be printed on the sales invoice is available for printing. The warehouse personnel can bring up the order on their screen and print the invoice immediately on the printer located in the warehouse office. A second printer is available which prints a shipping label for each item ordered. If 20 tires of a given size are ordered, 20 shipping labels with the invoice number, customer number, item number, and description will be printed.

The terminal can be used to call up a customer's order or to direct the computer to print an additional invoice or new set of stickers. Printing duplicates will not affect customer's balance or the quantity of items on hand.

One copy of the invoice is retained in the office while the other two parts are sent to the warehouse so that the order can be filled.

Filling the Order

Although other methods have been studied and rejected as not being as efficient or for being too costly, orders are filled as they are called out. The caller receives two copies of the invoice. An experienced caller can sometimes detect errors on the invoice. For example, the caller might question the quantity if the invoice indicates 100 large, expensive tractor tires had been ordered by a distributor.

Usually four individuals are assigned to assemble the orders and on an average 800-1000 items per hour are assembled for shipment.

Checking the Order

The checker receives the two copies of the invoice used to assemble the order. The order checker in the warehouse provides part of the control function. Each order is checked to make certain the correct number of items have been assembled. The checker also attaches the labels to each individual item. If there are too many or too few labels, an investigation must be made to determine the cause of the problem.

Ron Jones uses a data flow diagram to illustrate the additional features to be incorporated into the TASCO system.

The order is loaded on the truck that will deliver it to the assigned destination. The driver also checks to see that the correct number of items have been loaded before signing the invoice. One copy of the invoice goes with the shipment and the second copy is returned to the warehouse office. In the office one copy is filed and one copy is sent to TASCO. The invoice becomes part of the audit trail and is used to verify that the correct number of items were sent to each customer and that each customer is billed for the correct amount.

SYSTEM AUDIT

The system audit for Phase I of TASCO indicates the objectives established for the system have been achieved. The users in the warehouse feel the system is easy to work with and indicated that very seldom are sales orders processed incorrectly.

PHASE II

In Phase II of the TASCO project, emphasis will be placed on providing additional inventory control. The procedures used for recording the return of merchandise as well as recording new shipments of merchandise from suppliers will be revised. Also included in this phase of the project are the procedures used to record payments and returns made by customers. When the second phase is operational, there will be tighter control over inventory. Management will be able to make additional queries into the entire TASCO sales/inventory/receivable system.

SUMMARY

The TASCO sales order system is an excellent illustration of how analysts must involve users—members of the warehouse staff, members of the inventory, sales, and accounts receivable departments, and management. Once the second phase is completed, there will be one more link in the Texas Farm Bureau's extensive management information system.

By the time Phase II is implemented, all segments of the system will have been investigated to determine the most cost effective way of providing service to the users and information to management. The system will consist of traditional reports and will also include the means for obtaining answers to online queries.

12 PROGRAMMING CONSIDERATIONS

Looking Ahead

After reading the text and completing the learning activities you will be able to:

- Determine the four stages of growth that have occurred within electronic data processing.
- List the tasks that must be completed during the implementation phase of the project.
- List the steps needed to design, develop, and implement a program.
- Identify the reasons industry is concerned about the productivity of programmers.
- Identify the three constructs used to develop a structured program.
- Identify the conventions used in constructing a program hierarchy chart and the purpose of the chart.
- Identify the format used to illustrate the logic of a program when flowcharts, HIPOs, pseudocode, Nassi-Shneiderman charts, and Warnier-Orr diagrams are used.
- Explain why a design or code walkthrough should be conducted and the objectives to be achieved from this activity.
- When given the program specifications and detailed logic plan for a program, determine how the program should be tested.
- Identify the topics that might be covered in design and coding standards.
- Define and utilize the words and phrases listed in the end-of-chapter glossary.

INTRODUCTION

It is imperative that the implementation phase of a project be well planned and executed according to the plan. The plan must provide for the tasks listed in Figure 12-1.

Implementation Tasks

- site preparation

- procurement of the necessary hardware, software, and supplies

- program design, coding, testing

- file conversion

- training of personnel

- system testing

- conversion to the new system

- system audit

FIGURE 12.1 Tasks to be completed during the implementation phase

How long each task will take and when each task will be completed must be determined. Earlier in the discussion of system design concepts it was indicated that all of the activities should be divided into small units (or tasks) that can be completed in a given period of time (usually one week). Progress can then be measured by determining whether all of the stated objectives are being met.

With the increasing dependency on computers as a tool for management and the availability of analysts trained in planning, organizing, and controlling projects, it is no longer excusable to bring projects in weeks or months behind schedule. However, when planning the implementation, there must be some slack in the schedule to allow for unexpected occurrences or delays.

FOUR STAGES OF GROWTH IN ELECTRONIC DATA PROCESSING

In order to appreciate the changes that are occurring you should be familiar with the four stages of growth in electronic data processing cited in an article by Richard Noland in the *Harvard Business Review*.

1.	Initiation	Cost savings were realized as computerized systems replaced manual and unit-record systems. The objectives for EDP systems were limited and controls were inadequate.
2.	Expansion	EDP systems became larger and more powerful; more applications were added. Objectives were still limited and controls inadequate.
3.	Formalization	Steering committees (computer policy committees) were used for formulating objectives and making decisions. Systems were better planned and there was both quality and cost control.
4.	Maturity	Standards were developed in many areas of EDP and both the industry and the professionals came of age. Systems became better planned, designed, and controlled.

Managing EDP

Today it is clearly understood that the new technology must be *managed*. There must be adequate control in all areas of EDP. If the expansion of EDP and the development of new systems are not controlled at the corporate or top-management level there may be: a lack of standards, incompatibility between systems, an excessive amount of both hardware and software, lack of control, and unproductive EDP personnel.

In some organizations where the acquisition and use of microcomputer systems have not been included in the organization's master plan for EDP, the company may have taken two backward steps—from the "maturity stage" back to an uncontrolled "expansion" stage.

Developing Software

With the cost of hardware decreasing and personnel costs increasing, the cost of developing the application software could be twice as much as the combined total of the other costs. In one installation, a minicomputer was obtained for $30,000. However, the cost of developing the initial software was well over $100,000. With rising personnel costs, management has been increasingly concerned over the productivity of programmers. The industry is changing and more is expected of analysts and programmers.

System design and programming is one area within EDP that has matured. There are design standards, programming standards, programming aids, and new languages that make analysts and programmers more productive.

DEVELOPMENT OF STRUCTURED DESIGN AND PROGRAMMING CONCEPTS

What criteria should be used to evaluate a computer program? In the past the answer might have been, "If it works, it is a good program." Today the validity of output is not the only test used to determine the merits of a program. Why has the philosophy changed?

Since the lines of code written in a day were extremely low and personnel, costs were increasing, management became concerned with the productivity of programmers. A high percentage of the programming staff's time was spent in doing programming maintenance. Since many of the programs were poorly written, the programs were difficult to maintain.

Programmers wrote poor code for a number of reasons. Often they were self-taught or had very little formal training. There were no established guidelines for evaluating programs. When machine or assembler languages were used, the programmers were restricted by the characteristics of the language. Programmers seemed to have the attitude that it was impossible to write a program that would work on the first run, so why try.

Little time was spent on the design of the program. At the completion of the project, there was also a tendency to skip the documentation and go on to a new project. With the pressure of constant change and the increased demands being made on the CIS department, the attitude seemed to be "Who has time to design or document programs?"

More progressive CIS departments now have standards that are used in evaluating programs and often a structured approach is used in designing and writing programs. The development of structured programming will be presented first and then the design and programming standards for CPI will be reviewed.

STRUCTURED PROGRAMMING

Probably 50 percent or more of today's programmers design well-structured programs. Since the first paper regarding structured design was published in ACM's (Association of Computer Machinery) *Communications* in 1964, it seems odd that there are still so many poorly constructed programs being written. In that paper, written by Corrado Bohm and Guiseppe Jacopini, the authors indicated that all programming logic, no matter how complex, can be expressed using the following structures:

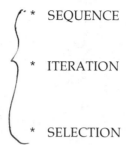

* SEQUENCE   —   Unless there is a branching command, computers always execute commands in the order in which they are stored within memory.

* ITERATION   —   Loops are constructed. The commands within the loop will continue to be executed until a given condition is fulfilled. WHILE or UNTIL are often used to construct the loop.

* SELECTION   —   One of two commands will be executed. IF/THEN/ELSE statements are used to evaluate a condition and to determine which command will be executed.

A second phase of structured programming was entered when Edsger W. Dijkstra advanced the theory of GOTO-less programming. In his letter published in ACM's *Communications*, Dijkstra indicated that GOTO statements should be eliminated from programs coded in high-level languages. The use of GOTO statements make programs difficult to understand and maintain.

After Dijkstra's letter was published, a number of theories were developed around structured programming concepts. Unfortunately some programmers thought that if they eliminated the GOTOs their programs would be structured! Others felt that as long as **modules** supported the main logic of the program, their programs were structured. A module contains code that accomplishes one major function and minor tasks related to the major function. The use of modules is illustrated under Designing a Program.

Many success stories concerning the use of structured programming concepts have been related. Industry has been turning to structured programming in order to: increase the productivity of programmers; increase the readability of programs; decrease the time needed to design, implement, and maintain programs; and as a way of establishing guidelines.

Guidelines are now established that can be used in developing structured programs. Under the guidelines, programmers no longer start by writing code. Time must be spent in developing and testing the design (or logic) of the program before any coding is done. While more time is now spent in the design of programs, less time is required to code and implement programs. Although as a general rule, GOTOs should be eliminated, there are situations where a well-thought-out GOTO *used within a module* simplifies the problem and makes the coding easier to understand. Also when some versions of BASIC are used, GOTOs must be used to construct a loop. However, many programming languages make it possible to design easy-to-follow and easy-to-maintain programs that only use the sequence, iteration, and selection constructs.

Constructs Used   **Sequence**   Unless there are conditional or unconditional branches in the program, any program will be executed in the order in which it is stored within the computer's memory. This also is the order in which the code has been written. Figure 12-2 illustrates a simple sequence that might be used to calculate a student's grade point average. If you refer to the program flowcharting symbols in Appendix II, you will note that a rectangle is used to depict a processing event. The computer will execute the program in the sequence it is written. In the simple sequence control structure, one command follows another.

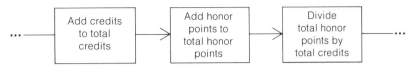

FIGURE 12.2 Logic expressed by SEQUENCE

Iteration Conditional loops are often established by using WHILE or UNTIL. The format of the statement is determined by the language being used. Unconditional loops, or count-controlled loops, are also used and can be constructed in BASIC by using FOR/NEXT or in COBOL by using a version of the PERFORM statements.

When a conditional loop is established, the commands within a loop will be executed as long as a given condition is true. Figure 12-3 illustrates a simple WHILE structure. Code written in BASIC is also provided to illustrate the instructions needed to execute the logic of the program. The BASIC written could be executed on a DEC or IBM PC computer. Some versions of BASIC do not support WHILE and do not permit data names to be as long as those used in the illustration. When a GOSUB is executed, a branch occurs to the module that is coded at line 1500. After the commands in the read record module are executed, control returns to the process records module. In some versions of BASIC, WEND is used to indicate the end of the WHILE loop; in other versions NEXT is used. RETURN is used to indicate the end of the module.

Selection Most selection is accomplished by using IF/THEN/ELSE statements. Figure 12-4 illustrates the section of the flowchart that would be needed to illustrate the use of an IF/THEN/ELSE statement. In the illustration, each sales representative gets a 10 percent commission on the first 50,000 of sales. Commission on all sales over 50,000 is based on 12 percent. The BASIC statement needed to calculate the commission under each condition is also provided.

CHECKPOINT

1. Why is it reasonable to assume that a project will be completed according to schedule?
2. Are all CIS departments in the "maturity" stage of their development?
3. What might occur in regard to EDP if the growth in EDP within an organization is not controlled?
4. In writing structured code, what three constructs are used to execute the logic of a program?
5. Why was management concerned about the productivity of programmers?
6. What major advantages have been attributed to the use of structured programing?

DEVELOPING AND IMPLEMENTING PROGRAMS

In the detailed design phase, the program specifications and the macro logic of the program required to implement the system were developed. The documentation completed during the design phase should be the first three items listed under the heading Steps in Programming. Using the documentation provided, a programmer or a programming team should complete the remainder of the tasks.

Steps in Programming

1. Define the problem.
2. Determine the output needs and formats.

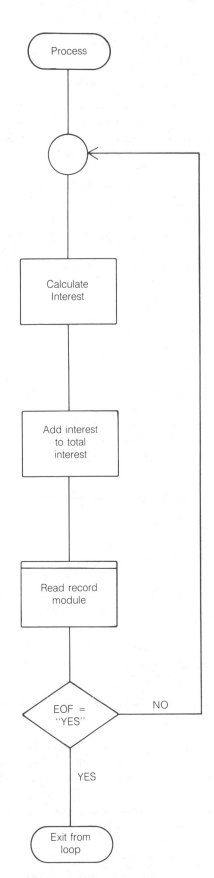

```
2000  REM   PROCESS RECORDS MODULE
2010  REM   CONTROL REMAINS WITHIN THE LOOP
2020  REM   UNTIL ALL RECORDS ARE PROCESSED.
2030  WHILE EOF = ''NO''
2040      INTEREST = BALANCE * RATE / 12
2050      TOTAL = TOTAL + INTEREST
2060      GOSUB 1500
2070  WEND
2080  RETURN       ! END OF LOOP
```

FIGURE 12.3 A loop is created by using WHILE and WEND

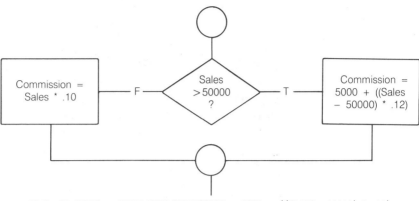

FIGURE 12.4 Selection is accomplished by using an IF/THEN/ELSE statement

```
2010   IF SALES > 50000 THEN COMMISSION = 5000 + ((SALES - 50000) * .12)
       ELSE COMMISSION = SALES * .10
```

3. Determine the input needs and formats.
4. Design the program. *Eflor chart / psudo code*
5. Test the design of the program (walkthrough).
6. Code the program.
7. Test the code (walkthrough).
8. Punch the code into cards, enter it on diskettes, or enter it into the system by using a terminal.
9. Compile the program.
10. Test the program.
11. Document the program.
12. Evaluate the program.

Defining the Problems

The problem should have been well defined in the detailed design phase. However, if the programmer has any question about the objectives and scope of the program, the systems analyst should be consulted.

Output Needs and Formats

During the detailed design phase, the output needed should have been determined. The file layouts, report layouts, and screen designs should have been completed and approved by the user. During the implementation phase, there should be no misunderstanding regarding the content or format of reports or screens.

Input Needs and Formats

The most effective way of entering variable data into the system should have been determined. The input specifications should also indicate how the data will be verified by the operator and by the program that processes the data. The decision should also have been made for each application regarding where the data will be entered and who will enter the data.

Source documents, input records, and screens should be designed simultaneously. The operator will be more productive and make fewer errors if the source document, screens, and input records all have the data fields in the same sequence.

Designing the Program

Within an installation there should be established guidelines regarding the use of program modules and structured, top-down programing. The programmer completing the detailed logic plan should be familiar with the programming aids available that can be used to simplify coding.

Students are sometimes heard to say, "I do the logic plan after I code the program." Those who make such comments are the same students who spend hours at a terminal trying to get their programs to work. Students who design and *test* a logic plan *before* coding the program often have valid output after one or two computer runs. Although good programmers know they must test their programs, they do not expect to have to debug their programs!

How do you design a program? If you were to ask five programmers the question, you might get five different answers. Figure 12-5 illustrates what might seem to be a logical approach to designing a program.

Logical Approach to Designing a Program

- Understand the problem completely.

- List the major functions to be performed.

- Under each major function identify the closely related minor functions that must be performed.

- Identify all the decisions that must be made and the exception routines that will be needed.

- Identify the modules that will be needed.

- Develop a hierarchy chart that identifies the relationships that exist between the various modules.

- Use a top-down approach and develop a detailed logic plan for each module.

- Test the main logic of the program.

- Test the detailed logic for each individual module.

- Walk through the logic using test data that will cause control of the program to exit to the exception routines.

FIGURE 12.5 A realistic approach to developing a detailed logic plan for a program

Understand the Problem   If programming specifications were prepared during the detailed design phase, the programmer should understand the problem after studying the specifications. However, there may be some tasks specified with which the programmer is unfamiliar. An example in the payroll system might be the calculation of federal income tax or the calculation of overtime. In the first example, the programmer will need to review the federal laws pertaining to income tax. In the second example, union contracts and company policy need to be studied.

Listing the Major Functions   Although most people accept the one-major-function-per-module concept, what constitutes a function is not always agreed upon. For example, printing a detail line is a major function. The print-detail module usually consists of the coding needed to: determine whether new headings should be printed; move the data to the print lines; print the detail line; and add one to the counter that keeps track of the number of detail lines printed on a page. A separate module would be developed for printing the headings on the top of a new page.

Common sense must be used in constructing modules. Certain basic functions are needed in every program. One such function is *initialization*—the let's-get-going activities. In the initialization module, the one-time-only tasks that must be performed before an input record can be processed are executed. Another common function is *termination*—wrapping up on the job. In the termination module, totals are printed, files are closed, and the job is ended. If only four or five lines of code are needed to end the job, it is probably better not to have a separate print-total module.

Once the major modules are determined, the closely related tasks that should be included within the module are outlined. Any decisions that need to be made should also be determined. The programmer must decide whether separate modules are needed for exception routines. Exception routines are needed whenever an abnormal occurrence is detected, such as alphanumeric data in a numeric field or a number out of range.

Developing a Hierarchy Chart   A program hierarchy chart is also called a **VTOC— visual table of contents**. By studying a hierarchy chart, such as the one shown in Figure 12-6, the relationships that exist between the modules can be determined. There is only one "boss" module—often identified as 1000. The main control module, module 1000, controls the execution of the 2000 level modules. The 2000 level modules control the execution of the 3000 level modules; and the 3000 level modules control the execution of the 4000 level modules. Unless a large, complex program is being developed, it is seldom that there are more than four levels.

Once the hierarchy chart is completed, detailed logic plans can be constructed for each module. However, as the detailed logic plans are developed, it may become necessary to add additional modules. Sometimes one module becomes too complex and difficult to understand. By dividing the modules into two or more modules, the logic is easier to develop and to understand and excessive nesting can be avoided.

Developing Detailed Logic Plans for Modules   The logic plan should be constructed first for the 1000 modules, then the 2000 level modules, and so forth. The man logic of the program should be tested before the detailed logic plans for the supporting modules are determined. The 3000 and above modules are often considered to be the supporting modules.

There are many different tools and techniques used in developing detailed logic plans. Flowcharts, **HIPO charts, Nassi-Shneiderman charts, Warnier-Orr diagrams**, and **pseudocode** are among the more popular tools that are used to depict the logic of a program. In the inventory change program illustration, a flowchart is used to illustrate how a structured logic plan should be developed. Each of the methods listed that can be used to develop a logic plan will also be illustrated. Within a CIS department a decision should be made as to what method will be used. The detailed logic plan will become part of the program documentation. Programming maintenance is easier if only one type of logic plan is used. Also, the logic plan may be used in explaining to a user or to an auditor how data is processed.

Conducting a Design Walkthrough

After the programmer has tested the main control logic of the program and the design of the individual modules, a walkthrough may be conducted. If a team approach is used to programming, members of the team will be given the program specifications, input and output formats, and the detailed logic plan. At this point enough test data should have been designed to test the normal execution of the program—when good data is read—and all of the exception routines.

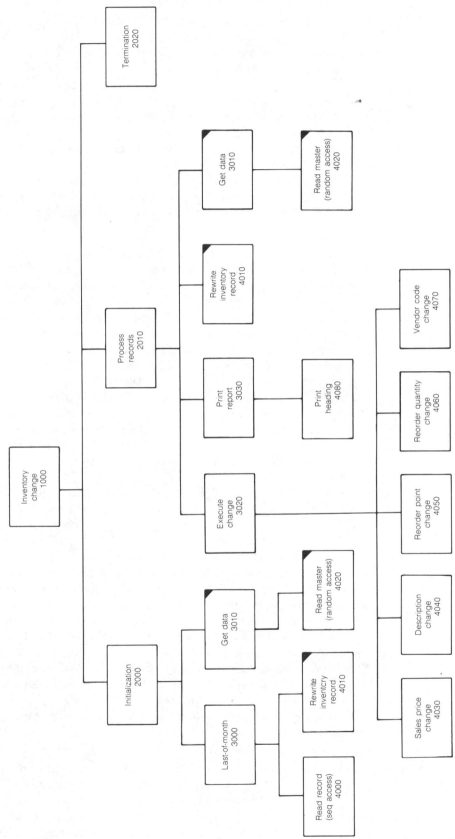

FIGURE 12.6   Hierarchy chart for the inventory change program

The programmer who developed the logic plan will go through the logic plan as if he or she were the computer processing the test data. This process is also referred to as *desk checking* or ''playing computer.'' Other members of the team should help to identify omissions or errors in the logic. A walkthrough is not an evaluation of the programmer. The purpose is to catch errors before a program is coded and implemented. If errors or omissions can be found prior to implementing the program, valuable resources will be saved and costly mistakes will be avoided.

Coding the Program

Although different compilers may be available for the computer being used, within a company it is wise to select one language that will be used for programming all business-oriented applications. If a number of different languages are used, it is more difficult to maintain the programs.

Companies with large mainframes use COBOL for coding business-oriented programs more often than any other language. The trend will probably continue because so many programs are already written in COBOL. Since COBOL is standardized and is considered a universal language, a change to another vendor's computer system does not present a major problem.

More software for microcomputers is written in BASIC than in any other language. This trend will also probably continue because so much software is already written in BASIC and because a BASIC compiler is often a standard item that is included as part of the purchase price of a microcomputer system.

Both large and small companies would do well to investigate the use of languages other than the traditional languages such as BASIC, COBOL, or PL/I. Some companies have reported their programmers are anywhere from 5 to 15 times more productive when a fourth-generation language such as FOCUS is used. FOCUS is a nonprocedural language that is used to generate reports and to manage databases.

Testing the Source Code

When a microcomputer or a terminal is used, the source code is often keyed in by the programmer. If an **interactive BASIC interpreter** is used, most syntactical and clerical errors are detected as soon as a line of code is entered. When COBOL or other types of languages are used, a listing of the source code may be obtained that identifies the clerical and **syntactical errors**.

A team walkthrough should be conducted to determine whether there are omissions or errors in the source code. The walkthrough is conducted using a clean listing of the program—one that is obtained after routine clerical and syntactical errors have been cleaned up. The team should again play computer and trace test data through the various decisions that must be made as different types of data are processed.

Testing Programs

The design of test data is important. Every condition that might occur should be tested before the program is considered operational. Usually the following test procedure will be used.

1. Test the program with valid data. Each condition should be tested.
2. Test each exception routine. Test data that will cause each of the error routines to be invoked must be used.
3. Test the program without any data. A message should be displayed indicating that data was not available.

Before testing the program, the test data should be listed in the sequence that it will be entered into the computer. Knowing the sequence in which the data enters the computer is important. For each test record the programmer must determine how the computer will process the data and how record counts, totals, and printed reports will be influenced. If a file is to be updated, the contents of the records to be changed should be displayed or printed before and after the program is run. A utility program can generally be used to display the contents of selected records.

Although the programmer is usually responsible for the design of the test data, the analyst generally checks to see that all possible conditions have been tested. When a program that has been operational for two years suddenly aborts at 3:00 A.M., the problem can usually be attributed to a combination of circumstances—a combination that had never been tested!

Records of test runs which include the data and the results should be kept as part of the programming documentation. The test documentation can be used to assist in locating errors that might occur at some later date or used when changes are to be made in the program.

Stub Testing   Testing or programming standards might include coverage of **stub testing**. When a top-down modular approach is used in coding a program, it is possible to test various modules of the program before the entire program is completed. This is referred to as stub testing. The main module of the program is written and tested. Dummy supporting modules, or stubs, are written and may contain statements that will show how the logic of the program is being executed. For example, the following statements might have been printed:

```
CONTROL PASSED TO PRINT-DETAIL MODULE.
CONTROL PASSED TO PRINT-HEADING MODULE.
CONTROL PASSED TO THE PROCESS-RECORDS MODULE.
```

As each module is developed, it replaces the dummy statements used previously. When complex programs are written, the main logic of the program should be tested without the programmer being concerned about minor details such as moving data to the print line, clearing the line counter to zero, and other such menial, *but very necessary*, tasks.

Documenting the Program

Standards that indicate how both a program and a system should be documented should be developed and followed. The documentation serves a variety of purposes and is essential to operations personnel and maintenance programmers who must work with the system. Chapter 14 provides an in-depth coverage of documentation.

Within an organization, individuals have many different views regarding documentation. Sometimes, fortunately not often, management views documentation as an unproductive use of a programmer's or an analyst's time. Analysts and programmers in turn look upon documenting as one of the less desirable features of their job. Unless programs are not considered operational until all phases of the documentation are completed, the CIS manager may find that the installation's documentation is totally inadequate.

Evaluating the Program

When a single program is being evaluated, the criteria used are simple. The analyst, programmer, user, and operations manager must be convinced of the following factors:

1. Error-free output was produced. Adequate controls were built into the program to validate the output.
2. The output met the objectives specified by the user.
3. Menu and help screens were provided so that operators entering the data could do so easily and comfortably.
4. Execution-time problems were identified and reported to the CIS manager, and corrective action was taken.
5. The cost of entering the input and producing the desired output is not significantly greater than the projected costs. If the costs exceed those projected during the detailed design phase by more than 10 or 15 percent, an investigation should be undertaken to determine where and why the problem occurred.

The elevation may show that there are problems or weaknesses that must be corrected. Well-designed programs developed by testing both the design and source code are likely to produce error-free output and to have few, if any, execution-time problems.

In addition to the initial evaluation of the program, there must be an ongoing evaluation to see whether problems occur at a later date. The ongoing evaluation and the process by which systems are evaluated will be covered in Chapter 14.

CHECKPOINT

7. When should the output and input needs and formats have been determined?
8. What should be done before a programmer attempts to develop a detailed logic plan for a program?
9. What information can be determined by studying a program hierarchy chart?
10. What relationship exists between the 1000, 2000, and 3000 level modules?
11. Why is a design walkthrough conducted?
12. What is an interactive BASIC interpreter?
13. What is a syntactical error?
14. How should an individual program be tested?
15. How should managers view documentation?
16. What criteria should be used to evaluate a program?

THE INVENTORY CHANGE PROGRAM

In the case study it was indicated that normal changes would include changing the special sale price code, the sale price, the reorder point, the reorder quantity, the description, and the vendor codes. Since changing the sale price code and sale price is considered as one type of change, only five change codes are needed. Let's assume that the following change codes have been assigned:

1 = Item placed on sale
2 = Description changes
3 = Reorder point changes
4 = Reorder quantity changes
5 = Vendor code changes

Changes will be entered in a batch and a confirmation report printed. The confirmation report will show the record being changed, the field being changed, and both the old value and the new value of the field. The report will be compared with the source documents to make certain that all changes were made correctly. Except for the last day of the month when the sale items for that month are reset to the normal selling price and new items are placed on sale, relatively few changes are made to the inventory records.

Detailed Flowcharts

Flowcharts graphically illustrate how the data will be processed. The symbols used to develop a flowchart are explained in Appendix II. When a top-down structured approach is used, a flowchart is developed for each module. The VTOC for the inventory change program was illustrated in Figure 12-6. Figures 12-7 through 12-10 on pages 435 to 438 illustrate the detailed flowchart for the inventory change program. Note that within the symbol, ordinary English statements are used rather than writing some type of source code.

Execution starts in the main control module—Inventory Change (1000). Immediately control branches to Initialization (2000). If it is the last day of the month, control passes to the Last-of-Month module (3000). Control remains within the module until all records are processed. In the module, the inventory file is opened for sequential access. Each logical record is read and the question is asked ''Is the sale code = N? If the answer is yes, processing is bypassed and another record is read. When files are processed sequentially, after the last record is read that contains data, a record with an end-of-file marker will be read. Each time a record is read from a sequential file, the operating system checks to see whether the end-of-file marker has been found. When this condition exists, the programmer can cause whatever action is needed to occur. In the Last-of-Month module, the file is closed and control of the program returns to the Initialization module.

After the instructions in the Initialization module are executed, control returns to the Inventory Change module (main control module) and control branches to Process Records. Each time the predetermined process symbol is used (the rectangle with the bar across the top), control branches to another module, the commands within that module are executed, and control returns to the calling module. A module may call, or invoke, other modules as needed.

In developing the logic of a program, the programmer should think about: how easy it will be to maintain the program; how easy it will be for others to understand how the data is being processed; and how efficiently the computer will be able to execute the program. Flowcharts, or any type of logic plan, have three basic functions. They serve as:

- *a guide to be used in coding the program.*   Think of the logic plan as a road map. Would you start out on a trip from New York to California without a map? Should you start to code a problem without a well-defined map?
- *a means of illustrating to management and to auditors how data is processed.*   Sometimes a single picture is better than a thousand words; the same is true of a good logic plan.
- *a guide for maintaining the program.*   Without a logic plan it may take the maintenance programmer a long time to understand the logic and to ''get into the program.''

The major advantage of flowcharts over other types of logic plans is that a better graphic picture is provided. Also if each major module is on a separate page, the old complaint that flowcharts are difficult to maintain is no longer true. When structured concepts are used, most changes affect only one module.

There is software available that can be used to prepare the final version of a flowchart. The flowcharting program is stored in the memory of the computer and the source code is read. Output, rather than being a listing of the program, is a hierarchy chart and detailed flowcharts. The printed chart should be identical with the original, manual design.

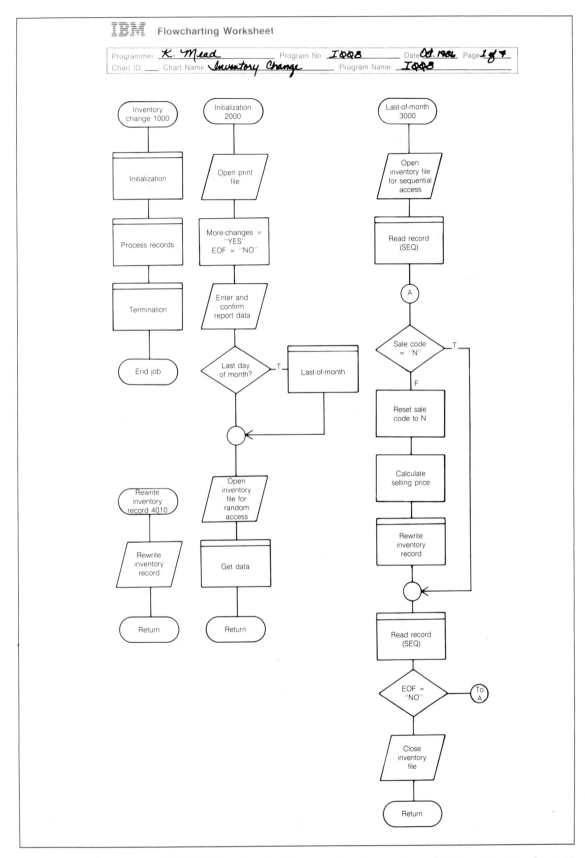

FIGURE 12.7 Detailed flowchart for the inventory change program (Page 1)

436 CHAPTER TWELVE

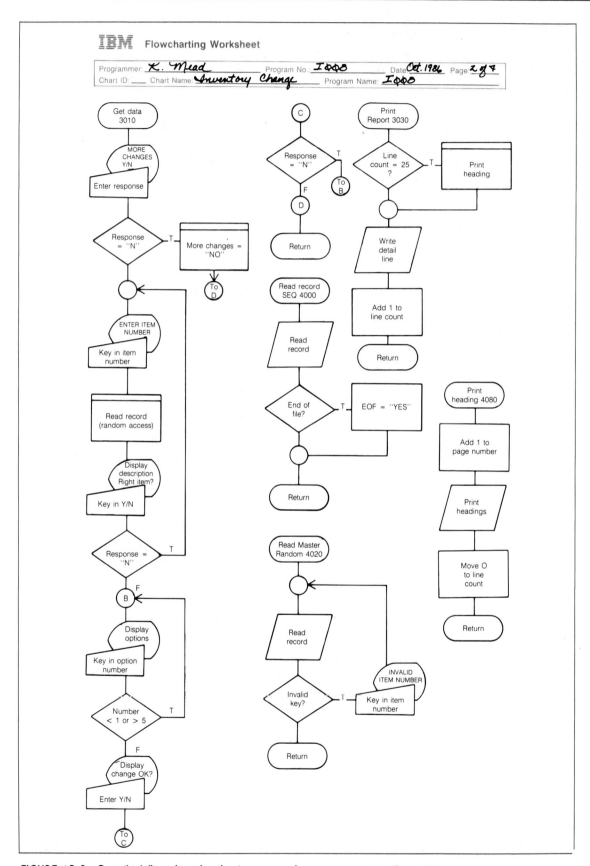

FIGURE 12.8   Detailed flowchart for the inventory change program   (Page 2)

© 1985 SRA

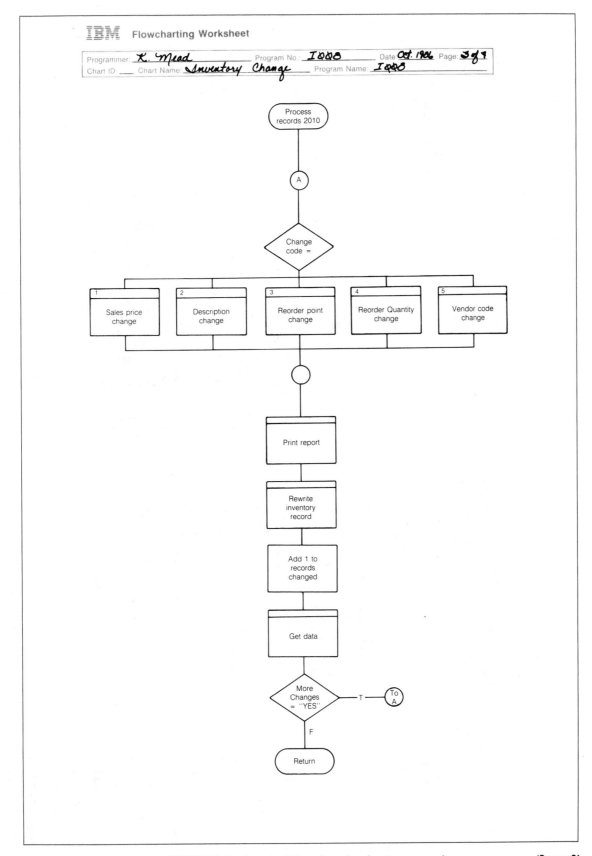

FIGURE 12.9 Detailed flowchart for the inventory change program (Page 3)

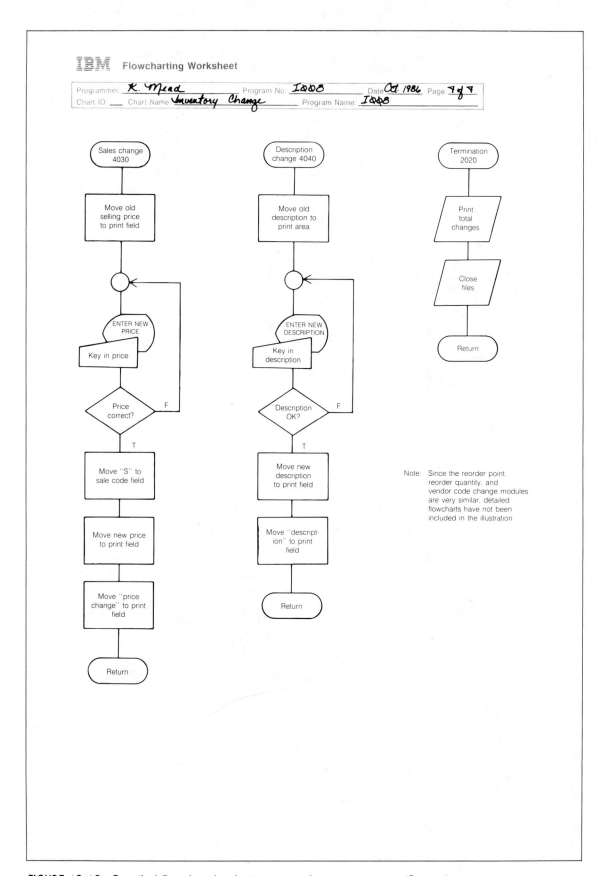

FIGURE 12.10 Detailed flowchart for the inventory change program (Page 4)

HIPO Charts IBM developed **HIPO (hierarchy/input/processing/output)** charts as an alternative to using flowcharts. Each module has its own HIPO chart. The major advantage of using HIPO charts is that the programmer is forced to think about the basic data processing cycle—input, processing, and output. Symbols may be used to show the input and output media used. In Figure 12-11 on page 440 the input indicates that the print and inventory files must be available. Shown as output are the individual fields of data that are available as the result of the initialization module commands being executed.

The arrow at the top of the chart indicates the module being executed when the Initialization module was invoked. The arrow at the bottom indicates where control will branch to when all instructions within the module are executed. When structured programs are written, the module branched from and to should be the same one.

The number used in the reference column relates the HIPO chart back to the hierarchy chart. When a branch occurs because it is the last day of the month, the reference number indicates that module 3000 will be executed.

Within the Process block of the worksheet, ordinary English is used to express the logic and to show what must be done. However, most programmers use key words such as PERFORM. PERFORM means to branch down to the module identified, execute the commands within the module, and return to the statement following the PERFORM.

Regardless of the language used to code the program, internal documentation will relate the coding back to the hierarchy chart and to the detailed logic plan. Figure 12-11 illustrates the HIPO chart that would be developed for the Initialization module of the inventory change program. Illustrated below are a few lines of code that show how the internal documentation for the Initialization module of a program written in BASIC would be coded.

```
1000   REM   INITIALIZATION MODULE-2000
1010   REM   CONTROL FIELDS ARE SET TO THEIR INITIAL VALUE AND
1020   REM   THE DATE FOR THE REPORT IS ENTERED. IF THE END-OF-MONTH
1030   REM   ROUTINE IS INVOKED THE MASTER FILE RECORDS ARE PROCESSED
1040   REM   SEQUENTIALLY. THE SALE CODE IS RESET TO N AND THE NEW
1050   REM   SALES PRICE IS CALCULATED. AFTER ALL RECORDS ARE PROCESSED
1060   REM   THE FILE IS THEN CLOSED AND REOPENED SO THAT RECORDS CAN
1070   REM   BE ACCESSED RANDOMLY. AFTER THE GET DATA MODULE IS
1080   REM   INVOKED, CONTROL RETURNS TO THE MAIN CONTROL MODULE.
```

Nassi-Shneiderman Charts In 1973, I. Nassi and Ben Shneiderman developed a charting technique that is considered compatible with structured programming. Symbols are provided for process, decisions, iteration, and **case entry**. Process is shown by using a rectangle. A rectangle divided into three triangles is used to show a decision. Iteration and case entry are shown as illustrated in Figure 12-12. In the illustration, NULL is used to indicate there is no action required if the condition is true.

Figure 12-13 illustrates the Initialization module of the inventory change program. Unless a module has a great many nested IF statements, the Nassi-Shneiderman chart is easy to construct and a module generally fits on a page. The scope of a loop is well defined and only four basic symbols are used. The programmer can leave the amount of space needed to record a meaningful statement about the section of the program being defined.

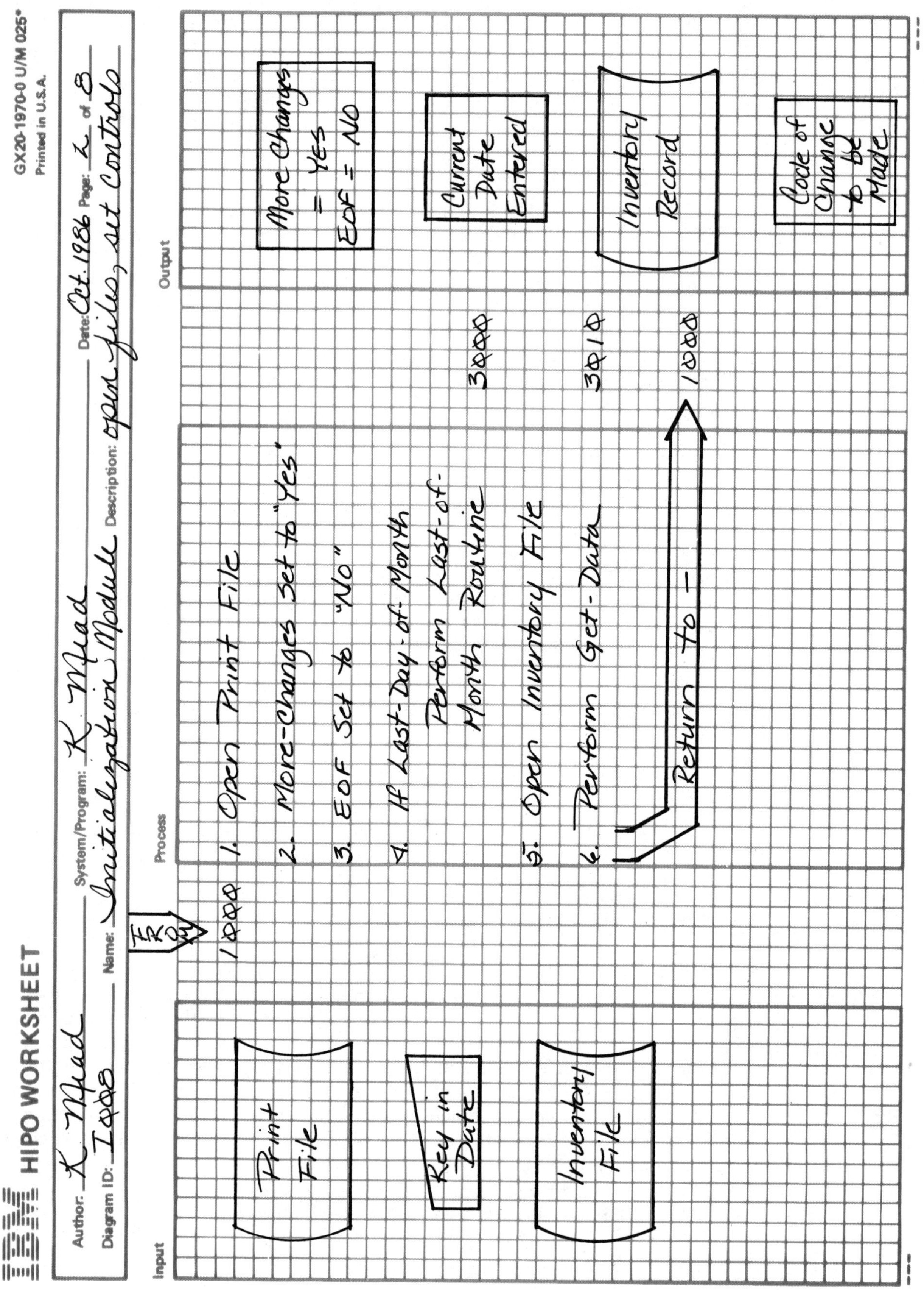

FIGURE 12.11 Hipo chart for initialization module

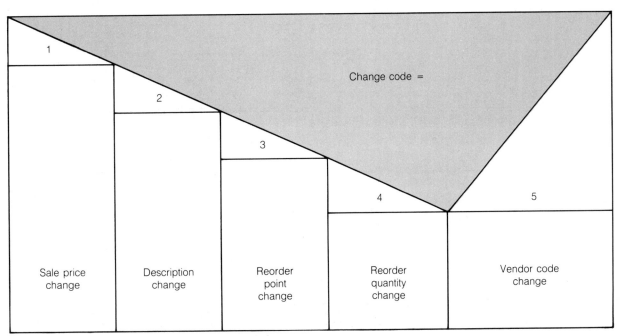

Case entry symbol

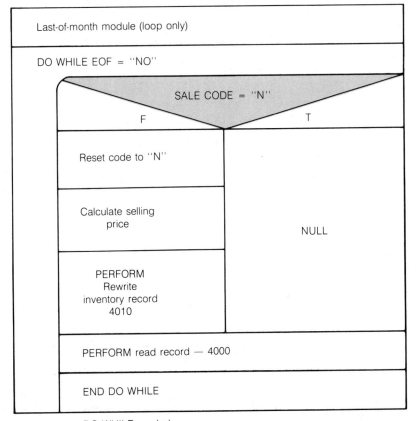

DO WHILE symbol

FIGURE 12.12 Case entry and **DO WHILE** symbols used to construct a Nassi-Shneiderman chart

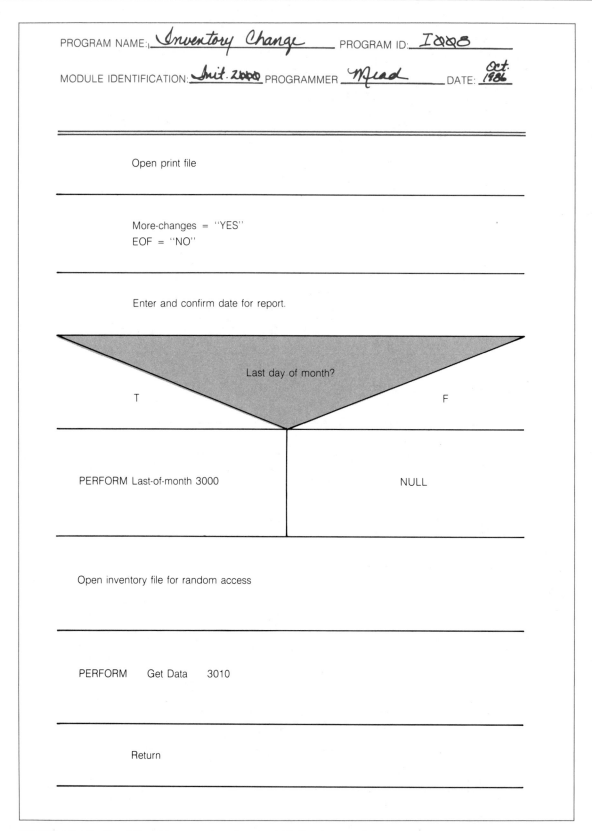

FIGURE 12.13   Nassi-Shneiderman chart for the initialization module

Since standardized forms are not available for use in constructing Nassi-Shneiderman charts, one has been developed for use in this textbook. When a form is used, the programmer is more inclined to record the identifying information needed.

Pseudocode

Pseudocode consists of simple English statements (not dependent upon any programming language) that illustrate how the program will be coded and executed. Although some keywords are used such as READ, WRITE, IF, and PERFORM, actual file and data names need not be used. Some programmers object to pseudocode since it is wordy and much like COBOL. Others who use pseudocode feel it is easier to use than a flowchart.

In some installations, only pseudocode is used to develop the logic. As with program flowcharts, pseudocode can be formatted in many different ways. Figure 12-14 on page 444 illustrates the pseudocode for the Initialization module of the inventory change program. Note that the key words are in uppercase letters.

Warnier-Orr Diagrams

The basic concept for the use of the diagrams in program design was developed by Jean-Dominique Warnier. Later Ken Orr extended the basic concept to include system design. The diagrams have been widely used for some time in France and are gaining acceptance in the United States. Some individuals describe the diagrams as a hierarchy chart turned on its side while others describe the diagrams as a structured pseudocode. Figure 12-15 illustrates the main control, initialization, process records, and termination module. When a (1) is placed under a statement, it is only executed once. When an (N) is used, it indicates that the statement will be executed a number of times until a given condition is true. When (0,1) is used, it indicates that the function being performed is mutually exclusive—if that function is performed none of the other functions coded with an (0,1) will be performed during that particular cycle. PERFORM is again used to indicate a branch to, and back from, a specified module.

DESIGN PROGRAMMING STANDARDS

In the age of maturity, standards were developed in many areas of EDP. For a given company, guidelines should be established for developing the logic of a program, conducting walkthroughs, coding problems, testing programs, documenting, and evaluating a program. The ones illustrated in Figures 12-16 through 12-18 can be considered typical of what might be found in a standards manual for developing program flowcharts and writing COBOL source code.

Design Standards

The hierarchy chart illustrated in Figure 12-16 might be considered a model. Although more modules may be required, seldom will a report-producing program have fewer modules. Each file requires a separate module for both read and write operations. In programming, any I/O device is referred to as a file.

Separate modules are generally used to print detail lines and heading lines. Some standards might require that a separate module be developed that would be used for nothing more than causing the print line stored in a buffer to be transmitted to the printer and printed.

Programming Standards

The concern over programming productivity and the need for maintaining programs is what led to the development of programming standards. Structured,

Author: K. Mead	Program: Inventory Change	Date: Oct 1986
VTOC ID: 2000	Module Name: Initialization	
Function: Open files, set initial values, get first record		Page 1 of 1

PSEUDOCODE	REF	FILES, RECORDS, FIELDS
Open print file		Print file
More-changes = yes		More-changes
EOF = No		EOF
Enter and confirm report date		
IF last day of month		
PERFORM last-of-month	3000	
OPEN inventory file		Inventory file
PERFORM get-data	5010	

FIGURE 12.14 Pseudocode for the initialization module

© 1985 SRA

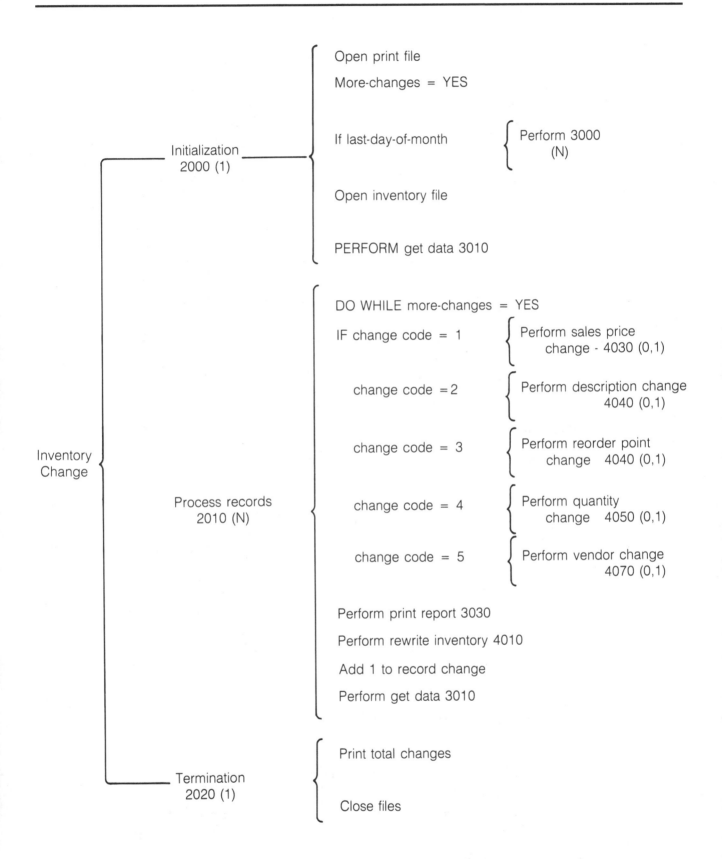

FIGURE 12.15 Warnier-Orr diagram for the inventory change, process records, initialization, and termination modules

C^PI **computer products inc.** DATA PROCESSING STANDARDS MANUAL

SECTION	Computer Programming	SECTION	CHAPTER	SUBJECT
CHAPTER	Programming Design	5	3	1
SUBJECT	Developing and testing the design	DATE August 1986		PAGE 1 of 2

1. The macro logic will be developed by an analyst. The analyst will provide the programmer assigned to the project a macro flowchart and written specifications.

2. During the detailed design phase, file layouts, print layouts, screens, and specifications will be developed.

3. Complex programs will be developed by a programming team consisting of three members. Less complex programs may be developed by an individual programmer and reviewed by other team members.

4. All programs will be modular. Each module will contain one major function and tasks closely related to the function.

5. A system hierarchy chart will be constructed for the system and a program hierarchy chart will be constructed for each program within the system. The program hierarchy chart must be constructed using the conventions illustrated below:

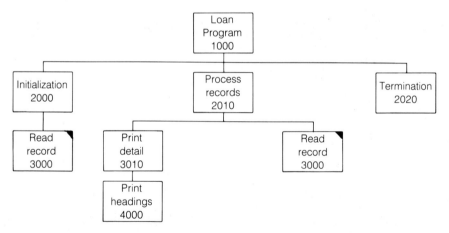

A. The main control module will be numbered 1000.

B. Modules invoked by the main control module will be numbered 2nnn.

C. Modules invoked by the 2000 level modules will be numbered 3nnn; those invoked by the 3000 level modules will be numbered 4nnn.

D. The module names will be used on the flowcharts and in coding the program.

FIGURE 12.16   Standards manual illustrating the standards used in developing the design of a program   (Page 1)

computer products inc. DATA PROCESSING STANDARDS MANUAL

SECTION	Computer Programming	SECTION	CHAPTER	SUBJECT
CHAPTER	Programming Design	5	3	1
SUBJECT	Developing and testing the design	DATE August 1986		PAGE 2 of 2

6. Program flowcharts will be used to develop program logic. The following guidelines are to be used in developing flowcharts:

   A. Standard ANSI symbols are to be used.

   B. Each module illustrated on the hierarchy chart must be supported by its own flowchart.

   C. The module names and reference numbers used on the hierarchy chart will correspond to those used on the flowcharts.

   D. Nontechnical and meaningful terminology will be used within each symbol.

7. A design walkthrough will be conducted according to the following guidelines:

   A. Prior to the walkthrough, the programmer responsible for the program will supply each team member with the following items:

      (1) definition of the problem
      (2) program specification
      (3) file layouts
      (4) system and program hierarchy charts
      (5) program flowcharts
      (6) test data

   B. Prior to the walkthrough, each team member will study the documents and note any errors or ommissions.

   C. The walkthrough will start with a general discussion of the system and the program. Errors that have been noted will be discussed and noted on the flowcharts.

   D. The project leader will walk through the design using the test data provided.

   E. Corrections and suggestions regarding how to improve the program will be discussed, approved, and then incorporated into the design of the program. The required changes will be made by the programmer who developed the logic plan.

8. In coding the problem, the standards listed under the subject CODING STANDARDS, SECTION 5, CHAPTER 3, SUBJECT 2 will be followed.

FIGURE 12.17 Standards manual illustrating the standards used in developing the design of a program (Page 2)

well-formatted programs are easier to understand, implement, and maintain than unstructured programs. Once standards are developed, management should insist that the standards be followed. If the programmers find that the standards are unrealistic, the standards should be changed and the manual updated. However, the standards should not be ignored. Figures 12-18 through 12-20 on pages 449 to 451 illustrate what might be considered typical standards for COBOL programmers.

After a program has been compiled, tested, and stored in a library, if modification is necessary, the program must often be copied into a text library. Code can be changed, new instructions added, or old instructions deleted. Each proposed change should be evaluated *before* it is made. After the changes have been entered, tested, and documented, a presentation regarding the changes should again be made to a subcommittee of the computer policy committee. Often approval for minor changes can be obtained from the project team.

DOCUMENTATION STANDARDS

Perhaps one of the most common problems in EDP today is the obsolescence of documentation. When changes are made to programs, the documentation is frequently not changed. Soon there is a difference between the formal (documented) procedure and the informal (undocumented) procedure. When this occurs, the job of everyone directly involved with the system or program becomes more difficult. When new personnel are to be trained, the user's job is much more difficult. The documented procedures might be totally different from the steps that are actually followed in preparing the input and working with the output generated from the system.

Standards must provide for the currency of documentation. When a procedure or program is changed it should not leave the test status and be considered operational until the documentation is updated.

INCREASING PROGRAMMING PRODUCTIVITY

Programming Support Products

Programming support products are available that increase the productivity of programmers. Most installations today have database software that supports the use of data dictionaries and query languages. **Tutorial languages** are often available that make it possible for an end user or inexperienced programmer to write a program by carrying on a dialogue with the computer. Software is also available which converts a logic plan into COBOL source code.

Programmers and analysts must have a working knowledge of all programming-support software. Whenever a new software package is obtained, the CIS manager should provide a hands-on training session for the programmers and analysts. If the programming product is designed for use by the end user, hands-on training should also be provided for those who are interested in learning to use the software.

Analysts and programmers must also be aware of the types of products that are available. When a new product is announced, a feasibility study should be conducted to determine whether the product should be obtained. The product should be evaluated in terms of the needs of the installation. Use of product should be cost-effective and make it easier to satisfy the needs of the users.

When selecting a new computer system, the individual making the investigation must determine the language supported by the computer system and the type of programming products available.

Using a Building-Block Approach

Far too many programmers do not use a **building-block** approach to programming. They tend to write the same routines each time the routines are used in a different program. Experienced programmers know that certain routines will

computer products inc.   CIS STANDARDS MANUAL

SECTION	Computer Programming	SECTION	CHAPTER	SUBJECT
CHAPTER	Programming Design	5	3	2
SUBJECT	Coding Standards	DATE August 1986		PAGE 1 of 3

1. All business-oriented applications will be coded in COBOL 81.

2. The standard CPI COBOL coding form will be used by all programmers.

3. The form will be completed in pencil. Each character will be printed in a column box. Corrections will be made be erasing the incorrect code and writing in the correct code.

4. The following conventions will be used for characters that might be misread by data entry operators:

    Letter I   Number 1   Number 7

    Letter O   Letter Q   Number Ø

    Letter Z   Number 2

   A space not otherwise apparent will be noted by a ɓ (Small b with a slash).

5. Coding for each division, file, record identification, and module should begin on a new page.

6. Sequence numbers must be entered in columns 1-6 of the coding form.
                1 - 3 Page identification
                4 - 6 Line identification

   To allow for insertions, line numbers should be in increments of 10.

7. A program identification code should be coded in columns 73 - 80 of the coding form. For example, in the accounts receivable programs, AR is used to represent the series and the numbers identify the program. The program to create new records would be coded as AR 001.

8. The COBOL rules for the use of the A and B margins should be followed.

9. Level numbers in the data division and file section should start at column 8 for level 01, at column 12 for the next level. A four-space indentation should be used for each additional level. The various levels should be incremented by 10. For example:

```
Ø1  MASTER-AR-RECORD-ARMASTER.
    1Ø  NAME-AND-ADDRESS-ARMASTER.
        2Ø  NAME-ARMASTER               PIC X(25)
        2Ø  STREET-ARMASTER             PIC X(25)
        2Ø  CITY-AND-STATE-ARMASTER     PIC X(25)
        2Ø  ZIP-CODE-ARMASTER           PIC X(25)
```

FIGURE 12.18 Standards for writing COBOL source code (Page 1)

$C^P I$ **computer products inc.** DATA PROCESSING STANDARDS MANUAL

SECTION	Computer Programming	SECTION	CHAPTER	SUBJECT
CHAPTER	Programming Design	5	3	2
SUBJECT	Coding Standards	DATE August 1986		PAGE 2 of 3

10. The PICTURE clause should begin in column 43.

11. Statements in a nested set of conditions must be aligned to emphasize where alternatives or clauses begin. Each identation will be four spaces. For example:

```
IF ERROR-TYPE=WS = Ø  THEN
    PERFORM COMPUTE-GROSS-ROUTINE
    IF GROSS-PAY >    5ØØ THEN
        MOVE ''***''  TO FLAG-FIELD-REPORT
    ELSE
        MOVE '' '' TO FLAG-FIELD-REPORT
ELSE
    PERFORM INVALID-RECORD-ROUTINE.

READ DISKIN
    INVALID KEY
        MOVE ''NO'' TO VALID-SEEK-WS
        MOVE 2 TO ERROR-TYPE-WS
        PERFORM ERROR-ROUTINE.
```

12. Data and file names must be meaningful and add to the internal documentation of the program. A suffix will be used to denote the source of the information. Examples:

    File names: TRANSACTION-DISKIN

                       MASTER-INVENTORY-DISKIN

    Data names: DESCRIPTION-TRANS

                       DESCRIPTION-MASTER-IVT

                       DESCRIPTION-REPORT

                       DESCRIPTION-WS

13. Once file and data names for transaction and master file have been established, the names must be used in all programs that utilize the files. File FDs and 01 structures will be copies from the library.

14. The reference number should be used as part of the module name. For example: 2ØØØ-INITIALIZATION-MODULE.

FIGURE 12.19 Standards for writing COBOL source code   (Page 2)

computer products inc. DATA PROCESSING STANDARDS MANUAL

SECTION	Computer Programming	SECTION	CHAPTER	SUBJECT
CHAPTER	Programming Design	5	3	2
SUBJECT	Coding Standards	DATE	August 1985	PAGE 3 of 3

15. At the beginning of each division or module a brief description shall be provided. The overview of the program as presented in the Identification Division might be as follows:

```
**************************************************************************
*     INVENTORY CHANGE PROGRAM.  ON THE LAST DAY OF THE MONTH THE FILE  *
*     IS OPENED AS A SEQUENTIAL FILE AND ALL PRICE CHANGES FOR THAT MONTH *
*     ARE RESET TO THE NORMAL SELLING PRICE.  OTHER CHANGES ARE ENTERED  *
*     AND CONFIRMED BY THE OPERATOR.  BOTH THE FORMER AND CURRENT VALUE  *
*     STORED IN THE FIELD ARE PRINTED.                                   *
**************************************************************************
```

The first asterisk must be in column 7 and the last asterisk in 72.

16. Comments should be used throughout the program. However, excessive use of comments should be avoided. For readability a blank space should be left on either side of the comment. Comments must be written in non-technical English.
```
*
* Employees with a status code of W, V, or R are exempt from
* contributing to the state retirement fund.
*
```

17. Constants to be used in a program should be defined and given a value in the WORKING-STORAGE section. The data name should then be used in the calculation. Whenever a change is required, only the statement in WORKING-STORAGE will need to be changed.

```
01  CONSTANTS.
    10  STATE-TAX-RATE-WS              PIC  V999  VALUE .046.
    10  STATE-TAX-DEDUCTION-WS         PIC  99V99 VALUE 57.43.
```

18. Library functions must be used whenever possible.

19. Whenever possible, input must be tested for validity. All data entered from a terminal must be edited as well as confirmed visually.

20. Data and program conditions must be tested in decreasing order of probability of occurrence.

21. Programs will be written using the three basic constructs—sequence, selection, and iteration. A GOTO can only be used to transfer control within a module.

22. A code walkthrough will be conducted. The guidelines used for the design walkthrough (Section 5, Chapter 3, Chapter 1) also apply to the code walkthrough.

FIGURE 12.20 Standards for writing COBOL source code (Page 3)

be used in almost every program they write. For example, whenever a record is to be randomly retrieved from file, provision must be made for an invalid key. The same invalid key routine might be used in 20 or more different programs. If the routine is placed in a library, it can be copied into each of the programs. A programmer/analyst working for a large organization indicated that when he writes a new program, usually less than 10 percent of the program is new code. Ninety percent is either code copied from a **source library** or subroutines called in from an online library.

CHECKPOINT

17. Why should uniform symbols be used when constructing a flowchart?
18. What might be considered the major advantage of using a flowchart rather than some other means of depicting the logic of a program?
19. What three basic functions should be provided by the detailed logic plan?
20. What are the three major sections of an HIPO chart?
21. How is the relationship to other modules shown on an HIPO chart?
22. Why do some programmers and analysts object to the use of pseudocode?
23. In pseudocode, what is a key word?
24. How might a Warnier-Orr diagram be described?
25. Why should design and programming standards be developed?
26. Why should the documentation of a program be changed whenever a program is modified?
27. When a new program is to be written, why might the programmer only be required to write 20 or 30 percent of the code?
28. If a new programming product is obtained, what should occur within the organization?

SUMMARY

Many CIS departments in both large and small organizations have reached the age of maturity. Standards have been developed and *are being followed*. Standards and guidelines are often available for: designing systems, designing programs, coding programs, testing programs and systems, and documentation.

Due to increasing personnel costs and the unproductivity of programmers, attention was focused on structured design concepts. It was proven that if more time was spent on designing a program and testing the design, less time would be spent on coding, testing, and debugging a program. When structured programs are written, each major function has its own module. Each module has its own detailed logic plan. The most common ways of depicting the logic of a program are to use flowcharts, HIPO charts, pseudocode, Warnier-Orr diagrams, or Nassi-Shneiderman charts. Within an installation, all programmers should adhere to the programming standards which should include the method to be used for developing the logic of a program.

Programming productivity can also be increased by the effective use of programming prodducts, copying from source libraries, and by using subroutines. Analysts and programmers must have a general knowledge of the programming prodducts available and they must have a thorough knowledge of the functions and applications of inhouse prodducts.

As conditions change, existing standards and guidelines must be reviewed and revised to reflect the hardware and software available and the current state-of-the-art.

DISCUSSION QUESTIONS

1. What functions must be performed during the implementation phase of the project?

2. What are Noland's four stages of growth in EDP? Are all organizations in the maturity stage? What might occur within an organization that would cause it to reenter the expansion stage?

3. Could what is occurring in office automation be compared to Noland's four stages of growth?

4. What are the major concepts of structured programming? Why was management concerned about the productivity of programmers? Did the implementation of structured programming concepts increase productivity?

5. What must be done prior to the development of a logic plan? If the programmer does not completely understand the problem, what should be done? Might the programmer need to do some research and consult external sources rather than conferring only with the user?

6. Contrast flowcharts, HIPO, pseudocode, Nassi-Shneiderman charts, and Warnier-Orr diagrams on the following points:
   a. number of symbols used
   b. ease of construction
   c. ease of maintenance
   d. ease of use—how easy is it to understand the logic being presented?

7. Defend or refute the following statements: Each programmer within an installation should be free to elect to use whatever technique he or she desires to construct detailed logic plans. If they so desire, they can elect to code the problem without constructing a logic plan.

8. What is a design walkthrough? What should be done prior to the walkthrough? Is the purpose of the design walkthrough to evaluate the programmer?

9. What type of topics should be covered in the design standards?

10. What type of topics should be covered in the programming standards? Would the programming standards developed for a company using BASIC as their major programming language be the same as those presented in the text?

11. In the programming standards, it was stated that once the file name and data names for a file had been established, all programmers had to use those names. Why should this be done? Why not let programmers use whatever names they wish?

12. How should an individual program be tested? Why is it important that test data be carefully constructed?

13. You are the CIS manager and have recently hired a new programmer. The new programmer is intelligent, knows how to write really complex programs, and knows COBOL. However, his total work experience has been in a department that let everyone do their own thing! What would you do to introduce the new programmer to your philosophy regarding programming and your installations standards?

14. As manager of the CIS department, you have decided to insist on design walkthroughs. You made this decision because you have observed that several of your programmers spend a great deal of time debugging programs. Their attitude seems to be "Let the computer catch the errors for me." Operations personnel also complain about the number of jobs that abort. Skip Pittsley, one of your programmers, has refused to subject his programs to a design walkthrough. What would you do to convince Skip about the merits of design walkthroughs?

15. Your CIS department has been a traditional COBOL shop and well-defined standards have been developed for designing and implementing COBOL programs. Recently microcomputers were obtained. Two programmers were hired and the language to be used is BASIC. The programmers feel their environment is totally different from the ''world of COBOL,'' and they feel they do not need to adhere to any of the standards. As manager, how would you handle the situation?

Team or Individual Projects

1. One of the programs to be developed is to produce a report that will list items: out of stock, below the reorder point, and within 10 percent of the reorder point. The specification form for the program indicated:

   a. The inventory master file was to be accessed sequentially. After all records have been processed, the job will be terminated.
   b. The date for the report is to be entered from the operator's console.
   c. Separate columns are to be used for indicating which situation exists. The items to be listed are either out of stock, below the reorder point, or approaching the reorder point. If none of the three situations exists, the item will not appear on the report. In checking the inventory status, checking for out-of-stock items will be done first, then for items below the reorder point, and finally for those items that are within 10 percent of the reorder point.
   d. Totals are to be maintained for each situation. At the end of the job, a second report will be printed that indicates what percentage of the total inventory each situation represents. For example:

	Total Items	Percentage
Items out of stock	180	.9

   e. The job will be run at the close of business each day. The report must be available for the inventory manager by 8:00 A.M. of the following morning.
   f. The report will list the names of the first two vendors identified by the vendor code. The vendor code can be used to randomly retrieve the vendor's record. Only the code number and the vendor's name will be printed. The vendor's records are to be accessed randomly.
   g. Neither file (the vendor or inventory) will be updated.

Directions:

(1) Review the material regarding the steps in programming and the construction of logic plans.
(2) Prepare a print layout form for the two reports. The detailed report is printed as the records are processed. The report is to be double-spaced. There should be 25 detail lines plus the headings on each page. The summary report is printed after all records have been processed.
(3) Prepare a hierarchy chart for the program.
(4) Develop a detailed logic plan for the program.
(5) Walk through the design of the program. As you do, check off on the report when each item became available for printing. Make sure that your logic plan provides the information that is needed and that details such as opening files, printing headings when needed, closing files, and so forth, are accomplished.
(6) Determine how the program would be tested. In this case, a copy of the actual file would be used to test the program. How would the programmer know the results were accurate?

2. Make an appointment to talk to an experienced programmer who writes software for microcomputers and ask the following questions:
   a. What technique do you use to develop the logic for a program?
   b. How would you describe a well-designed, structured program?
   c. In developing a new program, do you use a building-block approach?
   d. Are external subroutines and functions available?
   e. What occurs when the program must be modified?
   f. What design, programming, and documentation standards that you must follow are available in your installation? If formal standards are not available, what are your standards regarding the items listed?
   g. In developing a fairly complex program, what percentage of your time is spent in each of the following areas:
      (1) developing a detailed logic plan;
      (2) coding the program and eliminating syntactical errors;
      (3) testing the program; and
      (4) completing the documentation?

GLOSSARY OF WORDS AND PHRASES

building block A section of code that can be used in one or more programs. For example, the code used to read from a master file randomly might be used in several programs. The code can be stored in the source library and called into programs that require the file to be accessed randomly.

case entry A convenient way to use logic structure that describes multitest (more than two) conditional branching.

HIPO (Hierarchy/Input/Processing/Output) charts HIPO charts illustrate the relationships between various programming modules as well as showing what input is needed, what processing occurs, and what output is generated.

interactive BASIC interpreter A program that checks the source code entered for syntactical and clerical errors. As each line of code is entered, the interpreter checks for errors. If none is found, the code is translated into machine language.

module A section of code or part of a program dedicated to one function. Usually a module has 50 or fewer lines of code. One module might contain code required to print detail lines and a second code required to print headings.

Nassi-Shneiderman charts Simple diagrams used to graphically represent a structured solution to a problem. Symbols are used to represent the structured programming constructs.

pseudocode English statements used to illustrate the logic of a program.

source library A library used to store structures that describe data or files and other types of source coding. The use of prewritten code stored in the library saves coding and reduces the time needed to get a program to the execution stage.

stub testing The main logic of a program is tested and "dummy" modules are used to represent control passing to the supporting modules. Module by module, the dummy modules are replaced with actual code.

syntactical errors Errors that violate the rules of the language. For example, in BASIC the variables in a READ statement are separated by commas. The code "READ A B X" will cause an error message to be displayed. The code should be written "READ A, B, X".

tutorial language Software that enables the user to write programs in a conversational mode. When a tutorial language is used, the programmer or user is guided through the steps necessary in developing a program to solve a particular problem.

VTOC (visual table of contents) Another name for a program hierarchy chart.

Warnier-Orr diagram A graphic representation of the logic of a program that is often described as structured pseudocode.

STUDY GUIDE 12

Name _____

Class _____ Hour _____

A.   Indicate whether the following statements are true (T) or false (F). If false, indicate in the margin how the statement should be changed to make it true.

_____   1.   Since the programmers and analysts have expertise in developing plans for implementing a system, there is no need to have any slack in the schedule.

_____   2.   According to Noland, the last stage of growth in EDP is formalization. Computer policy committees are formulated and there is better control over the development of systems.

_____   3.   Management of the expansion of EDP and office automation should be at the middle-management level.

_____   4.   Programmers are completely at fault for not documenting and testing programs completely.

_____   5.   The three constructs used to develop structure programs are sequence, iteration, and the use of GOTOs.

_____   6.   Structured design concepts were first introduced in *Communications* in 1975.

_____   7.   Selection is usually implemented by the use of IF/THEN/ELSE statements.

_____   8.   Management became interested in structured design concepts due to the low productivity of programmers and the high cost of developing programs.

_____   9.   The source documents should be designed prior to the design of the screens and file layouts.

_____   10. Programmers who start coding prior to developing logic plans spend less time in testing and debugging programs.

_____   11. Within a CIS department, each programmer should be able to elect the type of logic plan to be developed.

_____   12. The test data used during the design walkthrough should only test the normal routines. It is not necessary to test the exception routines.

_____   13. Most mainframes installations use BASIC to code business-oriented programs.

_____   14. When an interactive BASIC interpreter is used, all the errors detected are printed after all of the source code has been entered.

_____   15. A code walkthrough should be conducted before the syntactical and clerical errors are detected.

_____   16. If a program is completely tested, only valid data and data that contains errors need to be used.

_____   17. Today programs are evaluated solely on the criteria that valid output is produced.

_____   18. The cost of entering data is not a consideration in determining the effectiveness of a program.

_____   19. In the inventory change program, changes were entered as they were approved by the inventory manager.

____ 20. One flowchart should be developed that will illustrate the entire logic of a program.

____ 21. When a file is processed sequentially, it is always possible to determine when the end-of-file marker has been detected.

____ 22. Any good logic plan should serve as a guide to be used in coding the program, a means of illustrating what the program does, and as a guide for maintaining the program.

____ 23. HIPO is an acronym for Hierarchy/Input/Processing/Output.

____ 24. When the word PERFORM is used, it means to branch down to the statements specified. Within the module will be a branching command that tells the computer where to branch to after the commands within the module have been executed.

____ 25. There is normally no way to relate the coding to the hierarchy chart and to the detailed logic plan for a module.

____ 26. In programming, an I/O device is referred to as a file.

____ 27. Unstructured programs are easier to maintain than structured programs.

____ 28. One of the most common problems cited for EDP today is the lack of valid, up-to-date documentation.

____ 29. Programmers and analysts only need to be familiar with the programming products used within their installation.

____ 30. A building block is code used within a program to cause an internal subroutine to be executed.

B.   Multiple choice. Indicate which method of constructing a logic plan is described by entering the letter of the method in the blank provided.

         a. flowcharts
         b. HIPO charts
         c. Nassi-Shneiderman charts
         d. pseudocode
         e. Warnier-Orr diagrams

____ 1. Uses only four basic symbols for showing processing, a decision, a loop, and case entry.

____ 2. Are sometimes described as structured pseudocode.

____ 3. Sections are available for identifying the required input, processing, and output.

____ 4. Key words may be used such as PERFORM, READ, or WRITE.

____ 5. Is sometimes considered the best way of graphically showing the logic of a program.

____ 6. An N is used to indicate that the step or series of steps is to be repeated until a given condition is true.

____ 7. Was developed by IBM as an alternative for flowcharting.

____ 8. An arrow is used to indicate where (which module) program control transferred from and where control will return after commands within the module are executed.

____ 9. Some programmers object to its use because it resembles COBOL.

____ 10. Its symbols have been standardized by ANSI.

C.   Multiple choice. Record the letter of the best answer in the space provided.

_____   1.   A section of code or part of a program dedicated to one function.
a. subroutine   b. function   c. module   d. routine

_____   2.   Used to store structures that describe data or files that are written in a language such as COBOL.
a. subroutines   b. source library   c. functions

_____   3.   A convenient way to use logic structure that describes multitest conditional branching.
a. case entry   b. subroutine   c. iteration   d. selection

_____   4.   A program that checks source code for errors as the code is keyed in by the programmer.
a. subroutine   b. interactive interpreter   c. function

_____   5.   A section of code that is normally stored in a source library that is used in one or more programs.
a. subroutine   b. building block   c. function

_____   6.   A 2000 level module invokes the
a. 1000 level modules.   b. 3000 level modules.   c. 4000 level modules.

_____   7.   The module names used on the hierarchy chart
a. are not used on the flowchart or coding.
b. are used as the module names on the flowcharts.
c. are used as a comment or remark on the coding.
d. should be used on both the coding and the flowcharts.

_____   8.   How code is formatted is important because well-informatted code
a. is easier to maintain.
b. is easier to understand.
c. increases the readability of a program.
d. provides the advantages specified in a, b, and c.

_____   9.   In CIS departments that have reached maturity, the major problem that still exists is
a. the lack of programming standards.
b. the lack of design standards.
c. documentation is not updated.
d. coding is started prior to the development of detailed logic.

_____   10.   One of its major strengths is that it forces the programmer to think in terms of input, processing, and output.
a. flowchart
b. pseudocode
c. HIPO
d. Nassi-Shneiderman chart

13 PREPARING FOR THE NEW SYSTEM

Looking Ahead

After reading the text and completing the learning activities you will be able to:

- Determine the tasks required to implement a new system.
- Identify the importance of determining a specific plan for converting existing files to the new format or medium.
- Identify some of the ways that output from a single program can be tested for validity.
- Explain the difference between program, link, and systems testing.
- Explain why the users should be directly involved in systems testing.
- Identify some of the ways that the personnel who will be working directly with the new system can be trained.
- Identify for the situation described the type of training materials that might be used.
- Identify the type of information regarding a procedure that should be available in the standards manual.
- List the items that should be included as part of the final report for the project.
- Explain why users should prepare their own documentation.
- Identify the difference between direct, partial, and parallel conversions.
- Define and utilize the words and phrases listed in the end-of-chapter glossary.

INTRODUCTION

A good motto for individuals involved in systems design and analysis is:

Plan the implementation, implement the plan.

In developing the plan, attention must be given to the sequence in which the various events are executed. Throughout the entire implementation phase, the necessary programs are designed, coded, tested, and documented. These are the most time-consuming tasks that must be completed during the implementation phase of the project.

While the programming activities are occurring, the site must be prepared. The hardware and control software must be ordered, obtained, and tested. Before the final testing can be done, files must be converted to the new format and the necessary supplies must be available.

How easily and successfully the conversion from the old system to the new system proceeds depends on a number of factors. The conversion will go more smoothly if the users and management have been directly involved in, and informed about, the new system. Also the analysts must consider the reaction of the employees who will be affected because of the changes and must work with them. Some employees may need to be retrained while others may be assigned to a different position. Others, even if informed of what will occur, may be apprehensive and will resist change.

If both the individual programs and the entire system have been completely tested, the conversion will be easier than if "bugs" have to be fixed during the actual conversion. The conversion will also be easier if the new computer system is fairly compatible with the old one.

Long before the actual conversion date, the documentation should have been completed. Everyone who will work directly with the system should have been trained in his or her part in the procedures prior to the change from the old to the new system.

FILE CONVERSION

Planning for the File Conversion

When a conversion to a new system is being planned, all factors must be considered. The plan must indicate: how the data in the present files will be converted; the amount of new data that must be keyed in; and when the files must be converted. Once the amount of data that is needed and how it will be converted are determined, the personnel requirements can be ascertained. The *exact* time additional personnel will be required must be calculated. If the company's own personnel will be unable to handle the conversion, the job can be assigned to a service bureau or personnel can be obtained from a concern that provides temporary help.

Creating New Master Files or Databases

Often master files must be converted to a new format or to a different medium before a new system can be considered operational. The time needed to complete the conversion depends on a number of factors. If the conversion is from a computerized system to a more sophisticated computerized system, programs can usually be written that reformat the data and create the new file. The program may also need to check for invalid data and convert existing codes to a different format. One of the problems which may complicate the conversion is that the documentation pertaining to the old files is obsolete and does not adequately define all of the codes.

If several independent files are to be converted to a database, a conversion program can be written that will read the old files as input and transfer the data into the database. When a batch system is being converted to a timesharing system, it may be necessary to transfer the data from cards or magnetic tape into the new system's database.

If a manual system is being converted to a computerized system, operators may be required to key a great deal of data that will be recorded on diskettes, disk, or magnetic tape. During this phase of the conversion, it may be necessary to obtain additional employees from a temporary job placement service or to obtain the services of an organization that specializes in data entry.

TESTING PROGRAMS AND SYSTEMS

The programmer must follow the analyst's specifications and design test data that will test all possible conditions. Every error routine used to generate error messages for invalid data must be tested.

Designing Test Data

The test data should be designed *before* the program is written. The programmer should prepare good data that tests all of the normal routines and bad data that will test each exception routine. The test data should be listed in the sequence that it will be entered into the computer. It is usually easier to test a program if each transaction record is specifically designed to test one condition.

Programmers must test for multiple errors. For example, a program might function properly if one bad record was encounted that was followed by a record that contained valid data. However, invalid results might be obtained if two records in sequence both contained one or more fields of invalid data.

Test Files

When programs are being tested, files are created that contain controlled information. In the sales/accounts receivable/inventory system, accounts receivable, sales, and inventory test files would be created that contain all the conditions that would exist if "live" data were used. The programmer can control the test data and make certain all conditions are tested. As long as each condition that could exist is tested, the volume of records contained in the file is not a consideration.

When major changes must be made to an operational program, the test files should be used rather than the actual files. After the changes have been tested on the controlled test data, *copies* of the actual files should be used to further test the changes made in the programs.

Testing an Individual Program

Once a program has been written and compiled, it must be tested. The test data prepared by the programmer is used to test all normal and exception routines. The testing procedures used are generally included in the documentation.

The validity of the output is determined by using one or more of the following methods:

- The output is compared with mathematical calculations made by the programmer.
- Output written on tape or on disk is displayed on the printer and checked by the programmers.
- Printed batch totals are checked against the totals prepared from the source documents.
- Printed record counts are compared against the number of test records processed.
- Totals are **crossfooted** to determine whether they balance correctly.
- Error messages are checked against the test data to determine whether all errors have been identified correctly.

Link or String Testing

Usually a series of programs is run to edit data and to update the master files. Although each program is tested individually, the entire sequence of programs executed in a series should be tested. After each program is tested individually, the jobs are entered onto the queue for execution in the order in which they will normally be run.

Assume that you wish to test the payroll system and that the payroll database is already created. The string of programs that you wish to test might include the programs that: add new employees; flag employees for deletion; change data in the employees' records; correct data stored in records; copy the file to a backup file; edit the data stored in the current transaction file; write the payroll register; update the database; and write the payroll summary. The first time the string is run, the payroll register must agree with the payroll summary.

A limited number of records should have been transferred to a test database. The contents of the databases should be displayed before and after running each program that updates any of the records. After all the jobs are run in a string, the programmer must determine that everything is in balance. The documentation should indicate how the validity of the output is checked.

Systems Testing

Once all the programs have been individually tested, and tested as part of a series of related programs, the programmer is ready to begin testing the entire system.

In program and link testing, the programmer controls the input. Test files are carefully designed so that it is possible to test all conditions—normal and abnormal. The programmer controlled the preparation of the input. Although a limited number of records are used, all conditions are tested in an environment carefully controlled by either the programmer or the analyst. Now it is time to test the system.

Systems testing is usually deferred until the portion of the documentation employed by the user and operators in preparing the source documents and the input is completed. The user and operations personnel should have received their documentation and instructions regarding how the system will function. The required training session should also have been completed.

Systems testing might be compared to the shakedown cruise of a ship or the dress rehearsal of a theater production. Everything should be live! Copies of the actual master files are used in systems testing. The preparation of the input or transaction files is no longer rigidly controlled by the programmer or analyst. The input is prepared by the user's personnel or by data entry clerks.

Source documents are prepared from actual data by the user's personnel. Batch totals and record counts are recorded on the transmittal forms. The documents are sent to the data control clerk (in a small installation, to the operations manager). The data control clerk logs in the documents and transmits the forms to data entry. The data entry clerks process the documents and return them to data control where they are logged.

The computer operations personnel prepare the necessary job control language statements and proceed to execute the jobs by following their documentation (run sheets). The output is sent to a data control clerk, who follows the procedures outlined for verifying the output. The output and the source documents are then sent to the user. The user follows documentation in his or her department in the verification and utilization of the output.

The analysts and programmers should observe the entire procedure. If they have done their jobs correctly, there should be no problems. They are interested in the workflow as well as in the validity of the output. Did the work proceed smoothly, without interruptions or were there obvious bottlenecks and problems?

Everyone—from the individuals who recorded the data on the source documents to the user who worked with the output—should have known what to do without asking questions. If any individuals have a question, they should be able to refer to their documentation to determine the answer. If not, either the employees were not properly trained or the documentation was incomplete or poorly written.

The complete cycle should simulate all of the daily, monthly, and yearly procedures. All procedures should be executed in the proper order to see that all of the reports are in balance and check with any totals that might have been prepared by accountants or other non-data processing personnel.

CHECKPOINT

1. Why should test files be used rather than actual files for testing new programs or testing changes made in operational programs?
2. What must be ascertained before the personnel requirements for converting a present file to a new format or medium can be determined?
3. What is the difference between testing a program, link or string testing, and testing a system?
4. What weaknesses might the system test detect?
5. What are the analysts and programmers looking for when they evaluate the results of a system test?

TRAINING PERSONNEL

Unless there is a division or department within the company that conducts all on-the-job training, the project team or user will need to develop and implement the training program. In the detailed investigation report, the personnel requirements for the new system were identified. An estimate was also made of the amount of retraining employees working with the old system would need to enable them to work effectively with the new system.

One of the constraints often imposed upon the design of a new system is that new employees cannot be hired. It must be possible to implement and run the new system with existing personnel.

Task Identification

Each new task or existing tasks that will be changed when the new system is operational must be identified. Each task must then be analyzed to determine what subtasks are involved. The next step is to determine the most effective way of performing each of the subtasks. The length of time needed to perform each subtask under varying conditions must also be determined. After this is done, standards can be developed to evaluate the effectiveness of employees who perform the tasks. Until the subtasks are evaluated and standards developed, the personnel requirements for the new system cannot be identified accurately.

If new jobs are being created, job descriptions similar to the ones illustrated on pages 61 and 64 should be prepared. Usually when new positions are created they are posted and made available to qualified individuals from within the company before they are advertised outside the company.

After the personnel requirements have been identified and it has been decided whether the new positions will be filled from within or from outside the company, the amount and type of training needed prior to converting to the new system can be determined.

Identifying the Type of Training Needed

The sales/accounts receivable/inventory system will be an online system. Although some of the individual programs are batch jobs, 90 percent of the sales data will be keyed in by sales representatives. New documents have been designed that will be used to record the initial transaction. After the document is filled in, the sales representatives transmit the data into the system. Sales representatives, as well as the order entry clerks, will need instruction regarding how to fill in the source document and how to key in the data.

Besides the new sales-order form there are a number of internal forms that will be used to record the data needed to update and maintain the master files. Transmittal forms must be filled out that contain the necessary control information.

The tasks involved in completing the necessary source documents and transmittal forms are analyzed and each new task is documented. Next, the analyst must determine how often it will be necessary to train people to perform the required task and how many people will need to be trained. If the tasks are ongoing, such as instructing sales representatives in the use of the source documents and terminals, different types of training materials will be developed than if a single training session is needed.

The location of employees needing training is another key factor. If all the employees are located in the corporate office, it is easier to provide hands-on training than if the affected personnel are located in different sections of the country. Should the employees be brought to the corporate headquarters for training or should the training be done in division offices?

Training the Sales Managers

Assume that the decision has been made to bring the district sales managers into the corporate headquarters for briefing sessions regarding the new sales/accounts receivable/inventory system. The session might be divided into three major areas: background information regarding the system; use of terminals for entering sales data; and instruction regarding the use of the query language.

Background Information The sessions might start with a presentation regarding the problems identified with the old system and the objectives of the new system. The presentor should explain the features that are incorporated into the new system that will provide better, more reliable information. Examples of the reports that will be available on a regular basis as well as those available upon demand should be provided. Specific information regarding new source documents and forms should be also be covered. The sales managers must understand what is expected of each sales representative.

Use of Terminals for Entering Sales Data In working with the district sales manager, instruction regarding the use of terminals for entering sales information can present a problem. Since this group of individuals may be unfamiliar both with terminals and with using keyboards, they may feel very uncomfortable unless the training session is well planned. Some companies have started off such sessions by using interesting and informative games. The trainees become involved in the material being presented and forget about the monster—the terminal—that they must master.

The documentation that the sales representatives will follow in entering sales data should be used in the later part of the training session. The sessions should end with the managers feeling comfortable in using the terminals. They should also understand how the data is entered, what errors could be made, and how each type of error will be detected. Regardless of their keyboarding ability, they must feel at ease in entering sales data. Although the sales managers will not normally enter data, they may want to demonstrate the system to a new employee or upon occasion, enter data for an order.

Using the Query Language   Before the sales managers can use the query language, they must understand the commands that can be used and the information stored in the database. The manager should receive instruction regarding how to extract information based upon different criteria, how to sequence the information extracted, and how to format the report or display. The mathematical capabilities that are available as part of the query language should also be explained.

Demonstrations should be given of the typical queries that might be made. Following the demonstration, worksheets might be used that provide more typical queries. This time the managers will use the terminals to generate their own output.

The training sessions must be well planned and the instructions given in a clear manner. If the managers understand the benefits they will receive by being able to use the query language, half of the battle will be won. Highly motivated people are more responsive to learning than individuals who see no value in the material being presented.

Training the Sales Representatives

Once the sales managers are familiar with the new system, training sessions should be provided for the sales representatives. The sessions should probably be held in the district offices. Because it is critical that the sales representatives have a complete understanding of the new system, someone who relates well to others and who likes to provide hands-on sessions should provide the instruction. Essentially the same instructional materials can be used to train the sales representatives as were used for the sales managers.

If any of the sales representatives cannot keyboard reasonably well, the trainer might suggest a tutorial program designed to teach keyboarding. Representatives also can obtain a basic keyboarding text and practice. The sales representatives will be entering data from the offices of customers and should look proficient and feel comfortable. A potential customer may become upset if it takes a sales representative 15 minutes to determine that an item is not in stock.

PREPARING AUDIOVISUAL MATERIALS

Use of Films

If a substantial number of sales representatives are hired throughout the year, it may be advisable to prepare some type of audiovisual materials or a training film. Since films are expensive to make, their use is limited. If large numbers of people are to be trained and if the material presented in the film will not become obsolete in a short period of time, the use of films should be investigated. Also it may be possible to obtain training films on specific topics from hardware or software vendors.

Slide/Tape Presentations

If a regular slide projector is used, the slide/tape presentation can be shown to a large group of people. When the presentation is viewed individually, the viewers may be asked to stop the tape, participate in some type of learning activity, and then restart the tape. The learners may also be asked to check their work with the solutions projected.

Since the sales representatives who need to view the materials are located in different sections of the country, it is not feasible to bring them into the corporate office for training. In addition, when new sales representatives are hired, the slide/tape presentation can be used as part of their orientation to the company and to the sales system. If the materials will not be used on an ongoing basis, it might be unwise to invest the time and talent needed to develop a presentation of this nature.

Videotape

Although commercially prepared videotapes are often expensive, videotapes can be made inhouse and used very effectively to train employees. CIS employees can be taped as they complete the necessary tasks. Once a presentation is prepared with visual aids, such as slides, overhead transparencies, and demonstrations, the entire presentation can be taped. Knowing that the presentation will be taped often encourages the presentor to prepare better visual aids and to be better organized.

Other Types of Visual Aids

If an instructor or lecturer will be making the entire presentation, slides, overhead transparencies, computer output displayed on a large monitor, or flip charts can be used effectively. In large organizations, a media or graphic arts department is usually available that will assist with the preparation of visual aids. If not, the analysts in charge of the project will have to be creative and develop their own materials.

PROGRAMMED LEARNING MATERIALS

When ongoing training is needed for individuals in different sections of the country, various kinds of programmed instruction materials may be used. When programmed instruction materials are used, the learner may work individually at his or her own pace. Slides, filmstrips, and videotapes may be incorporated into the programmed instruction materials.

The computer is a very effective device to use for individualized instructions. Text can be displayed, the learner can be asked to read material from outside sources, and a series of tests can be generated. The computer keeps track of the progress of the learner, and if programmed to do so, can make suggestions when a wrong answer is given. With the graphic capabilities that are now available, excellent illustrations can be incorporated into computer-assisted instruction programs. One company has their complete service manual for repairing locomotives computerized. When a part needs to be repaired, directions, including excellent diagrams, are displayed on the VDT.

TESTING TRAINING MATERIALS

Regardless of who prepares the training materials, the materials should be tested before being put to actual use. Individuals without any prior knowledge of the topics presented in the training materials should be selected for the test. In completing the tasks assigned in the training materials, the only source available during the test will be the documentation. If the individuals testing the materials cannot complete the assignments without asking questions, the training materials should be examined since changes may need to be made.

In writing instructions regarding the use of computers or terminals, the omission of a comma or period can be critical! One individual spent hours trying to get out of a loop created by a programming error merely because the documentation did not explain CTRL/C. The documentation did not state it was two keys (CTRL and C) that had to be held down simultaneously.

In-Service Training Programs

Vendor's Schools

Most of the large vendors offer educational programs designed to provide both specific training and general education regarding new products and services. Before IBM unbundled its computer services into different areas such as software support, maintenance, sales, and education, most of the IBM schools or training programs were free to users of IBM systems. Certain types of short courses, such as one-or two-day orientations to a new system such as the System/36 are still free. Longer courses and ones that provide specific training in a certain area, such as IMS (a two-week course), or DL/I, usually must be paid for by the participant's company. A firm that elects to send its employees to a vendor's school will need to pay the tuition, living expenses of the individuals attending the school, and transportation costs. The general-education-type programs are usually offered in a number of different education centers, while the more specific training programs tend to be offered in the larger cities such as Detroit, Chicago, or Dallas.

The material is usually well presented and good audiovisual aids are used. If the course covers specific training on hardware, hands-on laboratory experience as well as lectures will be provided.

Inhouse Training

If a number of people must be trained in the use of a software package or on a particular piece of equipment, it may be more economical to bring the classroom to the students. With today's technology, this could be done by providing a teleconference which would allow two-way communication between the learners and the presentors. Often, however, in submitting their response to an RFB, vendors indicate a willingness to provide in-service training for the customer's employees at a site selected by the customer.

If the vendor is unable to provide inhouse training, local colleges or consultants should be contacted to see whether an inhouse training program can be developed. The training sessions must be tailored to the precise needs of the company making the request.

With the high cost of travel, more companies may turn to some type of inhouse training rather than sending employees to conferences, seminars, workshops, and formal classes.

The data entry operators, computer operators, sales representatives, and sales managers should be trained in how to execute procedures and use the necessary equipment long before the new system is considered operational. This is another example of why an implementation plan must be developed. The site must be prepared and the equipment *operational* before the organization's employees can be trained.

Programmed Instructional Material

In many areas, commercially prepared **programmed instructional materials** can be used to provide in-service training. The materials are usually available in areas where there is sufficient demand to warrant the cost of developing the materials. Often a large company will carry a contract for a given number of courses each year with a vendor that specializes in individualized learning programs. Companies that provide such courses include Deltak, Inc., Learning Corporation of America, and Professional Development, Inc. Deltak, for example, produces a series of videotape presentations on such topics as distributed data processing, database systems, and telecommunications. IBM and other computer vendors publish excellent programmed instruction materials that can be used as a supplement to other learning activities.

Formal Courses

Most academic institutions provide training in different areas of EDP. When specific needs are identified, such as providing keyboard for management or learning about a relational database, there may be an academic course offered that meets those needs. If a course is available within the geographic area of the company whose employees need to take the course, the company usually pays for the employees' tuition, materials, and sometimes their transportation. If enough people are involved, the course may be taught inhouse rather than at the college or university. Employees taking the courses are often given released time from their normal duties or responsibilities.

Some academic institutions also specialize in responding to the needs of industry and provide intensified training or educational offerings. For example, if CPI were to obtain microcomputers for their sales representatives, a special course in the software packages available on the systems might be initiated. Perhaps the microcomputer course designed for CPI would include training in Symphony software (a database package developed by Lotus) rather than dBASE III (database software developed by Ashton Tate) which might normally be included in the regular college course.

Which Method Should Be Used

If there is a training department within the organization, it is the responsibility of the department to determine which alternative can be used most effectively. If there is no training department, it is up to the CIS manager and the user to decide how to obtain the necessary training for new systems. The decision will be reached by determining the resources that exist within the organization and those that are available within the community.

USING DOCUMENTATION AND STANDARDS MANUALS

When a new system is being installed, a great many tasks are eliminated while new ones are added. A section in the standards manual should be developed for each of the new tasks. Although the task may seem simple to the analyst, clear, explicit instructions must be provided. To the employee who has been doing a task a given way for a number of years, the change may be awesome and frightening.

Keeping all the people who will be involved in the new system informed will help to dispel some of their worries. Equally important is providing instructions that are:

- well written
- easy to find (well indexed)
- easy to understand, and
- comprehensive

When well-documented instructions are provided, many of the employees' apprehensions may be resolved. The instructions in the standards manual can be used to train present or new personnel or as a reference.

Sections of the Sales/ Accounts Receivable/ Inventory Standards Manual

Figures 13-1 and 13-2 illustrate how the procedure for adding new customers, deleting customers, and changing some of the fields within a customer's record might be documented. A clerk within the sales department will be assigned the responsibility for keying in the data. A menu is displayed and the clerk selects the program to be run. Figure 13-3 illustrates the form that will be used. Although one form is used for all three types of changes, a separate program is used for each function. Figure 13-4 illustrates the transmittal form. The two documents, along with the computer printouts that are referenced, should be included in the standards manual.

computer products inc. CIS STANDARDS MANUAL

SECTION	Sales/Accounts Receivable/ Inventory	SECTION 10	CHAPTER 4	SUBJECT 1
CHAPTER	Completing Forms			
SUBJECT	Adding, deleting, and changing records in the accounts receivable file	DATE: October 1986		PAGE 1 of 2

ORIGIN OF FORM DP 077:

    The form may originate either in the credit department or the sales department. The credit department is responsible for determining the credit limit and the discount code; all other information is provided by the sales department.

UTILIZATION OF THE FORM:

    The form is used to add new customers to the database, to delete records from the database, or to change a portion of the data stored within a record.

Function

Add new customers:
    The N is checked at the top of the form. The form is initiated within the sales department. It is then submitted to the credit department for the determination of the credit limit code and the discount code. After the credit department completes its portion of the form, the document is returned to the sales department.

Delete customers:
    The D is checked on the top of the form. The customer's account number and name are recorded. The customers to be deleted are supplied by the accounts receivable department. The list is reviewed by the sales manager.

Changing data in existing records:
    The C is checked at the top of the form. All changes other than credit or discount code changes are authorized by the sales manager. When a credit or discount code change occurs, the document is initiated in the credit department and then sent to the sales department.

COMPLETING THE FORM:

    All information on the form is to be printed. When boxes are supplied, the appropriate box is checked. The data recorded on the document will be keyed into the system by a data entry operator located within the sales department.

Account number
    Sequential numbers are assigned to new customers.

Name
    If the customer's name exceeds 25 characters, standard abbreviations should be used.

Address 1
    Either a firm name, street address, or box number may be recorded in the area provided on the form.

Address 2
    Either the street address or city and state may be recorded.

FIGURE 13.1   Documentation for completing form DP 077 and form DP 978   (Page 1)

C^PI **computer products inc.** CIS STANDARDS MANUAL

SECTION	Sales/Accounts Receivable/ Inventory	SECTION	CHAPTER	SUBJECT
CHAPTER	Completing Forms	10	4	1
SUBJECT	Adding, deleting, and changing records in the accounts receivable files	DATE: October 1986		PAGE 2 of 2

Address 3

When a four-line name and address is needed, the city and state are recorded. When only three lines are needed , address 3 is left blank.

Zip code

If the zip code is not supplied by the person submitting the information for new customers, the zip code directory must be used to determine the customer's correct zip code.

Credit code

Determined by the credit manager in accordance with the policies set by the corporate manager.

Business classification

Determined by the sales manager.

Discount code

Determined by the credit department manager in accordance with the policies set by management.

Date record is entered

Only the month and year are to be used. The date for a form initiated on February 10, 1986 would be recorded as 0286.

Authorized by

The credit manager or the person delegated by the credit manager must authorize credit and discount code changes. All other changes are authorized by the assistant sales manager.

TRANSMITTAL FORM:

Form DP 078, the transmittal form, must be completed and attached to the forms. The transmittal form must indicate when the documents were submitted for processing and the number of documents in each group. The documents are to be grouped as follows: new customers, deletions, and changes.

UTILIZATION OF THE REPORT:

After the documents are processed, they are returned to the assistant sales manager along with the change verification reports. The printout will be used to determine the accuracy of the changes. The person who authorized the changes, additions, or deletions must confirm the accuracy of the transactions. The rekport indicates the value stored in the field before and after the change is made. When new accounts are added to the file, all of the data is listed and should be confirmed.

ERROR PROCEDURE:

If errors are detected as the results of the operator keying in data incorrectly, a notation should be made on the document. The document should then be resubmitted to the operator with the next batch of documents. If the error is the result of inaccurate data on the document, a new document should be prepared.

FORM RETENTION:

Forms DP 077 and DP 078 as well as the printouts should be retained for six months. Forms and reports over six months old can be shredded for recycling.

FIGURE 13.2 Documentation for completing forms DP 077 and DP 978 (Page 2)

C^P_I **computer products inc.** CUSTOMER MASTER CHANGE FORM

[N] New customer. Fill in all sections of the form.

[D] Delete customer. Fill in customer's number and name.

[C] Change data in existing record. Enter the customer's number and
name as well as the information for the fields that must be changed.

— — — — — — — —
Account Number

— — — — — — — — — — — — — — — — — — — —
Name

— — — — — — — — — — — — — — — — — — — —
Address 1

— — — — — — — — — — — — — — — — — — — —
Address 2

— — — — — — — — — — — — — — — — — — — —
Address 3

— — — — —
Zip code

CREDIT CODE:

[1]   5,000.00 credit limit   [3] 50,000.00 credit limit
[2] 10,000.00 credit limit   [4] Unlimited credit

BUSINESS CLASSIFICATION

[1] Wholesale   [3] Retail
[2] Jobber   [4] Consumer

DISCOUNT CODE

[1] 5 percent   [3] 10 percent
[2] 7 percent   [4] 12 percent

Date record is created: _ _   _ _
                         Month   Year

Change authorized by: _____

Form: DP 077
       10/86

FIGURE 13.3 Customer master file change form

C P I   **computer products inc.**   TRANSMITTAL FORM FOR
ACCOUNTS RECEIVABLE CHANGE FORMS

NUMBER OF FORMS SUBMITTED:

New customers   _____

Deletions   _____

Changes   _____

Total   _____

TRANSMITTAL OF DOCUMENTS:	Date	Time	By
Transmitted to operator	_____	_____	_____
Returned to assistant manager	_____	_____	_____

The individual transmitting or receiving the documents must initial the
form in the space provided in the ''By'' column.

ADDITIONAL DIRECTIONS:

COMMENTS FROM OPERATOR:

Form No   DP 078
10/15/86

FIGURE 13.4   Transmittal form to accompany the accounts receivable change forms

The printout included in the documentation should provide examples of invalid data. The data might have been entered incorrectly and detected by an edit routine or might have been recorded on the source document incorrectly. Notations as to how the problem should be handled should be made next to the examples of invalid data listed on the printout.

Information that Should Be Included

As you study the information provided in the standards manual, you will observe that the following questions regarding the document being described are answered:

Where does the form originate? Essentaily this is the beginning of the audit trail that will be established regarding the changes that are to be made in the accounts receivable database.

Who completes the form? The position of the person completing the form should be indicated rather than the name of the individual. If it is critical that the form be completed by a particular time, an alternate should be named.

How is the data recorded on the form? Specific instructions should be given. Frequently errors are entered into a system because handwritten material was misread. Individuals completing forms should be required to print or type, and boxes that can be checked should be supplied for codes.

Who authorizes the changes? A change must not be made in records stored in databases without authorization. The person who can authorize the changes, and an alternate, must be specified.

How should the transmittal form be completed? In a distributed data processing environment, transmittal forms are less important than when documents flow from department to department. However, the transmittal form can provide useful information such as when the documents were prepared, entered, and returned for final inspection. The form also contains control information that is needed to make sure all data was processed.

How should the printout returned with the form be used? If a report is printed it is for a specific purpose. The report in the illustration is to be used to confirm the accuracy of the changes made to the records stored in the database. Each change is critical. Assume that the discount code was changed incorrectly. Think of how unhappy the customer would be if 5 percent rather than 12 percent were used.

What procedure is to be used to correct errors? Regardless of how well-trained employees are or how carefully they do their work, on rare occasions errors do occur. The important factor is that there is an established audit or checking procedure that will detect the errors before a disaster occurs. There must also be an established way of correcting the error.

How long should the documents and reports be retained? The length of time documents are retained depends upon factors such as the type of document, company policy regarding certain types of information, and federal and state requirements.

How are the documents and reports to be destroyed? Many documents are highly sensitive and contain confidential information. With our increased concern over ecology, fewer documents are burned. Many are shredded and bundled for recycling.

As you refer to the documentation, you might feel it is unnecessary to indicate what information is to be recorded in each address area. However, consider what would occur if the person completing the form decides that since four lines are provided, a three-line address must be spread out over four lines. Or the person might decide that address line 1 or 2 is to be left blank since the city and state must always be recorded on the third line. The instructions provided for the terminal operator should also indicate how the data is to be recorded.

CHECKPOINT

6. Within a company, who might train new employees or train present employees to work with a new system or procedure?
7. What is the first step in designing a training program?
8. What are performance standards?
9. Why should it be determined whether a training program will be needed on an ongoing basis?
10. Why should the makeup of the audience be determined before the training materials are developed?
11. When would you recommend that some type of programmed instruction materials be developed or obtained from an outside source?
12. In regard to entering sales data, would the same level of performance be expected of sales representatives and managers?
13. How important is it that the managers understand why they are to learn about the database query language and the databases available?
14. Within the general area of EDP, why is in-service training a vital issue?

CONVERTING TO THE NEW SYSTEM

Figure 13-5 illustrates what must be done prior to converting from the old to the new system. If analysts are working on a complex problem and have submitted detailed plans for conversion, when the day finally comes it is very rewarding when everything has proceeded according to schedule.

Tasks Completed Prior To Conversion

- New equipment must be ordered, installed, and tested.

- Necessary supplies and forms must be ordered and on hand.

- Each program in the system must meet its objectives and have been completely tested.

- Valid results must have been achieved from string and system testing.

- All documentation must be prepared and tested.

- Employees preparing data, entering data, operating the computer, or working with the output must be trained to perform their particular tasks.

- A detailed conversion plan must be prepared and approved.

- A detailed report should have been given to management regarding the implementation of the system.

FIGURE 13.5 Tasks that must be completed prior to the conversion to a new system

The outline illustrated in Figures 13-6 and 13-7 might be used in preparing the final report and will contain the items illustrated in Chapter 14 under the heading Final System Documentation. The analyst in charge of the project should make the verbal presentation and be prepared to answer any questions that arise. There should be few questions since some individuals are only interested in obtaining an overview of the system and in the cost-benefit analysis. Other individuals who will be working with the system should have been directly involved in its development and should understand the system.

Checklist for Final System Documentation and Report

1. Cover letter. The letter should provide an adequate introduction to the history and scope of the project.

2. Table of contents.

3. Old system. Although the material should be presented as concisely as possible, the following points should be covered:

   A. Description of the old system.

   B. Systems flowcharts. The flowcharts may be placed in an appendix.

   C. Operating costs.

   D. Problems. All major problems cited should be clearly identified.

4. New system.

   A. Overview of the system. A brief presentation of other alternatives studied should be included.

   B. Specific objectives for the new system. This section should also include how the new system addresses itself to the problems cited for the old system.

   C. Cost benefits to be derived from the new system.

   D. Constraints imposed by management and/or budget.

   E. Systems flowcharts.

   F. Master file or database requirements. An overview of the data stored in the file and the reports available should be presented.

   G. Controls built into the system.

   H. Training programs developed for ongoing training and to perform specific tasks.

   I. Conversion plan.

5. Documentation for each procedure. The following items must be completed and approved. Although referenced in the report, not all items will be included in the report submitted to management.

   A. Cover form. The form identifies the procedure and is used to record all changes authorized and implemented after the procedure is operational.

   B. Systems flowchart and overview of the procedure.

   C. Written overview of the input/output requirements.

FIGURE 13.6 Final documentation and report checklist (Page 1)

D.   Transaction file specifications and layouts.

E.   Master file layout. Master file information can be placed in the front of the manual and referenced.

F.   Output specifications and layouts.

G.   Written description of the procedure which highlights the controls built into the procedure.

H.   Program flowchart. The established guidelines for preparing flowcharts must be followed.

I.   Run sheets. The run sheet must be completed according to the established guidelines.

J.   Job control language requirements. The JCL should be tested and then listed for inclusion in the documentation.

K.   Directions for data entry personnel. The documentation should include (if applicable) the record layout and specific instructions for recording the data.

L.   Testing procedures used. The same procedures should be followed when a procedure is modified.

6.   User documentation. The user's documentation will consist primarily of the material prepared for the standards manual, source documents and forms, and computer printouts.

7.   Evaluation procedures. The ongoing evaluation that will be made to determine if the system meets its stated objectives and to determine the actual costs are equal to or less than the projected costs should be described. It is assumed that interviews, observations, abend reports, and error logs will be used in preparing the evaluation.

FIGURE 13.7   Final documentation and report checklist   (Page 2)

The Final Report

The outline for the report may seem to represent a staggering amount of work. However, most of the items listed were prepared during the implementation phase of the project. The final report presented to the computer policy committee might be compared to the final checkout of a missile on a launching pad. Before the missile is launched, every precaution must be taken to make certain its flight will be successful. Before conversion from the old to the new system is begun, every precaution must be taken to make certain the new system will produce accurate, timely results that meet the stated objectives.

Most of the material listed under item 3 of the checklist is available from earlier reports. An abstract should be prepared for material that contains a great deal of detail. The abstract can be included in the final report and the detailed information can be included in an appendix to the report.

The items listed under 4 are also available. There might be some modification to the material since the cost factors may need to be revised or the objectives may not be as definitive as the ones presented earlier. While projections regarding personnel requirements had been made in earlier reports, at this point a summary of the required training programs should also be included.

The verbal report will probably stress the conversion plan more than any other item. It is important that the plan be developed in minute detail, approved by management, and then implemented according to the plan.

Documentation

The documentation described under item 5 of the checklist is completed as the procedures are being developed. Working copies should be put into a more acceptable final form. Support personnel should assist with the more mechanical aspects of preparing the final documentation. Word processing should be used so that revisions and corrections can be made easily.

The majority of the documentation described under item 5 is designed for the use of CIS personnel. The first few items, the systems flowchart and the written overviews, are less technical and can be used by anyone who wants to understand how the data flows through the system and is processed. Auditors and top management are often interested in reviewing this type of documentation about a new system.

User's Documentation

Often users will develop their own documentation. If the users have been directly involved in *all phases of the project* they will understand the system and may welcome the chance to once again become directly involved. After the documentation has been prepared, it should be reviewed by the analyst and tested on one or two employees before it is put in its final form.

Personnel who are unfamiliar with either the present or proposed system should be asked to follow the steps outlined in the standards manual. If they have to ask questions, or if they make mistakes, the material may be poorly written. The time involved in making the necessary clarifications so that all instructions are understandable and easy to follow is time well spent. The direct payoff will be fewer errors—which in turn, means a saving in dollars.

Online Documentation

There are many advantages of putting the documentation online in a text database that is addressable by using a terminal. The major advantages are:

- *A single copy of the documentation exists.* The user will always be referencing the final, updated version.

- *Ease in updating the documentation.* Word processing can be used to change the text material. The old problem of updating a number of different manuals does not exist.
- *Easier to use than printed manuals.* The material is available when and where the employee working with the system needs the information. Often help menus can be displayed that provide a list of the available documentation.

Conversion Methods

The degree of difficulty in converting from one computer system to another depends on several factors. If the same hardware is being used without any additional control software, the conversion will be easier than when a totally new computer system is being used.

If a conversion is being made that requires the use of a new computer system, a new procedural language, and a totally new approach such as changing from a centralized batch system to a distributed transaction processing system, the conversion may be difficult.

The conversion must be well planned. All subsystems must be up and ready when conversion day arrives. Although there are several approaches that can be used in converting from an old to a new system, the three most widely used conversion methods are: direct, partial, and parallel.

Direct Conversion Although a **direct conversion** is the least expensive method and the one that will produce the fastest results, it is seldom used for major conversions. Direct conversion is simply too risky!

A direct conversion is easy to explain—one day the company processes data by using the old system and the next day it processes data using the new system. If new hardware is being used, the old hardware could be returned to the vendor or sold if not needed for other systems. The company has become totally committed to the new system.

In order to accomplish a direct conversion, everything must be ready to go at the same time. Why is this risky? The point of no return has been reached—there is usually no way of going back to the old system once the initial conversion is made. Although the system has been tested, were all the combinations that could result from the thousands of variables that must be considered actually tested? In theory, yes—in actual practice, it is questionable.

After the conversion, new files are updated; old ones soon become obsolete. The transaction files produced as output from the new system are not in the format, or on the right medium, to be processed by the old system. If for any reason it becomes necessary to return to the old system, there would usually be an insufficient number of personnel available to reprocess the data according to the old method. If the old system were used to reprocess the data and to process some of the new data, the new system's files would become obsolete. Additional programs would be needed to transfer data from the old system into the new system.

When months of study, work, and preparation have gone into the development of a new system, is it worth the risk of failure in order to do a direct conversion? In some cases, where the new computer system is totally incompatible with the system being used, it may seem as if there is no other choice. Usually, however, there are other options, and the new system can be implemented in stages providing the company is willing to keep the old hardware until the total conversion is completed. Although this approach might seem more expensive, it may actually be cheaper since a smoother, easier-on-everyone conversion to the new system can be undertaken.

Partial Conversion   There are several ways to accomplish a **partial conversion**. One approach is to break the conversion into phases. For example, in the sales/accounts receivable/inventory system it might be possible in the first phase to convert the accounts receivable files and to use the new files in the old programs. All of the new programs that provide for the maintenance and security of the files could be implemented and tested under live conditions. The next phase of the conversion might be to implement the changes that affect the inventory system. The third phase would be to implement the changes required for the sales-order processing procedures.

When a large organization is converting to a new system, it might allow one region or section of the company to serve as a pilot study to test the new system. If the pilot study proves successful and when the fine tuning of the system is accomplished, other regions can convert to the new system.

Parallel Conversion   When **parallel conversion** methods are used, the transaction data is input into two systems—the old system and the new system. It is often necessary to obtain additional personnel to perform the tasks needed under both systems.

Advantages of Parallel Conversion   An entire fiscal period (one day, one week, or one month) may be used to test the new methods before a firm commitment is made to the system. All of the testing is done under live conditions. Under live conditions, the documentation and the effectiveness of the training program can be tested. If the new system is defective, data can be processed in the old way until the necessary changes are made in the new system.

Disadvantages of Parallel Conversion   Parallel conversions are costly. The company must pay for operating both systems. Additional personnel may be needed and it is sometimes difficult to obtain qualified temporary help. Unless a different computer system is used for the new system, there may not be enough computer time to run both the old and new system. It may also be difficult to compare the results of the old and the new system since additional features were incorporated into the new system.

Time Required for Parallel Runs   The length of time a company decides to run parallel is determined by a number of factors. At one extreme, when a savings and loan company converted from a batch to an online system, parallel runs were made for seven months. At the other extreme, a company might elect to run parallel payroll systems for one pay period. This means that none of the monthly, quarterly, or yearly procedures would have been tested with live data. However, the more critical aspects of the payroll system would have been tested if all the weekly procedures were run in parallel.

Combining Partial and Parallel Conversions   Test or pilot locations can run parallel until management becomes convinced that the new system produced accurate results. The ideal approach is to run parallel at a typical branch or regional location. This permits all portions of the new system to be tested under live conditions before a major commitment is made to the new system.

Selecting the Method   No single formula can be used to decide exactly how a new system should be phased in. Each situation is different and, based upon the resources available and the magnitude of the change, the best possible conversion plan should be developed.

The type of conversion plan to be used must be considered early in the project's development. Both partial and parallel conversions require a commitment to both the old and new systems until the conversion is 100 percent complete. The conversion plan must contain a timetable stating when each phase of the new system will become operational. It is imperative to the success of the project that a reasonable and accurate timetable is established than can, and must, be followed.

There should also be a "fallback plan." If the system has been well designed and tested, everyone anticipates that the conversion will go smoothly and that there will be no major problems. However, a plan should be developed if it becomes necessary to revert to the old system. If prior thought has not been given to this issue, it can be a traumatic, costly, and dangerous to the reputation of the project team and to the CIS department.

CHECKPOINT

15. Why might there be few questions during the verbal presentation of the final report?

16. Should the employees be trained before or after the final report is given?

17. What one item of the final report will probably be discussed the most and is still subject to the approval of the computer policy committee?

18. When is most of the documentation listed under item 5 of the final report checklist in Figures 13-6 and 13-7 prepared?

19. Which method of conversion is probably the safest?

20. Which is the cheapest and fastest way to convert from one system to another? Would you recommend this type of conversion?

SUMMARY

As with any other phase of the project, the key to a successful conversion is planning. The type of conversion should be determined before the implementation of the system begins. All efforts are directed toward achieving a painless and smooth conversion from one system to another. There is no margin for error.

Before the conversion to the new system can begin, equipment must be ordered, installed, and tested by using the procedures that will make up the system. The string and systems testing must also be done on the hardware that will be used when the system is operational. Throughout the development of the many procedures that make up the system, the documentation should be established and tested.

The users of the system must be trained to prepare the source documents and to verify the accuracy of procedures and various types of output. Exception routines must be available to recover from errors that should not but sometimes do occur.

While the final oral report is a recap of many items already presented, it provides control since actual costs and achievements can be compared to projected costs and timetables. The conversion plan must be presented and approved.

The written report contains a condensed version of information that has already been presented. It also brings together into a logical sequence a great many items that have been developed over a long period of time. From the lengthy final report and documentation, abstracts can be made that meet the needs of management, auditors, users of the system, maintenance programmers, and operations personnel.

DISCUSSION QUESTIONS

1. When should the conversion plan be developed? What type of information should be provided in the final report regarding the conversion plan?

2. What type of string testing might you want to do regarding the programs needed for the inventory subsystem?

3. You are the project leader and it is now time to do the system testing for the sales/accounts receivable/inventory system. What should have been done prior to the system test (be specific)? How would you involve the user? Besides providing a final test for the programs, what else is tested during the system test?

4. Your company is getting a new computer system for the implementation of a complex new payroll/personnel system. Although many of the jobs will be scheduled and run in a batch mode, the maintenance of the database is distributed. The query language will enable authorized managers to extract a great deal of information from the database. Three employees within the personnel department will be doing most of the database maintenance. Explain the type of training program you would provide for management personnel, the individuals who will do the data entry function, and the operations personnel (computer operators and the operation's manager). What type of visual aids would you use? With which groups would you use documentation similar to what is illustrated in this chapter and in Chapter 14?

5. Why are the availability of good training materials and both formal and informal classes regarding current issues in EDP important?

6. If microcomputers had been obtained for all of the sales representatives, what type of training do you feel should initially be provided? If you were the sales manager, what additional type of training might you authorize for the sales representatives who are interested in learning additional ways to use their microcomputers in the performance of their jobs?

7. As the project leader you have been asked to explain to the sales manager the different ways in which the conversion to the new online sales transaction system can be accomplished. Identify the different methods that can be used and recommend the one that you feel CPI should use in implementing their new sales system. Give the rationale for your decision.

8. The sales manager is very negative about having his personnel do the documentation for the sales system that will be used within the department and is also very negative regarding the use of word processing and electronic mail. He stated: "Why document? You'll change the system and then of what value will our documentation be? If you want documentation, your people can do the work!" How would you respond to the sales manager's statements?

9. A friend of yours, Jean Riley, has a small business which employs four people who do the office functions manually. By the first of January, a microcomputer capable of multitasking will be installed and only three employees will be needed. This really doesn't present a problem since one of the employees has indicated she will be leaving because her husband has been transferred to Memphis. Jane has not informed her employees of the change and the company from whom she is obtaining the system (hardware and software) has indicated no training is necessary. The software is all being developed locally and has not been field tested. Yet in some way all or part of each employee's job will change and each employee will be working directly with the system. What advice would you give Jean?

10. The president of CPI has indicated that there is a severe cutback in funds for the CIS department. Although a new computer that has several new software packages will still be installed, all money allocated in the budget for

training will be cut. The president feels that the managers can learn to use the new system ''on their own.'' He also has cut the funds for completing the documentation and has indicated that there will be a direct conversion to the new system. Very shortly both the new online sales system and the payroll/personnel system are scheduled for implementation. Also one of the new software packages that is scheduled for delivery is a rather complex office automation package. If you were the CIS manager, how would you respond to the president's budget cuts?

Team or Individual Projects

1. A new type of data collection device is to be installed throughout the plant as part of an online job cost accounting system. The data entered will affect both the payroll system and the job cost accounting system. Although the payroll manager and assistant manager have been quite enthusiastic about the project, personnel in the job cost accounting area have been very negative about the new system. It is critical that every employee learn to use the data collection device correctly. Also management needs an orientation to the new system. At the present time there are approximately 5,000 hourly employees who will be using the data collection devices. The project leader has been asked to research the type of data collection device that should be obtained.

   Directions:

   a. Research the subject of data collection devices and select one that might be used in a factory environment for recording the raw materials, subassemblies, and labor put into production. Pick a data collection device that you feel will be appropriate and obtain detailed information about the project.
   b. Prepare a detailed outline of the type of orientation and training that you would provide for:
      (1) management,
      (2) supervisors in the plant,
      (3) hourly employees.
      Indicate what type of materials you would prepare or have prepared for each group and what outside resources you might use.
   c. Include in your report to management:
      (1) a report on the environmental requirements for the data collection device;
      (2) the rational for selecting the device; and
      (3) design considerations that will help provide for the security and integrity of data entering the system at the remote locations.

2. You are the owner and president of a small company that will very shortly convert from a manual accounts receivable and billing system to a computerized system. This will be the first application that will be installed on an IBM System/36. Although at the present time you will only install two workstations, you will increase the number of workstations and add additional applications including payroll, general ledger accounts, accounts payable, and word processing. Your office staff consists of an office manager, two accountants, three clerks who work under the accountants, two secretaries, and three clerk/typists. All 11 of the individuals listed will be directly affected by the arrival of the System/36. Although a few of the office employees have taken a course or two in data processing, none of the employees is familiar with the System/36 or has used terminals or other types

of EDP equipment. The equipment including the first application software, will be arriving in three months. In addition to the staff specified, you have three managers that are assigned various responsibilities.

Directions:

a. Determine the capabilities of the System/36 and what type of training programs are available.
b. Outline the type of general orientation that you would provide for your staff regarding the System/36.
c. The systems to be installed will be menu driven, online systems. Outline the type of training you would provide for each group of individuals, who should provide the training, and what problems, if any, might occur.
d. Assume that your secretary, who has worked for you for 15 years and has done an excellent job, has indicated that she "will not learn word processing or have anything to do with the new computer." You feel she is extremely valuable to the organization. How would you handle the situation?

GLOSSARY OF WORDS AND PHRASES

crossfooted Column totals are added to a single total which represents the sum of the individual column totals.

direct conversion The entire new system is implemented and considered operational. The old system is no longer used.

programmed instructional material Self-paced materials that a learner can use without additional instructions.

parallel conversion Both the old and new systems are operational. The output from the two systems is compared and should produce the same results.

partial conversion The new system becomes operational in segments. As each portion is proven to produce valid results, another segment of the new system becomes operational.

systems testing All programs for the procedures that make up the system are tested in a series. Normally the users follow their documentation and prepare the data for the systems test.

STUDY GUIDE 13

Name _____

Class _____ Hour _____

A.   Indicate whether the statements below are true (T) or false (F). If false, indicate in the margin how the statement can be corrected to make it true.

_____   1.   A good motto for an analyst is ''Plan the implementation, and implement the plan.''

_____   2.   A conversion will usually progress more smoothly if the users have been directly involved in the design and implementation of the system.

_____   3.   The quality of the documentation does not affect the ease with which the conversion from an old to a new system can be made.

_____   4.   As part of the conversion process, two or three files may be merged to form a database.

_____   5.   The conversion plan only indicates how the data in the existing files will be reformatted and the amount of new data that will be entered.

_____   6.   When an individual program is tested, live data is used.

_____   7.   Link testing involves the users. Users follow the documentation to prepare documents and enter data.

_____   8.   During the system test, the analyst and programmers are only concerned with the validity of the output.

_____   9.   The first step in preparing training materials or training seminars is to identify the tasks that must be performed.

_____ 10.   The district managers and the sales representatives would receive the same type of training regarding the use of the terminals.

_____ 11.   In making a presentation on the conversion to a new system to the district sales managers, the two topics covered were a general orientation to the sales system and the use of the query language.

_____ 12.   The CIS department is always responsible for training individuals to perform the tasks identified for the new system.

_____ 13.   Colleges, universities, and private organizations that specialize in developing training programs are usually unwilling to tailor the programs to the needs of a particular organization.

_____ 14.   Documentation and specialized training materials are developed after the system is considered operational.

_____ 15.   Direct conversion is recommended over the other conversion methods because it is cheaper and there is less risk involved.

_____ 16.   A partial conversion is only made when the entire system is not yet operational.

_____ 17.   If a parallel conversion is scheduled, it may be necessary to obtain personnel from a placement service that provides temporary employees.

_____ 18.   The conversion plan is prepared after the final report is presented and approved.

_____ 19.   When the final report is given, it is assumed that there will always be a large number of questions and points that need clarification.

_____ 20. The only information shown on a transmittal form is where the documents originated, where and when the documents were transferred to other individuals, and when they were returned to the individual responsible for checking the output.

B.   Matching. For the situation described, indicate what type of training materials or programs would be most appropriate. Record the letter of the most appropriate answer in the space provided.

a. training film	c. slides	e. slide/tape
b. video tape from a source such as Deltak materials	d. transparencies	f. programmed learning

_____ 1. You wish to show how the forms are to be filled out. As you give the instruction to 25 people, you will fill out the form.

_____ 2. One person within your organization is to receive extensive training on database systems.

_____ 3. Everyone within the organization is to receive a general orientation to the new time and attendance system. As new people are hired, they will also receive the same general orientation. The president of the company has set aside a considerable sum for the development of the materials.

_____ 4. Detailed instruction is to be provided to each sales representative. As new sales representatives are hired, often one at a time, the same training will be provided.

_____ 5. Individualized training is to be established for current employees and employees who will be hired. Manuals will be used, but you also want to verbally present some material that will be interspersed with visual images.

_____ 6. Technical material regarding the new job control language is to be taught to two individuals within the CIS department.

C.   Multiple choice. Record the answer that is most correct in the space provided.

_____ 1. The documentation for completing the forms used to add, delete, and change accounts receivable records describes: the utilization of the form; how to complete the form; the transmittal form; utilization of the report; form retention; and
   a. how to operate the terminal.
   b. how to make a correction if the operator keyed in incorrect data.
   c. the error procedure to be followed if the report did not agree with the source documents.
   d. the responsibilities of the sales manager and credit manager.

_____ 2. In designing the customer master file change form, dashes were used rather than solid lines in order to
   a. define how much data could be recorded on the form.
   b. encourage the person completing the form to print.
   c. make the data recorded on the form more legible.
   d. accomplish the objectives specified in a, b, and c.

_____ 3. In designing instructional materials the analyst
   a. only considers who will be using the materials.
   b. only considers how often the materials will be used.
   c. must consider who will use the materials and how often they will be used.

d. must consider the task to be taught, who must learn the task, and whether the materials will be used on an ongoing basis.

_____ 4. In regard to the final report,

a. only the users will be interested.

b. management will be interested in the general concepts of the new system and how projected costs and actual costs compare.

c. there is no need to attach additional documentation as most of the materials have already been presented.

d. most of the report and its attachments are prepared after the rest of the implementation tasks have been completed.

_____ 5. Training materials

a. are always prepared by the analyst.

b. are usually prepared by the user.

c. can be prepared by the analyst, the user, or obtained from outside sources.

d. should only be developed for large audiences.

_____ 6. The conversion plan that is most expensive and involves the least amount of risk is a

a. direct conversion.

b. partial conversion.

c. parallel conversion.

d. combined partial and parallel conversion.

_____ 7. In the final report,

a. only the new system is described.

b. both the old and new systems are described.

c. only the documentation for the user is attached.

d. only the documentation for the CIS department is included.

_____ 8. Today the general feeling is that

a. all documentation should be prepared by the analyst.

b. all documentation should be prepared by the programmers and analysts.

c. operations personnel should not do any of the documentation.

d. users should be directly involved in the design and implementation phase and should be responsible for their own documentation.

_____ 9. Vendor schools are

a. always free.

b. sometimes available both for a general orientation to a computer system and for in-depth coverage of topics such as databases or job control language.

c. always provided at a cost to the user.

d. available only at the vendor's corporate educational center.

_____ 10. The first step in training is to

a. develop the materials.

b. contact the vendor to see what types of programs are available.

c. select the employees who will be retrained.

d. identify the tasks that must be performed.

14 IMPLEMENTATION: DOCUMENTATION AND SYSTEM AUDIT

Looking Ahead

After reading the text and completing the learning activities you will be able to:

- Identify the differences that exist between the three types of documentation that should be available for a computer system or a major application.
- Determine the four major functions served by documentation.
- Identify the documentation used by management, auditors, users, analysts, programmers, and operators.
- Explain the major purposes of internal program documentation.
- Identify the information provided when the playscript method of writing documentation is used.
- Identify the guidelines and standards that should be followed in documenting a procedure or a system.
- Explain the purpose of the system audit and determine how it should be conducted.
- Identify how management, users, the operations manager, and the project leader might evaluate a system.
- Identify the factors that should be included in a comprehensive audit conducted by an EDP auditor.
- Explain how the documentation checklist, cover sheet, and abort form are used.

INTRODUCTION

Why document? The arguments against documentation are usually that it is never accurate and it takes time away from the more productive work such as designing procedures and programming. Who looks at documentation?

There is only one answer to the first question: Documentation is essential to the effective development, implementation, modification, operation, and utilization of any system. In answer to the second question, users, management, analysts, programmers, operations personnel, and auditors all benefit when a system is well documented.

Documentation is designed to serve different individuals and different functions. Once a system is completely documented, different users can abstract from the total documentation the material that meets their particular needs.

How agonizing it must be for a manager of an installation without documentation! What if one of the key people in the system leaves? Do others within the department know how to fill out the forms and work with the output? What about operations? Can a new operator successfully run jobs if there is no documentation? What happens when a job aborts? When a program must be maintained, how does the programmer know what impact a change in one program will have on the total system?

Regardless of the size of the shop, there should be a hard and fast rule which states that *all jobs must be completely documented before they can be considered operational*. Realistic standards must be developed for completing the documentation. The user's documentation, which is often overlooked, must provide more information than merely how to fill in the input form and how to interpret the output.

FUNCTIONS PERFORMED BY DOCUMENTATION

In the past, the worst documentation conceivable was often supplied by mainframe vendors. It took a six-foot (or more) library of manuals to describe a computer and its operating system. Seldom was a good index or glossary available, and the manuals were written using technical terms. The manuals were barely off the printing press when one or more addendums were printed. One IBM lecturer said that if the computer industry was compared to the automotive industry on the basis of manuals, when you drove your new car home the backseat would be full of addendums to the owner's manual in the glove compartment.

Mainframe manufacturers and companies that specialize in developing large software packages have taken a lesson from suppliers of software for microcomputers. Companies such as Peachtree Software, Inc. that develop a wide variety of software for microcomputers have developed documentation that is easy to use and to understand by those who are not highly skilled or knowledgeable in the use of computers.

Today vendors who develop large software packages or computer systems often provide three different types of manuals: user guides, reference guides, and technical guides.

User Guides

The first type is nontechnical and designed for the end user. Documentation for the user must be written in clear, concise English and nontechnical terms should describe the task to be performed. Also, with menu driven software that is now considered a standard, more of the documentation is internal to the program. Once the operator sits down at the terminal to perform a task, directions are available on the VDT.

Reference Guides   Although not task oriented, the **reference guide** should also assume no EDP expertise. The guide is designed to provide nontechnical material about a number of topics and different options that exist within the software or as part of the hardware. One reference guide for an operating system starts out by explaining to the reader that the book is intended for the user who is working with the computer and who needs detailed information about a particular command. The guide provides detailed information about the operating system commands that are *available to the user*. The dictionary of commands is followed by a detailed description of how to use each of the commands defined in the dictionary.

Technical Guides   The **technical guide** is designed for application programmers, system programmers, and analysts. The descriptions of the features included in the hardware or software are written in technical terms. The guide often includes an overview of the hardware or software, the design philosophy, and detailed, technical information regarding the features built into the hardware or the software.

FUNCTIONS SERVED BY DOCUMENTATION   Documentation for systems designed and developed inhouse means different things to different people. Regardless of how people view documentation, they are in agreement that it is used for: instruction, communication, establishing performance criteria, and historical reference.

Instruction   Figure 14-1 illustrates how documentation can be used to provide instruction to users and to operators. It would be difficult to determine which of the functions listed is most critical. If the source documents are inaccurate, the whole system could fail. To ensure the validity of the output, controls must be built into the system. The user's personnel must know how to initiate the controls and to check the final results of processing with the predetermined totals or check figures.

> Documentation tells:
>
> Users how to:
>
> 1. Fill in source documents.
> 2. Complete transmittal forms.
> 3. Establish controls.
> 4. Key information into online screens.
> 5. Check the validity of output.
> 6. Utilize reports.
>
> Operators how to:
>
> 1. Prepare JCL and system commands.
> 2. Set up the job stream so that programs will be executed in the most efficient manner and initiate jobs.
> 3. Interpret and respond to error messages.
> 4. Restart jobs.
> 5. Check the validity of the output.

FIGURE 14.1 A great deal of documentation provides instruction

Unless the documentation used by data entry operators is current, the data recorded on the source documents may be entered into the system incorrectly. If the documentation used by operations personnel is obsolete, invalid control information may be entered and invalid results will be produced.

Any documentation designed to provide instructions should be stated as clearly and simply as possible. With the increased graphic capabilities of the computer, documentation including illustrations can be stored online. When a request is made for information regarding a procedure, well-illustrated directions can be displayed on the VDT. The graphic and voice capabilities of the computer can be used to provide better instruction.

Communication

A number of people, such as managers, auditors, users and job trainees, may reference a portion of the documentation to obtain an understanding of the system and how it was developed. If the documentation is up-to-date, studying it is a good way to learn about the individual procedures that make up the system. Portions of the documentation, such as the systems flowchart, general overview, I/O requirements, file specification, processing controls, and error messages that might occur, provide a broad overview both of the individual procedures and of the system.

Establishing Performance Criteria

The expected results should always be included in the documentation. If the job is to produce a printed report, a sample report should be included. The documentation should also detail all of the various ways that the output is to be checked against predetermined totals. Any control used in checking the accuracy of data processing functions should be well documented.

Often the data control clerk, operations supervisor, or someone within the accounting department will post the current totals to a post-run sheet and maintain a running balance. For example, each pay period, the totals printed on the payroll register are posted and added to the post-run sheet totals. When the payroll summary is printed, the year-to-date totals must agree with the post-run sheet totals. The documentation should describe the procedures to be used for posting the figures to the sheet and how the totals are used. The procedure to be used when corrections are made to the file should also be illustrated.

The documentation should include an explanation and examples of all possible occurrences—normal reports, exception reports, and error messages. By studying the documentation, the operator and user should know exactly what might occur when a procedure is run. If something other than what is indicated in the documentation occurs, a standardized method should have been established for recording the problem. Usually an abend (abnormal job ending) or error report of some type will be filled out.

Historical Reference

Part of the documentation merely serves as a historical reference of what occurred. Much earlier in the text, memos were illustrated that summarized the results of an interview or a conference with the user. It may be necessary at some time in the future to determine the results of an interview, an observation, a phone call, or a survey. Conclusions reached verbally should always be documented to provide a reference as to why a particular decision was made. Anything worth doing is worth documenting!

When a systems study is conducted, a great many decisions are made that are the direct result of a discussion, observation, review of the literature, visits to another installation, interview, or conference. Whenever any activity resulted in implementation of some type of action, there should be sufficient documentation to support the rationale for the decision.

TYPES OF DOCUMENTATION

There are four major types of documentation, each of which is designed to serve a particular need. The four types are systems, program, operations, and user documentation.

Systems Documentation

The **systems documentation** includes the detailed specifications and objectives of the study. In addition, it normally includes memos, letters of authorization, reports, a narrative of the system, file descriptions, input/output specifications, controlling and testing information, and systems flowcharts. An analyst or someone on the analyst's staff is responsible for completing the systems documentations.

Efficient systems can only be designed if well-defined documentation specifications and objectives are established. Yet the systems documentation is generally the type of documentation most often missing.

When systems are being developed, problems frequently arise because of a lack of good systems documentation. Each step in the growth of a system is dependent upon the accuracy and effectiveness of the work and documentation of the present system.

After a new system is implemented, a review of the system is conducted to determine whether the end results of the weeks or months of planning and implementation have produced a system that meets the specifications. If the user's and analyst's specifications have not been documented, management will not have an accurate gauge to evaluate the effectiveness of the new system.

If changes are to be made to the system, the systems documentation should be consulted. If systems documentation is not available, the work of the programmer or analyst assigned to make the changes becomes far more complex.

If a project is postponed midway in its development, the programmers or analysts later assigned to reinitiate the project must start at the beginning if the necessary systems documentation is not available. Time taken to relearn what should have been documented increases the time needed to complete the necessary changes. Valuable computer and operations staff time can be lost if there is poor or nonexistent documentation regarding the relationship between scheduling maintenance runs and the availability of current, up-to-date files (and databases).

Generally the time spent in completing the systems documentation results in better systems being implemented that are far easier to maintain than systems developed in an installation that did not take, or make, the time to complete the systems documentation.

Program Documentation

Programmers generally prefer not to do the programming documentation. Although most programmers would agree that good documentation is necessary, time spent to complete documentation seems less productive than time spent in designing applications or writing source code.

Management is partially at fault for the negative view that programmers have toward documentation since documenting is given a low priority. Programmers are often reassigned to another project before they have completed the documentation for the first project. When programmers are not given enough time to document a program according to accepted standards, they may do just enough to get by and then promise to complete it "when things settle down a little." The problem is often most acute in some of the smaller installations since the time when things settle down a little never seems to arrive.

In shops where programmers are not given time to document according to even a minimal standard, the maintenance programmers and operators soon learn not to have much confidence in the documentation. Instead of using the documentation to provide insight into the program, they ignore the documentation and try to figure out what the programmer was doing from the coding. This takes time, and consequently the cost of maintaining programs is higher than necessary. Since no one uses the documentation when it is inaccurate and poorly done, why bother to conform to standards? Thus the cycle continues. Bad documentation fosters more bad documentation.

Needs of the Maintenance Programmer   Maintenance programmers assume responsibility for programs written by someone else and are expected to understand the functions and logic of the programs. Unless maintenance programmers understand the logic of a program, they will be unable to debug or modify it. While a detailed examination of the code can provide the maintenance programmer with an understanding of the logic, this is a tedious and time-consuming way to determine what the program should do—and is doing.

Even if the authors modify their own programs, what happens if a programmer leaves or becomes seriously ill? Furthermore, anyone who programs much knows that after a few months, or a few years, it is difficult to get back into the logic of a program—especially one that is poorly written and documented.

Programming documentation need not be a burden. If standards that require good programming practices are developed and enforced—including designing programs before attempting to write the code and a reasonable amount of internal documentation—documentation should not be a major problem.

The programming documentation should be completed as the various steps associated with designing and writing a program are accomplished. If this is done, at the end of the project there will not be a huge amount of tedious documentation to complete.

Items Included as Program Documentation   Often the documentation needed for program maintenance is duplicated and put into a separate notebook or folder. If a large notebook is used, a general overview of the system and the major databases should be provided. Then each program that is part of the system has the following items: **cover sheet**, systems flowchart, narrative description of the job, record and report layouts, detailed logic plan, source listings of the programs, description of the test plan used, run instructions, and a sample run.

The cover sheet should be similar to the one illustrated in Figure 14-2. The cover sheet identifies the job, the programmer, and who should be contacted in case a problem occurs. A very important part of the cover sheet is the portion that shows when a revision to a program was made, who authorized the change, who made the change, and why the change was made.

The narrative description of the program describes in nontechnical terms the objectives of the program, the input used, the controls built into the program, and the output produced. Anyone wanting a brief overview of the program and

$C^P I$ **computer products inc.** RUN MANUAL COVER SHEET

PROGRAM NAME	Accounts Receivable Change	LANGUAGE USED	COBOL

JOB NUMBER	AR002	PROGRAMMER M. Aguilar	DATE	OCTOBER 1986	

EXECUTION PROBLEMS CONTACT: M. Aguilar

REVISIONS:

By	Date	Authorized by	Reason for change
Walczak	Sept 1986	Dennis Paulson	Ship to code field and key field added to the customer master file records.

FIGURE 14.2 Cover sheet for the accounts receivable program documentation

what it does should be able to find the information they need by reading the narrative. This page would also be included in the documentation provided for managers, analysts, and users.

The importance of a well-developed detailed logic plan has already been discussed. It is probably worth stating one more time that *whenever the program is changed, the detailed logic plan must also be changed*.

Usually several versions of the source listing of a program are maintained. Assume that you are a maintenance programmer and receive a memo indicating that you are to change the payroll register program. You would obtain the documentation including the source listing, code the required changes, and *test the changes made*. You might elect to highlight the changes made on the source listing and attach the memo authorizing the changes to the new listing. Both lists would be filed with the documentation for the program. If a problem occurs when the revised program is put into an operational status, the highlighted changes would be the place to start looking for the cause of the problem.

Whenever changes are made to a program, the original test data should be used. A few more records may need to be added that will specifically test the changes made. What must also be considered is the impact that changing one program might have on other programs.

The job control language of system commands needed to run the program should also be included in the documentation. Often the job control language needed to run a job is stored in a special library under a unique name. To run the job, the operator enters the name the JCL is stored under. A listing of the required JCL should be part of the documentation on file.

Operations Documentation

Operations documentation is more widely accepted than any other type. Management, the CIS manager, analysts, programmers, and operations personnel all agree there is a need for up-to-date operations documentation.

In a small shop the operations staff might be able to manage the scheduling, data control functions, job setups, operation of the systems, and report distribution without documentation. But what if someone leaves? What if the installation is run on a two-shift basis and both operators leave? Does the manager understand how to run all of the jobs—especially those that have been modified?

What about the large shop where thousands of different jobs run? Very often the programmers and analysts are unaware of the problems created in operations when poorly written programs abort due to data exceptions caused by a lack of editing or by not providing for all of the exceptions that can occur. Often the programming standards developed do not include coverage of various programming techniques that would assist the operations personnel.

There is some duplication between what should be included in the operations documentation and what must be included in the documentation designed for programmers. Most of the documentation listed for programmers should be included in the manuals available within the computer center. However, since most operators do not change source code, the source listings and detailed logic plans are usually omitted. An exception occurs when the operator is in a very small shop that has only one or two EDP employees. The operator might be assigned to make minor modifications to programs, enter data, and do the routine functions assigned to an operator.

In addition to the items listed for the programmers, the documentation for operators has a **run sheet** that includes detailed information regarding: the control procedures, report distribution procedures, **restart procedures**, and error messages that might be displayed during the execution of a program. A typical run sheet is illustrated in Figure 14-3. Operators are normally told not to initiate a job without consulting the run sheet. An installation may have several thousand jobs, and it is impossible for the operator to remember how each is to be initiated, what files are to be used, and how the output is to be tested for validity unless the run sheet is consulted. The operations supervisor may be responsible for seeing that the run sheet is completed for each program.

PROGRAM RUN SHEET

PROGRAM NAME	Sales order processing for mail in orders	FREQUENCY:	Daily
JOB NAME	SR005	TYPE OF RUN	Maintenance Report

INPUT

1. Control record containing the date for the invoices and the beginning and ending invoice numbers are entered from the operator's console.

2. Sales transaction file is read from disk.

3. Accounts receivable and inventory master files. Both files will be updated.

OUTPUT

1. Sales invoices printed on six-part preprinted forms.

2. Control report printed on stock paper lists the total number of invoices printed, the beginning and ending invoice numbers, the total amount added to the accounts receivable master file records, and the total quantity of all items sold.

SPECIAL INSTRUCTIONS	ACTION

1. The job control statements required to execute the job and to access the files are stored in the procedure library under SR005.

2. In case of a job abort, the restart procedures outlined on the attached page are to be followed. The cause of the abort should be recorded in the abort log.

PROGRAMMED MESSAGES:

INSERT SALES INVOICES   Mount sales invoices in printer, align forms, and then enter a 1 on the console.

INSERT STOCK PAPER   Mount single-part stock paper and then enter a 1 on the console.

CONTROLS

1. Total number of invoices printed and the total number of items sold must agree with the totals specified on the transmittal form.

2. The beginning and ending invoice numbers must agree with those specified on the transmittal form.

REPORT DISTRIBUTION:

1. Burst and decollate the sales invoices.

2. All invoices, the control report, and the transmittal form are to be delivered immediately to a sales-order clerk. Before distributing the invoices, the sales-order clerk will visually compare each invoice to its corresponding order.

FIGURE 14.3 A typical run sheet used to provide instructions for the computer operator

Restart procedures are used when a job aborts. Often checkpoint restarts can be included in a program so that if a job aborts it is not necessary to go back to the beginning and rerun the entire program. If programs and systems are completely tested, why are restart procedures needed? In small shops there often is not a backup power supply. If the power fails, the computer needs to be restarted (IPLed—initial program loaded or booted) and the application programs restarted. Sometimes computers do "crash." There can be an unsolved problem with the operating system or a user does something so bizarre that the computer literally throws up its hands and stops! How often have you been at a bank or some-where and been told your transaction cannot be processed because the computer is down?

User Documentation

In an installation that does not have complete and accurate documentation, the likelihood of the user's documentation being current is fairly remote. The documentation in the user's standards manual should provide step-by-step in-structions on how to complete source documents, fill in transmittal forms, and work with the output produced from processing the input. When distributed data entry or data processing concepts are used, it is even more important that the user's documentation be complete. The documentation must be completed before the conversion to the new system. It is vital to the continued success of any system to have the user's documentation updated when changes are made to procedures that the affect way source documents are prepared or the way data is entered into the system.

Examples should be provided that illustrate how any type of form is filled out. The documentation should also provide information on how to solve prob-lems that might arise. For example, what should the payroll clerk do if some of the data needed is not available on the forms being used to fill out the new employee source document? Who should be contacted to obtain the required information?

INTERNAL PROGRAM DOCUMENTATION

With the cost of online storage decreasing and an increasing number of online applications, more emphasis must be placed on internal documentation. The two major purposes of internal program documentation are to:
- make it easier for programmers modifying the program to understand how the data is being processed; and
- assist terminal operators to enter data correctly.

The documentation should be action oriented and written in a clear, concise man-ner. Although documentation will be printed on the source listing, the efficiency of the program will not be affected by the amount of internal documentation. However, a well-documented program is easier and cheaper to maintain than a poorly documented program.

Information About the Program

At the beginning of the program, the internal documentation should summarize what is accomplished by the program, what input is used, and what output is generated by the program.

Each module should include the purpose of the module, how control passes to the module, and where the control returns after the module is executed.

Instructions

Some of the internal documentation may be displayed on the VDT at execution time. Often the operator is told to enter HELP if additional instructions or in-formation is needed.

The programmer who writes the documentation for the operator must be absolutely certain the instructions are clear. For example, one computer-assisted instruction program displayed the following statement:

```
WHEN YOU SEE THE PROMPT, ENTER THE LOAN AMOUNT AND INTEREST RATE.
```

What is wrong with the message? Not everyone knows what a prompt is. Also does everyone know that a comma should separate the amount of the loan from the interest rate? Also, what about the interest rate? Should a decimal be entered?

Experienced operators can bypass the required instructions by not entering HELP. However, the internal documentation should be written as if all operators are *inexperienced*. Assumptions should never be made in EDP. The internal documentation must tell the operator what a prompt is and exactly how the two fields of data should be entered.

PLAYSCRIPT METHOD OF WRITING DOCUMENTATION

Today as you review the documentation provided by manufacturers, software houses, and by various CIS departments, you will see many different formats and innovative ideas. It is not uncommon to see cartoons and color used to make the manuals more attractive. Sometimes the final documentation is written by technical writers rather than by analysts or by someone within the user's department. One innovative idea regarding documentation was developed by Leslie Matthies and published in his book *The Playscript Procedure: A Tool of Administration*. In his book, the playscript method of writing instructions is described. After the task is identified, each instruction provides the name of the actor, the logical step sequence, the key action word, and the action sentence. Although a form with an appropriate heading should be used, the following illustration shows how effectively directions can be written using the playscript method.

Procedure:	*Transmittal of Sales Order to Data Entry*
Sales-order clerk:	1. Stamp the date and a sequential number on the processable documents received from the assistant manager.
	2. Establish a batch total on quantity. Verify the tape.
	3. Fill in the transmittal form. A sample form is attached.
	4. Log out the documents in the sales-order log. The entry must include:
	a. the time and date the documents were sent to data control.,
	b. the batch total on quantity.,
	c. the numbers of the documents transmitted.,
	d. the total number of documents transmitted., and
	e. the signature of the person transmitting the forms to data control
	5. Transmit the documents and the transmittal form to data control. Since interoffice mail causes unnecessary delays, it should not be used.
Data control clerk:	6. Log in documents. The date, time, batch total, numbers of the documents, and the number of documents received are recorded.
	7. Log out the documents to data entry.
	8. Transmit the documents to data entry.

Data entry supervisor:	9. Log in the documents. Record the date, time, batch total, and numbers of the documents received.
	10. Assign the job to an operator.
Operator:	11. Key the data into punched cards.
	12. Accumulate totals on quantity and total number of cards punched.
	13. Confirm the quantity total with the predetermined batch total.
	14. Find errors keypunched into the quantity field.
	15. Correct cards with the wrong quantity.
Verifier:	16. Verify all data punched into the cards.
	17. Stamp the documents to indicate the date and time they were processed in the data entry department.
Supervisor:	18. Confirm the batch total.
	19. Log out the documents.
	20. Transmit the documents and the cards to data control.
Data control clerk:	21. Log in the documents and cards.
	22. Log out the cards to computer services.
Computer service operator:	23. Process cards.
	24. File cards.
	25. Transmit invoices and reports to data control.
Data control clerk:	26. Confirm the batch total and record count.
	27. Audit the sales invoices by comparing the invoice to the order.
	28. Note discrepancies on the transmittal form, and record the date and time the documents were transmitted to the sales department.
	29. Log out the sales invoices, source documents, and transmittal form.
Sales department clerk:	30. Log in the documents and transmittal form.
	31. Confirm the batch totals and record count.
	32. Transmit documents and reports to the assistant manager.

In the format used, the task (transmittal of sales orders to data entry) and each actor is easily identified. Each task is numbered and starts with an action word that is followed by a short, simple action statement.

When any type of documentation designed for instruction is written, testing of the documentation should occur prior to putting it in final form. If the person performing the tasks finds it necessary to ask numerous questions, the documentation is not clear and concise enough.

GUIDELINES AND STANDARDS USED IN WRITING DOCUMENTATION

Written instructions should either be developed by the user or with the direct involvement of the user. The guidelines illustrated in Figure 14-4 should be used.

The checklist illustrated in Figure 14-5 can be used to evaluate documentation or by the individuals responsible for developing the documentation. Obviously not all documentation needs to include tutorials, nor is a glossary or index required for documentation of a single task. However, when developing the documentation for an entire system, the checklist questions will cause the person responsible for the documentation to consider some of the options.

Guidelines for Writing Instructions

- Identify the task or procedure.

- Use forms that provide the necessary heading.

- Include as little stage setting as possible. Start with action steps rather than with long policy statements.

- Identify the individual or group assigned to each task.

- Write the material for the individual who will perform the task.

- Use simple, nontechnical terms.

- Use simple sentences rather than complex sentences.

FIGURE 14.4 Guidelines for writing documentation to be used for instruction

MAINTAINING DOCUMENTATION

The most effective way to develop some of the documentation is to use word processing. The text will be stored in online files and can be displayed on the VDT or printed. Since many computerized systems have been in existence for several years and are constantly being updated, the use of word processing is almost mandatory.

When word processing is used, the text (narrative that describes the system and the instructions for tasks that must be performed) is stored in a file. When changes are to be made, the text is recalled from the online file and displayed on the VDT. The word processing technician can update the text by changing some of the narrative, deleting obsolete material, and inserting new material into the text. If a number of insertions are made, the software will repage the material and insert the new page number references where needed. After the updated information is reviewed on the VDT, the operator can print as many copies as are needed.

When word processing is available, management, analysts, programmers, and users are encouraged to keep documentation current.

FORMAT AND COMPLETION OF DOCUMENTATION

The format of the documentation and how it is packaged may vary since the guidelines used differ from one company to another. Within any one company, however, it is important that a standard format be used. The documentation must also meet the needs of the individual using it.

Once all the documentation is completed, separate reports can be abstracted. The final documentation should be completed prior to the final report on the new system. A checklist similar to the one illustrated in Figure 14-6 should be used to make certain that everything is completed prior to turning the programs (or system) over to operations for scheduling production runs. In Figure 14-6 note

Checklist For Developing Documentation

1. *What is the purpose of the documentation?* Will the documentation be used to provide instruction, for communication, to establish performance criteria, or for historical reference?

2. *For whom is the documentation intended?* Is the documentation to be used by management, by the technical support staff, or by the users? Make sure the material is geared towards the audience for whom it is intended.

3. *How much of the required documentation should be online?* While some of the documentation is well suited for use as HELP screens, other, less-used portions of the documentation or portions that require special forms should be maintained in a printed form.

4. *Can a tutorial be used effectively?* an action-packed tutorial often provides the most effective way of providing instruction. The most effective way to present a tutorial may be as an online, hands-on learning experience.

5. *Should a task-oriented approach be used?* Each task should be explained briefly and then the steps needed to perform the task identified and illustrated.

6. *Is a glossary that defines technical terms needed?* The technical terms that cannot be avoided should be defined in a glossary at the front of the manual.

7. *Is an index needed?* The manual that contains the complete documentation for a system should be indexed.

8. *Is the manual well organized?* The standards for the installation should provide specific guidelines regarding how manuals are to be organized.

9. *Are all questions regarding the system answered?* If the various segments of the documentation are tested by the audience for whom they are intended, omissions should be detected.

10. *Is the size of the manual convenient to work with?* Rather than develop one large, oversized manual, it is sometimes better to divide the material into two more manageable manuals.

11. *What material should be extracted and repackaged as documentation designed for the users, management, or programmers?* Each individual, such as a maintenance programmer, should receive the specific documentation designed for his or her use.

12. *Who is responsible for updating the manuals when changes are made?* When pages or sections are reprinted, someone must be specifically assigned to insert the pages into all existing manuals.

FIGURE 14.5 Checklist for developing documentation

that occasionally all of the major parties (users, operations personnel, programmers, and analysts) must approve certain portions such as the initial design and the final systems test. You will also note that when two or more individuals are responsible for completing portions of the documentation, one person will be listed in the "person responsible" column. There must always be just one person responsible for seeing that a specific task is completed.

CHECKPOINT

1. What three types of documentation are available from vendors of computer systems or major software packages?

Systems Documentation Checklist

SYSTEM NAME:	SYSTEM NUMBER:		
PROJECT NAME:	PROJECT NUMBER:		
PROJECT LEADER:	Identifier: S = Systems P = Programming U = Users O = Operations A = All		
TASK	BY	DATE	PERSON RESPONSIBLE
1. Design acceptance	A		Project leader
2. Procedural specifications completed	U/S		Project leader
3. Program specifications completed	S		Analyst
4. Detailed logic plan developed	P		Programmer
5. Design for program and system testing completed	A/P		Project leader
6. Test files created	P		Programmer
7. Program testing completed	P		Programmer
8. Source document and record retention determined	S/U		Analyst
9. Run sheet and restart procedures completed	P/O		Operations supervisor
10. Data entry documentation completed	P/U		Analyst
11. File layouts completed and cross referenced	P		Programmer
12. Data dictionary entries completed	P		Programmer
13. Operations schedule updated to include new program	O/S		Project leader
14. Training schedule completed	A		Project leader
15. Documentation completed according to installation standards	A		Project leader
16. Systems (acceptance) test completed	A		Project leader

FIGURE 14.6 Checklist for assignment and completion of documentation

2. How does a user guide differ from a technical guide?

3. When documentation is developed for a computerized system, what four functions are served by the documentation?

4. Is the documentation provided for systems personnel the same as the documentation provided for programmers or for operators?

5. What type of documentation must be available for a maintenance programmer?

6. Are source listings and detailed logic plans usually included in the documentation provided operations personnel?

7. Why should most of the documentation designed for the terminal operator be part of the program used to enter the data?

8. When the playscript method is used for providing instructions, what information is provided?

9. What information is provided on a run sheet?

10. According to the checklist illustrated in Figure 14-5, what should be done as the first and second step in developing any portion of the documentation?

11. When two people were assigned the dual responsibility for developing some portion of the documentation, why was a third person such as the project leader listed in the person responsible column in Figure 14-6?

12. Why should the manual that contains the complete documentation for a computerized system contain an index and glossary of terms and be well organized?

SYSTEMS AUDIT

The project should not be considered complete until a system audit is performed. In many ways the system audit is similar to the initial investigation. Someone is assigned the task of evaluating the system. To do this, the key people involved with the system are interviewed to determine their viewpoint regarding the success of the system. The project must be evaluated in terms of the objectives defined for the system and for each procedure.

What constitutes a successful project? There could be many different answers depending upon the type of system being evaluated and who is being interviewed. Management, users, the operations manager, auditors, and the project leader need to be interviewed regarding their concept of the operational system. Often the audit is not completed until the system has been operational for anywhere from six months to a year.

System Reviews

Systems should be reviewed periodically. There should be a stated policy regarding when the reviews should occur. After a new system is operational, a review of the system should probably occur sometime within the following four to six months. At that time the problems (if any) can be discussed. The people working with the system will have a much better idea of the effectiveness of the system than they had when it was being implemented.

A CIS manager of a small company might say, ''Who can afford time to evaluate a system that is up and running with all that needs to be done?'' Management should respond, ''How can be afford *not* to assess where we are and where we are going?''

If statistical data is maintained by operations, users, and sytems personnel, the ongoing evaluation should not take much time. Yet the time spent may result in very cost-effective changes being made.

Management Evaluations   After receiving the final report, reviewing the system documentation, and seeing the results of the system, management will consider the project successful if:

- *Costs were controlled.*   The actual cost of the project's development and implementation, as well as its ongoing cost, should be reasonably close to the projected costs.
- *Target dates were met.*   A PERT or Gantt chart, or similar tool was used to project the completion dates of the various phases of the project. Unless the planning chart was constructed poorly, the work should have been completed as scheduled.
- *The project was well managed.*   Users and management should have been involved and well informed during all phases of the project. Their input should have been received and considered before any major decisions were made. The project should have progressed according to schedule with few conflicts or problems. Everyone affected by the system should have had a clear understanding of the objectives and the resources needed to achieve the stated objectives.
- *Adequate controls and checkpoints are built into the system.*   A clear audit trail must be provided.
- *The project was not oversold.*   The system should accomplish the stated objectives. Users and management should not have been promised features that would not be incorporated into the initial design of the system.

User Evaluations   The user's initial response to a new system may be very favorable. There still must be an ongoing evaluation to provide the mechanics necessary to keep the system updated in terms of the business and economic environment as well as technological changes in the EDP industry. Users will consider the system successful if it meets the following criteria:

- *Easy to understand and to use.*   Systems that have task-oriented, online documentation are often easiest to use. Tutorials should have been available during the training sessions and everyone working directly with the system should have received adequate training. Learning by the trial-and-error method should be avoided.
- *Users were directly involved.*   If users have been directly involved during all phases of the project, their specific needs should have been identified and fulfilled.
- *Solves the problems identified for the old system.*   One of the primary reasons for studying the old system is to determine the cause of each problem so that an effective solution can be developed.
- *Reports contain the proper amount of detail.*   Reports must be easy to work with, contain the proper amount of detail, and be available when needed.
- *Queries can easily be made to obtain additional information.*   Today database systems are affordable and users can learn to abstract needed information. Many CIS departments provide a "help desk" that can respond to requests for additonal assistance.
- *Provision is made for growth and for change.*   Routinely the system should be reviewed to see what changes are needed. There must be enough capacity, in terms of CPU time, terminals, and file space, so that the system can be expanded.

- *The actual cost is close to the projected cost.*   Since it is the user's budget that the developmental and ongoing costs are charged to, the user is usually satisfied if the actual costs are within 5 to 10 percent of the projected costs.

The user must assume responsibility for improving and maintaining the effectiveness of the new system. If reports contain errors, an analysis should be made to determine what created the problem. If additional features should be incorporated into the system to provide management with the type of information that is needed, a request for additional systems work should be initiated.

People should not rely on their memories or remarks made while engaged in casual conversation. A log should be retained, similar to the one used in the computer center, for recording the problems that occur. While it might be easy for the CIS manager to ignore a casual remark such as, "You know, Ben, that report you give me each month seems to contain some errors," it is not easy for the CIS manager to ignore a written summary, based upon the error log, of problems that have occurred.

Conditions change. Management should not feel that a system was poorly designed if modifications must be made once the system is operational. A law might have changed or a new accounting system might be initiated within the company that requires a reformatting of the output produced by the system.

The user is in the best position to determine, on an ongoing basis, the effectiveness of the system. If the system is not yielding a return on the investment made in the development and continuing use of the system, it should be either modified or abandoned.

Operations Manager Evaluations

The operations manager views the system from a completely different perspective than any of the other people who have been involved in the system's development. Operations managers will consider the system successful if:

- *Complete documentation is available.*   Operators must know how to set up batch jobs, respond to messages, recover from errors, restart a job that has aborted, and distribute reports. When transaction processing systems are implemented, the documentation is even more vital to the success of the systems than when a batch system is implemented.

- *Procedures and programs have been well tested.*   Procedures regarding the flow of information must be established, documented, and easy to follow. Each program must be completely tested. Programs are not considered operational until the system testing is completed. Transaction processing programs simply cannot abort.

- *State-of-the-art technology was included in the design of the system.*   Although there will always be some batch jobs, many applications today should be designed as online realtime systems. When query languages are available, many of the reports required "on demand" should be generated by the user rather than be initiated by operations personnel.

- *Backup and disaster recovery procedures have been clearly defined.*   Factors such as "What would happen if a flood or tornado destroyed the CIS department?" must be answered and plans developed to recover from such disasters.

Since it is necessary to schedule operational programs and because operations plays a vital role in a successful project, the operations manager should have been involved in all phases of the project. The operations manager and the personnel directly involved in operations are concerned when jobs must be restarted due to abends that occur because of poor programming. Rerunning jobs because there was a large number of unprocessed records is also an unproductive, and expensive, use of personnel and computer time.

An analysis should be made of the type of errors listed in the error report. Once it is determined what type of errors occur on a regular basis, an analysis can be made to determine how the errors can be prevented.

Figure 14-7 illustrates a form that can be used to record abends. If a check mark is placed in the circle denoting that an applications program created the problem, the CIS manager should have a maintenance programmer determine how the program abort could have been prevented. In a poorly managed installation, job cancellations may occur without being analyzed to determine the true nature of the problem.

When the maintenance programmer changes the applications program, all the documentation should be updated. While it would be unrealistic to assume that a job will never cancel due to some type of programming error, it is just as unrealistic to assume that nothing can be done to prevent the aborts from occurring.

Project Leader Evaluation

The project leader or analyst in charge of the project uses different criteria in evaluating the success of the design and implementation of a system than do managers and users. The project leader will consider the system successful if:

- *The needs of the users were defined correctly.* Users were aware of the problems with the old system and knew what additional features should be incorporated into the design of the new system. Users must define their own needs.
- *Specifications were approved without major changes.* If the project teams followed good procedures in conducting the initial investigation and the feasibility study, the specifications for the new system should have been approved without major changes. When the specifications are presented, the analyst should feel confident that only minor modifications will be requested.
- *Sufficient resources were committed to the project.* The hardware, software, and personnel needed to design and implement a state-of-the-art system were committed to the project. If the specifications or design had to be altered because of limited resources, the original objectives might not be met.
- *The schedule for designing, implementing, and converting to the new system was realistic.* Management did not expect instantaneous results and enough time was budgeted so that users could be directly involved and the documentation, training, and system testing could be done prior to the conversion date.
- *Management was actively involved.* If management is directly involved, there will be fewer problems due to administrative red tape. Not only is the support of management needed but their input will help to develop a better system.
- *Changes were controlled.* After the detailed design report was accepted, very few changes were required.
- *Management and users have expressed confidence in the ability of the EDP professional staff.* As a result of the project, guidelines and procedures were refined so that future studies will go even more smoothly.

$C^{P}I$ computer products inc. ABEND REPORT

PROBLEM LOG:

DATE:

RUN NUMBER:

SHIFT:

TIME: OPERATOR:

APPLICATION: PROGRAMMER:

◯ HARDWARE PROBLEM ◯ APPLICATION PROGRAM PROBLEM ◯ OTHER SOFTWARE PROBLEM ◯ OTHER

BRIEFLY DESCRIBE THE PROBLEM:

Include console messages, program status word, and other pertinent information.

CONSOLE LOG NUMBER: _____

DOWNTIME: _____

RERUN TIME: _____

IF TAPE OR DISK TROUBLES:

INPUT _____ OUTPUT _____

WAS ANOTHER DRIVE TRIED?

YES _____ NO _____

DID THE JOB RUN SUCCESSFULLY ON ANOTHER DRIVE?

YES _____ NO _____

WAS ANOTHER PACK OR TAPE TRIED?

YES _____ NO _____

DID THE JOB RUN SUCCESSFULLY ON ANOTHER PACK OR TAPE?

YES _____ NO _____

ADDRESS OF FAILING DRIVE:

_____

VOLUME I.D. _____ REEL NO. _____

RESOLUTION OF THE PROBLEM:

CE called? Yes _____ No _____

CE received call: _____

CE corrected: _____

Signed by: _____

FOLLOW-UP:

Responsible area: _____

Area charged: _____

Party contacted: _____ By whom: _____

Category code: _____

FIGURE 14.7 A form used for recording program aborts or other problems

The analysts or project team members who were involved in the design and implementation of the system should be responsible for evaluating it on an ongoing basis. When a new enhancement is made to the computer system (either hardware or software), its utilization by the users of some of the present systems should be carefully considered.

When CPI adds the necessary software and hardware to its system so that terminals and telecommunications can be supported, some of the procedures used for entering data into the system will need to be reevaluated to determine whether using terminals will be more efficient and cost-effective than the methods originally incorporated into the system.

If a modular approach was used to design the system, it should be possible to look at any one module to determine whether terminals or realtime concepts can be incorporated into that module. If the system was poorly designed, it may be virtually impossible to modify one module or procedure without creating serious problems.

Auditor's Evaluation of the System

As systems become more complex, management is more concerned over the number of computer-related crimes reported. Today auditors who represent large accounting firms audit through the computer and conduct their audits according to well-established guidelines. A growing number of auditors are specializing in EDP auditing. Although computerized systems are becoming increasingly complex, the use of better-designed auditing techniques and better documentation standards have made it possible for auditors to function more effectively.

If internal EDP auditors are not involved in the design of the system, they should be involved during the testing and evaluation of the system. CIS managers have indicated that the involvement of auditors helps to provide better system controls, reduce fraud, and increase user confidence and satisfaction.

Many auditors still believe they should not be involved until after the system is operational. If they are involved prior to that time, they feel they may lose their objectivity regarding the system. This viewpoint seems to be giving way to the belief that early participation is the key to ensuring adequate controls. Management personnel often support this concept since they feel it is more costly to modify the system to add controls than to design the controls into the system.

A COMPREHENSIVE EDP AUDIT

When a comprehensive EDP audit is conducted it should include evaluation of the following five items:

1. The controls built into the system.
2. Input/output scheduling and controls.
3. Procedures used to provide for the physical security of the computer system—both hardware and software.
4. Controls used to protect the program and data libraries.
5. Backup and disaster-recovery procedures.

In conducting the audit, the auditor will make use of the documentation. Auditors also employ the same techniques used by the analyst to conduct the detailed investigation and design the system. By observing individuals perform the tasks that make up the system, by interviewing key personnel, by studying documents and reports, and by testing the system, auditors are able to provide an objective evaluation.

Auditors should be aware of the difference between the formal and informal organization and should note any variation between the documentation for a procedure and the manner in which the procedure is performed. Also of concern will be the abend reports, and reports showing that check totals and other forms of batch totals were not in balance.

CHECKPOINT

13. Is the project considered complete after the final report is accepted and the system is implemented?
14. What criteria are used by management to evaluate a system?
15. Should an ongoing evaluation of an existing system take much time?
16. What might be the most important consideration when a system or procedure is being evaluated by a user?
17. Why should users maintain a log of the problems they encounter while using different programs or procedures?
18. How does the project leader's evaluation differ from that of management?
19. What three factors does the operations manager consider when evaluating a system?
20. What information is recorded on the form used for program aborts?
21. What does an EDP auditor evaluate when conducting an audit of a system or a procedure?

SUMMARY

Different portions of the documentation will be used by different individuals who will be working either directly or indirectly with the new system. Unfortunately, in some installations there is almost a complete lack of documentation. In other installations, the documentation is inadequate and obsolete.

There should be well-developed standards regarding the amount and type of documentation that must be available. Standards must also be established and followed regarding using, storing, and maintaining the documentation.

A new system must be evaluated to determine whether the stated objectives were met. If qualified analysts and programmers worked with the user and management from the system's conception to its completion, the system probably will meet its objectives.

If adequate resources were committed to the project and good controls were built into the system there should be few, if any, execution-time problems. However, because of the inevitability of change, there should be an ongoing evaluation of all operational systems.

Management cannot ignore the impact of change upon systems and must strive to be aware of environmental and technological changes. It is not always necessary to be the first to utilize a new technology, but no progressive company wants to be the last to put aside old, and perhaps, obsolete methods.

DISCUSSION QUESTIONS

1. What seems to be occurring in regard to the documentation available for computer systems and applications software? What impact did the microcomputer software industry have on documentation?
2. Explain the four major purposes for which documentation is used.
3. How does the documentation used by a maintenance programmer differ from the documentation used by a computer operator?

4. What type of documentation is well suited to the use of tutorials and to being developed as internal program documentation? How should the documentation be prepared and tested?

5. The manager of a very small company that employs two people in its EDP department has stated that "since both employees know their jobs so well, it is unnecessary to document new systems." How would you convince the manager that documentation is needed?

6. In reviewing the guidelines for writing instructions, which ones seem to be the most important? Give the rationale for your answer.

7. In reviewing the checklist for developing documentation, which ones do you agree with? Are there any points listed with which you disagree?

8. How do the criteria used by a manager for evaluating a system differ from the criteria used by the project leader?

9. Why should a system audit be conducted after the system is operational? If the initial system audit is satisfactory, is there any need for an ongoing evaluation?

10. What functions are performed by an EDP auditor? Why is the EDP audit field likely to expand? How does the EDP auditor go about auditing a computerized system?

11. A CIS manager with whom you are acquainted has indicated that all of the documentation should be prepared and tested by an analyst. Tell why you either agree or disagree with the CIS manager.

12. A CIS manager of a department that includes a staff of 12 EDP professionals has just published guidelines regarding documentation. In the guidelines it is stated that for each system there will be one comprehensive manual that contains all of the documentation for that system. The manual will be stored in the office which is shared by two analysts. Whoever is interested in using the documentation can get it from the office. The manager feels it is not necessary to make additional copies of the comprehensive manual. Would you support or reject the manager's policy? If you were a programmer working for the organization, how would you respond to the policy?

13. In reviewing standards to be used for developing documentation, you note there is no provision made for maintaining the currency of the documentation. What policies would you suggest be reviewed for possible implementation?

14. You are the manager of a CIS department and one of your programmers refuses to complete his part of the documentation. The programmer defends his position by stating that "my time is too valuable to waste completing documentation. After all, the programs work and I understand them. What more is needed?" How would you handle the situation?

15. The president of your company has called you in for a conference. You were the leader of a project team assigned to develop a new payroll/personnel system. The system is operational and you feel you followed good procedures in all phases of the project. The payroll manager has complained to the president, but not to you, that the system has more problems and shortcomings than the old system. What points would you bring out in your conference with the president?

Team or Individual Products

The run manual cover sheet, illustrated in Figure 14-2, indicates that a modification was made to the AROO2 program. A ship-to-code field and a field for the key that identifies the record of the ship-to-address were added. The ship-to-address is in a record stored in a VSAM file.

If an S is stored in the ship-to-code, the key in the next field will be used to retrieve the shipping address from the file. A ship-to-address is needed when the billing is to be sent to one address and the merchandise is to be sent to a different location. When this occurred in the past, the sales representatives merely called in the ship-to-address which was then handwritten on the sales invoices. Several shipments have gone to the wrong location and it is obvious that the sales invoicing procedure must be changed.

Directions:

    a. Consider each of the programs required to maintain the accounts receivable file and to print the sales invoices. What changes, if any, will have to be made to each program?

    b. What new programs, if any, will have to be developed?

    c. What source documents, reports, and file layouts will require change?

    d. What documentation will need to be changed?

    e. Outline the procedures that should be followed to: authorize the changes, make the required changes to the programs and documentation, and restore the programs to an operational status.

2. Use several textbooks and current publications to prepare a report on the role of the EDP auditor. Include the type of background the auditor should have, why the demand for EDP auditors is increasing, and the type of activities in which auditors will be involved.

3. If possible, obtain a very old user's guide for a computer system such as the IBM 360 and contrast it with a user's guide obtained for a new computer system. Contrast the two manuals on readability, organization, ease of finding material, and general appearance. Ideally the same type of manual should be used for each. For example, the user's guides for two different text editors might be contrasted.

GLOSSARY OF WORDS AND PHRASES

cover sheet Used to identify the job, the programmer who should be notified in case of a problem, and the reasons why the program was modified.

reference guide Non-technical material covering a number of topics and the options that are available when using the hardware or software specified.

restart procedures Detailed instructions for recovering from a job cancellation. The documentation should indicate if it is necessary to run another job prior to restarting the cancelled job.

systems documentation Contains information regarding the detailed specifications and objectives for the system. Also included as part of the documentation are memos, letters, file descriptions, I/O specifications, controlling and testing information, and systems flowcharts.

technical guide The features of the hardware or software specified are described in technical terms. The guide is designed for the use of EDP professionals.

STUDY GUIDE 14

Name _____

Class _____ Hour _____

A. Indicate whether the following statements are true (T) or false (F). If false, indicate in the margin how the statement should be changed to make it true.

_____ 1. EDP managers sometimes feel that documenting is an unproductive use of an analyst's or programmer's time.

_____ 2. Regardless of the size of the EDP shop, all jobs must be completely documented before they are considered operational.

_____ 3. Often the three types of documentation supplied by vendors are user's guides, reference guides, and standards manuals.

_____ 4. A reference guide should be written for individuals who have expertise in EDP.

_____ 5. Well-developed documentation provides instruction and information, serves as historical reference, and is also used to establish performance criteria.

_____ 6. The expected results are never included in the documentation.

_____ 7. Although not often referenced, the final documentation includes copies of memos and the results of interviews, phone calls, and memos.

_____ 8. The systems documentation is omitted more often than any other type of documentation.

_____ 9. The systems documentation contains memos, letters of authorization, reports, a narrative of the system, file and database descriptions, and input/output specifications.

_____ 10. The maintenance programmer must have documentation that provides an overview of the program, the file and report layouts, a detailed logic plan, and the methods used to test the program.

_____ 11. Since operations personnel sometimes modify programs, source listing and detailed logic plans should be included in their documentation.

_____ 12. The playscript method of documentation only shows the logical sequence of the tasks, the key action word, and the action sentence.

_____ 13. Run sheets are only used to train new operators.

_____ 14. A run sheet usually shows the messages that might be displayed and the action that the operator should follow.

_____ 15. Job control language statements are often stored in a file and called in by the use of a single command that includes the name of the file.

_____ 16. Internal documentation is provided to make it easier for the person working with the program to understand what will occur and also to assist the data entry operator.

_____ 17. If internal documentation is included in a program, it decreases the efficiency of the program.

_____ 18. When word processing is used, changes are more difficult to make than when the documentation is typed and then copied.

_____ 19. Management, users, and the project leader use the same criteria to evaluate a system.

_____ 20. The operations manager is most concerned about the quality and completeness of the documentation, how well the procedures have been tested, and how effectively state-of-the-art technology is utilized.

_____ 21. The abort form only shows the type of problem that occurred and how the problem was solved.

_____ 22. It is easier to maintain written documentation than documentation stored in online files.

_____ 23. When EDP auditors conduct an audit, they are only concerned with the controls that are built into the system.

_____ 24. Auditors are never involved until after the system is operational.

_____ 25. Auditors investigate a system only by interviewing management and observing the workflow.

B. Matching. Documentation is designed to meet the needs of four groups of people. Match the items with the individuals who would find the documentation most useful.

       a. operations personnel   c. systems analysts
       b. programmers   d. users

_____ 1. detailed logic plan

_____ 2. run sheet

_____ 3. system flowchart showing the data flow throughout the system

_____ 4. internal documentation showing how to enter data

_____ 5. abort form or log

_____ 6. record and print layout forms

_____ 7. source listing of the program

_____ 8. objectives for the system and procedures

_____ 9. detailed explanation of how a source document is to be filled out

_____ 10. cover sheet for the documentation for a single program

C. Multiple choice. Record the letter of the best answer in the space provided.

_____ 1. The documentation designed for microcomputer software
     a. had no impact on the mainframe vendors.
     b. was more technical than the documentation created for mainframe users.
     c. paved the way for a trend to make user's documentation less technical.

_____ 2. A reference guide should
     a. be technical and designed for professional EDP personnel.
     b. be task oriented.
     c. provide nontechnical material about a number of topics.
     d. be written in technical language and designed for users.

_____ 3. Documentation serves four functions. It provides communication, a historical reference, performance criteria, and
     a. information.
     b. internal documentation.
     c. instruction.
     d. detailed information for mangement regarding how data is processed by each program.

_____ 4. Documentation is
   a. entirely developed by EDP personnel.
   b. often developed as the system is in the design and implementation phase.
   c. seldom written by technical writers.
   d. always updated when a program or procedure is revised.

_____ 5. A restart procedure
   a. is only used in the event of a power failure.
   b. is developed by the programmer and documented for the use of computer operators.
   c. is only used if the job must be started again from the beginning.
   d. is seldom documented since computers rarely crash.

_____ 6. Once all parts of the system documentation are completed
   a. they are indexed and used primarily by analysts
   b. sections may be extracted and "packaged" for users, management, operations personnel, and maintenance programmers.
   c. they should not be changed.

_____ 7. Management will consider a system successful if costs were controlled, target dates were met, the project was well managed, and
   a. the documentation was completed on time.
   b. adequate controls were built into the system and the project was not oversold.
   c. no new equipment had to be obtained.
   d. no additional suggestions were made when the final report was presented.

_____ 8. In the checklist or guidelines to be used for developing documentation, the first two major considerations are
   a. the amount of internal documentation that should be provided and the most effective way of using tutorials.
   b. how the material is written and what type of terminology is used.
   c. the function the documentation is to serve and for whom it is intended.
   d. the type of glossary and index that should be provided.

_____ 9. Should be consulted by the operator before initiating a program.
   a. abort log   c. run sheet
   b. cover sheet   d. overview of the program.

_____ 10. Ongoing evaluations should be conducted because
   a. the system may not have been completely tested and "bugs" might still remain.
   b. the initial system audit was unsatisfactory.
   c. the objectives for the system were not met.
   d. there are dynamic changes being made in computer technology, environmental conditions, and within the organization.

Appendix I

CPI CASE STUDY

CHAPTER 1: Background Information on CPI and Its Computer System

You will be asked to perform some tasks and make decisions for Computer Products Incorporated (CPI). To do this you will need to understand how the company and its data processing department developed. Although CPI has been involved with computers for over 30 years and is a relatively large corporation, its computerized applications are batch rather than online systems.

EARLY HISTORY

The company was established in the early 1950s by Mr. Greenlee. Office equipment, furniture, and supplies were sold to retail stores. They also sold directly to large corporations such as General Motors and the Wickes Corporation. At that time, the name of the firm was Greenlee Office Equipment and Supplies. The company expanded very rapidly and began to investigate better ways of processing its financial data. In the 1950s some of its data was processed by a service bureau.

Punched-Card Equipment is Obtained

In 1961 punched-card equipment was obtained. Data punched into standard 80-column cards was processed by electromechanical equipment controlled by wired control panels. This was referred to as *unit-record equipment*. Machines were available that could record, sort, condense, and calculate data. Jobs required a great deal of card handling and the equipment's calculating and printing capabilities were limited.

Applications Are Transferred to a Computer

In 1963 a second-generation computer was acquired that had a CPU with 8000 words of memory, two tape drives, a card read/punch unit, and a printer. It was programmed in a low-level assembler language and did not have an operating system. Only batch jobs could be run and direct-access storage (magnetic drum or disk) was not available. Since direct-access storage devices (magnetic drum or disk) were very expensive, their use could not be cost-justified. The vendor provided training for the data processing manager, programmers, and operators. From within the firm, two individuals were selected to be trained as programmers. Although both individuals expressed an interest in programming, they had no prior experience with electronic data processing (EDP). Neither the data processing manager nor the operators had any prior experience with computers.

The applications were not redesigned. Punched-card applications run on the unit-record equipment were transferred to the computer as fast as the programs could be written. In the years that followed, the 1963 computer system was updated and expanded by adding disk drives and more memory. In 1969 it was necessary to secure a new system. A third-generation system was obtained which had an operating system, utilities, and more powerful, high-level programming languages. Once again the applications were transferred from the old to the new equipment without being redesigned to take advantage of the new technology.

THE COMPANY IS REORGANIZED

In the early '70s the founder of the corporation died and a new president, Matthew Dwan, was hired from outside the company. Dwan immediately implemented a number of changes. Perhaps the most dramatic change was to increase the products to include an extensive line of computer-related items. Since a large number of firms were obtaining minicomputer systems and developing data processing departments, Dwan could not have picked a better time to introduce the new products.

The sales representatives received inhouse training regarding the new products. The company name was changed to CPI so that there would be a greater emphasis on the new line. Since most of CPI's customers were also expanding their product line to include computer furniture, equipment, and supplies, the new products met with instant success.

During the '70s CPI's third-generation computer equipment was updated by adding more memory, faster I/O devices, and more online storage. It was apparent that the equipment no longer met CPI's data processing needs. All jobs were run in a batch mode. The variable data was recorded into punched cards, verified, and then processed. Because of the length of time CPI had been involved with computers, Dwan felt its EDP system should be far more advanced.

A TASK FORCE IS APPOINTED

In 1979 Dwan appointed a special task force that was assigned the following responsibilities:

- Data processing applications were to be studied to determine what methods and procedures were being used.
- A study was to be made of alternate ways of performing the typical EDP recordkeeping functions. The president was especially concerned about sales-order entry processing.
- Long-terms goals and objectives for EDP were to be established.
- Recommendations, based on the long-term objectives, were to be made regarding the type of new equipment that should be acquired.

Although the task force reported its recommendations in 1981, it was not until 1982 that equipment specifications were prepared and distributed to the computer vendors. Each vendor was asked to submit bids based on the needs identified in the specifications. Since the only individuals within the organization who were knowledgeable regarding computers were the members of the EDP department, the EDP manager and his staff were to make recommendations regarding the equipment to be obtained. The EDP department had a reputation for being disorganized and slow to respond to requests for system studies or for assistance. Therefore, the new equipment was not selected until late in 1982 and was not delivered until the early part of 1983. Once again old systems were implemented on new hardware.

CPI'S EDP EQUIPMENT   As illustrated in Figure CS1-1 a state-of-the-art computer system is available that can be used to support online processing. A comprehensive operating system was obtained with the computer that supports batch processing, timesharing applications, and teleprocessing. If application software were developed, sales representatives could submit their orders directly into the system for processing. The long-range goals and objectives indicated that some of the applications should be developed as online systems while others should be executed in a batch mode.

Hardware Obtained   CPI obtained the following equipment that is now functional:

CPU	The CPU currently has four megabytes of memory. A megabyte is one million bytes of memory. As the need develops, four additional megabytes of memory can be added.
Display console	A CRT/keyboard console is available. On the CRT a maximum of 24 lines of 80 characters each can be displayed. The keyboard is used to communicate with the computer.
Direct-access storage	Three disk drives with removable packs are available. Each pack can store 300 megabytes of data. If online applications are developed and additional online storage is needed, five more disk drives can be added to the disk controller.
Printer	One 1100 line-per-minute printer is being used. However, as new applications are developed it may be necessary to obtain additional letter-quality printers that will be installed at remote workstations.
Magnetic tape drives	Two tape drives are available for backing up files.
Card reader	The model selected can read 1000 cards per minute. As soon as applications are converted from batch to online applications and the jobs that will still run in a batch mode are redesigned, the card reader will be phased out.
Card punch	Cards are punched row by row at the rate of 300 cards per minute. As soon as possible, the card punch will be phased out.
Terminals	At the present time there are 14 **hardwired terminals** that are located in or near the computer center. From each terminal, immediate access can be gained to the computer. Unfortunately the terminals are rarely used.

If the sales-order system is redesigned, it may be necessary to obtain 150 dial-up terminals or microcomputers that are capable of being used as terminals. Each of the 150 sales representatives in the field will have a terminal.

The system can only support 64 **ports**. Since 14 are dedicated to handling the hardwired terminals, less than 50 ports would be available for use by the **dial-up terminals**. If an individual dials the computer and all ports are in use, a busy signal will be received. However, it is unlikely that all of the ports will be busy when a sales representative wants to place an order.

Controller   Because the system is oriented to teleprocessing, a **controller**, or **frontend processor**, that has 64 ports is already available. It is possible to add an additional controller with 32 ports. However, 96 terminals is the maximum number that can be supported by the present computer system.

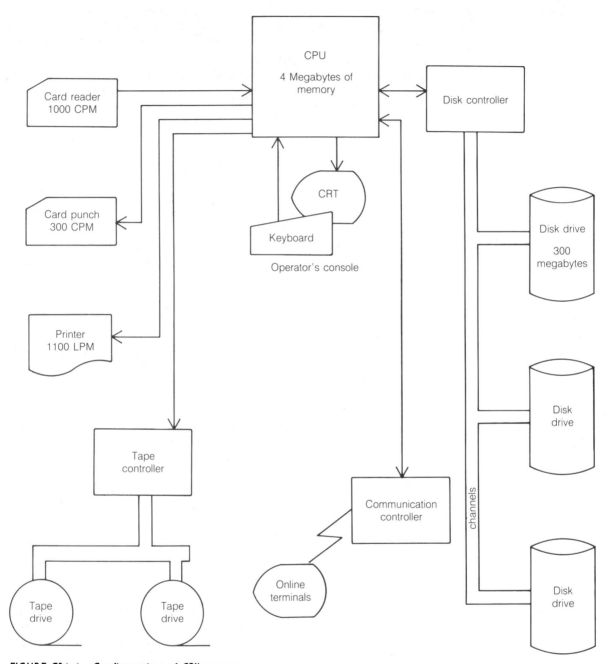

FIGURE CS1-1   Configuration of CPI's system.

Some of the details presented regarding the hardware may not be meaningful until you have covered more of the material presented in the text. You should realize that CPI has state-of-the-art hardware that is not being fully utilized. At the present time, **keypunchers** and **verifiers** are being used to transcribe the data recorded on sales orders and other source documents into a machine readable format. As applications are redesigned, the keypunchers and verifiers should be phased out.

CPI's computer is a high-performance computer ideal for routine EDP applications, office automation, CAD/CAM, and realtime applications. Whenever the need arises, the system can be upgraded to a larger model. This would increase the size of the memory, allow the addition of more disk drives, and increase the number of ports available. Since the 9250 is a more advanced model, the internal calculation speeds are faster than those reported for the model CPI is now using.

Control Software The computer's operating system supports both transaction and batch processing. The operating system also supports the communication controllers, and additional software is available that allows a network of computers to be established. The basic operating system provides good file handling and protection of users' files. A job-account system is also included and both multiprogramming and timesharing are supported. Since jobs running in a multiprogramming environment must share the resources of the system, an internal priority system was established.

When batch applications are run, jobs requiring punched-card input are normally loaded in from the card reader. Jobs that do not require punched-card input are called in by using a terminal to enter the appropriate system commands. Jobs requiring punched-card input, punched-card output, or printed output are executed in three phases:

1. The job stream (job control statements and data cards) is read in and stored in a queue (file). A queue is an area set aside on disk for the job stream.
2. According to a predetermined priority, the jobs are loaded from the queue into the computer's memory for execution. Output to be punched into cards or printed is stored in the punch or print queue.
3. After the job execution has ended, the output stored in the print or punch queue is printed or punched.

The operating system has programs that maintain and service the libraries. By using the correct job control language (JCL) statements or system commands, programs and source statements can be catalogued into or deleted from the libraries. Elements can be copied from one library to another and the location of the libraries changed. The directory of any library can be printed. Stored in the libraries are application programs, subroutines, commands that establish job streams, compilers, error routines, and other control software.

Utility programs are available that make it possible to do various functions such as copying or displaying data stored in files. A sort/merge program can be executed by supplying the necessary information such as which files are to be merged and the fields to be used in sequencing the records. Numerous routines are available for the detection, analysis, and recovery of the machine and system functions. System/operator communication is provided through the use of job control language and system commands. Messages that provide the operator with information regarding the status of the jobs and the computer system are displayed on the system's console.

The text editor software provides for creating new files, editing (changing) data stored in existing files, adding records to an established file, and deleting records from a file. Office automation software is included that makes it possible to do many of the tasks normally associated with word processing. However, other than the EDP staff, no one uses the text editor and the program that prints or displays the data.

Files stored on disk can be organized and accessed according to several different methods. File protection and security are achieved by using JCL and system commands. In addition, a system of account numbers and passwords protects each user's files from unauthorized use. The operating system provides a number of routines that relate to file handling such as unblocking of physical records and determining whether the right file of data is being processed. The operating system determines where each file is stored on disk. When files are backed up, the files are reorganized so data that is used most often can be accessed more rapidly than data that is seldom used.

Additional Software Obtained for the System

Although CPI has not yet developed transaction processing systems, additional software was obtained that would provide for office automation and database management. Dwan is concerned that no one seems interested in learning to use the following available packages:

- Database software. The database software has been used by many different companies and includes programs to define the database, process the data stored in the database, and retrieve data from the database.
- Networking software. Various options are available which include networking mainframes together or interfacing minicomputers or microcomputers with the network.
- The office automation software combines word processing, management communication and support, and advanced text management with EDP.

SUMMARY

Although CPI has a computer system with a great deal of potential, the applications being run still resemble those run on the unit-record equipment. As the company upgraded from a second-generation computer with 8K (K is 1000) of memory to a third-generation computer with an operating system, more memory, and faster I/O devices, the systems were not redesigned. Again when the third-generation computer was phased out and a fourth-generation computer system obtained, the old applications were merely transferred to the new computer.

Although long-term goals and objectives were developed by the task force, little progress has been made. Few individuals within the organization seem to be interested in investigating the possibility of developing online systems or in using the computer for anything other than simple recordkeeping functions.

While the company has had outstanding growth in sales and has developed an excellent reputation for service, its EDP leaves a great deal to be desired. The president of the company, Matthew Dwan, is very concerned that nothing has been done regarding the implementation of the long-term goals and objectives.

Case Study Discussion Questions

1. Do you think that the development of data processing at CPI might be similar to what has occurred in other companies? What information are you using as the basis of your decision?

2. At the present time, what are some of the problems that seem to exist within CPI's data processing department?

1. Contrast the computer system that you have on campus with the one described in the case study. Determine for your system:
   a. the amount of memory being used and the amount that can be added.
   b. the amount of online storage that is available.
   c. the number of tape drives and functions for which they are used.
   d. the lines per minute capabilities of your printer. (Also find out if there are other printers located in work areas.)
   e. the number of terminals being used and how many more can be added to the computer system. (Also find out which terminals are hardwired and which are dial-ups.)
   f. the major features included in the operating system and the other major software packages being used such as database software.
   g. whether the mainframe is networked with other computers or not.
2. Review the material presented in both the chapter and the case study and then brainstorm the questions listed below under a and b.
   a. Why did a dynamic company like CPI that was so successful in sales and marketing make so little advancement in the utilization of computers in processing data and in providing more useful information for management?
   b. If you were a consultant hired to suggest solutions for some of the problems identified, what recommendations might you have for the company?

Directions

1. Brainstorm both of the questions. When you brainstorm a given topic you are not concerned about how practical your ideas are.
2. List the ideas that come forth as the result of brainstorming the questions.
3. Evaluate your list and indicate which reasons for CPI not developing good application software seem the most realistic.
4. Evaluate your recommendations and indicate which ones should be implemented.
5. Keep both lists. As you progress through the text and the case study material, see how you might add to, delete from, or change the problems you identified and the recommendations you would make.

Note: Words in boldface type are defined in the end-of-chapter glossary.

CHAPTER 2: Background Information on CPI's Computer Service Department

Figure CS2-1 is the organizational chart for CPI's computer information services department. Although not indicated on the chart, the CIS manager is also the systems development and maintenance manager. Since a team approach is used and the individuals assigned to the team do the investigation, design, and implementation, members of the system's development are classified as programmer/analysts. The programmer/analysts also do the maintenance programming.

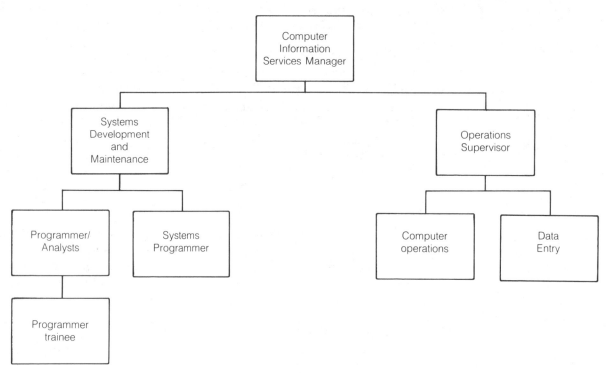

FIGURE CS2-1 CPI's Computer Information Services Department.

The operations supervisor does the tasks usually assigned to the data control clerk and to the scheduler. The computer operators perform the functions normally assigned to a librarian.

If anything, the department is understaffed. Perhaps this is why the ''fire fighting'' continues long after other CIS departments in comparable organizations have settled into a more relaxed and productive schedule. The staff consists of:

CIS/Systems development manager	Dennis Paulson
Operations supervisor	June Mitchell
Programmer/analysts	Kay Walzcak, Glen Nichols, Martha Aguilar, Mike Arnold
Data entry	Cindy Page, Peter Gomez, Nancy Green, Jan Gorney
Computer operators	James St Clair, Sue Henderson
Programmer trainee	Kim Mead

THE PROGRAMMING STAFF Kay Walzcak does the systems programming while the other three programmers are concerned with the development and maintenance of applications programs. Kay has been with the company for a short time. She has a degree in computer science but has a very limited background in business or in any phase of management. In brainstorming sessions, Kay usually comes up with some good ideas that are rarely implemented. There seems to be resentment regarding Kay's background and her being hired as a programmer/analyst.

Arnold and Aguilar have each been with the company for 15 years. They were hired as accountants. Just before Greenlee, the former president of CIS, died, they transferred into data processing. Greenlee was more concerned that employees were given an opportunity for advancement than he was with how well qualified they were for the positions. Although Arnold and Aguilar have bachelor's degrees, they have had very little course work in data processing, computer science, or management.

Glen Nichols has been with the company for five years. Prior to that time he worked for an old, established company that was still using a second-generation computer. Nichols had obtained a BS degree in Information Science. However, since obtaining his degree, he has done very little to keep current regarding changes in computer technology. Glen might be described as a good, solid programmer who is very careful to test and document his programs. However, he doesn't seem to be creative or comfortable in working with the other programmer/analysts.

Kim Mead is the newest member of the department and has an excellent background in computer technology, accounting, management, and statistics. Kim also belongs to several professional organizations and often comes back with ideas that were presented at meetings or conferences. Others in the department are not receptive to Kim's ideas or suggestions. They seem to have the idea "why change?" As far as most members of the department are concerned, things are going well. Users make few requests and there has been little pressure to develop online systems or to help users develop additional applications.

None of the present staff has had experience in working with databases, computer networks, microcomputers, or areas relating to office automation. Kim and Sue have expressed an interest in attending seminars or going to some of the schools provided by the vendor to become more knowledgeable regarding the software packages available on their computer system. However, the previous manager would not provide the necessary release time or any financial assistance.

OPERATORS

Both computer operators have associate's degrees in computer science and are working on bachelor's degrees. Dennis Paulson, the new CIS manager, feels they are intelligent, hard working, and capable of accepting more responsibility. They seem to have a good understanding of the operating system, utilities available, and system commands, and can program in the languages currently available on the system. In contrast to the programmers, they seem to be more energetic and motivated.

DATA ENTRY OPERATORS

The data entry operators are responsible for keying in data. All four operators can operate the data recorders that punch and verify cards. When the new system was installed, they were trained to use the terminals. At the present time only a limited amount of data is entered into the system by using terminals. The operators have excellent motor skills and are good operators. However, they have expressed no desire to learn additional skills. Cindy did enroll in some courses but dropped the courses because of personal problems at home.

DENNIS PAULSON— THE NEW MANAGER

The manager, Dennis Paulson, has been with the company for two months. He came to CPI because he felt there was an opportunity to expand the department and to develop a number of new applications. Dennis Paulson has an excellent background and previously worked with a large company that had a well-developed management information system. His former company stressed the use of standards and documentation. Paulson has a degree in Computer Information Systems and has worked in a number of supervisory positions. He has good managerial and leadership skills. However, he has been told that he should develop better interpersonal skills and a different leadership style. His leadership style has been very dogmatic. When he becomes enthusiastic about a new project, he cannot understand why others do not share his enthusiasm.

During the past two months, Paulson had been learning about CPI, its product line, organization, goals, and objectives. He has reviewed the personnel files of each of his employees and has also conducted in-depth interviews in order to determine how the employees view their position, relationships with others in the department, and objectives concerning EDP. He has become aware of some of the conflicts that exist between various staff members.

THE PREVIOUS MANAGER

The previous manager of the CIS department retired after 35 years of service with the company. He had been an accountant until CPI acquired unit-record equipment. Since he expressed an interest in the equipment, he was sent to school and was asked to develop a data processing department. Each time new equipment was acquired, he was sent to school. However, he rarely passed any of the information he received regarding the features incorporated in the design of the new computer system on to other employees.

The employees within the department were very loyal to him and enjoyed the relaxed conditions under which they worked. The users felt he did a fairly good job of designing systems. When problems occurred, most users felt they were hardware related rather than software related. In looking back on his 35 years with the company, the former manager felt satisfied. Although he was aware that much more should be done in the area of EDP, for the past several years he hesitated to make major changes since he might not be around to see the completion of the projects.

MAJOR CONCERNS

Although many positive things could be said about the department, Paulson has some major concerns:

- Standard manuals are not available for the CIS department.
- The documentation consists of source listings, obsolete run sheets, and a few handwritten notes.
- Although there is a schedule board that shows when jobs are to be run, the operators seldom follow the schedule since there is a constant demand for "got-to-have-it-now" reports.
- A few managers have indicated that they want additional computer services, but there does not seem to be a systematic way of determining priorities.
- Although no error log is available, from observing the operators, it is obvious that many of the jobs abort.
- Second-generation applications are run on a fourth-generation computer.
- Adequate controls have *not* been built into the system. Often the accountants complain about errors they find in the output.
- Users have not been involved in the development of their applications.
- The terminals that are available are seldom used. Too much input is entered into the system by using punched cards.
- Databases have not been developed. The users seem to be totally unaware of the capabilities of the database management software.
- Top management is unaware of the capabilities provided by the office automation software that is available on the system.
- Most of the programmer/analysts are working with second-generation skills in a fourth-generation environment.

DENNIS PAULSON'S OBJECTIVES

The president of the company, Matthew Dwan, is well aware of Paulson's capabilities and has indicated that Paulson can hire three new staff members and reorganize the department. Dwan already approved the name change to CIS. He would also like to have Paulson organize a computer policy committee. Dwan feels this will help Paulson develop a good working relationship with top and middle management. Paulson and Dwan worked out the following objectives for the CIS department:

- Develop standards for designing systems, implementing the design, and evaluating systems.
- Reorganize the department in such a way that it functions more effectively.
- Upgrade the skills of the CIS staff.
- Provide in-service training for top and middle management.
- Develop a plan for decentralizing EDP and developing transaction processing applications.
- Develop a plan for studying current office procedures in order to determine which functions can be automated. Dwan is very concerned that so few of the office support staff utilize word processing.

The president realizes that it will take Paulson a year or two to accomplish all of the objectives. What he wants Paulson to do is to develop a tentative plan for accomplishing the objectives and to indicate the problems that Paulson feels must be solved first. He has also cautioned Dennis about making changes too rapidly. The president feels that since there is so little documentation it would be unfortunate if any of the existing staff decided to leave.

Discussion Questions

1. For each of the concerns, determine the possible causes.
2. For each of the concerns, what recommendations would you make that might help to solve the problem?
3. Considering the various factors that motivate individuals, if you were the employees, what concerns would you have?
4. Considering the various factors that motivate individuals, if you were the CIS manager, what concerns would you have?

Team or Individual Projects

Based on information presented in Chapters 1 and 2 and information found in other textbooks, current publications, and from interviewing individuals involved in EDP or in management, develop solutions for the projects listed below. Keep in mind that there is no one right answer for any of the projects. If you were to contact five EDP managers, you would probably have five different solutions for each of the projects.

1. Considering Paulson's concerns and the objectives he was given, develop a list indicating which problems are the most critical and explain briefly how they might be solved. You may assume that the president indicated your plan of action should be an 18-month plan. In developing the list of problems and their solutions, consider:
   a. Which of the problems are most critical and need an immediate solution.
   b. The steps that would need to be taken to solve the problem. What additional information or support would you need from the organization in order to implement the solution?
   c. The points that would need to be considered in order not to compound, rather than to solve, the problem.

2. At the present time CPI has vice-presidents for marketing, production, research and development, finance, personnel, public relations, and environmental control. The vice-presidents are line officers who function as an advisory committee to Dwan. Dwan believes in structure and feels there should be little difference between the informal and the formal organization. Therefore, as changes are made, job descriptions, manuals, and organizational charts are changed.

   a. Develop brief descriptions for the three new people you would like to hire. In developing the descriptions, determine what voids you feel exist in the CIS department and keep in mind that no one individual can be an expert in all areas of EDP.

   b. Based on the descriptions for the three individuals, indicate how you would reorganize the CIS department.

   c. Indicate what additional changes in staff you would recommend as EDP becomes more decentralized and online applications are developed. In preparing for the changes you would recommend, what type of in-service training programs would you begin to develop? What problems concerning staff might develop? What steps would you take to prevent those problems from occurring?

CHAPTER 3: Developing Gantt and PERT Charts

Review the material in the CPI case study for the first two chapters. If you were assigned Project 1 in Chapter 2, review your solution. Now assume that you are Dennis Paulson. You have had your interview with Matthew Dwan regarding your objectives. Since you would like to impress the president of your company, you would like to achieve the objectives in one year. During that year, you feel the major thrusts should be to reorganize the CIS department, to provide in-service education, and to work more closely with the other departments.

Paulson is hopeful that one small application could be redesigned as an online system and serve as a model for other systems. You are very concerned about the lack of controls built into the system, documentation, and standards.

As you complete the projects, keep in mind that in systems work there is no one source of data that you can refer to in order to determine a solution to a problem. Analysts must learn to gather data and then analyze the data they obtain in order to arrive at a solution.

Team or Individual Projects

1. Using the material presented in Chapters 1 and 2 of the CIS case study and the material provided in Chapter 3 of the text:

   a. Develop a Gantt chart which shows your one-year plan for meeting the objectives defined in Chapter 2 of the case study. Determine what activities can occur concurrently and which ones would need to be completed prior to a new phase of the projected plan. Before doing the Gantt chart, make certain you have listed all tasks in a manner similar to that used in Figure 3-1 on page 90. You may also wish to look in the index and see where other examples of Gantt charts are illustrated so that you have a better understanding of how they are developed and used.

   b. Select someone from the present CIS staff to work as your assistant. You feel you should work closely with someone so that the success of the project is not dependent on one individual.

    c. Develop a PERT chart using the same tasks and time elements.

2. If you were a CIS manager, would you recommend that your staff use PERT charts, Gantt charts, or a planning chart developed on a weekly planning calendar like the ones available in most office supply stores? Give the rationale for your decision.

3. One of the major tasks that you identified as a high-priority item is the development of standards. Refer to standards in the index and read the material presented in the text.

    a. Develop a key question and exit criteria for standards.

    b. List the tasks that you feel must be completed to meet the exit criteria. For each of the tasks you identify, develop a key question and the exit criteria.

    c. List all of the resources that you would use in developing the standards.

CHAPTER 4: Preparing a Report on the Requirements for a Decision Support System

At the present time, all of CPI's systems are batch oriented and both data entry and data processing functions are centralized. Paulson feels strongly that once standards have been developed and both the users and CIS staff have received in-service training, online systems must be developed. Paulson also feels it is better to establish long-range goals that might not be obtainable in order to reach the objectives established each year.

In order to make the staff more aware of the types of systems that are being designed for some of their competitors, Dennis asked Glen Nichols to prepare a report on decision support systems. Dennis has asked Nichols to do some independent research by reading some of the literature available. In addition, Nichols is attending a conference designed to provide a better understanding of DSSs.

Paulson has asked Nichols to prepare a brief report on DSSs. The report is to be prepared prior to attending the conference. Paulson felt that if Nichols has a better understanding of DSSs, he will get more out of the conference. When Nichols returns, he is to put the final touches on his report and make a presentation to the CIS staff as well as to some of the users.

Directions:

Assume that you are Nichols and research DSSs by reading articles printed in current publications.

1. Prepare a three- to five-page summary of your findings. Use at least five references in preparing your report.

2. Prepare an outline of the presentation you would make to the CIS staff.

CHAPTER 5: A Survey of Office Procedures at CPI

Three months ago, Dennis Paulson developed a Gantt chart to show how he would achieve his goals of reorganizing the CIS department and developing standards that can be used in all phases of the systems development life cycle.

The plan included a detailed outline of the in-service training program provided for the EDP staff and users.

Matthew Dwan was very impressed with the ideas that were incorporated into the plan and the way that Dennis Paulson was working with his staff. Each staff member was given certain areas of responsibility and made to feel an integral part of the EDP department. The implementation of Paulson's plan is on schedule.

The in-service training provided to date has been for EDP staff members. In addition, three months from now, seminars and workshops will be provided for the users. The individuals included as users are the executives and some of the support personnel who utilize electronic data processing. The objective of the in-service training for the users is to make them aware of the company's computer system. Paulson also wants to incorporate into the in-service training program information regarding the necessity for input validation, internal programming controls, and better documentation.

Hands-on workshops are scheduled to train some of the executives to use the system's query language. Paulson hopes that as online transaction processing systems are developed, some of the reports now printed can be eliminated. Since databases will be online and updated as transactions occur, more meaningful information can be obtained by using the query language than is available by reviewing weekly and end-of-month reports.

Dwan feels that the areas typically associated with EDP are coming along fine and that within two or three years his company will have a fairly well-developed MIS and in addition will be able to answer many of the "what if?" questions. However, he is very concerned about the way executives spend their time. Dwan hired a consultant to conduct a survey to determine what percentage of executive time was spent on tasks typically performed by executives. Dwan was amazed at the results. The consultant, Robert Scott, indicated that the results were almost identical to those obtained from a national survey. The results indicated:

Activity	Percentage
Paper handling	6.4
Writing	8.9
Searching information	15.0
Proofreading	2.0
Handling incoming mail	4.9
Reading technical materials	3.0
Dictation	6.0
Telephone—individual and conference calls	13.0
Calculating—ratios, percentages, budgets	6.0
Attending ad hoc and informal meetings	6.2
Attending scheduled meetings and seminars	5.0
Travel	9.0
All other activities	14.6

Dwan has felt for a long time that executive personnel—both upper and middle management—are spending too much time doing tasks that are essentially nonproductive and should be done in a more cost-effective manner by support personnel. The report also included recommendations regarding changes that the consultant felt should be made. Scott was shocked to find that office automation software was available and not being used. The office automation software provides for word processing, interoffice electronic mail, scheduling of meetings, and handling phone messages.

The financial vice-president, Norman Black, is responsible for most of the office functions and personnel. The organization of the office if very traditional and each manager has his or her own private secretary. The secretaries that work for middle- and lower-level managers seem to have a great deal of free time. Some of the stenographers and clerks that report to the secretaries never seem to be able to get their work done.

Norm has been with the company a long time, is well liked, and plans to work four more years and then retire. In regard to office automation, Norm has refused to become involved. He seems to feel that if the office procedures were streamlined to include new concepts and technologies, a major reorganization would be required. Although some members of the top management team feel the letters, reports, and other documents could be more professional, they hate to complain to Norm. Dwan views this as another example of the type of problem that sometimes develops when people have worked together for a long period of time and when members of the top management team are too complacent.

Some of the support personnel Black supervises indicated an interest in having the personnel they supervise use word processing and some of the other available software. Black implied the company's system is not dependable and should be changed. He also brings in newspaper articles such as the one that stated "using terminals is hazardous to the health of the operator." However, he never seems to mention the medical reports that indicate many of the facts reported regarding hazards associated with the use of terminals were incorrect. It has become obvious to Dwan that Black does not intend to do anything about office automation. Because of Black's attitude, some of the other managers and support personnel also have a negative attitude regarding OA.

Dwan has called Paulson in and has discussed the situation with him. Dwan feels there will be some major changes resulting from the implementation of Paulson's long-range plans regarding EDP. Dwan feels office automation should either be:

1. integrated into a master plan that would coordinate the development in both EDP and OA; or
2. developed as a separate functional area in a manner that would parallel Paulson's plan to achieve his long-range goals and objectives.

Dwan asked Paulson to think about coordinating the development of the two areas and to prepare a report regarding his recommendations. Paulson indicated that there would be advantages and disadvantages to developing either an integrated or a parallel plan. In response to Paulson's comment, Dwan qualified his request and asked Paulson to prepare a report that would identify what he felt were the major problems that would be encountered in reorganizing the office and using new methods in an effort to increase productivity and the quality of the work produced. Dwan also asked him to include in his report the advantages and disadvantages of both approaches (integrated and parallel) to the development of a more automated office.

Team or Individual Projects

Before completing the assignment, review Chapter 5 and look up additional material regarding the impact of office automation on organizational structure within the office. What new job descriptions are emerging and how will the change affect executive and support personnel? Before preparing the report to be given to Dwan, you should brainstorm the advantages and disadvantages of both an integrated and a parallel plan for the development of OA. You may also wish to find out what progress some of the companies in your area have made regarding OA and what organizational changes were made. To complete the assignment, do the following:

1. Prepare a list of questions that you would like to ask Dwan prior to preparing your report. Also state any assumptions that you make regarding the computer system, acquisition of additional equipment, development of workstations, and the physical surroundings of the executives and support personnel.

2. Prepare a list of the problems that you feel will need to be solved before any progress regarding office automation can occur.

3. Based on the information presented in the case study and the research you have done, list the major advantages and disadvantages of each approach (integrated versus parallel) to developing and implementing a plan for OA.

4. Indicate which approach you would recommend and the rationale for making that choice.

CHAPTER 6: Problems Involving the Current Sales/Accounts Receivable/Inventory System Are Identified

CPI has grown at a constant rate. Its major source of revenue is from sales, which are growing at a fairly steady pace. CPI has five major product lines: office supplies, office equipment, office furniture, stationery products, and certain types of computer-related products. Sales in the computer product and furniture lines have increased faster than in the other areas.

The board of directors was satisfied that the business has been well managed. Lately, however, one or two of the board members have expressed concern regarding the average age of the individuals in management. Little thought seems to have been given to hiring younger, less settled executives who would inject new ideas into the company. If any one fault is to be cited, it would be that management has become content and complacent. Then one day their bubble of complacency burst.

One of the line officers read an article on the growth rate of sales over the past 20 years for companies that sold the same type of products. CPI's growth rate, in both sales and assets, was considerably less than average.

The stir the article caused at the line officers' meeting was similar to some of the more exciting meetings that occurred when Greenlee, the former president, was a young man! An immediate investigation was started to determine why CPI's growth rate was less than that of companies with similar product lines. Was CPI's market research ineffective so that the wrong products were being promoted or was too little money being spent on advertising? Were customers dissatisfied with CPI's products or service?

The informal investigation indicated that CPI obtained a large number of new customers each year. However, the total number of customers listed on the schedule of accounts receivable remained constant. Although new names were listed, many former customers' names were not on the list.

A list was printed that showed the names of one-time customers who were no longer placing orders. Each sales representative was given a list of these former customers, by territory, and was asked to find out why they were no longer placing orders. Were CPI's prices too high or were their products inferior to those of their competitors?

Although the survey was conducted on an informal basis, the reasons given by former customers usually fell into one of the following categories:

1. A considerable amount of time elapsed between when an order was placed and when the merchandise was shipped. This occurred even when the order was marked "rush."
2. Although the sales representatives seemed to understand what merchandise was needed, the wrong items were shipped.
3. Customers were billed for items they had not received.
4. The sales representatives often did not inform customers of the "specials" that were available.
5. Trade discounts were not given to customers who qualified.
6. Sales tax was charged when the customer qualified for an exemption.
7. Several days after an order was placed, customers were notified that the item had to be backordered. Often this meant that a sale was lost because the item was not in stock and an alternate item was not suggested to the customer. Customers were upset when they found out after the order was placed that the item was not available for immediate shipment.

Since most of their former customers liked CPI's product line, they seemed sorry they were no longer dealing with CPI. The major problem seemed to be the result of an inadequate sales/accounts receivable/inventory system. Certainly management should have been aware of the large number of customers that were being lost each year!

Since one of the board members had read the article mentioned above, the problem was discussed at the next board meeting. When the board members read the report on the reasons some former customers were no longer dealing with CPI, they asked the president to submit a request for a formal study. The major issue was "Why could other companies process orders faster and give better service?"

There were many possible answers to the questions posed by the board of directors. Were the sales representatives not performing their jobs satisfactorily? About 90 percent of the sales orders received are those submitted by sales representatives; 10 percent are either phoned in by the customer or are on the customer's order forms and mailed directly to CPI. It usually takes CPI four or five days to process an order. During peak sales periods, it takes longer.

Discussion Questions

1. What do you feel is the true nature and scope of the problem?
2. Do you feel management is at fault or is the problem due to uncontrollable external forces?
3. Should management have had an exception report that listed inactive customers? An inactive customer might be one that has not placed an order in the past 12 months.
4. Can the problem be solved by designing a new sales system that processes orders in less time or are there other problems that must be solved?
5. In which categories might the sales representatives be at fault? What type of information must the sales representatives have regarding customers and CPI products so they can provide better service and avoid making errors?

6. If you were the analyst assigned to the initial investigation, whom would you interview?

7. As an analyst investigating the problem, what documents would you study? What observations would you make?

Team or Individual Projects

1. Fill out a Request for Systems Analysis. On the back of the form explain any of the statements you made that you feel need to be justified. For example, if you assigned a priority of 1 to the study, tell why you feel it should receive such a high priority.

2. Using the form provided, write a memo to the sales manager confirming your appointment. List at least five specific topics you wish to discuss during your meeting.

3. Write a final report that you would submit to the computer policy committee. At the present time, the format of the report is probably more important than the content. Unless you had access to a great deal of information and had interviewed a number of people, you could not be expected to know the extent of the problems that exist. In preparing your report, you will have to make some assumptions.

CHAPTER 7: Sales System Feasibility Study

A team has already been assigned to the project since the study was expanded to include the subsystems directly related to sales. The subsystems closely related to sales are inventory, purchasing, and accounts receivable. Management has indicated that funds are available for terminals, additional workstations, and communication equipment.

The project team started by reviewing the initial investigation report and documentation. A brainstorming session was held to determine what must be done. The tasks were assigned to the analysts and support personnel. A color-coded chart was prepared that illustrated the tasks to be performed, who was assigned to each task, and when each task should be started and completed.

In making the study, the project team proceded as follows:

1. Interviewed a number of people in the sales, inventory, purchasing, accounts receivable, and shipping departments.

2. Observations were made regarding how:
   a. sales orders were processed.
   b. telephone orders were handled.
   c. complaints from customers were processed.
   d. backorders were processed.
   e. merchandise was prepared for shipment.
   f. payments from customers were processed.
   g. merchandise was reordered.
   h. customers' statements were prepared.

In making the observations, the analysts were concerned with what type of controls were built into the procedures.

3. Numerous documents, reports, and articles were studied. Some of the materials reviewed were:

    a. articles regarding the advantages of online sales systems.

    b. articles regarding the advantages of distributed data entry and data processing.

    c. documentation regarding a software package that would provide a comprehensive sales system.

    d. documentation regarding CPI's telecommunication system and database software.

    e. articles on inventory control systems.

    f. the results of three questionnaires.

    g. a summary of the types of complaints that were most frequently received from customers.

    h. a report on the percentage of orders that could not be processed due to incomplete information on the order, invalid codes, items requested that were not in stock, or orders from customers who had exceeded their credit limit.

    i. reports submitted by sales representatives to regional managers regarding the number of calls they had made and what occurred during each call.

    j. sections of the standards manual that covered how the sales orders were processed, how items were to be reordered, and how customers were billed.

    k. forms or reports such as the sales order, sales invoice, accounts receivable statements, and letters sent when items were backordered.

4. Questionnaires were sent to a random sample of individuals:

    a. responsible for placing orders.

    b. involved in accounts receivable, accounts payable, and in the direct placement of sales orders.

5. Two installations were visited. Both of the firms had recently developed online sales-order processing systems. The objectives of the visits were to determine the advantages and disadvantages of the company's new online sales system. While attending a recent DPMA meeting, CPI's CIS manager had heard about both of the systems.

After all of the data was gathered, the problems of the present system were redefined. The objectives of the new system or a revised system were also determined in more detail. In order to solve the problems and to meet the objectives, four alternatives were studied:

1. Modify the present system. Additional controls would be built into the present system and some of the manual procedures would be automated. This would be considered a temporary solution until such time as the system could be redesigned.

2. Define the requirements for an online sales system and submit **requests for bids** (RFBs) from independent software houses. The software must meet the needs specified for CPI.

3. Develop an online sales system inhouse. Sales would be entered from remote locations. However, many of the data processing functions would be centralized.

4. Develop an online sales system and distribute many of the data processing functions to the sales, inventory, or accounts receivable departments.

In both the written and oral presentations, the first two solutions were discouraged and the disadvantages listed. The project team strongly recommended the third solution. The rationale for their decision was included in their report.

The team's presentation was well received and the computer policy committee requested that the documentation and the formal report be made available to all committee members. The members of the committee were asked to study the report. A second meeting was scheduled and at that time the computer policy committee would make a decision regarding which alternative they felt should be pursued.

Discussion Questions

Before answering some of the questions you may wish to reread the material that has been presented regarding CPI's CIS department, their computer system, and their sales system. Remember, additional information was also included in some of the chapters.

1. If the sales system is being investigated, why did the project team also investigate the inventory subsystem?
2. Why did the project team send a questionnaire to the customers' accounts payable managers?
3. Why did the team observe the procedures used in shipping merchandise to customers?
4. Why were the reports submitted by the sales representatives to their sales managers studied?
5. What type of visual aids would you suggest that the team prepare? What type of film might be available from a vendor that could be used to explain some of the concepts presented in the third and fourth alternatives?
6. Why did the project team feel that it was unwise to modify the present system?
7. Based upon the information presented regarding CPI's CIS personnel, do you see any problems in trying to implement an online system? Give the rationale for your answer.
8. Based upon the information presented regarding CPI's management personnel, would you support the third or fourth alternative? Give the rationale for your answer.
9. If software could be purchased or leased for less money than it would cost to develop the system inhouse, why did the project team feel the software should be developed inhouse?
10. Why did the computer policy committee schedule a second meeting rather than making a decision regarding which option to pursue? Did this mean they were dissatisfied with the report and the work of the team?
11. In looking over the summary of activities performed by the project team, do you see any that were overlooked?
12. If you were the CIS manager for CPI, would you feel that a consultant should be employed to work with the project team? In looking over the backgrounds of the staff members, who would you make the project team leader? If three people are to be on the project team, whom would you select? Would you have anyone from outside of the CIS department on the team? Why might it be better to have a three- or five-person team rather than a four-person team?

Team or Individual Projects

1. Research the advantages and disadvantages of pursuing the fourth alternative and distributing more of the EDP functions. In completing the assignment, read four or five articles regarding distributed data processing and write a brief report on your findings.

   After you have read the articles and written your report, answer the questions which follow. You are to assume that many of the EDP functions will be distributed. However, there will still be a centralized mainframe and database.

   a. Why is it necessary to have a centralized database?

   b. What did Dwan have Paulson do that might indicate he wanted EDP to be more distributed?

   c. Based on the information presented in the other segments of the case study, do you see any problem with developing a more distributed sales system?

   d. Since the sales representatives take orders from the customers and then send them to the company, what additional hardware would need to be obtained to permit the representatives to enter the orders directly into the system?

   e. Do you feel that the sales department should do all of their own file maintenance (enter data into the database when changes are made such as when customers get a new telephone number or changes their name)?

   f. Are there any ways microcomputers might be used in the revised sales system?

2. If you were on the project team and asked to visit a company that is similar to CPI:

   a. What company in your area would you pick?

   b. If you were not familiar with any company that had the same kind of business as CPI, what resources could you use to find out if such a company exists in your immediate area or close to where you are located? (You might wish to check with your librarian to see what he or she would recommend).

   c. What would you do to prepare for your visit to the companies with whom you have appointments?

   d. List the most important questions you would ask the CIS or sales manager. List the major objectives for your visit. If the opportunity were available, would you want to talk to a systems analyst or progammer as well as to the CIS manager?

3. Visit a CIS department and interview the manager or an analyst and obtain the following information:

   a. What guidelines are available regarding what should be done when conducting a feasibility study?

   b. For a major project, is a team assigned to the project or is one analyst responsible for the study?

   c. What other techniques are used for gathering data other than those listed in the text?

   d. What types of reports must be submitted at the end of the feasibility study? What information is included in the reports?

   e. To what extent are surveys used?

   f. Has the company employed a consultant? If so, for what type of projects?

   g. To what extent is EDP distributed? Is the control centralized or decentralized?

4. Review the description of CPI's present computer system that is found on page 000. If either option 3 or 4 is selected, does it seem likely that the present computer system can support an online sales system? What if other systems were also converted from card-oriented batch systems to an online system? How might the system need to be upgraded?

CHAPTER 8: A New Sales System Is Designed for CPI

The recommendations of the project team to develop a transaction processing sales-order system and to distribute the database maintenance procedures to the sales department were also approved.

Although Dennis Paulson was pleased that the team's recommendations were approved, he was concerned that all of the standards and guidelines for the CIS department had not been developed. Since the staff training sessions regarding the operating system, office automation package, and database management system had been completed successfully by all staff members, he felt somewhat more comfortable about having his staff develop an online system.

Glen Nichols has become the inhouse expert on the office automation software and will begin to work with the sales department's personnel on use of the software. Glen will train the word processing technicians and also work with some of the other staff members who have expressed an interest in learning to use some of the features of office automation. Paulson is delighted at the interest and involvement shown by the sales department.

Kim Mead, who has been designated as the resource person for the database management system (DBMS), will start working with some of the sales department's management personnel to show them how to use the system's query language. Eventually the entire sales force will be trained to use the DBMS software. In redesigning the system, some of the seldom-used reports were eliminated. Management will be able to generate more meaningful ad hoc reports by using the query language.

Workshops had been held for the entire CIS staff on the principles involved in developing online systems. Martha Aguilar is the only staff member who still seems to have a "why change?" attitude. For this reason, Paulson appointed her the project leader. During the data gathering stages of the investigation and general design phases of the project, Martha and Kim visited several companies that had sales transaction systems. The sales managers of the companies they visited were extremely enthusiastic about their transaction processing systems and Martha did learn a number of "DOs and DON'Ts" from the visits.

The general design project team consisted of Mead, Aguilar, and Nancy Lopez. Nancy is the administrative assistant to the sales manager and very knowledgeable about the company, its products, and the problems with the existing sales system. During the feasibility study, the following objectives had been determined for the general design:

1. automate all manual procedures where this can be cost-justified;

2. build additional controls into the system;

3. integrate the sales-order system with the accounts receivable and inventory systems;

4. determine which procedures should be distributed to the sales department;

5. determine which procedures should be interactive and which should be run in a batch environment;

6. provide management with the necessary exception reports that will enable them to make needed decisions;

7. develop databases that can be used to generate ad hoc reports; and

8. investigate the security of the system.

Management is concerned that unauthorized individuals could obtain information from the online databases. Kay Walzcak was asked to investigate the security provided by the operating system and database management software and recommend any additional measures that should be taken. Kay was also appointed as the security officer responsible for developing security guidelines for the development of online systems.

Portions of the sales-order design report are illustrated in Figures CS8-1 through CS8-3 on pages 542 through 544. Although not illustrated, the sales-order design report would also include:

• an overview of each subsystem illustrated on the hierarchy chart;
• the estimated cost and time needed to complete the detailed design and implementation phases of the project;
• the operational cost of the system;
• tentative recommendations on how the new system can be phased in; and
• a brief explanation of the changes to be made in the manual procedures.

The items referenced in the report as Exhibits 2 and 3 are not included in the example.

The documentation for the design phase would include all of the working papers reviewed by the project team and informally by management and the users. The planning chart and budget report for the general design phase would also be included. The budget report should include a comparison of the projected costs with the actual costs.

Discussion Questions

1. After reading the material presented in this section of the case study, do you feel that Dennis Paulson is achieving his objectives in regard to staff development?

2. What advantages are provided in having the sales department personnel maintain the sales, inventory, and accounts receivable databases?

3. Review the general guidelines for developing specifications illustrated in Figure 8-1 on page 257. Give specific examples of how the guidelines have been followed in determining the general design for the sales-order system.

4. Explain the advantages and disadvantages of appointing Martha Aguilar as the project leader. If you were the manager of the CIS department, would you have made the same decision? Give the rationale for your answer.

Team or Individual Projects

1. Prepare a report indicating the advantages and disadvantages of having the sales department do the routine database maintenance, submit orders directly

computer products inc.   SALES-ORDER DESIGN REPORT

Page 1 of 3

PROJECT NUMBER   1051

REPPORT PREPARED BY: Martha Aguilar

Date: July 9, 1986

OVERVIEW OF THE SALES ORDER SYSTEM:

The sales order system is divided into the five major subsystems illustrated in the hierarchy chart identified as Exhibit 1. The subsystems are identified as maintenance, sales order, query backup and report.

Each of the subsystems is divided into a number of procedures and each procedure is divided into a number of tasks. While many of the tasks are automated and involve the use of the computer, some tasks involve manual operations. In developing the design specifications, each task was evaluated to determine whether it could be automated or a more efficient manual method could be designed. The objectives identified in the Feasibility Study Report for the sales-order system will be met in the followind manner:

1. The credit and inventory checking procedures will be automated. As a sales order is entered, the customer's credit is automatically checked. If the customer has exceeded his or her credit, an exception routine will be invoked. If there are too few of a given item in stock, a backorder exception routine will be initiated.

2. The security of the system will be determined. Procedures will be such that only authorized personnel can gain access to information stored in the online databases. In addition, internal and external controls will be included in the design of all programs.

3. When a sales order is successfully processed, a record containing the date, transaction number, transaction code, and amount is added to the accounts receivable online database. The record in the online inventory database for each item sold are updated to reflect the decrease in items on hand.

4. Sales department personnel will do the routine maintenance for the accounts receivable, inventory, and sales databases. When the inventory control department obtains terminals and has personnel trained to use the DBMS software, the inventory database will be maintained by inventory department personnel.

5. The management staff of the sales department will have terminals available and will be able to generate ad hoc sales reports and to obtain information from the accounts receivable and inventory databases.

FIGURE CS8-1   Sales system design report   (Page 1)

computer products inc.   SALES ORDER DESIGN REPORT

Page 2 of 3

PROJECT NUMBER: 1051

REPORT PREPARED BY: Martha Aguilar

DATE: July 9, 1986

6. The file maintenance and sales-order systems will be interactive-there will
    be two-way communication between the terminal operator and the computer.
    All sales representatives will have portable, intelligent terminals that
    will be utilized in placing sales orders. Ninety percent of all orders
    will be placed by the sales representatives. Rationale for obtaining
    intelligent terminals is illustrated in Exhibit 2.

7. Each day two exception reports will be printed. The first report will list
    all items that must be reordered and those that are close to the reorder point.
    The second report will list customers who are within 10 percent of their
    credit limit. The credit manager can review their credit histories and
    determine whether their limits should be extended.

8. Upon demand, two additional exception reports will be printed. One report
    lists customers with inactive accounts and the second lists items that are
    below last year's sales volume.

MAJOR ADVANTAGES OF THE PROPOSED SYSTEM

1. Better inventory control. When the recommendations detailed in Exhibit 3
    are implemented, fewer items will be carried in stock and the number of
    items that must be backordered will decrease.

2. Accounts receivable will be better controlled. Recommendations have been
    made to the accounts receivable manager that will be implemented immediately.
    There should be fewer uncollectable accounts and the average age of
    receivables should decrease. Since the credit limit can be increased for
    customers who have paid their accounts promptly, fewer orders will be delayed
    pendng approval by the credit manager.

3. Managers will have immediate access to information stored in the databases.
    Therefore, many of the reports that had been printed and were seldom used will be
    eliminated.

4. The number of manual operations will be decreased.

5. Approximately 98 percent of the sales orders will be processed on the same
    day as they are received. Ninety percent will be entered into the system
    from remote locations by the sales representatives; 10 percent will be
    received over the phone or as mail-in orders from customers.

FIGURE CS8-2 Sales system design report (Page 2)

computer products inc. SALES-ORDER DESIGN REPORT

PROJECT NUMBER: 1051

REPORT PREPARED BY: Martha Aguilar

DATE: July 9, 1986

6. The information retained in the sales databases will provide management with far more information than has been available. The information can be used for forecasting sales, determining areas where there should be a concentration of advertising, determining product lines to either phase out or expand, and to provide additional information as input to the inventory control system.

7. Since orders will be processed online, all prices quoted will be those stored in the inventory database.

SYSTEM OBJECTIVES:

The objectives of the proposed sales system are to:

1. Provide better services to customers. Since online transaction processing will be used for 90 percent of the orders, orders should be processed the same day they are received.

2. Decrease the number of errors. Since the sales representative will be able to confirm that the right customer's record is retrieved and that the record for the correct item is accessed, fewer errors will be made.

3. Provide better internal and external controls.

4. Provide a well-defined audit trail. The written sales order, terminal transaction file, invoices, and the sales register report will all be part of the audit trail.

5. Provide management with additional sales information that can be used to make better decisions regarding the marketing of products and the buying trends that will affect the product line.

6. Interrelate sales with the accounts receivable, inventory control, and backorder systems.

7. Decrease the number of items that will be backordered.

8. Provide sales representatives with complete and accurate information regarding the product line. Since each sales representative will have a terminal, electronic mail can be used to keep the sales representatives informed. Once the sales representatives learn to use the DBMS query software, needed information can be extracted from the databases.

FIGURE CS8-3 Sales system design report (Page 3)

into the system, and use the database query language to generate ad hoc reports. In order to prepare your report, you will need to research the advantages and disadvantages of both distributed and transaction processing.

a. Based on the material presented in the case study, does it appear that the sales department personnel will get the full cooperation of the CIS manager and department? Include the rationale for your opinion.

b. Are there any other functions associated with the sales-order entry system that you would assign to the sales department? Please list and explain any other functions associated with the sales-order entry system you would assign to the sales department.

CHAPTER 9: Designing Reports and Source Documents

In Chapter 8 a hierarchy chart was illustrated that indicated some of the procedures needed for the sales/accounts receivable/inventory system. Although the entire report was not given, from the data provided, the following information was stated or implied:

1. Databases are to be created for accounts receivable, sales, and inventory records. Programs must be provided that add, delete, and change records stored in the databases. The databases must be online when sales orders are entered in order to do the following:

   a. Retrieve the customer's record and determine whether the new sales order can be processed.

   b. Retrieve the records for the items ordered and determine whether the quantity ordered is available or the item must be backordered.

   c. Update the databases. Information regarding the transaction is added to the accounts receivable database and the appropriate records in the inventory and sales databases are updated. The inventory record is updated by subtracting the items sold from the items available for sale, and in the sales database, the amount of the sale for each item is recorded in the appropriate record.

   d. Print the sales invoices, backorder report, and overlimit report. The invoices are printed as soon as all the data is entered by the sales representative. Confirmation of the order is also printed on the sales representative's terminal. The data for the backorder report and overlimit report is stored in files and printed in a batch mode.

2. Interactive processing was to be used to maintain the databases; transaction processing would be used for processing sales orders.

Team or Individual Projects

1. Prepare a specification form for the sales invoice. Before preparing the form you should determine what information is usually printed on sales invoices. You may want to refer to invoices you have received or have seen in an accounting or management text.

2. Prepare a print layout for the sales invoice. Follow the rules provided in the text such as printing the information that will be preprinted on the form, using red Xs to indicated serially numbered forms, and Xs for the variable data that will be printed by CPI's printer. Make sure the form is a convenient size to handle, meets postal regulations, and contains all of the required information.

3. Prepare a print layout for the sales register. The data for the sales register is generated as the sales orders are processed, stored in a file, and after all orders are placed for the day, used to print the register. Refer to your accounting text to determine what information is normally printed on a sales register (sometimes called a sales journal).

4. The sales transaction records require the following fields to be entered:

	Field Size	Data Type
Account number	10	N
Shipping Code	4	A
Sold by	3	N
Item number	5	AN
Quantity	5	N
Order Number	6	N

In reviewing your notes, you will find that at the present time CPI has 150 sales representatives that have numbers from 1 to 150. Also it is very unusual that anyone orders more than 100 of any one single item. On the transmittal form, both the first and last order (reference) number is recorded. On an average when an order is placed, five items are ordered. After the control information is entered, the operator enters the customer's account number. The computer retrieves the record from the database and checks the customer's credit. If the order can be processed, the reference number, shipping code, and sales representative's number are entered. Those three fields are entered once per order. Next the sales representative enters the item number and the corresponding record is retrieved from the inventory database. If the item is available, the quantity ordered is entered. The operator is then asked whether another item is to be entered.

Directions:

1. Design the screens that will be needed to enter the control information and the data recorded on the sales orders received through the mail or from phoned-in orders. As you design the screens, consider how you want the operator to verify the data. Also determine what will occur when the customer's credit limit is exceeded or an item is out of stock. Remember the computer is to print control totals that list the total items ordered, items backordered, quantity of items ordered by customers who exceeded their credit limit, and the items to be shipped.

2. Answer the following question: Based on the information presented in the text, what additional ways might the computer be programmed to verify the data? Be specific and identify the ways in which each field might be edited.

3. After the system has been operational for some time, statistical data indicates that approximately 2 percent of the orders cannot be processed because customers are over their credit limit or because items are out of stock. Under the old system approximately 10 percent of the orders called or mailed in could not be processed. What procedures were implemented that caused the percentage to decrease from 10 percent to 2 percent? Would it be easier to decrease the items that had to be backordered or cause fewer customers who have exceeded their credit limit to place orders?

4. Refer back to the description of the products that are sold by Computer Products Incorporated and the information regarding codes presented in your text. You might also wish to review some catalogues or brochures from office and computer supply companies. Based on your findings, determine what sort of codes might be used in developing a new system of item numbers.

What factors should you keep in mind as you develop a coding system? If you make any assumptions about the product line as you develop a coding system, state the assumptions you have made. You may increase the size of the field so that a more definitive code can be developed.

CHAPTER 10: The Inventory Analysis Report

Martha Aguilar has been working very closely with the sales manager who seems very happy with the suggested reports and the type of information that will be available by making online queries. Since Martha consulted with the sales manager, she did not feel it was necessary to work closely with the inventory manager, Pete Greenway. Greenway was very upset when Martha Aguilar discussed the tentative design for the VSAM file, which was illustrated in Figure 10-9 on page 362, and the list of reports that would be available. He told Martha that he would have less information under the new system than he had at the present time.

Greenway was also concerned that Martha had not taken the time to find out what information was presently available in the old ISAM file. He did indicate that enough information was available in the new individual order files which contained the item number, order number, data, quantity, and cost. He suggested that the year followed by the Julian date should have been used for the date of the orders. Past experience has shown that some slow-moving items are retained in stock several years before they are finally written off as a loss or are sold at a reduced price.

The inventory manager provided Martha with the following information regarding the inventory. The item number is a coded number. The first letter denotes the major category, the next five digits are the item number, and the last two digits are the model number. Record numbers are sequentially assigned, in increments of ten, to all items and are used to randomly retrieve the master file records. Greenway indicated that he liked the idea of separate records being recorded for each inventory shipment received. Under the present system, the detailed information regarding each lot is stored in a table within the master file record. Since enough space for ten shipments is provided, unused space exists in most of the records. On an average, the inventory for an item consists of three shipments. At the present time, CPI stocks approximately 20,000 different items. Some of the items have as many as ten different models or sizes.

The lowest cost price is .05275 per unit and the highest is 4,789.00. The cost price in the master record is a weighted average cost that is only used to compute the selling price. The selling price is based on a 110 percent markup of the weighted average. On the first of each month, the weighted average and selling price are calculated. If the selling price has changed from the previous month, the item is listed out. Each sales representative receives a copy of the price changes. On an average, each month approximately 4 percent of the items in stock have a price change.

Seldom are more than 10,000 units of an item stocked. The reorder point and the reorder quantity are calculated using a mathematical formula that is based on the average amount sold per month and the length of time that it takes to get the item from the current supplier.

The data stored in the vendor code field is the key used to randomly retrieve the vendor's records from a second ISAM file. The keys are in order of preference.

The vendor's record contains needed data such as name, address, current balance, as so forth. In addition, the record contains the history of the last five orders—when the order was placed, when the order was received, the condition of the order, and a field that contains a code which is used to indicate price increases or decreases. The inventory manager would like to be able to display this information prior to placing a new order. If the history shows unusual delays in shipments, merchandise arriving in poor condition, or a continuous increase in price, a new vendor may be selected. Usually the vendor that is listed first is the one with whom the company normally places the order.

The information needed to produce the inventory analysis report that Greenway wants is currently available either in the inventory master ISAM file or in other files. Greenway would like an inventory analysis report printed each month that contains the following information:

Item Number

Description

Quantity on Hand

Average amount sold per month. This is calculated by taking the units sold during the last 12 months and dividing by 12. Each month the units-sold tables for the current and previous year are updated.

Quantity sold during the same month of the previous year.

Quantity sold during the current month.

Percentage of increase or decrease over the same month of the previous year.

Percentage of the annual total that was sold during the current month. If more than 15 percent of the annual total for an item was sold during the past month, the items should be flagged for attention.

Quantity sold during the previous year.

Quantity sold during the current year.

Annual percentage of increase or decrease. All increases greater than 10 percent are flagged as well as all decreases of more than 5 percent.

Aguilar informed Greenway that it would not be possible under the new system to obtain the type of report that he wanted. She also implied that he certainly didn't have time to go through all 20,000 items that were listed on the report. Greenway told Aguilar that the report was absolutely necessary. His market research staff used the report to determine trends and to decide whether the reorder point and reorder quantity should be adjusted. If the sales of one item dropped substantially during a month, Greenway was not concerned unless the trend continued. The same was true when an unexpected increase in sales of a certain product occurred.

Greenway called Dennis Paulson and complained about having less information under the proposed online sales/accounts receivable/inventory system than under the old batch method. After talking to Greenway, Paulson called Aguilar for an appointment. During the appointment, he informed her that she must revise the contents of the inventory master file to provide the required information. He also asked her to design the report needed and to fill out the required report specification form. Although she was given a specific assignment to make certain that the information was available in the files for the inventory analysis report, Paulson plans to review the current inventory reports to see what else should be included in the files.

Inventory Analysis Report

The inventory analysis report must be run after all sales transactions are
entered for the month and before any sales are recorded for the new month. The
inventory analysis report program prints the report and updates the two tables
that are illustrated below. In addition, before the record is rewritten, the
quantity sold field is set to zero. Normally the file is backed up, the program
run, and the control totals checked. After the totals are checked, the report
is queued to the printer. If necessary, the report can be written to tape and
printed when time is available on the line printer. However, the report should
be distributed to the inventory manager prior to the fifth of the month.

   The tables illustrated below are part of the master inventory record
maintained for each item.

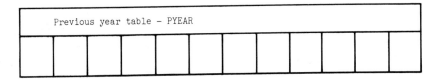

   In the following illustration, assume that the inventory analysis report
for April is being run. As part of the control information, a ''4'' (for April)
is entered. The four will be used to access the data stored in the previous-
year table and the current-year table. The 4 is entered into a field called
month. Also entered is the date for the report.

   In order to obtain the quantity sold during the previous year—from April
through March—the quantity sold for last April is moved into the previous year
table by using the following instruction:

                    PYEAR (MONTH) = CYEAR (MONTH)
   Next the table for the current year is updated by moving the amount from
the quantity sold field (QSOLD) into the current-year table.

                    CYEAR (MONTH) = QSOLD

   Data is now available in the two tables to accurately determine the total
quantity sold for the current year and the preceding year. The counters for
the count-controlled loops are used as the subscript to access each area of the
two tables. If BASIC is used to write the program, the totals would be
calculated as follows:

```
              FOR J = 1 TO 12
                  CTOTAL = CTOTAL + CYEAR (J)
                  PTOTAL = PTOTAL = PYEAR (J)
              NEXT J
```

   Just prior to rewriting the record, zero is moved into the quantity sold
field for the current month.

FIGURE CS10-1 Documentation for the inventory analysis report.

Aguilar indicated that she was not certain how the sales quantity tables could be used to supply the required information. Paulson referred her to the documentation for the inventory analysis report that is illustrated in Figure CS10-1 on page 549. Paulson also reminded her that in determining field sizes, provision should be made for expansion. However, an excessive amount of extra space should not be provided. Paulson indicated that prices and quantities tended not to vary more than 10 percent a year. He also indicated that it was only necessary to project the required field sizes for the next five years.

Team or Individual Projects

Review the material illustrated in Figure C10-1. If you are still uncertain as to how data stored in tables is accessed and updated, refer to the section on tables which is in Chapter 11. Answer the following questions and complete the tasks that were specified by Paulson:

1. If you were Paulson, what would you tell Aguilar regarding the way she designed the VSAM inventory master file?

2. Should a complete inventory analysis report be printed? Or would it be better to only list the deviations specified as "being flagged"?

3. Explain how the redesign of the file illustrates the point that "files are the backbone of an EDP system."

4. Explain how the problem created by Aguilar not gathering all of the available data caused a portion of the decisions made in the general design phase to be reconsidered. In regard to the inventory subsystem, will additional data have to be gathered and analyzed in order to complete both the general design and detailed design?

5. Indicate what information will need to be in each record stored in the vendor's file. How would you organize the file? Would you recommend the use of variable length records or continue to store the data for the past five orders in a table within the master record?

6. Fill out the report specification sheet and indicate how each field of data needed on the report will be obtained. Be sure to indicate what information would be considered as "control information" and would be entered by the operator.

7. Complete a file layout form for the inventory VSAM files that includes the portion referred to as the master record and the records that provide the information for the individual shipments of merchandise. You should be able to determine the size for each of the required fields.

8. Do a print layout form that includes headings, detail lines, and any total lines that you feel might be required.

CHAPTER 11: Preparing Programming Specifications

In Chapter 10 you were asked to design the file layout forms for the master inventory record and for the record that would be used to record the individual shipments of merchandise. At the present time the master file contains the information identified in Figure 10-9 and the additional information identified under Inventory Analysis Report in Case Study 10. In Project 2 (page 408) at the end of Chapter 11 the reports identified by the inventory manager and the analyst as being needed were listed.

At this time you will do a report layout form and a program specification form for two of the programs. Additional information will be supplied for each of the programs. You may need to reconsider what should be included in the inventory master file.

Team or Individual Projects

1. New items are not added to the master file when the purchase invoice is received. At that time, a new item form is completed by the inventory department. The form provides space for recording the following fields of data: item number, record number, product description, quantity on hand, reorder point, reorder quantity, cost price, date of first shipment, and vendor codes. The inventory manager is responsible for determining the reorder point and quantity. Since it is a new product, there is no historical information upon which to make a decision. Therefore, the decision is based on past experience with similar products. The cost of the first shipment is used as the cost price. The selling price is computed. The date of the first shipment is recorded as the year (86) followed by the Julian date. The inventory manager feels it is important to be able to identify when additional items were added to CPI's product line.

Attached to the new item form is a copy of the purchase invoice received from the vendor. A terminal within the inventory department is used to enter the data into the system. The assistant inventory manager, Mary Beth Hughes, is responsible for filling out the new item forms and for determining the batch totals. Hughes functions as a data control clerk and reviews the confirmation report and batch totals. If any errors are detected, either a change form or the form for the correct program will need to be filled out to make the necessary changes in the master record and in the individual shipment record. Since the purchase invoices arrive several days ahead of the actual merchandise, new item records are "batched" and entered on Tuesday and Friday.

The program that will be used to enter the data will be designed to display a formatted screen that permits the operator to enter all of the data. After all data is entered, the message ENTER NUMBER OF INVALID FIELD OR "OK" IF ALL DATA WAS ENTERED CORRECTLY will appear. For instance, if the operator enters a 4, the cursor moves to the location where the fourth field of data is to be entered.

If a record number or item number is entered that is already used for an existing record, a message will be displayed. The operator will need to go on to the next record until a new number can be assigned by Hughes. The operating system will not permit a duplicate record number or item number to be entered during the execution of the ADDNEW program.

Directions:

a. Design the form that will be filled in by Hughes and used as the source document.

b. Design the transmittal form that will be attached to the source documents.

c. Design the formatted screen. Assume that your screen has 25 80- character lines. Be sure to include a screen that provides an opportunity for the operator to exit from the program and return to the main menu for the inventory programs.

d. Prepare the print layout form for the confirmation report and the control totals.

e. List each field of data to be entered and determine how the computer might be programmed to edit the data. For some fields, such as the product description, it will only be possible to visually confirm the data.

However, the computer can check to see whether too many or too few digits or characters are entered.

  f. Prepare the programming specification form. Use the format illustrated in Figures 11-6 and 11-7.

  g. How would the records stored in the file be accessed?

2. In the change program, the following fields of data could be changed due to decisions made by the inventory manager: product description, reorder point, reorder quantity, selling price, and vendor codes. If a special sale price is to be established, the selling price is changed and an S placed in a field that will either contain an S or an N. An N is originally stored in the field. When a sale occurs, an S is placed in the field and the sale price is recorded in the selling price field. After the sale, the change program is used to replace the N in the field and the normal selling price will be calculated by the computer. However, during the sale, when new shipments are received, the cost price is adjusted but not the selling price.

The sale price control field is used to provide another important function. When a sales representative places an order for a sale item a message is flashed on the screen. The message SALE ITEM. INFORM CUSTOMER. is displayed as a reminder that the customer might wish to increase the quantity ordered. The inventory manager feels that this will help to solve the problem of customers not being informed when items are on sale. Sale prices are entered after the close of business on the last working day of the month and the changes are normally in effect for exactly one month. The last working day of the month the sale prices are reset to the normal selling prices. Relatively few changes are made other than for the last day of the month.

The general approach to be used in designing the program is to have the operator enter a number from 1 to 5 (there are only five items that can be changed). If a 1 is entered, the current description will be displayed and the operator will be asked to enter the new description. Of course, the operator must make certain that the right record was retrieved from the master file. Also for any inventory item, more than one change might be made.

If any changes are detected by the person who functions as the data control clerk, the change form will need to be resubmitted and the record with the incorrect information changed immediately.

Directions:

  a. Answer the following questions:

    (1) Who should be responsible for authorizing the changes?

    (2) When should the change program be run?

    (3) Should the change program be run before the inventory analysis report is run? Should the inventory analysis report show which items have been on sale for that particular month? Why might this information be of value to the sales manager and to the inventory manager? Would there be any value in creating a second report that would list only the sale items? (Assume that at any one time there are never more than 300 items on sale.)

    (4) Would it be necessary to fill out individual change forms for the items when the sale is over? Remember, in a program it is possible to open a file, indicate that records are to be accessed sequentially, close the file, and then reopen the file and indicate that records are to be accessed randomly. A control character could be entered by the operator to indicate when the end-of-month sale price changes are to be made. This would have to occur *before* new price changes are entered.

(5) Why is the exact sequence in which jobs are run sometimes very important? What would have to be done if the jobs were run in the wrong sequence? Is it sometimes critical that the file backup be run exactly when specified in the job run schedule?

b. Design the source document that will be used to authorize the changes.

c. Design the screens that will be needed. The analyst has suggested that a help menu should be designed that explains the options available. Each type of change should have its own screen.

d. List the various ways each of the first fields of data can be edited or verified.

e. Prepare the program specifications.

f. Prepare the print layout form for the confirmation report and the batch total report.

g. How would the records stored in the file be accessed?

CHAPTER 12: Developing a Hierarchy Chart and a Detailed Logic Plan

Dennis Paulson feels that a great deal of progress has been made in developing standards and guidelines. Most of the staff members have been fairly cooperative and seem willing to try new methods of solving problems. Paulson recently attended an ACM conference and his interest was renewed in using design and code walkthroughs as a means of detecting errors before the system was considered operational.

Paulson wasn't sure that everyone on the staff would feel comfortable, so he asked Martha Aguilar to research the topics of structured walkthroughs and stub testing. Aguilar would attend a structured design conference where both topics would be covered in depth. Paulson hopes that Martha will become enthusiastic about the concepts and will want to put the theory into practice. By having Martha report on the two topics and develop standards for their implementation, Paulson hopes the other staff members will be more receptive to the ideas than if he were to make the presentation.

Team or Individual Projects

1. Research the topic of structured walkthroughs using materials from five sources other than your textbook. Prepare both a written and an oral presentation that includes: the advantages, the disadvantages, and suggestions for implementing walkthroughs. Use appropriate visual aids.

2. Research the topic of stub testing using materials from five sources other than your textbook. Prepare both a written and an oral presentation that includes: the advantages, the disadvantages, and the suggestions for implementing stub testing. Use appropriate visual aids.

3. The inventory manager would like a report printed upon demand that lists all of the items that are currently on sale. For each item the report should list the item number, description, normal selling price, and the sale price. He would also like the average monthly sales for the current year and the number of items sold so far this month. The program specifications include the following:

a. The report is to be double-spaced. Headings plus 25 detail lines are to be printed on each page.

b. The inventory file is opened for sequential access. Unless the item is on sale, all processing is bypassed.

c. The average number sold per month is to be calculated by creating a loop that will be executed 12 times. The loop's counter will be used to access the individual areas within the current-year table so that the 12 monthly totals can be added. The total is then divided by 12 to obtain the average. A separate module should be used to determine the average.

d. At the end of the job, a total is to be printed that states: PERCENT OF NORMAL MONTHLY SALES REACHED IS xx.xx PERCENT. To obtain the percentage figure, the total of all sale items sold to date is divided by the total of all the individual item averages.

Directions:

a. Prepare a list of the major functions that are to be performed.
b. List the tasks that are associated with each major function.
c. Prepare a print layout form for the report.
d. Prepare a hierarchy chart. Use as a guide the one provided in Figure 12-16. You may need to add a few more modules that will be invoked by the process records modules.
e. Develop detailed logic plans for each module.

CHAPTER 13: Developing Job Descriptions and Sections of the Standards Manual

The analysts, programmers, computer operations personnel, and the users at Computer Products Incorporated have been working toward a common goal—to get the sales system implemented as soon as possible. Although new equipment other than additional terminals was not required, the CIS manager is continuing to get his house in order. To date, the following policies and procedures have been implemented:

1. A security officer has been appointed who is assigned the task of preparing a report on the security of the hardware, software, and files. Recommendations for improvements are to be submitted along with the report. The security officer has been instructed to study the present security measures, to design minimum standards based upon research and observations of other systems, and to follow through to see that the recommendations are enforced after they are approved.

2. The standards manual is being revised to reflect the informal structure and procedures that exist rather than the somewhat obsolete formal structure. Although some of the staff feel this is wasted effort, the users of the various systems are cooperative and their staffs are helping to update the manuals. The manuals will be useful when new systems are designed or additional employees are hired. Procedures have been developed for ensuring that the manuals are updated.

3. Guidelines and standards have been developed for:
   a. designing systems
   b. designing procedures
   c. conducting design walkthroughs

     d. writing source code
     e. conducting code walkthroughs
     f. developing test files
     g. testing procedures
     h. string and systems testing
     i. maintaining test files for use in testing authorized changes to operational programs
     j. developing training programs
     k. documenting systems and procedures
     l. preparing documentation for the users

For the most part, the CIS staff has been cooperative since staff members feel most of the changes have resulted in the development of more efficient and reliable systems. As the time arrives to complete the final tasks necessary to convert to a new sales system, the morale of the members of the data processing department is high and there is a great deal of cooperation among them. Members of the department have a feeling of belonging to the team.

Team or Individual Projects

1. Assume you have been appointed the security officer for your installation. Research the responsibilities of a security officer and write a job description for the position.

2. Dennis Paulson has not formulated a code of ethics for CIS personnel. Do some research to see what code of ethics is suggested for professionals who work in EDP. Both the Data Processing Management Association and the Association of Computer Machinery have a code of ethics. You may wish to interview individuals involved in EDP to determine their concept of the ethics that apply to EDP. Based on your findings, develop a code of ethics for CPI's CIS department.

3. Refer to page 328 and study the sales-order form. Using the material presented in Figures 13-1 and 13-2 as a model, prepare the section of the standards manual that describes the preparation and handling of the sales-order form by the sales-order clerks. If you find it necessary to make an assumption, state in a footnote how you would determine the information necessary to prepare the sales-order section of the standards manual. Indicate how the sales department clerk who submitted the documents would determine the validity of the sales invoices and the control report. Provide an exception routine for handling any errors that are detected.

4. Prepare a training plan for the members of the sales department who will be trained to enter the sales-order information into the system. The three employees selected have keyboarding skill but have not used terminals.

     a. Indicate how you would train the terminal operators and what material you would use in the training program.

     b. How would you develop a standard that could be used to measure the productivity of the employees?

CHAPTER 14: Determining if the CIS Department's Objectives Have Been Achieved

Now that the sales/accounts receivable/inventory system is operational, Dennis Paulson is taking stock of the current status of the CIS department. He once again reviews the objectives given to him by the president and those that he developed for his own personal use. Although many of the objectives seem to have been accomplished, there are still areas of concern.

Team or Individual Projects

1. Review the objectives given to Paulson in the different segments of the case study. Answer the following questions:
   a. Which objectives seem to have been achieved? Be sure to include the rationale for your answer.
   b. Which objectives have not been addressed and still represent a major problem?
   c. What additional problems, if any, seem to have developed that should be solved? State the problems that seem to have developed and how each should be solved.

2. The procedure to be followed in initiating and making programming changes had not yet been formalized at CPI. Paulson is concerned that changes are being made without authorization and in a somewhat disorganized manner. Paulson has asked you to design a request form to be used when a program must be modified. In addition, he would like a one-page description of the procedure that should be followed to initiate the request, make the change, and alter the documentation.

Appendix II

PROGRAM FLOWCHARTING SYMBOLS

The program flowchart symbols illustrated are recommended by ANSI. A program flowchart shows the way data is processed within the computer.

Before flowcharting a program, keep the following points in mind:

1. Know the meaning of the various symbols.
2. A flowchart is a way of describing a task.
3. Before starting make sure you:
   a. understand exactly what is to be done;
   b. identify what data is to be used as input and what information must be produced as output;
   c. list any facts, formulas, and data relationships that apply to the procedure being flowcharted;
   d. list all information that must be calculated either as an intermediate result or as part of the output; and
   e. determine what modules you will need. Each module should perform one major function, such as processing records, printing a report line, or printing totals. Submodules can be invoked from your major modules.
4. Develop the mainline logic first, then proceed to develop each supporting module.
5. Use enough processing blocks so that your flowchart can be easily understood.
6. Within the symbols, keep the explanation clear and simple.
7. Test your flowchart. Walk through it, using sample data, and make sure it will process the date correctly. In selecting sample data, make sure all pathways through the flowchart are tested.
8. In the walkthrough, if a logic error or omission is detected, correct your flowchart and retest.

A flowchart is used for three functions. Its first function is to provide a guide for coding the problem. Second, when included as part of the documentation, it is used by management and auditors in determining how data is processed. The third function it serves is in programming maintenance. Often the programmer who wrote the program is not the one who maintains the program. A well-developed flowchart makes the maintenance programmer's job much easier.

Seldom will two programmers develop identical flowcharts. However, the key to success is to think in terms of modules. In a top-down approach, each module is developed as needed. Usually mainline logic modules will be very similar.

The ANSI symbols used in constructing a program flowchart are as follows:

Symbol	*Explanation*
Termination	Usually START, STOP, or END JOB is printed within the symbol.
Input/output	Used for opening files, closing files, reading records, and writing output.
Process	Any computational step.
Decision	Used when question is asked, such as whether gross pay is greater than or equal to $1000.
Flowline	An arrowhead is used to show direction. The usual direction is from top to bottom and from left to right.
Connector	When the connector symbol has a letter or number within it, there must always be another with the same letter or number.
Preparation	Used to set an indicator to a given value. It is most frequently used in the initialization, termination, and error routines.
Predetermined process	Used to indicate a submodule that will be detailed in a separate flowchart.

The following symbols are not part of those recommended by ANSI, but are used by many programmers:

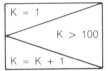

A rectangle divided into three triangles is used to illustrate a count-controlled loop. The counter (K) is initially set to 1. Each time the commands within the loop are executed, 1 is added to K. When K is greater than 100, control leaves the loop.

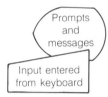

The display symbol is used to indicate a prompt, message, or formatted screen. The manual input symbol indicates that data is entered from the keyboard.

Appendix III

ANSWERS TO CHECKPOINT QUESTIONS

1. Words set in boldface type are defined in the glossary of words and phrases that is provided at the end of each chapter.

2. The answers to checkpoint questions are found in Appendix III. You knew the answer to this question, or you would not have read the previous sentence.

3. A system is an orderly means of accomplishing one or more procedures or tasks. Often a system, such as payroll, is composed of many subsystems. A procedure is a step-by-step method of solving a problem. Procedures are usually made up of tasks. A task is one step within a procedure. In a sales-order processing system, one task within the order-entry procedure is to fill out the order form. Another task might be to establish a batch total on quantity.

4. In a top-down approach to systems analysis and design, the system is divided into procedures and the procedures into tasks. Objectives are first defined for the total system and then for procedures and tasks.

5. In a batch system, data is accumulated and then processed. In contrast, in a transaction processing system, data is processed when transactions occur.

6. A system flowchart illustrates the input to be processed and the resulting output. However, a system flowchart does not illustrate how the data will be processed.

7. An analyst investigates a problem and then designs a solution.

8. Since the computer is being used to process a wider range of data and for many different types of applications, there is a greater demand for computer analysts.

9. Because there is a wider range of hardware and software from which to choose, an analyst must study many more alternatives to determine which should be used in the design of the system.

10. The six phases in the systems development life cycle are:

    a. Initial investigation. The nature and scope of the problem are determined.

    b. Feasibility study. A detailed investigation is made. The workflow of the present system is studied along with various alternatives.

    c. General design. A design to solve the problem is created that clearly shows the users' view of the application. During this phase, commitments are made regarding the type of system to be developed and the computer system (hardware and software) needed to implement the design.

    d. Detailed design. A detailed study is now made of how the system will work on the computer and specifications are developed for databases, files, reports, and displays.

    e. Implementation. Programs are written and tested, files are converted, documentation is put in its final form, and the users and EDP staff are trained to work with the procedures that make up the system.

    f. System audit. A study is made to see whether the objectives listed for the system and for each procedure and task have been achieved. The methods used in designing and implementing the system are also reviewed.

11. The word ergonomics comes from *ergo* (work) and *nomics* (law or management). The term is used to indicate the study of the relationship that exists between humans and machines.

12. An operating system is the control software that manages the computer and its resources. A comprehensive operating system makes it easier to communicate with a computer.

13. It may not be possible for an analyst to design an ideal solution to a problem due to constraints imposed by management. Typical constraints are limits imposed regarding time to develop and implement the system, personnel to be involved in the system's design and implementation, the hardware and software to be used, and the budget established for the design, implementation, and operation of the system.

14. As each phase is reviewed, a decision is made to either drop the project, schedule the next phase of the project, or have the task force submit a new report based on further study of the problem and its solution.

15. The recommendations of an analyst might be rejected because resources are not available to implement the

recommendations, new hardware or software has been announced that should be incorporated into the design of the system, or because the recommendation was disapproved for any one of many different reasons.

16. The primary objectives of the initial investigation are to determine the nature and scope of the problem.

17. The general design *must* reflect the user's view of the application.

18. A task force is two or more people, often with different skills and backgrounds, assigned to a project. Usually a better system will be designed when a task force is used than if one person is responsible for the entire project.

19. Laser beams are used to store data on videodisks.

20. The users might be working on some phases of the documentation, completing users' manuals, and designing test data.

21. A system audit is conducted in order to determine whether the stated objectives have been met and whether the operational system conforms to the specifications developed for the system. The system audit might also show that users or the operational staff are having problems with the system.

22. A system that met all its other stated objectives might be scrapped because the benefits received cannot be cost-justified.

23. A measureable objective for a payroll system would be "all employees will be paid the correct amount." A non-measurable objective would be "to provide employees with more meaningful information regarding their earnings and deductions."

24. Although the system is more costly, you could point out the additional information that would be available, how the information could be used in the decision-making process, and perhaps how queries could be made directly into the online databases. You could also point out how the new system and the services it provides compare to the system used by your leading competitors.

25. By determining how the system was developed, the reasons the system does not meet the needs of management and the users might be determined. Also, the task force and other analysts should benefit from the mistakes made in designing and implementing the unsuccessful system.

26. No. The most important factor in determining whether a system will be successful or not is often the amount of communication that occurs between the task force and the users. A successful system requires an in-depth study of the existing system and a careful analysis of the available alternatives.

27. When computers were first used for recordkeeping functions, the cost of the computer system was considered excessive unless large quantities of data could be processed and the computer could be kept running for more than one shift. For these reasons, data processing was centralized.

28. Control over acquisition and utilization of computers should be centralized. If this is not done, the computer systems acquired may not be able to communicate with one another and share common databases.

29. Due to the widespread use of microcomputers by all kinds of people, there is a greater concern regarding the security of information and computer crime.

30. A survey reported in *Computerworld* indicated that companies with a history of computer utilization have established guidelines and standards regarding the acquisition and use of microcomputers.

31. The informational needs of a small company are basically the same as those of a large corporation. However, because of the corporation's size and the volume of data, the systems developed for large corporations are more complex than those developed for small companies.

32. The communication needs of an organization must be considered when a system is being used. Today it is impossible to separate the informational needs of the organization from its need to communicate.

33. The five components of a computer system are hardware, software, procedures, data, and people.

34. The main components of a computer are memory, the central control unit which is responsible for executing instructions, and the arithmetic/logic unit.

35. The two types of software are control software (the components of the operating system) and application software (programs that process data).

36. Operating systems were developed to make computers more productive and to improve communication between the computer and the individuals who work with it.

37. The operating system is responsible for job management, system resource management, and data management.

38. IOCS (input output control systems) took care of the input/output operations. Programmers were no longer concerned with label checking, blocking of records, or many of the other tasks associated with I/O operations.

39. The four types of operating systems are single batch, multiprogramming, timesharing, and transaction processing (also called realtime).

40. New features and corrections must be incorporated into the existing operating system.

41. JCL and system commands are designed to instruct the computer on when and how a program is to be executed. Job streams (sequences in which jobs are executed) are set up by using either JCL or system commands.

42. To take full advantage of the capabilities of the computer system being used, an analyst must understand the operating system.

43. Computers are usually classified as microcomputers, minicomputers, or mainframes.

44. Although there are no set definitions, how a computer is classified is normally determined by the amount of real memory and online storage, word size, speed with which calculations are performed, and the complexity of its operating system.

45. No. One organization defined a minicomputer as costing less than $100,000 while a second organization defined a minicomputer as costing less than $50,000.

46. Microcomputers have affected the manner in which jobs are designed and the type of applications for which computers are used far more than minicomputers or mainframes.

47. A microcomputer can perform all of the functions performed by a mainframe. However, the size of the memory, amount of file space, and operating system being used impose limitations.

CHAPTER 2

1. A systems approach makes it easier to manage large, complex, and ever-changing organizations. The organization is divided into small, manageable parts that have measureable objectives.

2. When a systems approach is used, major goals are determined for the whole before the whole is divided into manageable subsystems. The interrelationships between the various subsystems are considered before defining objectives or determining the tasks that are required to solve the problem.

3. A systems approach must be considered creative since there seldom is an established formula or pattern that can be followed in solving the problem.

4. Yes. It is important to understand the relationship between the various subsystems. For example, it would be impossible to design a sales system and not consider the inventory or accounts receivable systems.

5. Yes. A systems approach is often used for solving personal problems. Assume that you are out of shape and wish to improve your overall physical condition. The major goal would be self-improvement. the problem can be divided into subsystems—diet, exercise, and an activity program. Specific objectives, achievable on a week-to-week basis, can be set in each area. Each week the results an be measured against the objectives and your progress can be charted.

6. Yes. A structured approach can be used in designing a home. The size and type of construction might be exterior and interior construction. The exterior system would consist of the foundation, framing, roof construction, windows and doors, and exterior finish. Once the components had been determined, a plan would need to be developed to determine when each component would need to be completed. In some instances, the work on two or more components could be in process at the same time. The design and construction of a home could not be accomplished without using a structured approach.

7. Responsibility for managing a complex organization is often divided into functional areas of responsibility such as marketing, finance, production, and so forth.

8. From studying the organizational chart you should be able to determine:
   a. how the business and the various subsystems are divided—by function, type of product, class of customer, geographic location, or project.
   b. the lines of authority and responsibility.

9. When a company undertakes a complete reorganization there *should* be little difference between the formal and informal organization. If no changes are made in the organizational structure, each year there will be a greater difference between the formal and informal organization.

10. The four levels of management are:
    a. strategic—top management.
    b. tactical—middle management.
    c. supervisory—lower management.
    d. functional—operational management.

11. Top management needs historical data that indicates trends, different types of financial reports, exception reports, "what if" considerations, and a great deal of external information. Lower management needs very detailed, current information, that shows day-by-day operations. Exception reports that show deviations from standards must be available so corrective action can be taken.

12. The five basic functions performed by management are directing, planning, staffing, controlling, and organizing.

13. a. directing
    b. planning
    c. staffing
    d. controlling
    e. organizing

14. The computer information service department provides services for other departments within the organization.

15. The three functional areas within the data processing are systems, programming, and operations.

16. Information services, or computer information services, is more descriptive than data processing since services are provided regarding the processing of data, storage and retrieval of information, and the communication of information to others. The term data processing conveys a more limited idea of what services are provided.

17. The computer information services manager must have good interpersonal, managerial, and technical skills.

18. An excellent programmer or analyst might make a poor manager because he or she lacks interpersonal and managerial skills.

19. The office systems coordinator should coordinate word processing, information (records) management, communication of data, teleconferencing, and telecommuting activities.

20. The database administrator is responsible for the selection of the database software, determining the contents and structure of the database, and providing for the security and integrity of the database.

21. The information center is established to provide assistance to users who wish to learn how to extract information from the databases or to develop applications for their particular area. Rather than information center, it might be called the help center.

22. The six major functional areas within systems work are: forms design and retention, work simplification, information (records) management, work measurement, analysis and design of new procedures and systems, and ergonomics.

23. A job family is a group or cluster of jobs that relate to the same activity. For example, over the years the functions performed by analysts have been identified as manager; lead; senior; A, B, C; and trainee.

24. Conceptual skills permit an individual to study a problem and visualize solutions for the problem that fall within certain parameters. Interpersonal skills are people oriented and allow an individual to work effectively with others.

25. The lowest priority of need is physiological—the basic need for food, clothing, shelter, medical services, and so forth. The highest need is for self-fulfillment—the necessity for realizing one's fullest potential in creativity and self-development.

26. The egotistic need is for recognition, status, and importance. After more basic needs are met, a new title may be more important that the increase in pay that goes with the promotion.

27. Since the only constant in the field of EDP is change, an analyst must continually update his or her skills and learn about new products, procedures, and concepts.

28. An analyst studies the problem, gathers facts, and comes up with a design for one or more solutions to the problem. When one of the solutions is selected, specifications are developed. The analyst may also prepare the overall logic plan. The programmer takes the specifications and general logic plan and prepares a detailed logic plan from which code can be written. Both the analyst and the programmer may be involved in the testing, documentation, and evaluation of the programs that make up the system.

29. The five functional areas within operations are data entry, data control, scheduling, computer and peripheral operation, and librarian functions.

30. Throughput is the amount of data that can, within a given period, be entered into the system, processed, and outputted in final form.

31. Batch totals and record counts were built into the time card edit program. The totals were established in the payroll department. However, the data control clerk checked the totals prepared by the payroll department with those printed on the report.

32. Controls are not eliminated when online systems are developed. Different types of controls may be included as part of the programs that process the data. Every effort must be made to prove that the data entered into the system is correct and that it is processed correctly.

33. Due to the high cost of entering data into the system, analysts are more concerned about when, where, and how the data will be entered into the system.

34. The trend today is to decentralize the data entry functions.

35. A librarian labels and store tapes, diskettes, and disks; determines when tapes and disk packs can be released; cleans tapes; and may also be responsible for maintaining documentation and standards manuals.

36. Computer operations is responsible for obtaining the most effective use of the equipment and getting as much throughput as possible.

37. If the job documentation is inaccurate or incomplete, jobs may abort or may produce invalid output.

38. An analyst working in a large organization has a more limited job description than does an analyst working in a small organization. Because the department that has minicomputers may have a limited staff, the analyst may perform tasks that in a large department may be performed by someone else.

39. Often a microcomputer is not used effectively because the individuals using the equipment have not learned to use the software efficiently. The individual may not have an in-depth understanding of what the system is capable of doing.

40. One advisory committee consists of top management and will meet infrequently to determine goals, objectives, and major policy changes. The second committee consists of middle management personnel and reviews recommendations made by project teams, establishes priorities, and makes decisions regarding day-to-day operations of the information service department.

CHAPTER 3

1. Projects have definite beginnings and endings, are unique, and often consist of one or more phases. Because each is unique, different talents and skills are needed for each project. By using standardized tools and techniques, the completion date and cost of developing a project can be determined.

2. A standard is a uniform method of accomplishing a task. Standards are available for developing systems and programs.

3. Before a plan can be developed, a list of required tasks must be developed. The time needed to complete each task should be estimated and the tasks that can be completed concurrently should be identified.

4. By studying a Gantt chart you can see the scheduled, in progress, and completed tasks.

5. A PERT chart illustrates the tasks that must be completed, how long each task should take, and the sequence in which the tasks must be executed. By using a mathematical formula, the shortest period of time needed to complete the project can be estimated.

6. In gathering facts, analysts interview people; use questionnaires; watch tasks being completed; study documents, manuals, and forms; watch demonstrations; and so forth.

7. By studying a detailed data flow diagram or a Diagram 0 you can determine the entities that provide input or receive output, the data stores that are developed or updated, and the process that transforms data into information. A data flow diagram does not show the logic used in the execution of the process, the controls built into the system, or exception routines that are initiated but not performed by one of the entities identified.

8. A context data flow diagram show the entities that supply data and receive information. The processes and data stores are not shown.

9. A logical model shows what the system does and the emphasis is placed upon the data and the processes. The physical model shows the processes in sequence and emphasis is placed on documents, entities, and forms.

10. Yes. Some analysts combine the characteristics of a logical and physical model into one diagram.

11. In developing a Diagram 0, the analyst should:
    a. Determine the data stores being used.
    b. List the major events that occur within the system.
    c. Draw a separate data flow diagram for each event.
    d. Assemble the separate data flow diagrams into a single data flow diagram.

12. A data flow diagram does not show the unique procedures used to start up a system or the ones needed to terminate a system.

13. A square is used for an external entity, data flow is represented by a line with an arrow, processes are represented by circles, and data stores are shown by a rectangle with an open right side.

14. An asterisk indicates that two data streams are required in order to produce a single output or that two outputs are produced from one process.

15. A ring-sum operator indicated an either/or situation. One of two data streams will be followed.

16. A system flowchart does not show how the computer processes the data. However, it does show the movement of data throughout the system.

17. System flowcharts are used by auditors, managers, users, and programmers.

18. An exception routine is a series of instructions or tasks that will be followed when a deviation occurs. No. Although a number of exception routines are built into procedures and systems, some of the routines may never be executed because abnormal occurrences do not take place.

19. When a file is updated, a flowline with arrows at both ends connects the rectangle that shows the program being executed with the symbol representing the file or database.

20. In developing a system flowchart, separate symbols are available for each input and output medium. The major storage media such as magnetic tape and disk are also represented by unique symbols.

21. Fewer symbols are used in developing data flow diagrams than are used in developing system flowcharts.

22. In developing system flowcharts uniform symbols should be used, a limited amount of detail should be included within the symbols, and the chart is developed from left to right and from top to bottom.

CHAPTER 4

1. Over half of today's work force is considered knowledge workers.

2. A knowledge worker does not produce a service or product and relys on information to perform his or her job.

3. In batch processing, data is collected and then processed in a batch.

4. In batch processing, or any other type of EDP, once the data is recorded on a medium that can be used to process the data electronically, multiple uses can be made of the data. This reduces the amount of human intervention and data handling.

5. With the development of more online systems, computers have become more user friendly, more data is online, fewer reports are printed, data entry and data processing have become more distributed, fewer source documents are used, and more controls are designed into systems. Also users are more involved in the design and development of applications and are less dependent on the EDP department for ad hoc reports. When a batch system is converted to an online system, the physical model of the application changes far more than the logical model.

6. When the physical fitness billing system was redesigned, the data entry function was distributed and some of the procedures were online rather than batch jobs.

7. A menu is a display on a VDT that indicates the options that can be executed. When an option is selected, a program will be loaded in memory and executed, a new menu will be displayed, or the main menu will be displayed. A *help* option is usually available that provides documentation regarding the use of the other options.

8. A formatted screen displays the names of the fields of data to be entered and also shows the size of the field. The cursor moves to the location where the variable data is to be entered. Often the operator is asked to review the material keyed in before it is entered into the system.

9. The file maintenance and charge procedures are online systems. As soon as a transaction occurs, the data enters the system and the appropriate files are updated.

10. The programs that produced reports were batch jobs. Also the procedure used to record deposits was a batch job. The total of all receipts entered had to agree with the total recorded on the deposit slip.

11. More ad hoc requests using the query capabilities of the system will probably occur because a more comprehensive database will be available. Because queries are made by users, fewer formal reports will be printed.

12. A file is often online only when the information is needed and is used for a limited number of programs. A database is usually online as long as transaction processing jobs are being run and is also used for a number of different application programs.

13. A batch oriented system often uses master and transaction files and after data is accumulated, both the data entry and data processing functions are centralized. An MIS is dependent upon the availability of online databases. Often both the data entry and data processing functions are decentralized. When an MIS is available, query languages are generally available and fewer reports are printed.

14. One of the most important ingredients of a successful DDS is an easy-to-use human interface which makes it easy for the user to get responses to "what if?" questions. In the future, more queries will be made by using voice input and the responses from the computer may also be provided using voice output.

15. In a large organization there may be great deal of hardware and software that does not have a standard interface. Also there is a lack of communication and cooperation between EDP and office automation personnel.

16. Artificial intelligence is the capability of a computer to simulate functions such as reasoning, learning, and self-improvement.

CHAPTER 5

1. Analysts should become involved in an office automation project in time to provide input regarding the procedures to be automated, the selection of the hardware and software, and the design of the office.

2. When a commitment is made to office automation, space must be provided for the equipment; wiring and cables must be provided for the new equipment; and provision must be made for the comfort of the personnel who will use the equipment.

3. Management has had the problems developing solutions to the problems occurring within the office due to the complexity of the solution, the multiple sources of products, and the lack of standardization.

4. Over half of today's work force is considered knowlege workers—those who work with information rather than producing goods or services.

5. The accounts receivable file would not be considered an up-to-the-minute database. If the resources are available, an analyst should suggest that an online system be developed. As a sales transaction occurs, the appropriate master file should be updated.

6. The environment of many offices has changed a great deal as more functions are automated. Wall systems are used rather than permanent walls, furniture is obtained that meets the needs of personnel and equipment, better lighting is provided, and more thought is given of the control of noise, heat, and acoustics.

7. When wall systems are used, various arrangements can be obtained. As the need for smaller or larger offices occurs, the wall systems can be moved to provide the space necessary. In some offices, people can elect to combine their wall system with others or to have a smaller, but private, office.

8. Predictions vary regarding how many VDTs will be available in a modern office. By 1990, one VDT may be shared by two or three individuals. However, some predictions are that 80 percent of the people who work in offices will have a VDT.

9. More VDTs will be used because of the availability of more online information and better query languages. Therefore, the amount of hardcopy (printed reports) should decrease.

10. Many of the office automation functions require computers, peripheral equipment, and software which supports multiprogramming, timesharing, teleprocessing, and the office automation functions. Most applications require a substantial amount of online file space.

11. When word processing systems were first developed, many people failed to recognize that the same technology was required for both applications. Files created under WP could not utilized for EDP applications.

12. Mainframe and minicomputer manufacturers developed software for word processing and other OA functions when they realized that the OA market was predicted to be twice as large as the EDP market.

13. Within large organizations either the mainframe is used for both applications or standards are developed that require WP equipment be able to interface with EDP equipment.

14. Word processing is cost-effective for applications in which the document is printed and used only once. When a typist or stenographer becomes familiar with word processing codes and techniques, he or she is more productive than when a typewriter is used to create letters, memos, and other documents.

15. Word processing software provides for creating text, editing text, storing text in files, formatting printed letters and reports, and backing up files.

16. The backup file is provided in case something happens to the original file of text material. It is possible that a portion of the disk becomes defective and the material stored on that area of the disk cannot be recovered.

17. When programmable terminals are used, some of the software as well as a portion of the text being created or edited is stored within the memory of the terminal. Many of the required commands will be executed without using either the logical or computational ability of the host computer.

18. When microcomputers are used for word processing, faster access to text is obtained when hard disks rather than diskettes are used. Also more data can be stored online when hard disks are used.

19. When decisions are being made regarding word processing, the analyst should help: determine what guidelines and standards should be used in obtaining equipment, establish criteria for evaluating hardware and software, and determine what features are considered essential.

20. Most word processing software includes, or has available as separate programs, software that can: check the spelling of words used in the text, check for grammatical errors, and combine two files of text. One file usually contains the text and the second file contains information that is to be merged with the text.

21. The most efficient way to produce the letters is to create two files. One file has the basic letter that will be sent to all clients. Stops are programmed so that whenever variable data is needed, the data is obtained from the second file which contains the names, addresses, and any other variable data that is to be merged with the information stored in the first file.

22. When a user signs on to the system, an account number and password are entered. The computer checks to see that the account number is valid and that the password is the correct one for the account number entered.

23. When a message is acknowledged, a message is sent to the sender indicating when the message was received. A log of messages sent and when they were received can be printed or reviewed.

24. Electronic mail is faster, more reliable, and less costly than using next-day mail services. Confirmation that the document was received is also provided.

25. A modem is a device that translates data back and forth from characters the computer can understand to the type of data that can be transmitted over a communication system.

26. If files aren't deleted when no longer needed, an excessive amount of online file space will be needed. Also it may cause delays in retrieving information from online files.

27. In order to schedule a meeting between several principals, each person's schedule must be stored in a file. The identification codes of the principals and the dates that are acceptable for the meeting are then entered.

28. Teleconference equipment can be justified by showing the time saved and the quality of meetings that are possible. The savings from eliminating the need to travel may offset the cost of the preparing materials and conducting a teleconference.

29. When a facsimile system is used, a copy machine reads the document and causes the information to be transmitted to a second copy machine that produces a hardcopy like the original. Electronic mail transmits information stored in a file into another user's file. The second user can display or print the information.

30. A computerized telephone can record messages, assign priorities to messages, sort the messages by priority, and print out the messages. A busy executive can review the messages to determine which one should be answered.

31. Surveys indicate there has been very little increase in the productivity of office workers.

CHAPTER 6

1. The objectives of the initial investigation are to: accurately define the problem; determine the scope of the problem; determine the resources needed to complete the feasibility study; estimate the resources needed to complete the project; determine the priority to assign to the project; and accurately estimate the resources needed to modify an existing system.

2. The five reasons investigations are made are: a modification to an existing system is needed; a new system is needed to solve a problem; management feels a system should be redesigned so that it will be more cost-effective; problems such as poor security or lack of controls have been detected within the old system; and new technological advances have occurred that should be investigated.

3. Users are more aware of problems that occurred in using the present system.

4. Management may request an investigation because there is not enough information available with which to make decisions. They may also request a study because problems with the existing system have been identified. Management may also have heard about an innovative new system that reportedly allows a more advanced decision support system to be designed.

5. No. General guidelines or standards should be available. However, many different methods may be used to complete the initial investigation.

6. There are no clear guidelines as to what should be done in the initial investigation and what should be done in the feasibility study.

7. A form should save time, avoid misunderstanding, be easier to use than writing a memo, and provide more information.

8. A form should be easy to fill out and should also provide complete information.

9. The computer policy committee determines policies and sets priorities.

10. If the CIS manager approves or disapproves all requests and assigns the priorities, the projects that the manager is most interested in may be approved and receive a higher priority than those that might be more important in regard to the organization's goals and objectives.

11. If all requests must be approved by the computer policy committee, requests for modifications of an existing system that should be implemented immediately may have to wait for the committee to meet.

12. An initial investigation is initiated by one or more individuals submitting a form stating the purpose of the investigation. The request is reviewed by an analyst and then submitted to the computer policy committee for action.

13. In the initial investigation, data is gathered and analyzed, the tasks performed are documented, a report is prepared that summarizes the project team activities and recommendations, and an estimate is made of the resources needed to complete the feasibility study.

14. During the initial investigation both internal and external documents are studied.

15. The three major purposes of the memo are to confirm the time and place of the appointment, list the points to be covered, and enable the person to be interviewed to prepare.

16. Finding out as much as possible about the person to be interviewed will help the interviewer to establish rapport with the individual and to determine the best way to proceed with the interview.

17. The form provides all the pertinent facts about the interview—who, when, and why—and also lists the questions that should be covered.

18. No. If the analyst takes extensive notes, his or her attention is not focused on the person being interviewed. Unless the person being interviewed suggests that a tape recorder be used, the interviewer should not use one.

19. The follow-up memo thanks the person interviewed for his or her time and lists the conclusions that were arrived at during the interview. The memo serves as documentation regarding the outcome of the meeting.

20. Observations are also made during initial investigations.

21. The standards manual should be written for the users. Nontechnical language should be used and the five C's—clear, concise, correct, complete, and courteous—should be put into practice.

22. The information in the standards manual is used by the individuals who must perform the tasks described, trainees, auditors, managers, analysts, and anyone who wishes detailed information regarding a particular task or topic.

23. If a standards manual is kept current, new sections will be added and obsolete sections deleted.

24. Damuth, during lunch with Paulson, learned about the word processing software that was part of the office automation package.

25. The office supervisor was interviewed to determine how the responses to complaints or requests for information were handled.

26. The job descriptions were studied to determine what tasks were being performed and to determine whether any standards regarding the employee's productivity were established. The rate of pay is needed in order to determine the cost of the present system.

27. When forms that establish a standard format are used, more complete information is usually included.

28. In order to determine the cost of the present system, the analyst must know the number of letters processed and the time that it takes to process each letter.

29. Yes. The second memo to Damuth indicates he was organized and that the meeting was very productive. The memo also indicates that Mary Green was receptive to the proposal and offered to supply any information that was needed.

30. All tasks regarding the processing of the letters need to be identified to determine the cost and to determine which tasks can be automated or performed manually in a more efficient manner.

31. Yes. The objectives of the new system were stated and the methods to be used to investigate the problem were listed. The issues identified were addressed in the conclusions and recommendations sections of the report.

32. Initial investigations must be documented so that when the preliminary investigation is begun the project team can determine exactly what had been done and the conclusions that were reached. Also if the study has a low priority, there might be a good deal of time between the two phases.

CHAPTER 7

1. The objectives of the feasibility study are to further define the problem, determine alternate solutions, ascertain the resources needed to complete the systems work, and estimate the cost of the proposed study.

2. Top management must provide support for the project, answer questions about the existing system, make materials available, make personnel available, and suggest possible solutions to some of the problems.

3. Before a consultant is hired, an investigation must be conducted to determine whether the consultant being considered has the required background. If the consultant is a member of a long-established firm that has a good reputation, the client need not conduct an investigation.

4. Large companies with their own staff sometimes hire a consultant who has in-depth knowledge in a highly specialized area. A consultant might also be hired when an impartial opinion is needed.

5. The disadvantages of hiring a consultant are: the consultant is not familiar with the organization; the organization's employees may resent the consultant and not cooperate; the consultant benefits most from the learning experiences derived from doing the study; and the consultant may remain with the firm and do tasks that should be assigned to the organization's employees.

6. If a consultant is not hired, the other options available are to: have someone within the organization released from other responsibilities so that he or she may do the study; work with vendors; provided in-service training for staff members; or hire a small-shop specialist.

7. By studying the documentation and the report, the analyst can determine what facts have already been gathered, what decisions have been reached, and what resources are committed to the feasibility study.

8. Usually routine decision-making processes such as when to reorder items that have fallen below the reorder level or when to send a collection letter to a customer with a past-due account can be automated.

9. Small tasks are more manageable and it is also easier to determine whether the project is being completed according to schedule.

10. Making a list is one way to divide the project into manageable tasks. Also, once the list is completed, it can be ordered according to the sequence in which the various tasks should be completed.

11. The six basic questions that must be answered are: What is being done? Why is it being done? Where is it being done? Who is performing the task? When are the tasks being done? How are the tasks being done?

12. During the feasibility study, operational personnel are interviewed. In the initial investigation usually only management personnel are interviewed.

13. The analyst studied the initial investigation report and documents, articles and technical material from vendors, cost studies, industry standards, letters and financial records, job classifications, and the standards manual.

14. By observation an analyst can determine the workflow, the locations and type of workstations, how the tasks are performed, and who performs each task.

15. A systems flowchart is easier to understand than a written or verbal explanation. Also in order to complete the flowchart, the analyst must understand the workflow.

16. A flowchart does not show the time that a task takes or the distance that a document travels between workstations.

17. Yes. People who must share material should be close to one another. Also individuals who must concentrate are more productive if their workstation is not in a traffic pattern or near a place where people tend to congregate. Another concern is the location of impact printers which tend to be distracting.

18. Yes. Since it has a direct impact upon productivity, analysts must be concerned with the physical environment of the worker.

19. The best way to determine how a task is performed is by observation. The documentation may be obsolete.

20. A questionnaire should be used to gather data when a limited amount of information is to be obtained from a large number of people.

21. The questionnaire should be short and have checklist types of questions. All questions must be clearly stated. The questionnaire should be tested on a small group of people before being sent out to the entire population.

22. If an invalid sample, one not representative of the total population, is used, the results will be invalid.

23. By visiting another installation, an analyst can obtain information regarding the performance of the system's hardware and software. Information might also be obtained regarding unique applications and the manager may share some of the problems encountered in using the computer system.

24. When listening to a sales representative's presentation, an analyst should remember the vendor's primary purpose is to sell products.

25. When investigating letter-quality printers, information published in current publications that evaluate different types of printers should be studied, material should be obtained from vendors, and a demonstration of some of the printers should be obtained.

26. To evaluate spreadsheet software, articles comparing different programs should be obtained, material should be obtained from the vendors, and a demonstration should be obtained. After the demonstration, a potential buyer should sit down and try using the program. The software should be internally documented and the external documentation should be complete, current, and easy to follow.

27. The constraints that might be imposed are time, money, personnel, and equipment.

28. After all data is gathered, it must be analyzed and decisions must be made. This phase of the study is more difficult than gathering the data.

29. A presentor should tailor the presentation to the audience. This cannot be done if the presentor has no prior information regarding the background of the audience.

30. A presentation on the word-processing system with an objective of training technicians would be very different from a presentation with an objective of providing an overview of the new system to the line officers of the organization.

31. An outline helps the presentor to plan the presentation. During the presentation, the outline helps the presentor to stay on target.

32. Technical terms must be defined so that the audience will understand the speaker.

33. Visual aids make the presentation clearer and help to keep the interest of the audience.

34. A program can be written that will cause "screens" of graphics or information to be displayed on a large monitor. Using the computer combines the features available when overheads and slide projectors are used.

35. Motion pictures are expensive and in a high-technology field tend to become obsolete. However, some of the major vendors do have some excellent films that can be used in making presentations.

36. A transparency master is far easier to use. The material can be prepared in advance, color and graphics can be used, and often the information is better displayed than when a chalkboard is used.

37. The presentor should check out the equipment to make sure everything functions properly.

38. If a different analyst is assigned to the project or if during the presentation questions are asked, the material is available. Also if a period of time elapses between when the feasibility study is finished and the next phase of the study begins, the project team will begin by reviewing the documentation.

39. The cost of the new system should be computed by using the ongoing cost of labor, equipment, and materials. In addition, the developmental costs should be prorated over the expected life of the system.

40. If standards and guidelines are not developed and used, costs for systems may be calculated differently. At best, some of the cost figures are only estimated; therefore, it is essential that costs are calculated in the same way for all projects.

CHAPTER 8

1. The feasibility study phase determines *what* must be done, and the design phase identifies *how* it will be done.

2. All major commitments are made during the design phase. After the design is accepted, there should be no major changes.

3. No. It is imperative that as the need arises, additional topics will be investigated.

4. Yes. Although the analyst is responsible for the design of the new system, many of the user's ideas will be incorporated into the design.

5. Before designing an online sales system, the analyst should make certain there is adequate CPU time to process the additional data and that there are enough ports available to accommodate additional terminals. The analyst must also find out if the additional terminals will make the response time adequate for all of the users.

6. The factor that was not considered in each of the situations described was:

   a. Designing "one large computer program" is not in accord with the guidelines to avoid undue complexity.

   b. If the sales manager was not consulted, it is doubtful that the needs of the user were a prime consideration.

   c. Having a clerk look up a customer's credit limit violates the principal that routine, repetitive manual functions should be automated.

   d. If the present computer system can only process the current volume of data, provision was not made for expansion and growth.

   e. If documentation is not developed and kept current, informal systems will develop.

   f. If the analyst did not consult with the accounts receivable or the inventory manager, it is doubtful that an integrated system was designed.

   g. If letters are being printed and mailed stating orders cannot be filled because a customer's credit limit was exceeded, the proper level of automation might not have been achieved. If letters are generated automatically, they should be reviewed by the credit manager.

   h. If the manager was not informed that three fewer data entry operators were needed, the impact of the system on its total environment might not have been considered.

   i. The inventory check should have been automated. Perhaps two guidelines were violated: routine, repetitive manual functions should be automated, and the systems and procedures with which the new system will interact should be considered. Sales processing has a direct impact on the inventory system.

7. If the departments have representation on the committee, there will be better communication between the departments and the project team. This may lessen the

resistance to change. In addition, the representatives from other departments contribute their expertise.

8. Even if the team consists of only two people, there should be a leader. The leader is responsible for the project, has been given a certain amount of authority, and is also designated as the contact person for individuals who needed information.

9. Before a PERT chart can be developed, the tasks to be performed must be identified and how long each task will take must be determined. The analyst must know the order in which the tasks will be performed.

10. Before the critical path can be determined, the relationships between the activities and the time to complete each activity must be determined. The critical path is determined by adding the times of the activities along the different paths. The critical path is the combination (or path) of activities that take the longest time to complete.

11. An activity is the application of time and resources to achieve an objective. An event is the point in time at which an activity begins or ends.

12. In structured analysis and design, the major goals and objectives are determined for the total system. The system is then divided into subsystems and the subsystems into procedures. Each procedure is divided into small, manageable tasks. Both the design of the system and the development of detailed logic plans for individual programs are developed from the top down.

13. A hierarchy chart shows the procedures that make up a system and the relationships that exist between the procedures.

14. The PERT chart and the hierarchy chart are reviewed to determine whether the time allocated to the various activities is realistic and to ascertain whether there are omissions or changes should be made in some of the major procedures. The specifications for all of the individual procedures must identify all of the major functions performed by the system.

15. Although the content is similar, the feasibility report is less detailed and less accurate than the general design report. Also the feasibility report tells what must be done and the general design report starts describing how various procedures will be done. Cost and time estimates are also more accurate in the general design report than in the feasibility report.

16. If good practices have been followed, users and managers were directly involved. If they have provided input and approved the design as it was being developed, users and managers will certainly approve the report.

17. A study might be conducted because management feels a new computer would be more cost-effective or because a new application is to be designed and there is not enough file space or CPU time to implement the new application. Users may want their own system that will become part of a network.

18. No. The membership of the project team for the computer selection study should have as members CIS staff, representatives from different areas of management, and members of the computer policy committee. The CIS staff members should provide the technical knowledge that is needed.

19. The first step in the selection of a new computer system is to determine the applications for which it will be used.

20. Before an RFB can be developed, the applications for the proposed system, the size and number of records to be stored, and the amount of online transactions to be processed must be determined.

21. In response to an RFB, a vendor usually provides a detailed description of the hardware and software being bid, the financial and maintenance agreements that are possible, the training programs for management and the CIS staff, a conversion plan, and bids on optional items.

22. No. Some vendors include the price of the operating system software in the basic price of the CPU, while others vendors list the price of each item separately.

23. A benchmark is an evaluation of the performance of a computer system. A typical mix of programs is run on several different systems and the results compared.

24. When a benchmark is done for a timesharing system, all systems being evaluated should have approximately the same number of users on the system as will the computer obtained for the designated applications.

25. If two computer systems can communicate effectively with each other, each backs up the other. In addition, files and programs can be shared.

26. If a computer that has been on the market for five years is obtained rather than one that is just announced, less capital outlay may be required and the ongoing monthly costs may be less. Several questions must be asked: "Does the system meet the needs identified. Is there room for growth and expansion?" "Will the computer continue to be supported by the vendor?"

27. The organization's tax consultant, accountant, and investment consultant should be directly involved in the decision to purchase or to lease a computer system.

28. It is sometimes possible to obtain an entire system, hardware devices, or software more cheaply from a third party than from the original vendor.

CHAPTER 9

1. During the general design phase, an analyst determines the medium to be used and what information will be needed for reports, displays, or files. During the detailed design phase, reports and displays are designed and file layouts are completed.

2. The statement is false. Sometimes the analyst finds that two reports are almost identical and can be combined into one report that serves the needs of the two users. Also users may ask that more information be added to

ANSWERS TO CHECKPOINT QUESTIONS

a report. Unless the changes are major (which is doubtful) and would prevent the project from being completed on schedule, the reports will be redesigned.

3. The analyst and user must determine: what output is needed; how the output will be used; if a turnaround document can be used; if the output will be used internally or externally; and how important it is that the data be timely.

4. Internal reports are usually printed on stock paper while external reports are printed on preprinted forms. Also, certain external reports may have to conform to the specifications of a government agency.

5. Several factors should be challenged. One of the major concerns is the timeliness of the data. Also the reports would be very large and costly to print. The sales manager should reject the design and insist that a transaction processing system be developed—databases are to be updated as the transactions occur. Also the updated databases must be accessed by the sales representative to determine the status of an account or of an inventory item.

6. When two-up forms are used, two identical copies are printed simultaneously. Two-up forms can be used to provide two originals, each of which might have as many as four or five carbon copies.

7. How the output is to be used determines which output medium should be used. If the output is seldom used and only serves as historical data, magnetic tape might be used as part of an audit trail or in the preparation of additional documents, a printed report may be needed.

8. A turnaround document is one that is printed as output and then a portion of the document is used to input the data back into the system. Utility bills and insurance statements are examples of turnaround documents.

9. Most people would probably answer as follows:

   (a) printer   (b) hard disk
   (c) magnetic tape (d) COM
   (e) voice output   (g) printer

10. You would probably obtain one or two diskette drives, and a letter-quality printer.

11. A company that has never worked with punched cards would probably elect to use terminals. If one of IBM's small-business systems was selected, diskettes might be used for recording and entering variable data.

12. Analysts should be aware of the advancements that are being made in the various types of I/O devices. For each I/O device, they should know the advantages, disadvantages, and applications for which it might be used.

13. The analyst should (1) make certain the form provides all of the required information; (2) create a form that is easy to handle; (3) make certain the form represents the company favorably; and (4) work with a representative of the company that will print the form.

14. The rules that must be followed in completing the design are: outline the form with a heavy line; print the data that is to be preprinted on the form; use Xs to indicate where the variable data is to be printed; and use dashes to indicate where the form is to be perforated.

15. If the user signs the layout for the form, it is an indication that the form is acceptable. at some time in the future, the user cannot say that the form is unsatisfactory because an opportunity to review it was not provided.

16. On a print layout form for an external report, the constant information that is printed will be preprinted on the forms.

17. On a print layout form for an internal report, the information lettered on the form is printed by the organization's printer.

18. When determining how long a printed report is to be retained, several factors must be considered: use of the report, company policy, and governmental regulations. After two or three weeks, the report might be microfilmed.

19. A decollator removes the carbon paper and separates the various parts of the form.

20. A burster separates individual pages of the report. Both decollators and bursters may have slitters that remove the pinholes from the sides of the paper. A number of optional features are also available on decollators and bursters.

21. After a report is printed, the documentation should be used to determine how the validity of the output is to be determined, whether the report is to be decollated or burst, and who should receive the report.

22. The basic rules to follow in designing screens are: clear the entire screen between displays; format the output so it is easy to read; prevent scrolling; don't over use color; be consistent; develop and use conventions; and test all screens.

23. The two conventions used in designing the screen for the sales-order system were: use a y for a positive reply to a question and display the instructions at the bottom of the screen.

24. Someone unfamiliar with the application should attempt to enter data and to follow instructions displayed on the screens.

25. Software is available that permits the user to design a screen by keying in the data and formatting it on the VDTs screen. Once the screen is designed, the format is given a name and stored in a file.

26. The costs associated with data entry are increasing while other data processing costs are decreasing. Also, there are many more feasible methods of handling data today than there were ten years ago.

27. The four major sources of input are transaction files, master files and databases, control data, and data entered from some type of terminal.

28. The amount of input that must be entered from transaction files, keyed in, or entered by using scanners, voice input, or some other method depends upon what output is needed and what information is already available in a master file or database.

29. Source documents, screens, and records for a particular application should be designed concurrently so that the flow of information is the same on all three. When the screen (or record layout for a data recorder) matches the source document, fewer errors are likely to occur and the operator will be more productive.

30. In designing the input format, the analyst must determine the record length needed, the length of each field, the codes that will be used, and the correlations that should exist between the source document and the input record.

31. Data entry operators may not be as productive as desired because source documents are poorly designed, inadequate instructions are provided, reference material is not available, or the quality of the data recorded on the document is very poor.

32. If invalid data enters a computerized system, the data may be part of several reports, master files, or a database. Once invalid data gets into a system, it is far more expensive, difficult, and time-consuming to make the necessary corrections than if the error can be detected before the data is processed.

33. An intelligent terminal can be programmed. For example, the terminal's microprocessor is programmed by the operator to check a particular code and see that it is a letter from A to E. Any other letter will cause a message to be displayed. The host computer is not involved in checking the data. If the same checking were done when a "dumb" terminal was used, the data would be transmitted into the host computer, checked, and if necessary, a message transmitted back to the operator.

34. External source documents should be easy to complete, have space to record all of the required data, provide adequate instructions, and represent the organization well.

35. Lines or squares are used to encourage the individual completing the form to print and to indicate the length of the fields.

36. Codes are identified on source documents to prevent errors. If a line were used rather than squares that contain the acceptable codes, an invalid code could be entered. Also the form is easier to complete if there are boxes that can be checked.

37. The data should be in the same sequence on all three items—the source document, display screen, and transaction record.

38. A transposition might have occurred and 45 was entered as 54, a zero might have been dropped and 100 was entered as 10, an extra 0 or digit was added, or the figures on the document were misread. Errors should be found so that the right quantity is shipped to the customer and the customer is billed for the correct amount.

39. The sales orders were prenumbered so that the number printed on the order could be used as a reference when discussing the order with the customer. The number that establishes the audit trail is the one stamped on the document.

40. By entering the first and last number, the computer can be programmed to check each reference number entered. The number must not be less than the first number or higher than the last number.

41. By using codes, less data needs to be keyed in and stored in online files or databases.

42. Zip codes are examples of group classification codes.

43. Libraries often use a decimal code.

44. A mnemonic code is one that makes use of letters or numbers that are easy to relate to the subject. For example, F for female is a mnemonic and easier to remember than if a 1 were used to indicate a female.

45. Yes. Forty-four used to represent Riverside High School is an example of a simple sequence code.

CHAPTER 10

1. A new era in data processing was created when magnetic disk storage became available. For the first time data could be accessed randomly.

2. During the general design phase, the content of files and the type of storage medium to be used are determined. During the detailed design phase, the field size and type of data stored in each field is determined.

3. Under each operating system, data is formatted on diskettes differently. Data stored on a diskette using an Apple computer may not be read by an IBM PC. However, if the same operating systems are used for both microcomputers, it *might be* possible.

4. No. The data is tranferred to magnetic disk for storage and the programmer is only concerned with how the records are formatted on disk. The records are automatically transferred from tape to magnetic disk by the MSS software.

5. Records stored on magnetic tape can only be accessed sequentially.

6. The major advantage of using magnetic tape is that per megabyte of storage it is far cheaper than storing data on magnetic disk.

7. The blocking factor is determined by the programmer or analyst. When tape is blocked an entire block (a physical record) is read or written. Therefore it takes far less time to read data from or write data on tape when physical records rather than logical records are read or written.

8. EBCDIC is an eight-bit code that can store 256 unique characters. ASCII is a seven-bit code that can store 128 unique characters. EBCDIC is used extensively for mainframes; ASCII is widely used by minicomputers and microcomputers.

9. When physical records are read or written, the data is stored in a buffer. The operating system software unblocks the records in an input buffer and moves one logical record at a time into an area where it can be processed. Information after it is processed is stored in a buffer (one logical record at a time) until the buffer is full. Physical records are written on tape or disk.

10. A parity check is a means of determining whether data is stored correctly on tape, disk, or in real memory. After data is stored, the operating system software checks to see whether either an odd or even number of bits is activated. Odd parity machines check for an odd number of bits; even parity machines check for an even number of bits. If data is stored incorrectly, a routine that will attempt to solve the problem will be executed by the operating system.

11. Header labels are written on tape so that when a tape is being used as input, the operating system software can determine whether the right file is being used.

12. Magnetic tape is used for backing up files, for transmitting reports, and for transmitting data between computer systems. A database would not be stored on magnetic tape since it would take too long to sequentially access the records stored on tape.

13. The five factors that should be considered when selecting disk drives are the capacity, organization and access methods supported, access time, cost of storing a megabyte of data, and the special features supported.

14. An analyst should know the capacity per track so that the maximum blocking factor can be determined.

15. Access time consists of seek time (positioning the read/write mechanism), rotational delay time (time that it takes for the records to be positioned so they can be read), and transfer time (the time that it takes to transfer the data from the disk into real memory).

16. The blocking factor can be determined by using the table supplied by the vendor. The ideal blocking factor when processing disk records is to store as many logical records on one track as possible. Therefore, the blocking factor should be the same as the maximum number of records that can be stored on a track.

17. An actuator is the access arm and the read/write mechanism.

18. When an operating system supports dynamic disk allocation, the operating system determines where each file will be stored on the online storage that is available.

19. The four most popular organizational methods are sequential, VSAM, ISAM, and direct.

20. VSAM files are preferred over ISAM because records can be randomly retrieved faster. Also less maintenance is required and both a primary key and an alternate key can be used. Therefore, within a single program, records can be accessed both randomly and sequentially.

21. An ISAM must be reorganized so that records stored in the overflow area can be positioned in the prime area. If this is not done on a periodic basis, the overflow area becomes full and the job will abort.

22. If a control interval becomes filled, a split occurs. The records stored in the one interval are rewritten onto two intervals and free space is available at the bottom of each interval.

23. The primary key is used to retrieve the master records; the alternate key is used to retrieve the first record of the group of records that has the same key. The rest of the records stored in the group are retrieved sequentially.

24. The primary key must have a unique value such as a social security number. The alternate key is unique for the group. A location number might be an alternate key. Everything stored in bin 428 can be accessed by using the B428 key.

25. Lasers are used to record or read data from optical disks.

26. Far more data can be stored on a single 14-inch optical disk than can be stored on a reel of magnetic tape or a 14-inch magnetic disk.

27. The cost of the 5 ¼-inch optical disk drive is expected to be only slightly more than the present cost of a Winchester drive.

28. The trends regarding databases have been to: increase accountability; have more centralized databases; increase the number of terminals used; develop more complex programming software; and develop sophisticated database software for small computer systems.

29. A data item is the small storage unit within a database (also called a field).

30. The relative record number is the number of the record within the file. If 9 records precede your master record, the relative record number of your record is 10.

31. The relative record number is used to randomly retrieve the first transaction record. The relative number stored in that record is used to retrieve the second transaction record and so forth.

32. When a zero is stored in the link field, it indicates that all of the transaction records for the group have been processed.

33. Records stored in a relational database should be visualized as being placed in a two-dimensional table. Each row contains a separate record; each column a unique field.

34. SELECT, FROM, and WHERE are keywords. Each is used to perform a unique function regarding the retrieval of data from the database.

35. Yes. Most query languages have instructions that can be used to format reports and print headings and total lines.

36. No. The records stored within the database are not sorted. The information regarding the selected records that will be used for the report are stored in a separate file. The records stored in the temporary file are sorted and the report is printed using the records from the sort program.

37. DDL is a data description language used to describe records and fields; DML is a data manipulation language used to add records, delete records, and change records stored within selected records.

38. A DML language is usually used by a programmer; query languages are used by management, support personnel, and end users.

39. Query languages are designed for the use of individuals who are not EDP professionals. Most people can learn to use a query language by referring to the documentation and following the instructions displayed on the VDT.

40. A. False. Information stored in a database can usually be accessed by using one or many different fields.

B. False. A database must be designed so that it can interface with future as well as present systems.

C. False. It must be possible to change an application program without changing the database. An exception would be if more data had to be added to the database.

D. False. A database should support multiple programs and applications.

E. False. A database can be fine tuned without changing the programs that use the information.

F. False. Often a database is designed so that some of the information can only be obtained by people who are authorized to use those fields of data.

G. False. Records stored in a database can be retrieved and processed randomly.

H. True.

I. True.

41. Stored within a data dictionary are the names of all fields used within a database, the programs in which the data is used, the field size, and the type of data stored within the field.

42. If database dictionary software is not available, a file or database utilization chart can be made that contains the same data.

43. On the record layout form, the relative positive of each field is shown along with the size of the field and the type of data stored within the field.

CHAPTER 11

1. The activities are interrelated. If another report is desired, it may be necessary to design a new source document and input record. Additional data stored in a master file or database may also be needed and a new procedure and program will be required. An analyst could start by determining the procedures and programs that are needed. However, the output required dictates the procedures and programs required.

2. The two statements are false. Most realtime systems do have less data handling than more traditional batch systems. However, errors can still result and controls must be established to insure the validity of the output.

3. Although there is a tendency to integrate systems, the complexity of a system should not be used as a criterion for evaluating the effectiveness of the system. A basic rule is still to "avoid complexity."

4. The five major factors to be considered when determining how to segment a system into procedures are: characteristics of the hardware to be used; characteristics of the software to be used; philosophy of the analyst designing the system; controls that must be included in the system; and the type of audit trail that is required.

5. The four major concerns that must be considered when standards are designed are: the integrity of the data; the maximum utilization of the system; the ease of maintaining and modifying the system; and the importance of developing a user-friendly system.

6. Information stored in files can be protected by: backing up the files according to a predetermined schedule; storing tapes and disks in a disasterproof area; using an unexpired retention date on disks and tapes; requiring authorization and documentation for changes; and establishing and enforcing standards.

7. Programs are needed to add new records to a file, delete records from a file, change data stored within a record, make the necessary corrections, and to backup files.

8. The analyst works with the operations manager to determine when jobs should and can be scheduled.

9. A utility program is one that is tested, debugged, and use to accomplish a common task such as copying or sorting a file.

10. When tables are used to store data, more efficient programs can be written and less storage space may be needed in files and databases.

11. In the examples, CLASS and LEVEL are called subscripts. By using subscripts, a unique address can be developed for each item stored within a table.

12. If programming specifications are too detailed, the programmer may feel that the only activity left is coding and may become uninterested in the program.

13. The logic of the program will be different if partial orders for an item can be filled than if the entire quantity must be available for shipment.

14. Included in the program specifications should be the origin of the data, the source documents required, a description of the input required, the program controls needed to validate the input, the output controls to be built into the program, how the reports should be distributed, and any special considerations that must be made regarding how the input is to be processed.

15. The design report gives a progress report, an evaluation of the effectiveness of the detailed design phase of the project, and a detailed plan that can be followed to implement the system.

16. The statement "the formal report becomes the plan of action" implies the implementation plan, which includes a management tool such as a PERT or Gantt chart, will be followed in the next phase of the project. If the plan is well managed, the resources needed to implement the design will be allocated to the project and available when needed.

17. When the formal report for the detailed design phase is given, it is not expected that many changes will be suggested. If the users have been directly involved in the general and detailed design, all of their suggestions should have been implemented. However, the project team should still be receptive to suggestions.

18. A planning tool is needed in order to develop and to adhere to an effective plan. Since more resources are committed to the implementation phase than to any other phase, it is imperative that a plan be developed and *followed*.

19. The documentation included in this portion of the project will be used in developing detailed logic plans, getting source documents and report forms printed, establishing formats for data stored in files and databases, in developing screens for VDTs, and for many other functions.

20. The actual cost and time needed to develop the detailed design should be compared to the projected costs. If there is too great a difference, a study should be made to determine what created the problem. By evaluating the reasons for the difference, analysts and management can benefit from past mistakes and hopefully learn to develop better estimates in the future. In some cases, the analysis may show it was due to users wanting additional features built into the system. When this occurs, the project becomes more expensive and more time is needed to complete the design phase.

CHAPTER 12

1. Today analysts are experienced in the design and implementation of projects. Therefore they can determine how long a new project should take based on their past experiences.

2. No. Not all CIS departments are in the maturity stage of their development. Companies that have obtained microcomputer systems as their first computer system might well be in the initiation stage of their development.

3. If the growth is not controlled at the top-management level, there may be a lack of standards, incompatibility between computer systems, an excessive amount of hardware and software, lack of corporate control, and unproductive EDP professionals.

4. The three constructs used to write structured code are sequence, selection, and iteration.

5. Management was concerned about the productivity of programmers due to the increased cost of developing and maintaining software. While the cost of hardware continues to decrease, the cost of developing software is increasing.

6. When structured programs are developed, the programmer is more productive. Also the programs are easier to understand and to maintain.

7. Input and output needs and formats should be determined during the detailed design phase of the project.

8. Before attempting to develop a detailed logic plan for a program the programmer must: completely understand the problem; list the major functions to be performed; list the tasks to be performed; determine the decisions that must be made and the exception routines that are needed; arrange the functions and tasks into modules; and develop a hierarchy chart.

9. By studying the hierarchy chart, the major functions to be performed by the program can be determined. Also the relationships that exist between the individual modules can be determined.

10. One-thousand level modules call the 2000 level modules; 2000 level modules call the 3000 level modules.

11. A design walkthrough is conducted to detect logical errors and omissions.

12. An interactive BASIC interpreter evaluates each line of code as it is entered by an operator.

13. A syntactical error is one that is in violation of a basic rule of the language being used.

14. Each program should be tested with good data, then data that contains errors, and finally with no data. Each exception routine must be tested.

15. Managers should view documentation as being necessary in order to have a successful CIS department. Unless documentation is available, there is a great deal of difference between formal and informal procedures.

16. A program should be evaluated on the basis of: structure, readability, ease of maintenance, and ability to produce error-free output. Well-written and tested programs should not abort during execution.

17. Uniform symbols should be used so that anyone reading the flowchart knows what is meant by the use of each symbol.

18. Some programmers and analysts feel that a flowchart provides a better visual image of the program than do other forms of detailed logic plans.

19. A detailed logic plan is a guide to be followed in writing source code, a guide to be followed in doing program maintenance, and is helpful to auditors and others who want to know how the program executes.

20. The three major sections described the required input, processing, and output.

21. The number in the arrow at the top tells where the module being described was invoked and the arrow at the left tells where the control will go after the commands within the module are executed.

22. Some programmers and analysts feel that pseudocode is too much like COBOL.

23. A keyword is one like PERFORM which means to branch to a given location, execute the commands within the module, and then return to the statement following the PERFORM.

24. A Warnier-Orr diagram is sometimes described as structured pseudocode.

25. Design and programming standards must be developed if design and program walkthroughs are to be conducted. Also system and program maintenance is easier if everyone within the department follows the same guidelines.

26. Unless the documentation is changed when a program is modified, the documentation does not agree with the program. It will be increasingly difficult to maintain the program.

27. If a programmer uses a building-block approach, a great deal of code stored in the source library can be copied into the new program. Also subroutines can be called into the program.

28. Whenever a new programming product is obtained, programmers and analysts should be provided in-service training regarding its use. If the product is also designed for users, those persons interested in using the software should receive in-service training.

CHAPTER 13

1. Test files can be controlled by programmers or analysts. When controlled data is used in the transaction file, the programmer is sure that all possible conditions are being tested. If live data were used, not all conditions might be tested.

2. Before the personnel requirements can be determined, the analyst must decide how the present files will be converted, the volume of new data that must be keyed into the system, and when the files will need to be converted.

3. When an individual program is tested, test data is created that will enable the programmer to test the conditions that exist in that particular program. In link or string testing, a series of programs is executed in the sequence the programs will normally be run. An undetected error in the first program in the series might cause the second program to abort. In both program and link testing, the programmer controls the data used in the transaction and master file. In a systems test, the entire system is tested, including the preparation of the source documents and the utilization of the output. Actual transactions are processed—not test data.

4. When the system is tested, any or all of the following situations might be detected:
   a. A bottleneck that hinders the workflow is occurring in one area.
   b. Source documents are made out incorrectly.
   c. A situation arises which the employee preparing the source documents does not know how to handle.
   d. The operator does not know the sequence in which jobs are to be run.
   e. The data entry personnel find it difficult to enter the data.
   f. The error messages generated when invalid data is processed are not understandable.
   g. Some invalid data was transferred into a file without being detected.
   h. A combination of circumstances occurred which created a situation that had not been tested, causing the job to cancel.

5. In evaluating the results of the systems test, analysts and programmers are looking for problems that resulted because a combination of circumstances existed that had not been adequately tested for, and problems created by employees who did not understand the documentation. Prior to the systems test, the environment in which the systems run had been completely controlled by the analyst. In the systems test, the programs must execute in a "real world" environment.

6. Training can be done by a department whose function is to train employees, by someone within the CIS department, or by the user.

7. The first step is to determine what tasks are required to implement and utilize a new system. Some of the tasks will represent new procedures for which employees need to be trained.

8. Performance standards determine the level of proficiency and amount of output that an employee is expected to achieve. In data entry, the performance standard could be stated as so many key strokes per hour with perhaps 99 percent accuracy.

9. Different types of materials will be developed for ongoing training programs than are developed for a training program that will be conducted one time. The number of people participating in the sessions also determines what types of materials will be used.

10. The makeup of the audience must be determine prior to preparing materials so that the visual aids and handouts can be directed specifically to the participants in the training sessions.

11. Programmed instruction materials should be developed when it is necessary to train one or two employees at a time. The employees can work at their own pace without the constant supervision of the training director.

12. No. Managers should be familiar with the way data is entered and be able to enter data on occasion. The sales representative should be proficient at entering the data.

13. If the managers realize the value of learning about the databases, the query language, and how it will benefit them directly, they will be more interested in learning the material.

14. Within the area of EDP, the only constant is change. Provision must be made for continued, in-service training for key employees.

15. During the verbal presentation of the final report, many of the people present have been directly involved with the development of the system. Other individuals only want an overview of the how the system functions. Most questions will have been answered prior to presentation.

16. Employees should be trained prior to the final report being given.

17. The item that will be discussed the most at the meeting is the conversion plan.

18. Most of the items listed under 5 were developed during the implementation phase of the project.

19. The safest, although the most costly, method is parallel conversion. Data is processed concurrently using both the new and old systems.

20. Direct conversion is the fastest and cheapest. Often direct conversion is done because of lack of personnel and equipment to do other types of conversions. However, most CIS managers would not consider it the recommended solution.

CHAPTER 14

1. Usually user, reference, and technical guides are available from vendors of computer systems or major software packages.

2. A user's guide is task-oriented and written in nontechnical terms. A technical guide is written for the EDP professional and contains information about the system and its maintenance that would not be of interest to the user.

3. Documentation provides instruction, information, performance criteria, and also serves as an historical reference.

4. The documentation provided for systems personnel is more general and contains information regarding objectives, specifications, data flow, and the descriptions of the input, output, controls, and test procedures. The documentation for the operators is more directly related to their role in initiating jobs and determining what procedure to follow in restarting jobs, responding to messages, and in distributing reports. The programmer's documentation contains the detailed logic plan, file and record layouts, and other items needed to develop or modify the programs.

5. A maintenance programmer must have the program overview, record layouts, report layouts, screen designs, and the detailed logic plan.

6. Since operations personnel usually do not modify programs, the source listing and detailed logic plans are not included in their documentation.

7. Much of the documentation designed for a terminal operator should be internal to the program used to enter the data. Once the operator has initiated the job, HELP screens should be available to answer any questions that might arise.

8. In playscript documentation the procedure is identified. Each instruction provides the name of the actor, the logical step sequence, the key action word, and the action sentence.

9. The run sheet provides the name of the program, the input required, the output produced, special instructions, programmed messages, controls to be checked, and how the report is to be distributed.

10. The first and second steps of developing documentation are to determine the purpose of the documentation and for whom it is intended.

11. When two people have dual responsibility for developing a portion of the documentation, either a third person or one of the first two must be responsible for seeing that the documentation is completed on time and in accordance with the installation's standards. Also other individuals must know whom to contact in case there is a problem or they have a question.

12. Since the manual may contain documentation for many different programs, an index should be provided. If any technical terms, unusual terms, or abbreviations are used, they should be defined in a glossary. Manuals must be well organized so that any material that is needed can be easily found.

13. No. the project is not considered complete until the system audit is completed.

14. Management considers a system successful if costs were controlled, targeted dates were met, adequate controls and checkpoints were built into the system, the project was not oversold, and the project was well managed.

15. If problems and required changes are documented, an ongoing evaluation should not take a great deal of time. Users and management will usually know what problems have occurred.

16. The most important factor to the user may be the ease with which data can be entered and the reports can be used.

17. A log of the problems encountered should be maintained rather than relying on the memory of an operator, user, or someone in management.

18. The projects leader's evaluation is concerned with the resources that were available and how well the users defined their needs. Management is more concerned that the projections regarding the resources and time needed to complete the project were accurate and that adequate controls were built into the system to safeguard the equipment and files. Management is also

concerned about the integrity of both the input and output.

19. The operations manager is concerned about the documentation available, how well the procedures and programs have been tested, and how effectively state-of-the-art equipment and techniques were utilized.

20. If a program aborts, the date, run number, and reason for the abort are recorded. How the program was resolved is also recorded on the form along with a space to record what follow-up action is neeeded.

21. An EDP auditor should evaluate the controls built into the system; I/O scheduling and controls; procedures used to provide security for hardware, software, and data; backup procedures; and the procedures to be followed in the event that a disaster occurs.

INDEX OF FORMS AND CHARTS

abend report 510

database change form 333

file utilization chart 374

Gantt chart 91, 214

interview form 177, 193

job description—analyst 61

job description—programmer 64

master file change form 473

observation report 195

office memo 176, 180

PERT chart 92, 268, 269

print layout form 307, 310

project reports
    final documentation and report checklist 477
    general design 542
    feasibility 236
    initial 196

questionnaire 223

record layout 331, 376

report specifications 306, 309, 326

request for systems analysis 189

run manual cover sheet 497

run sheet 499

sales order 328

standards manual 184, 446, 449, 471

systems documentation checklist 505

transmittal form 330, 474

INDEX

abend report 510
abort 73, 76
acceptance test 287, 269
account number 153, 146
activity 287, 266
actuator 379, 355
address 379, 356
American National Standard Institute
  (ANSI) 111, 26, 99
analyst 34, 9, 60
  characteristics of a successful 62
  functions of 10
  job description 61
  role of 9-11
applications 21-23
  online 24
  software 26
artificial intelligence (AI) 131, 129
ASCII 379, 350
assembler 34, 30
audiovisual aids 230-234, 467-469
audit trail 34, 10, 321, 392
automating functions 261

backordered 34, 18
batch jobs 34, 15, 118-119
batch processing system 131, 118
benchmark 287, 278
bit 34, 30
blocking factor 379
brainstorming 168
buffer 132, 121
building block 455, 448
burst mode 340, 325
burster 340, 308
byte 34, 30
bytes per inch (BPI) 379, 351
case entry 455, 439
centralized data processing 34, 21
characters per inch (CPI) 379, 351
COBOL 409, 396, 431, 443-451
code 131, 125
  types of 334-335
COM (computer output to
  microfilm) 340, 298, 304
compilers 34, 26
computer 30
  classifications 30
  languages 30
Computer Assisted Design/Computer
  Assisted Manufacturing
  (CAD/CAM) 34, 18

Computer Information Services
  (CIS) 54
Computer Information Services
  Manager 57-58
computer operations 65
Computer Policy Committee 69, 173
Computer Products Incorporated (CPI)
  achievement of objectives 556
  CIS manager 529
  CIS staff 526-528
  Computer Services Department 525
  hardware 521-522
  history of 519-521
  office automation survey 532
  problems identified 534
  see Sales/Accounts
    Receivable/Inventory
  software, control 523-524
computer systems 25
conditioning documents 340, 319
constraints 12, 212
consultants 207-208
continuous form 287, 271
control interval 379, 360
controller 34, 521-522
controls 34, 15, 390
  external 287, 66, 264
  internal 287, 66, 264
conversion to a new system 462-470
  checklist 477
  documentation 479-480
  file conversion 462-463
  final report 479
  methods used 480-481
  tasks identified 476
  testing programs 463-465
  training 466-475
cost-justified 34, 24
costs, estimating 227
cover sheet 515, 496-497
critical path 287, 267
crossfooted 485, 463
CRT 34
customer information control system
  (CICS) 242, 230
cylinder 379, 354

data description language (DDL) 380,
  368
data dictionary 111, 108, 379, 373
data entry 67, 323-324, 527
data exceptions 34, 18

data flow diagram (DFD) 111, 92
  developing 95
  examples of 96, 97, 99
  functions of 93
  logical model 111, 93
  physical model 111, 93
  symbols used 94
  types of 93
data item 379, 365
data manipulation language
  (DML) 380, 368
data module 379, 354
database 34, 10, 17
  dictionaries 373
  features 372
  languages 368-371
  software 371-372
  types of 365-368
  uses of 346-348
database administrator (DBA) 73, 59
database interrogation software 379,
  377
decision support system (DSS) 132,
  128-129
decollator 340, 308
design specifications 257-260
design standards 392, 443, 446-447
detailed design 14, 345-409
  report for 403-405
detailed logic plan 287, 267, 434-445
dial-up terminals 34, 521
direct conversion 485, 480
direct files 363
direct organization 380, 357
direct-access storage devices 34, 27,
  349-350, 353-364, 521
disk controller 34
disk storage 353-364
  access time 353
  capacity 353-355
  cost of storing data 353
  cylinders and tracks 354
  data modules 354
  drives 354
  files and databases 346-348
  organization methods 357-364
  RAMAC 346
diskette 34, 28
distributed data entry 323
distributed data processing 34, 30
documentation 35, 11, 479, 492-505
  checklist 504-505

format 503
functions of 492-495
guidelines and standards 503
internal program 500
maintaining 503
online 479-480
operators 498-499
Playscript method 501
program 496
system 495
use in training 470-475
users 479
documents, study of 216
download 287, 282
dynamic disk allocation 380, 356

EBCDIC 379, 350
editing 31, 15
electronic data processing
(EDP) 34, 11
electronic mail 146
services 147
trends 147
entry sequence 380, 360
ergonomics 35, 11
event 287, 266
exception reports 73, 51
exception routines 111, 99
exit criteria 111, 106
external control 287, 66, 264
external information 73, 51

facsimile system 153, 149
feasibility study 13
beginning the study 210
conducting the study 234
developing objectives 210-211
gathering data 215-224
involvement of management 206
team approach 206-207
use of consultants 207-208
field 132, 118
file maintenance procedures 124-126
files 346-364
maintenance of 395
security of 393-394
use of 346-348
first-in, first-out (FIFO) 380, 362
fixed-head drives 380, 354
flexible manufacturing systems
(FMS) 35, 22
flowcharts, detailed 434-438
formatted screen 132, 121, 124-125
forms 108
advantages of 170
directions for completing 172, 471
use of 108
frontend processor 35, 521
functions 35, 28

general design phase 13, 255-287
controlling the design 266
developing specifications 257-264
project organization 265
report 273-274
review 273
structured design 270
gigabyte 380, 357
Gantt chart 111, 89, 91, 214
goals 73, 46
growth in EDP 422-423

hardware 35, 8, 26, 388
selecting 274
hardwired terminals 35, 521
head-per-track 380, 354
hierarchy charts 287, 270
hierarchy-input-processing-output
charts (HIPO) 455, 429, 439-440
high-level programming
language 35, 22

implementation phase 15
inches per second (IPS) 380, 351
index area 380, 358
indexed-sequential access method
(ISAM) 380, 357, 358-360
organization of data 358
reorganization 358-359
use of 360
information 73, 46
information center 59
information processing 35, 22
information retrieval 73, 56
initial investigation 12, 165-199
conducting the
investigation 173-190
establishing priorities 173
final report 191-197
objectives 166
steps in initializing 172
input 294-340
devices available 296-297
format 319
major concerns 317
medium 320
needed 318
verification of 320-323
intelligent terminal 153, 143
interactive BASIC interpreter 455, 431
internal information 73, 51
interrecord gap (IRG) 380, 351
interrupt system 35, 28
interviews 174-181, 216
conducting 174
documenting the interview 181
follow-up memo 186
forms used 177-178
memo confirming appointment 176
responsibility of interviewee 181
responsibility of interviewer 179
types of questions 179-180

job accounting 29
job control language (JCL) 35, 27, 28
job family 73, 60

key questions 111, 106-107
key-sequence 380, 360
keypunch 35, 523
knowledge worker 132, 118, 137

last-in, first-out (LIFO) 380, 362
line officers 73, 49
logic plan 108
logical model 111, 93
logical record 380, 348
loop, example of 426

machine language 35, 30
magnetic disk 353-364
access time 353
capacity 353

cost of storing data 353
cylinder and tracks 353
magnetic ink character recognition
(MICR) 287, 277
magnetic tape storage 350-352
bytes per inch (BPI) 351
characters per inch (CPI) 351
inches per second (IPS) 351
interrecord gap (IRG) 351
labels 351-352
parity check 351
mainframe 35, 22, 30
maintenance program 409, 395
management 46-52
components 48
functions 52
levels of 50
systems approach 46
management by objectives
(MBO) 73, 53
management information system
(MIS) 132, 127
concepts of 128
goals for 23
mass storage system (MSS) 380, 349
master file 346, 462
mathematical model 73, 51
microcomputer 35, 8, 30
advantages in using 282
developing guidelines 281
disadvantages in using 282
environment 69
impact of 23
survey 282
microprocessor chip 35, 23
minicomputer 35, 22, 30
modem 153, 147
module 455, 424
monitor program 35, 26
motivation 62
multiprogramming 35, 27

Nassi-Shneiderman chart 455, 429,
439, 441-442
natural language 35, 17
networking 35, 10

objectives 73, 46
measurable 17
nonmeasurable 17
observations 181
office 73, 56
office automation (OA) 136
commitment to 137
facsimile systems 149
reasons for 136
requirements 141
survey results 352
teleconferences 148
office systems coordinator 58
offline files 35, 11
online application 36, 24
online master files 36, 11
online transaction processing
system 132, 119-120
operating system 36, 12
CPI's 523-524
defining 26
functions 26
maintaining 28
types of 27

operational 36, 15
optical character recognition
    (OCR) 340, 296
optical disk-memory system 380, 346,
    356-357
optical mark recognition (OMR) 340,
    296
output 294-340
    conclusions regarding 303
    considerations 298
    devices available 295, 298
overflow area 380, 358

parallel conversion 485, 481
parity check 380, 351
partial conversion 485, 481
password 153, 147
PBX system 73, 58
person/month 200, 168
physical fitness billing system 95-126
    batch system 95-98
    developing an online
        system 120-126
physical model 111, 93
physical record 380-348
planning 88
    developing a plan 89
    Gantt charts 89, 91
    PERT charts 89, 92
Playscript method 501-502
pointer 380, 356
population 242, 220
ports 36, 521
presentations
    audiovisual aids 230-232
    defining terms 230
    defining the audience 229
    hints 223
    organizing 229
    planning 229
prime area 380, 358
principal 153, 141
procedural flowchart 36, 15, 108
procedures 36, 8, 388-390
Program Evaluation Review Technique
    (PERT) 111, 89, 266-269
program specifications 399-402
programmable terminal 153, 143
programmer 63, 64
programmer/analyst 63
programs
    design standards 443-447
    developing 396
    documentation 432
    evaluating 423-433
    programming
        considerations 421-455
    specifications for 399-402
    steps in programming 425-428
    testing 431-432, 463-465
project 111, 88
    leader 265
    organization 265
pseudocode 455, 429, 444

query language 380, 468, 467
questionnaires 220-223
queues, report 523

real memory 287, 260
realtime system 36, 28

record 132, 118
    logical 380, 348
    physical 380, 348
    size 355-356
    unblocking 348-349
records management 73, 56
reference guide 515, 493
relational model 380, 367
relative record number 380, 360, 365
reports
    external 300-301, 305
    internal 300-301
    inventory analysis 547-549
    layout form 307, 310
    printing from queues 523
    retention of 310
    specifications 306, 309, 326-327
request for bids (RFB) 287, 276-280,
    537
request for systems studies 166-171
resources 200, 166
response time 287, 260
restart procedure 515, 498
retention date 394, 409
robotics 36, 22
run sheet 515, 498

Sales/Accounts Receivable/Inventory
    design for new system 541
    design phase report 542-545
    designing reports 545
    developing a hierarchy chart 553
    developing job descriptions 555
    feasibility study 536-538
    inventory analysis report 547-549
    preparing program
        specifications 549
    problems with current system 534
screens 121, 124, 125, 311, 313-316
scrolling 340, 312
security, file 393-395
sequential files 357-358
small shop specialist 209
software 36, 8, 26, 389
    selecting 225, 275-278
source document 36, 10, 319
    designing 325
    examples 328, 333
source library 455, 452
specifications, developing 263-264
spreadsheets 73, 69
standards 36, 11, 183
standards manual 200, 170
    definition of terms 188
    examples 184-186, 446-447, 449-451,
        471-474
stenographic pool 153, 142
stock paper 340, 301
streaming tape drives 380, 351
structured design 287, 270, 423
structured programming 423
    constructs used 424-425
    designing 428
    logic plans 429, 434-445
    walkthroughs 429-431
stub testing 455, 432
subscripts 409, 397
subroutines 36, 28
subsystems 8
supervisor 36, 26
syntactical errors 455, 431

system 36, 29, 263
    reason for failures 18
    unsuccessful systems 18
system audit 506-512
    comprehensive 511-512
    evaluation 507-512
    reviews 506
system commands 36, 29
system evaluation 17
system flowchart 36, 14
    guidelines in developing 99
    illustrations 102-106, 122-123, 219
    symbols used 100
system programmer 36, 28
systems approach to management 47
systems audit 15
systems department 60
systems development life cycle 36,
    11-15
systems documentation 515, 495
systems testing 485, 464-465

table 340, 318, 396-397
task 36, 8, 263
task force 36, 14, 520
task identification 213
technical guide 513, 493
telecommunications 73, 56
telecommuting 73, 56
teleconference 73, 56, 148-149
teleprocessing 36, 10
terminal 36, 17
testing programs 431-432
text editor 200, 187, 524
throughput 73, 65
timesharing system 36, 24, 27
top down 36, 8
track 380, 354
training materials and
    methods 466-474
transaction file 111, 95
transaction processing 36, 15, 28
transmittal form 474
tree structure 365-366
turnaround document 287, 277, 300
turnkey system 36, 31
tutorial language 455, 448
two-up forms 340, 300

unbundled 287-288
unit-record equipment 287, 519
universal product code (UPC) 340,
    296
user guide 492
utility programs 34, 26, 363, 396, 523

variable length record 381, 362
vendors 224
verification of data 321-323
video display tube (VDT) 11, 140
videodisk 37, 14
virtual storage access method
    (VSAM) 455, 429, 430
    accessing records 360-362
    control intervals 360
    keys 362-363
    types 360
virtual systems (VS) 287, 278
visual table of contents (VTOC) 455,
    429-430
volume table of contents (VTOC) 363

walkthrough 287, 269, 429
  design 429-431
Warnier-Orr diagrams 455, 443, 445
Watson, Thomas 20
Winchester pack 381, 354
word length 37, 30

word processing 37, 23, 142
  advantages of 143
  analysts' role in developing 144
  disadvantages of standalone
    systems 142
  examples of 145

  guidelines 142
  requirements 143
workflow 242, 211, 217
workstations 138, 218